Freeze Drying/ Lyophilization of Pharmaceutical and Biological Products

DRUGS AND THE PHARMACEUTICAL SCIENCES

A Series of Textbooks and Monographs

Executive Editor

James Swarbrick
PharmaceuTech, Inc.
Pinehurst, North Carolina

Advisory Board

Recent Titles in Series

Freeze Drying/ Lyophilization of Pharmaceutical and Biological Products

Third Edition

Edited by

Louis Rey
AERIAL, Illkirch, France
Cabinet d'Etudes, Lausanne, Switzerland

Joan C. May
United States Food and Drug Administration
Rockville, Maryland, U.S.A.

CRC Press
Taylor & Francis Group
Boca Raton London New York

CRC Press is an imprint of the
Taylor & Francis Group, an **informa** business

First published in paperback 2024

This edition published in 2010 by Informa Healthcare

First published 2010
by CRC Press
2385 NW Executive Center Drive, Suite 320, Boca Raton FL 33431

and by CRC Press
4 Park Square, Milton Park, Abingdon, Oxon, OX14 4RN

CRC Press is an imprint of Taylor & Francis Group, LLC

© 2010, 2024 Taylor & Francis Group, LLC

A CIP record for this book is available from the British Library.

Library of Congress Cataloging-in-Publication Data available on application

ISBN: 978-1-4398-2575-4 (hbk)
ISBN: 978-1-03-291845-7 (pbk)
ISBN: 978-0-429-15185-9 (ebk)

DOI: 10.3109/9781439825761

**Visit the Taylor & Francis Web site at
http://www.taylorandfrancis.com**

**and the CRC Press Web site at http://
www.crcpress.com**

Foreword

Preservation of perishable products has always been a major concern for humanity. Many processes have been used in the past, particularly natural refrigeration, then artificial refrigeration starting in the 19th century. Freeze-drying was developed in the 20th century (1906), not only for biological material and pharmaceuticals but also for certain foods, and not just instant coffee that is now widely produced.

Prof Louis Rey was a key pioneer of this technology, through both public and private research projects, and today is still one of the most eminent specialists in the field. The International Institute of Refrigeration (IIR) is honored to have had Prof Rey as honorary president of its commission devoted to cryobiology and cryomedicine, and honorary head of its Section C with ever broader scope encompassing biology and food technology.

When I became director of the IIR in 2004, through my studies and experience I was familiar only with freeze-drying of foodstuffs. I was extremely happy when I was contacted by Prof Rey and AERIAL at a later date concerning a scientific IIR conference on freeze-drying of pharmaceuticals: New Ventures in Freeze-drying. This milestone event was held in Strasbourg, France, on November 7–9, 2007. Thanks to this conference, I was able to grasp the lively, promising nature of freeze-drying technology used for biological products and pharmaceuticals. The preservation of these products involves huge stakes: the quality required by the processes used and the industrial production of such products require great skill and precise knowledge of the impact of the processes on the components of the products handled. Freeze-drying, used alone or in conjunction with other processes such as ionization, enables the quality to be maintained at a high level. It is in the biological products and pharmaceutical sectors that a wide range of refrigeration applications are not only needed but are also fast-evolving.

This is why it is important to monitor progress in these different research fields on a regular basis. Prof Rey and Dr. May's remarkable expertise networks have made it possible to gather contributions from all over the world to gain the greatest insight possible into the state of the art.

I hope you will enjoy this book and will become even more interested in this technology.

Didier Coulomb
Director, International Institute of Refrigeration
Institut International du Froid, Paris, France

Preface

Drying an aqueous system from the frozen state by direct sublimation of ice has been known for more than a 100 years. However, for a very long time it has remained confidential and restricted to a handful of specialists trying to stabilize delicate biochemicals in their laboratory. Then, during war time, it became a major process to provide human plasma to the Medical Corps on battlefields. A few years later, freeze-drying, now named lyophilization, experienced a rocketing development with the preparation of antibiotics and many delicate pharmaceuticals and biological products. Today it is a well-established technology with large applications under steady expansion.

Despite this very wide use, it remains, anyway, a challenging issue both because of the increasing sensitivity of the treated products, mainly proteins, and because of the more and more stringent regulations and constraints that impact the industrial process. In 1999, Joan May and myself already gathered in a first book some of the main issues linked to our basic understanding of the freeze-drying of pharmaceuticals and biological products. Five years later, in 2004, the situation had again evolved and it looked sensible to publish a second edition, revised and enlarged. That situation prevails once more today and this is why Informa Healthcare felt that it would be appropriate to launch a third edition of our book addressing both the basic science aspects and their industrial output. We have taken up that request with determination knowing that we would have the privilege to gather a large panel of renowned experts.

In that prospect we see that many fundamental issues are still in the limelight, such as water and ice "abnormal" properties, phase transition in frozen liquids, and more particularly glass behaviour. On another hand, more and more emphasis is given today to the product/container/closure interaction and this implies more knowledge on the properties of glass and elastomers and on their potential leaching out of undesirable components into the active substances. The structure of the dry plug is also an important factor in the long-time storage and final reconstitution ability of the freeze-dried product. Quite obviously the nature and amount of additives, bulking agents, and residual moisture are key issues for the preservation of large molecules as well as for living cells, hence formulation remains of major concern.

Last but not least, the industrial operations are more and more under close scrutiny and require continuous monitoring and control. Most manufacturing processes being now fully automated, developments such as process analytical technology (PAT) or quality by design (QbD) are becoming leading issues.

To cover such a diversified field of competence was not an easy task, and Joan May and I have been quite fortunate to be able to group together, in this third edition, a great number of distinguished specialists who are bringing gracefully their knowledge and experience. Altogether it has been a rewarding exercise for the editors, and we are pretty sure that it will equally be the case for our readers.

Needless to say that all these efforts would have been unsuccessful without the active support of our publisher, and it is our sincere pleasure to extend, here, our deep appreciation to Sandra Beberman who has been, since 1999, the driving force of that enterprise.

Louis Rey
Joan C. May

Contents

Contributors

Dalal Aoude-Werner Aerial-crt, Illkirch, France

Yukio Aso National Institute of Health Sciences, Tokyo, Japan

David S. Baker Pharmaceutical Sciences, QLT Plug Delivery Inc., Menlo Park, California, U.S.A.

Antonello A. Barresi Dipartimento di Scienza dei Materiali e Ingegneria Chimica, Politecnico di Torino, Torino, Italy

Marie-Claire Bellissent-Funel Laboratoire Léon-Brillouin (CEA-CNRS), CEA Saclay, Gif-sur-Yvette, France

John F. Carpenter University of Colorado Health Sciences Center, Denver, Colorado, U.S.A.

Fran L. DeGrazio West Pharmaceutical Services, Inc., Lionville, Pennsylvania, U.S.A.

Claudia Dietrich SCHOTT forma vitrum ag, St. Gallen, Switzerland

Davide Fissore Dipartimento di Scienza dei Materiali e Ingegneria Chimica, Politecnico di Torino, Torino, Italy

Fernanda Fonseca UMR 782 Génie et Microbiologie des Procédés Alimentaires, AgroParisTech, INRA, Thiverval-Grignon, France

Wolfgang Frieß Ludwig-Maximilians-Universitaet, Munich, Germany

Miquel Galan Telstar Technologies, S.L., Terrassa, Spain

Ken-ichi Izutsu University of Colorado Health Sciences Center, Denver, Colorado, U.S.A.

Tauno Jalanti Microscan Service SA, Chavannes-près-Renens, Switzerland

Paul Jefferson National Institute for Biological Standards and Control, Health Protection Agency, Potters Bar, U.K.

Florent Kuntz Aerial-crt, Illkirch, France

Daniele L. Marchisio   Dipartimento di Scienza dei Materiali e Ingegneria Chimica, Politecnico di Torino, Torino, Italy

Michèle Marin   UMR 782 Génie et Microbiologie des Procédés Alimentaires, AgroParisTech, INRA, Thiverval-Grignon, France

Susan W. H. Martin   Pfizer Biotherapeutics Pharmaceutical Sciences, Pfizer, Inc., Chesterfield, Missouri, U.S.A.

Paul Matejtschuk   National Institute for Biological Standards and Control, Health Protection Agency, Potters Bar, U.K.

Florian Maurer   SCHOTT AG, Mainz, Germany

Joan C. May   Center for Biologics Evaluation and Research, United States Food and Drug Administration, Rockville, Maryland, U.S.A.

Thomas Österberg   Pfizer Product and Process Development, Global Manufacturing, Pfizer Health AB, Strängnäs, Sweden

Adora M. Padilla   KBI Biopharma, Inc., Durham, North Carolina, U.S.A.

Diane Paskiet   West Pharmaceutical Services, Inc., Lionville, Pennsylvania, U.S.A.

Stéphanie Passot   UMR 782 Génie et Microbiologie des Procédés Alimentaires, AgroParisTech, INRA, Thiverval-Grignon, France

Michael J. Pikal   School of Pharmacy, University of Connecticut, Storrs, Connecticut, U.S.A.

Theodore W. Randolph   University of Colorado, Boulder, Colorado, U.S.A.

Louis Rey   Aerial-crt, Illkirch, France and Cabinet d'Etudes, Lausanne, Switzerland

Holger Roehl   SCHOTT AG, Mainz, Germany

Jim A. Searles   Aktiv-Dry LLC, Boulder, Colorado, U.S.A.

Stanley M. Speaker   Pfizer Biotherapeutics Pharmaceutical Sciences, Pfizer, Inc., Chesterfield, Missouri, U.S.A.

Michelle Stanley   National Institute for Biological Standards and Control, Health Protection Agency, Potters Bar, U.K.

Dirk L. Teagarden   Pfizer Biotherapeutics Pharmaceutical Sciences, Pfizer, Inc., Chesterfield, Missouri, U.S.A.

Jose Teixeira   Laboratoire Léon-Brillouin (CEA-CNRS), CEA Saclay, Gif-sur-Yvette, France

Ioan Cristian Tréléa   UMR 782 Génie et Microbiologie des Procédés Alimentaires, AgroParisTech, INRA, Thiverval-Grignon, France

D. Q. Wang   Bayer Schering Pharma, Beijing, China

Wei Wang   Pfizer Biotherapeutics Pharmaceutical Sciences, Pfizer, Inc., Chesterfield, Missouri, U.S.A.

Kevin R. Ward   Biopharma Technology Ltd., Hampshire, U.K.

Sumie Yoshioka   University of Connecticut, Storrs, Connecticut, U.S.A.

1   Glimpses into the Realm of Freeze-Drying: Classical Issues and New Ventures

Louis Rey

Aerial-crt, Illkirch, France and Cabinet d'Etudes, Lausanne, Switzerland

INTRODUCTION

In their 1906 paper to the Académie des Sciences in Paris, Bordas and d'Arsonval demonstrated for the first time that it was possible to dry a delicate product from the frozen state under moderate vacuum. In that state it would be stable at room temperature for a long time and the authors described, in a set of successive notes, that this technique could be applied to the preservation of sera and vaccines. Freeze-drying was then officially borne, despite its having been in use centuries ago by the Inca who dried their frozen meat in the radiant heat of the sun in the rarified atmosphere of the Altiplano.

However, we had to wait until 1935 to witness a major development in the field when Earl W. Flosdorf and his coworkers published some very important research on what they called, at the time, lyophilization (this name was derived from the term *lyophile* coming from the Greek λ υ ο ς and φ ι λ ε ι υ, which means "likes the solvent," describing the great ability of the dry product to rehydrate again). Freeze-drying had then received a new name, which has been in current use since then, together with cryodesiccation.

Many authors, in different comprehensive books dedicated to freeze-drying, have already described in full detail the scientific history of this method, and we will not attempt to do it again. Moreover, in these last 75 years, much research and substantial development have been devoted to freeze-drying, and it would be of little use to list papers that are well known and available to all the specialists concerned.

This, indeed, is the very reason why the present book has been designed to present to the readers essentially new experimental methods and data, as well as recent developments on our own basic understanding of the physical and chemical mechanisms involved in cryodesiccation.

Nevertheless, it would not be fair to skip the names of some of the great pioneers in the field. Earl Flosdorf, Ronald Greaves, and François Henaff fought the difficult battle of the mass production of freeze-dried human plasma, which was used extensively during World War II. To that end, they engineered the appropriate large-scale equipment. Sir Ernst Boris Chain, the Nobel Laureate for penicillin, introduced freeze-drying for the preparation of antibiotics and sensitive biochemicals. Isidore Gersh and, later on, Tokio Nei and Fritjof Sjostrand produced remarkable photographs of biological structures prepared by freeze-drying for electron microscopy. Charles Mérieux, on his side, opened wide new areas for the industrial production of sera and vaccines. In parallel, he developed a bone bank, a first move in a field where the U.S. Navy Medical Corps invested heavily a few years later under Captain Georges Hyatt.

At the same time, cryobiology was getting its credentials with many devoted and gifted scientists such as Basil Luyet, Alan Parkes, Audrey Smith,

Author contact information: Cabinet d'Etudes, Chemin de verdonnet 2, CH1010 Lausanne, Switzerland; email: louis.rey@bluewin.ch

Harry Meryman, Christopher Polge, Peter Mazur, and others. We had the privilege of living this exciting period together with all these people since 1954, and most of them were present in Lyon in 1958 when Charles Mérieux and I opened the first International Course on Lyophilization, with the sad exception of Earl Flosdorf who had agreed to deliver the opening address but died tragically a few weeks before the conference.

Today, 56 years later, we are pleased to see that freeze-drying still holds a remarkable place in our multiple panel of advanced technologies, more particularly in the pharmaceutical field.

BASIC FREEZE-DRYING

Lyophilization is a multistage operation in which, quite obviously, each step is critical. The main actors of this scenario are all well known and should be under strict control to achieve a successful operation.

The product, that is, the "active" substance, which needs to keep its prime properties.

The surrounding "medium" and its complex cohort of bulking agents, stabilizers, emulsifiers, antioxidants, cryoprotectors, and moisture-buffering agents.

The equipment, which needs to be flexible, fully reliable, and geared to the ultimate goal (mass production of sterile/nonsterile drugs or ingredients, experimental research, technical development).

The process, which has to be adapted to individual cases according to the specific requirements and low-temperature behavior of the different products under treatment.

The final conditioning and storage parameters of the finished product, which will vary not only from one substance to another but also in relationship with its "expected therapeutic life" and marketing conditions (i.e., vaccines for remote tropical countries, international biological standards, etc.). In other words, a freeze-dryer is not a conventional balance; it does not perform in the same way with different products. *There is no universal recipe for a successful freeze-drying operation*, and the repetitive claim that "this material cannot be freeze-dried" has no meaning until each successive step of the process has been duly challenged with the product in a systematic and professional way and not by the all-too-common "trial-and-error" game.

The Freeze-Drying Cycle

It is now well established that a freeze-drying operation includes the following:

The ad hoc *preparation of the material* (solid, liquid, paste, emulsion) to be processed, taking great care not to impede its fundamental properties.

The freezing step during which the material is hardened by low temperatures. During this very critical period all fluids present become solid bodies, either crystalline, amorphous, or glass. Most often, water gives rise to a complex ice network, but it might also be imbedded in glassy structures or remain more or less firmly bound within the interstitial structures. Solutes do concentrate and might finally crystallize out. At the same time, the volumetric expansion of the system might induce powerful mechanical stresses that combine with the osmotic shock given by the increasing concentration of interstitial fluids.

The *sublimation phase or primary drying* will follow when the frozen material, placed under vacuum, is progressively heated to deliver enough energy for the ice to sublimate. During this very critical period a correct balance has to be adjusted between heat input (heat transfer) and water sublimation (mass transfer) so that drying can proceed without inducing adverse reactions in the frozen material such as back melting, puffing, or collapse. A continuous and precise adjustment of the operating pressure is then compulsory to link the heat input to the "evaporative possibilities" of the frozen material.

The *desorption phase or secondary drying* starts when ice has been distilled away and that a higher vacuum allows the progressive extraction of bound water at above-zero temperatures. This, again, is not an easy task since overdrying might be as bad as underdrying. For each product an appropriate residual moisture has to be reached under given temperatures and pressures.

Final conditioning and storage begins with the extraction of the product from the equipment. During this operation great care has to be taken not to lose the refined qualities that have been achieved during the preceding steps. Thus, for vials, stoppering under vacuum or neutral gas within the chamber is of current practice. For products in bulk or in ampoules, extraction might be done in a tight gas chamber or an isolator by remote operation. Water, oxygen, light, and contaminants are all important threats and need to be monitored and controlled.

Ultimate storage has to be carried according to the specific "sensitivities" of the products (at room temperature, $+4°C$, $-20°C$). Again uncontrolled exposures to water vapor, oxygen (air), light, excess heat, or nonsterile environment are major factors to be considered. In that context the composition and quality of the container itself, (type of glass, elastomers of the stoppers, plastic, or organic membranes) have to be considered.

At the end, we find the *reconstitution phase*. This can be done in many different ways with water, balanced salt solutions, or solvents either to restore the concentration of the initial product or to reach a more concentrated or diluted product. For surgical grafts or wound dressings, special procedures might be requested. It is also possible to use the product as such, in its dry state, in a subsequent solvent extraction process when very dilute biochemicals have to be isolated from a large hydrated mass, as this is the case for marine invertebrates.

Figure 1 summarizes the freeze-drying cycle and indicates for each step the different limits that have to be taken into consideration. Figure 2 gives an example of a typical freeze-drying cycle.

INSIGHT INTO THE BEHAVIOR OF PRODUCTS AT LOW TEMPERATURE
A Sensitive Issue: The Freezing Step

This initial operation in the freeze-drying process is, of course, a critical one since, if the product is impaired ab initio, there is obviously no interest to go further. We will see in the following paragraph what are the governing parameters, thermal, electric, and structural, but before it is worthwhile to make a few comments on the way freezing is performed.

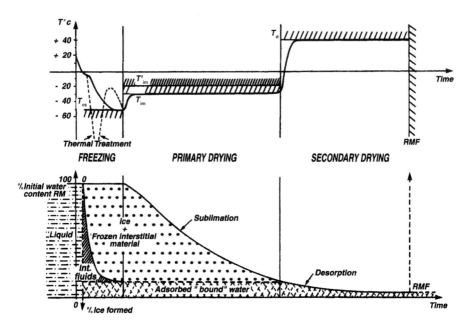

FIGURE 1 Schematic evolution of the freeze-drying process. Temperatures (*upper curve*) and water content (*lower curve*) versus time are indicated. *Abbreviations*: T_{cs}, maximum temperature of complete solidification; T_{im}, minimum temperature of incipient melting; T'_{im}, absolute limit for fast process; T_d, maximum allowed temperature for the dry product; RMF, final requested residual moisture.

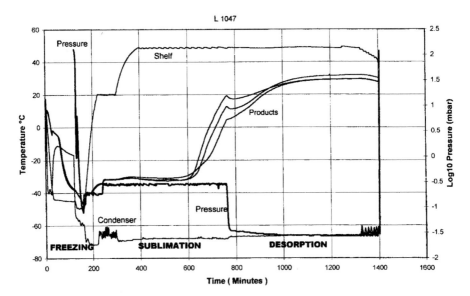

FIGURE 2 A typical freeze-drying cycle. Note that the pressure is raised to 0.3 mbar to increase the heat transfer during sublimation and then lowered to 0.02 mbar for desorption.

In diluted solutions, which is a current case in freeze-drying of pharmaceuticals, ice can develop in the course of cooling either as a well-defined front moving upward in the liquid from the cold supporting shelf or as a brusk cloud of individual germs appearing at the same time in the whole mass of a supercooled fluid. In the first case, and especially when the cooling velocity is low, there might be some cryoconcentration of the product, which provokes a solute gradient from bottom to top resulting in an increasing solid concentration in the upper layer. The result is, most often, the occurrence of a thin film, rather compact, at the surface of the dry plug at the end of the process, which might create problems at the reconstitution step and definitely impedes the water vapor (mass) transfer in the course of drying (Fig. 3). In the second case, when nucleation starts up all at once throughout the liquid the structure of the frozen mass is more homogenous, and this may lead to a more finely porous dry cake, but if the degree of supercooling is important and the ice development is pretty fast it may result also in the rupture of the vial. It is thus advisable in that particular case of freezing to measure ahead the crystallization velocity of the solution in function of the degree of supercooling (Fig. 4) to get a better control of the process.

Another concern is that supercooling might not be regular from one shelf to the other one and even all over one single shelf. As a result nucleation may occur at different temperatures and give different structures among the treated vials. Then it becomes advisable to control the structure of all products to equally control the nucleation step and make it simultaneous for the whole batch. The use of acoustic of ultrasound stimulators seems to give an elegant solution to that problem.

Another interesting issue is that if the freezing is delayed, for operational reasons, and the liquid stands for too long a time in contact with the inside surface of the vials, there might be an important interaction between the solution and the glass walls. Active substances might then be adsorbed, often in an irreversible way, or else glass chemicals might leak out into the liquid. These problems will be addressed further in this book but they have to be taken care of when handling solutions in which the active substances are in very small quantity (genetically engineered biochemicals) or as small individual particles (bacteria and virus vaccines).

Finally, the freezing process might be very delicate to carry when the initial liquid is an unstable emulsion and a fast processing, making use of cryogenics, might be then desirable.

An Interesting Development: Soft Ice Technology

Everybody is familiar with the concept of sherbets and soft ice-creams. They are basically frozen plastic pastes, which present a high viscosity and relatively moderate negative temperatures. They are not free flowing but can be stirred and mixed under moderate mechanical strength and can incorporate solid particles like chocolate crumbs, fruit, nuts, and others without losing their structure. Their rheological properties are rather complex and highly dependent on temperature. They are most generally commercialized as such but might be also the first step of a more elaborate process. For instance, in the freeze-drying of coffee, we start with the production of a viscous concentrated coffee extract, which is turned into a plastic icy foam by injection of carbon dioxide under pressure in a rotative scraped surface heat exchanger. There, the liquid is converted into a sherbet-like foamy paste, which is spread over a metallic belt

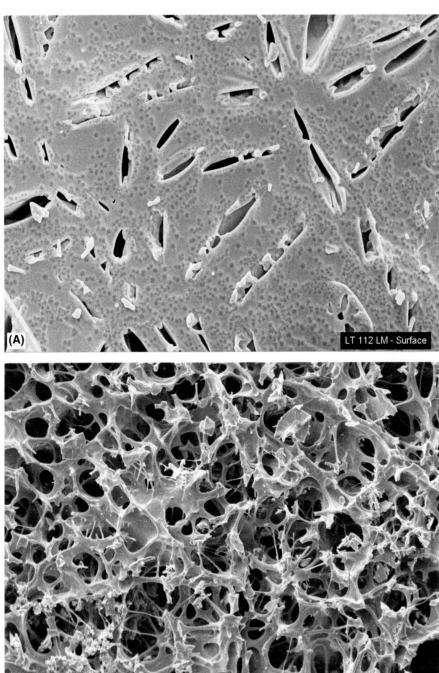

FIGURE 3 (**A**) Scanning EM of the upper surface of a freeze-dried pellet showing a membrane-like "skin" with elongated "vents" for the escape of water vapor. (**B**) Scanning EM of the inside middle part of a freeze-dried plug showing the extremely porous material resulting from the sublimation of ice-crystals. *Abbreviation*: EM, electron micrograph (Microscan).

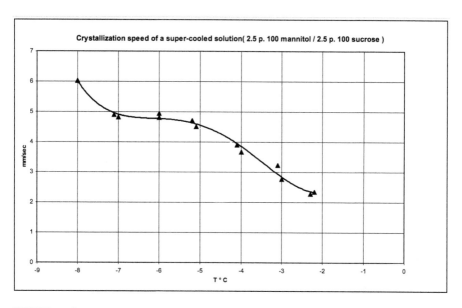

FIGURE 4 Crystallization velocity of a solution versus its degree of supercooling.

conveyor and frozen hard in a cold room. Then it is broken, ground or sliced, sifted, and finally fed into the freeze-drying plant.

In the preparation of many pharmaceuticals we know that some components in the formulation are not compatible together. For instance, if we mix acetyl salicylic acid with sodium bicarbonate in solution, there is an immediate reaction and a vigorous release of carbon dioxide. Nevertheless, it might be of interest to freeze-dry them together to have instant sparkling aspirin, but this is not possible, and the only way around this is to compact the products together in the dry state. The resulting tablet is not very stable and it takes a relatively long time to get back into solution.

Conversely, if such reactive substances are mixed together at low temperature after incorporation in a soft ice they will not react and the resulting paste can be molded in appropriate shapes and hardened by further cooling. The material can then be freeze-dried without difficulties. In that state, lyophilized aspirin, for instance, is perfectly stable and when water is added back it reconstitutes as a sparkling fluid in a few seconds because of its high porosity.

The "soft ice" technology is thus a very precious tool to prepare complex products, most often for oral route. Pure chemicals, drugs, vitamins, and mineral salts can be successfully freeze-dried in that way in rather elaborate formulations since it is possible to mix together "sherbets-lines" issued from different solutions and even add to the whole other solid ingredients as finely dispersed powders. The key to success in that process is a good control of the temperature of the icy paste, which is generally prepared in a cylindrical double-wall heat exchanger with a continuous scraped surface maintained at temperatures between −4°C and −20°C depending on the nature of the treated products. When the soft ice mixture is duly completed it can be molded by conventional equipment in plastic blisters and frozen hard in a blast tunnel.

Then this material can enter the freeze-dryer. Altogether, it is a rather simple online galenic process.

Thermal and Electric Properties

A fundamental paradigm of freeze-drying is to understand that, in almost every case, there is no direct correlation between the structure of a frozen product and its temperature since all structural features depend essentially on the "thermal history" of the material. In other words, the knowledge of the temperature of a frozen solution is not enough to allow the operator to know its structure since this latter depends, essentially, on the way this temperature has been reached, that is, on the freezing cycle.

For instance, we did show in 1960 that an aqueous solution of sodium chloride at $-25°C$ could be

either a sponge-like ice network soaked with highly concentrated fluids if the system has been cooled progressively from $+20°C$ to $-25°C$

or a totally frozen solid where all the interstitial fluids have crystallized as eutectics if the system has been cooled, first to $-40°C$ and then rewarmed to $-25°C$

Actually, when water separates as pure ice, as is the case for diluted solutions, there might be a considerable degree of supercooling in the remaining interstitial fluids. It is then compulsory to go to much lower temperatures to "rupture" these metastable states and provoke their separation as solid phases. This, indeed, has a great significance because it is precisely within those hypertonic concentrated fluids that the "active substances" soak, whether they are virus particles, bacteria, or delicate proteins, and there they can undergo serious alterations in this aggressive environment. This is the reason why we advised to cool the product at sufficiently low levels to reach what we have called T_{cs}, the maximum temperature of complete solidification. However, when frozen, the material will only start to melt when it reaches either eutectic temperature or what we called T_{im}, the minimum temperature of incipient melting.

Differential thermal analysis (DTA) and differential scanning calorimetry (DSC) are useful techniques to proceed to this determination. They can be very advantageously coupled with low frequency (LF) electric measurements since the impedance of the frozen system drops in a spectacular way when melting occurs (Figs. 5 and 6). In other terms, a high electric impedance is always related to a state of utter rigidity. Moreover, the electric measurements are more reliable than DTA/DSC alone. Indeed, when we are dealing with a complex system—more particularly when it contains high molecular weight compounds—the material hardens progressively during freezing and, often, in the course of rewarming shows incipient melting only at relatively high temperatures. Until that point the DTA curve remains "silent." Unfortunately, this is not always a sign of absolute stability since, quite often, a "softening" of the structure appears much earlier. It is our experience that a sharp decrease in electric impedance is a clear warning for an operator who should try and maintain the product during primary drying at temperatures below this limit. Figure 7 demonstrates this phenomenon in the case of U.S. standard for pertussis vaccine lot 9, and Figure 8 shows the same behavior for the Saizen mass 10 mg from Serono. In both cases it is quite obvious that sublimation has to be

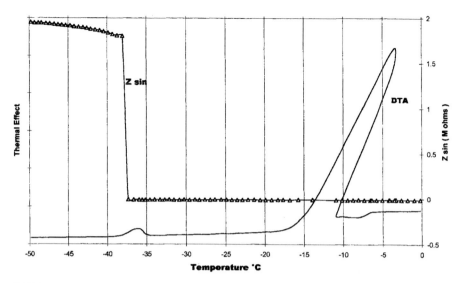

FIGURE 5 Differential thermal analysis (DTA) and impedance (1000 Hz − $Z\sin\varphi$) of a 2 p. 100 solution of Cl Na in water during controlled freezing.

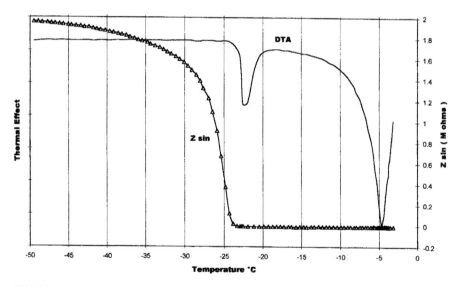

FIGURE 6 Differential thermal analysis (DTA) and impedance (1000 Hz − $Z\sin\varphi$) of a 2 p. 100 solution of Cl Na in water during controlled rewarming.

done at temperatures much lower than those, which could be derived from the DTA curves alone.

The evolution of electric properties is, thus, at least for us, a major parameter in the design and follow-up of a freeze-drying cycle. It can even be used for an automatic control of the whole operation, as we proposed in the past.

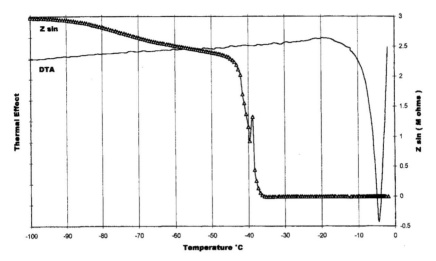

FIGURE 7   Differential thermal analysis (DTA) and impedance (1000 Hz − *Z* sin φ) of U.S. FDA standard pertussis vaccine lot 9.

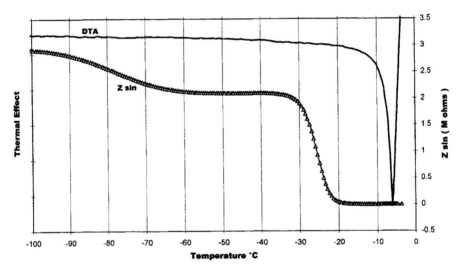

FIGURE 8   Differential thermal analysis (DTA) and impedance (1000 Hz − *Z* sin φ) of Serono Saizen mass 10 mg.

Glass and Vitreous Transformations

It is well known that certain solutions/systems do not crystallize when they are cooled down, especially when this is done in a rapid process (quenching in cryogenic fluids has long been in use by electron microscopy specialists to prepare their samples). Cooling results, in that particular case, in the formation of a hard material that has the properties of a glass. This vitreous body can prove to be stable, and when rewarmed it softens progressively and goes back to liquid

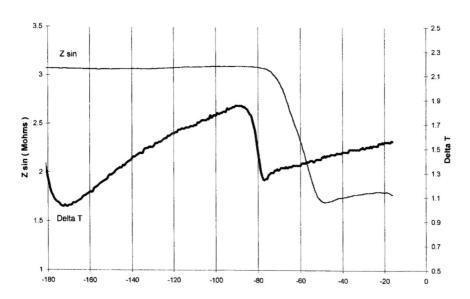

FIGURE 9 Differential thermal analysis and impedance (1000 Hz) of pure glycerol.

again. However, in the course of rewarming it generally undergoes a major structural change known as the *vitreous transformation*. During this process the glass, still in the solid phase, evolves from a solid-like state (with low specific heat and low specific volume) to a liquid-like state with a sharp increase in specific heat and specific volume. This occurs at a well-known temperature, the temperature of vitreous transformation, generally quoted as T_g. A typical example, pure glycerol, is shown in Figure 9. This phenomenon is fully reversible.

In some other cases the glass, when rewarmed, becomes highly unstable, and when the *vitreous transformation* is completed (or sometimes during this transformation itself) it crystallizes out abruptly showing a marked exothermic peak. This, to the contrary, is an irreversible process called *devitrification* (Fig. 10). All of the glass might then disappear or it might devitrify only partially. If the material is then cooled again and rewarmed a second time, the initial vitreous transformation disappears or is substantially reduced (and sometimes shifted to a higher temperature) and the exothermic peak no longer exists. During this cycle, which we called "thermal treatment" in 1960, the metastable system became unstable and has been "annealed." Figure 11 shows this type of evolution in a glycerol-water-Cl Na system. It is worthwhile mentioning that in both cases the vitreous transformation initiates a marked decrease of the electric impedance, though at first sight the system remains a compact solid.

The magnitude and temperature of these events are of course dependent on the system under investigation, as can be seen in Figure 12, which compares D_2O/glycerol and H_2O/glycerol mixtures.

This type of behavior is not uncommon in pharmaceutical preparations. Indeed, it is often required that a thermal treatment be applied to the material to

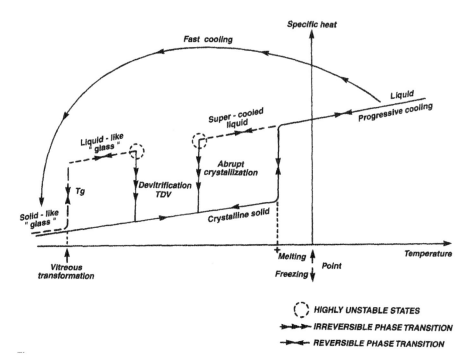

FIGURE 10 Theoretical diagram of the low-temperature behavior of a system susceptible to present glass formation. Specific heat has been taken as the "marker."

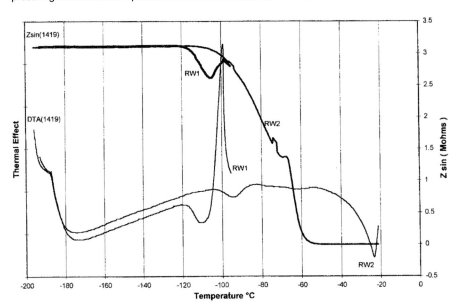

FIGURE 11 Low-temperature behavior of a 50:50 glycerol-H_2O system containing 10 per 1000 Cl Na. RW1: recordings of differential thermal analysis and $Z \sin \varphi$ during the first rewarming from $-196°C$ to $-95°C$. When the devitrification peak is completed ($-95°C$), the sample is immediately cooled back to $-196°C$. RW2: Second analysis from $-196°C$ to $-20°C$ after "thermal treatment."

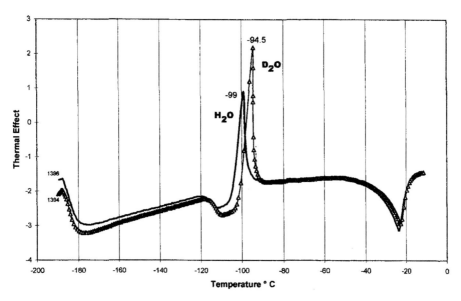

FIGURE 12 Differential thermal analysis of 50:50/100 mixtures of glycerol and D_2O/H_2O. Both normal and "heavy" water-glycerol mixtures demonstrate the same behavior but the transition temperatures for D_2O are shifted upward by 4.5°C.

secure a steady and successful freeze-drying operation. In the example depicted in Figure 13A and B, it was discovered that a refined biochemical (recombinant glycoprotein-Mol-2 from the Center for Molecular Immunology in Havana, Cuba) needed to be stabilized by thermal treatment to raise the softening temperature (as witnessed by the evolution of the electric impedance) from −32°C to −19°C and allow an easy freeze-drying.

In other instances—as this will be developed later in this book by several authors (Randolph, Searles, Carpenter, and Pikal)—it appears essential to safeguard the glassy state, which is indeed an absolute prerequisite to secure the tertiary structure of active proteins. Here again a precise knowledge of the boundaries and sensitivities of this glass is definitely needed.

Finally, let us mention the interesting case of the sucrose/water systems, which after drying gives a very sensitive porous cake susceptible to undergoing adverse transformations at *above-zero* temperatures and *in the dry state* because of the existence of a rather high T_g (57°C). It is, thus, compulsory to keep these products at temperatures low enough to prevent the shrinkage of the pellet by what has been called a "rubber behavior."

BULKING AGENTS: A MATTER OF CONCERN IN SUBLIMATION
This is, indeed, the heart of the lyophilization process: the step during which the frozen product is dried—from a solid phase—by direct sublimation of ice. During that phase the drying boundary sinks into the frozen material from top to bottom or else from outside to inside leaving an upper layer or surrounding shell of fine porous material. The resulting dry matrix is thus carved from the frozen mass under vacuum. To achieve this operation this frozen mass has to keep a structure when ice disappears and water vapor moves away. To that end,

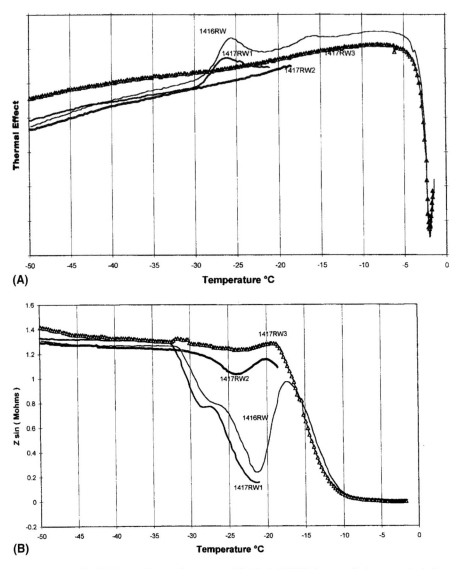

FIGURE 13 **(A, B)** Recombinant glycoprotein-Mol-2. 1416RW: In a preliminary analysis two slight exothermic phenomena were observed by DTA at −25.5°C and −16.4°C, followed by a progressive melting starting around −8°C. However, the electrical measurements already showed a marked decrease in rigidity starting at −32°. 1417RW1: "Thermal treatment" was applied by cooling the solution down from −20°C to −50°C, then rewarming to −21°C. The thermal curve is almost identical to the 1416RW and the first exothermic peak shows up. The system is then cooled back from −21°C to −50°C and rewarmed again to −18°C. 1417RW2 shows, during the second rewarming, that the first exothermic phenomenon has disappeared. As a consequence, the $Z\sin\varphi$ is by far higher 1417RW2 and "rigidity" good until −20°C. To improve, once more, the low-temperature stability of the system, it is cooled back a third time from −18°C to −50°C and then rewarmed for analysis. 1417RW3 shows that, in the course of the third rewarming, all "accidents" have disappeared on the DTA curve, which only indicates progressive melting from −8°C upward. Moreover, the stability of the frozen structure has been substantially improved since now $Z\sin\varphi$ remains almost constant during the whole period and only starts to drop at −19°C instead of −32°C without thermal treatment. It is then easy to freeze-dry the solution at −20°C or slightly below. *Abbreviation*: DTA, differential thermal analysis.

14

bulking agents have to be added to "build" that structure if the initial solution is too diluted. Lactose, sucrose, mannitol, PVP, dextrose, methyl cellulose, glycine, and many others have been used to fulfill this task.

New Carriers

In some cases, however, the introduction into the formulation of these "foreign compounds" might be undesirable and, ideally, it would be by far better to do without them and handle the active substance alone. An interesting option might then be to enclose the liquid to be dried into a porous matrix where it will be kept in the course of drying and thus prevented to fly away with the water vapor stream. Porous polymers, sintered metals, ceramics, porous glass, inorganic textiles, multilamellar pads, even small silica or glass beads, etc., could, at first glance, be resorted provided they demonstrate simultaneously a certain number of properties.

> They can adsorb the liquid solution easily as a hydrophilic material and withstand freezing and drying without mechanical rupture.
> They do not interact with any element of the formulation.
> They are clean and deprived of residues, particles, or contaminants.
> They hold enough liquid per unit volume, which means that they present at least a 30% to 50% porosity.
> They can be shaped as well-defined geometrical units: disks, rods, and spheres to be incorporated into the vial.
> Their pore structure is thin enough to hold the solution but wide enough to let the water vapor escape from the frozen liquid, which means an open-pore structure, a "sponge" with interconnecting interstitial channels.
> They release the adsorbed active product when they are flooded with the reconstitution fluid.

To our own experience, fulfilling these requirements is not an easy task and, actually, very few products are susceptible to provide this complete set of properties. While most of the structural issues can be solved, the most difficult, by far, remains the latter one: The "carrier" should release the totality or at least a major known amount of the active substance at the time of reconstitution. To that end, developments have been made to "exhaust" the carrier by percolating through it the dissolution fluid under pressure, as would be done with a conventional online filter. Special syringes have been manufactured where the original solution to be dried is pumped through the carrier, then allowed to dry there, and finally extracted by the reconstitution fluid using the same type of mechanism.

Today, the inert carrier issue is still under development, and it is more than likely that the enormous amount of research, which is currently devoted to new materials, whether glass and glass derivatives, polymers, fibers, etc., might help us to find new supports with almost nil adsorption properties and still an open porous structure.

OPERATING PRESSURE

Numerous books and papers have been written on the physics of primary and secondary drying. However, we would like to stress one single point that has not always been clearly understood—the central role of the operating pressure.

Indeed, as we have shown in Figures 1 and 2, freeze-drying deserves an accurate and continuous control of the pressure in the chamber during the whole operation.

With the exception of vacuum freezing (or "snap"-freezing generally used in the food industry), the *initial cooling* is always done at atmospheric pressure, sometimes in a separate cabinet.

Primary drying, to the contrary, is performed under vacuum. Whereas in the young days of the technique it was felt by most specialists that the higher the vacuum the better the process could be. It was shown by Neumann and Oetjen et al. that throttling the water vapor flow between the chamber and the condenser was increasing the speed and efficiency of the operations. Later on, Rieutord and I patented the air injection process, which could be applied to any equipment whether the condenser coils were placed in a separate chamber or in the drying cabinet itself. Since then, the "air bleed" has been used almost universally since a substantial rise in the pressure (0.1–0.5 mbar according to the T_{im} of the treated product) proved to increase considerably the heat transfer to the sublimation interface essentially by gas conduction-convection. As a consequence, the temperature of the heaters could be substantially reduced, which prevented melting of the still-frozen core and/or scorching of the already existing dry layer. Monitoring and control of this pressure is still an essential part of the process, and they can be geared to the intrinsic properties of the product such as its temperature or, better, its electric impedance. At any rate, during this whole sublimation period, the vacuum level is the master key to the heat transfer and can help to "rescue" a product that is becoming too hot and starting to soften dangerously. Indeed, in such a situation, pulling the vacuum down immediately works as a real thermal switch with an instantaneous result, whereas cooling the shelves requires a longer time.

Secondary drying is generally carried out under higher vacuum when the product has reached an above-zero temperature (or its electric impedance has reached the upper limit). Indeed, it has been shown by different authors that isothermal desorption was *faster* and allowed *lower final residual moistures* when the pressure lay in the level of 10^{-2} mbar but that higher vacuums (10^{-3} mbar) did not drastically improve the operation.

Final conditioning has always been much discussed and, in the last decades, vacuum and dry neutral gas both got their supporters. Today, stoppering the vials under a slightly reduced pressure of dry nitrogen gas looks to be the favorite option. Some experiments that we did in the past and that remained unpublished push us to think that stoppering under dry argon could give better results for long-term storage, as is the case, for instance, for international biological standards.

SOME CHALLENGING WAYS TO INVESTIGATE THE FINAL DRY PRODUCT

The dry freeze-dried cake undoubtedly has a very peculiar structure since, as we have already seen before, it has been "carved" under vacuum from a solid matrix (Fig. 3). At the end of this process and when it is still under vacuum its internal surface is quite "clean" and very reactive. It can be easily understood that its first contact with a foreign element, such as the gas used to rupture the vacuum, is determinant. This is, in fact, why the choice of this gas as well as the procedure to introduce it are critical. In that context the structure of the dry cake and the properties of its internal surface constitute a major element.

Another important factor is the level of the residual moisture in the pellet and its evolution, if any, during storage.

Finally, it is definitely of interest to know how this internal water is bound to the structure.

These are three recurrent problems in freeze-drying, among many others. We shall propose some selected experimental means for their investigation.

Internal Surface: BET

Applying the Brunner, Emmett, and Teller (BET) method to pharmaceuticals, whether ingredients or finished products, is of common use in powders for the determination of their specific area and pore sizes. However, generally, the operator has a reasonable amount of substance in hands, and most often this material is relatively resistant to water vapor and can be manipulated without too many risks in the open space of a laboratory or, if needed, within a conventional glove box.

The situation is quite different with sensitive freeze-dried specimens contained in sealed vials or ampoules, which present both a very reduced weight and, of course, a very low density as well as a high hygroscopicity. In that case, the prolonged contact with moist air can provoke a real "collapse" of the internal structure, which cannot be restored to its original state by prolonged pumping. Then a very precise methodology has to be followed to get a reliable measurement. We would like to explain how we proceed to that end.

> The sample to be checked (5–100 mg of dry product in a sealed vial or ampoule) is quickly opened with a diamond saw and the bottom part, containing the plug, placed in a special stainless steel BET cell that is immediately capped, weighted, and connected to the manifold of the pumping device. Vacuum is then pulled over the sample at room temperature.
>
> Two to three days later vacuum is broken with dry nitrogen gas and the BET cell connected to an automatic, computerized analyzer (Den-Ar-Mat 1000).
>
> Vacuum is pulled again and repetitively checked until the leak rate (combined with the gas release from the product) falls to the order of 2–3×10^{-4} mbar/sec.
>
> The BET cell is then flooded with helium for the determination of the sample volume, which allows the determination of its real density and porosity, since we already know both its weight and its apparent volume.
>
> The BET cell is again pumped and immersed in liquid nitrogen.
>
> When temperature and pressure are stable, known amounts of helium are introduced in the measuring chamber; thanks to special gas-tight microsyringes (7 µL for the smallest one) to perform the calibration of the cavity before adsorption.
>
> The BET cell is pumped again and then the measurement can start. Known amounts of the adsorption gas are then introduced in the specimen chamber and each time we wait for a steady equilibrium. The adsorption isotherm is, thus, constructed progressively throughout the BET range.
>
> When saturation is reached, the BET cell is slowly pumped down, stepwise, to ensure controlled desorption to measure pore size and distribution.

For absolute surfaces ranging from 50 to 5 m^2, the measuring gas to be used is nitrogen. For absolute surfaces of 5 to 0.5 m^2, the measuring gas is argon. For absolute surfaces below this level we use krypton. In the latter case, the only

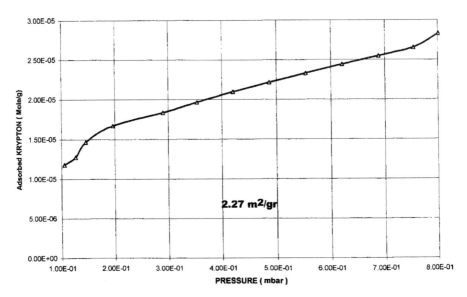

FIGURE 14 Determination of the specific area of freeze-dried Serono Gonal F75 UI by BET (sample weight 0.0325 g).

possible determination is that of the specific area since, with krypton being solid at liquid N_2 temperature, it is not feasible to perform a complete adsorption-desorption cycle for the evaluation of pore distribution and sizes. For obvious reasons, krypton measurement is bound to become the routine one for most freeze-dried pharmaceuticals.

Figure 14, for instance, represents the krypton adsorption curve for a sample of Gonal F75 UI from Serono, and we can see that it gives a pretty accurate reading despite the fact that the absolute surface measured is only 0.07 m^2.

It is beyond the scope of this chapter to indulge in more details about this technique, but it might be of interest to note that it allows the evaluation of the influence, on the product structure, of different factors such as the initial concentration of the starting fluid, the rate of freezing (macro or microcrystalline structures), the pertinence of the drying cycle, and of the final conditioning.

Since many functional properties of the dry cake derive from its internal surface and porosity such as solubility, oxygen, and water sensitivity, these measurements might be of great interest, especially for the assessment of reliability of a new process in view of its potential validation.

Some people will ask why we did not use penetration by mercury for the measurement of pores. Our answer is that for delicate, flexible products that can be easily crushed or simply distorted in their morphology by outside constraints, we found that the application on the cake of pressures ranging to a few thousand bars could be considered irrelevant.

Equilibrium Water Vapor Measurement in a Sealed Vial
For the reasons that we have just described, the measurement of residual moisture within an isolated vial or ampoule is a difficult undertaking. Joan May, in this book, also deals with this topic, and she explains how the FDA developed

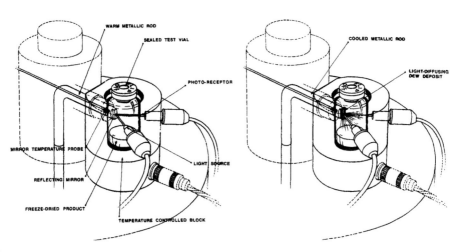

FIGURE 15 Operating principle of the device made to measure the equilibrium water vapor pressure inside a sealed vial.

very accurate standardized techniques to execute these determinations by chemical Karl Fischer titration or thermo gravimetric techniques. The only drawback is that, in both cases, the material is destroyed.

It is, in fact, because of that specific point that our approach has followed a different route.

Use a nonintrusive method leaving the product intact.
Be able, under those conditions, to do repetitive measurements on the *same vial* in the course of time to assess its evolution during storage.

Figure 15 shows the principle of the method we developed jointly with J. Mosnier (deceased). The selected vial or ampoule is placed in a thermostatic metallic block where temperature can be maintained constant from −10°C to +60°C. On one side of the block a metallic "finger," having at its end a tiny polished stainless steel cylindrical mirror (2 × 6 mm), is pushed in close contact with the outside wall of the glass vessel and the connection is made optically and thermally tight, thanks to a trace of silicon grease. On the other side is placed a near-infrared diode that beams light throughout the vial on the mirror, which, in turn, reflects that light on a phototransistor placed in a symmetrical position.

When the temperature of the mirror is equal to or higher than that of the metallic block (hence the temperature of the sample) the level of reflected light is steady. To start measurement the temperature of the mirror is progressively decreased, thanks to combined Peltier and cold nitrogen gas. Because of the good thermal contact on a very limited surface (12 mm^2) the internal wall of the flask is equally cooled on a small isolated spot. When, finally, the temperature of the mirror (and the temperature of the spot on the internal surface) reaches the value corresponding to the saturated water vapor pressure within the flask, some dew (or ice) is deposited on the internal wall that immediately becomes diffusive. The reflected light drops sharply. The measurement is done. The mirror can thus be warmed again, dew evaporates, light is restored, and the sample can be placed back into the storage cabinet for further determination. Figure 16 gives an

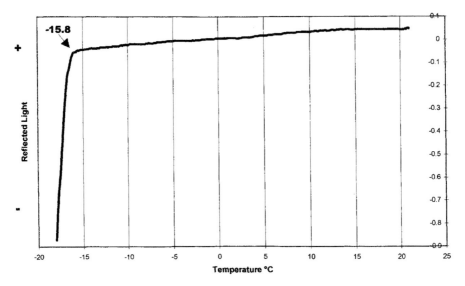

FIGURE 16 Novartis calcitonin dry substance in ampoules.

example of this determination for a sample of Novartis calcitonin dry substance in a sealed ampoule measured at +25°C. At the corresponding temperature of −15.8°C the water vapor pressure inside the ampoule is 1.53 mbar.

The sensitivity of this method is very high since measurements are still possible for condensation temperatures of −60°C, which, in a vial of 10 cm³, means that less than 0.1 µg of water is present in the overhead space above the pellet.

Over the last 15 years systematic determinations have been done in collaboration with Dr. May from FDA and with several drug companies on a vast number of products. Most of them have been followed during storage (at −20°C, +4°C, or room temperature) for more than two years. Dr May reports on some of her findings later in this book.

Without going into too much detail, let us mention some of the key issues of this particular work.

In sealed glass ampoules and for the same product we found substantial differences in the equilibrium water vapor pressures according to batches and, within a given batch, according to the position of the ampoule on the shelves of the freeze-dryer. The so-called wall, door, condenser, and corner effects could then be analyzed and amended.

In sealed glass ampoules we also witnessed an evolution in the course of storage showing that often the remaining water was "restructured" in the course of time going from a rather mobile form to a more firmly bound form, displaying a lower water vapor equilibrium pressure (Fig. 17).

In stoppered vials we could quite easily follow the transfer of moisture from the stopper toward the plug, which acted as a "getter pump" until a new equilibrium was reached (Fig. 17). We could establish, then, a direct correlation between the water vapor pressure measurements and the Karl Fischer titration (Fig. 18).

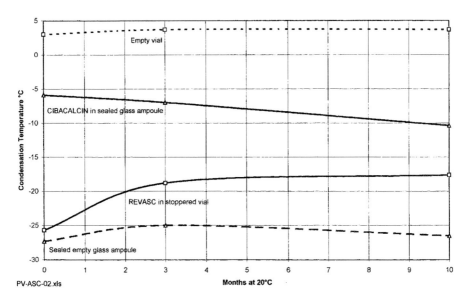

FIGURE 17 Evolution of the condensation temperatures of freeze-dried Cibacalcin and Revasc (Novartis) in the course of storage at 20°C.

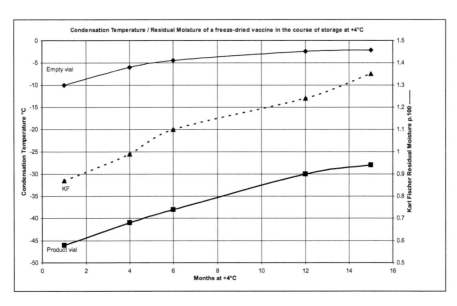

FIGURE 18 Evolution with time of the condensation temperature of the vial gas phase and of the Karl Fisher residual moisture (measured on an alternate sample) of a freeze-dried vaccine.

Sometimes our method helped the pharmaceutical technicians to modify the formula of the excipients and stabilize the water content. They could also better understand the as-yet-unknown behavior of certain additives like heavy polyalcohols, glycerol, or tensioactive agents.

Today we are convinced that there are numerous applications for this new analytical tool, especially when controls have to be performed on rare, high-cost, or infectious material for which a nonintrusive remote technique is compulsory or that repetitive measurements need to be done on the same sample over prolonged periods.

Low-Temperature Thermoluminescence of Dry Material

Thermo luminescence analysis has been applied to solids for quite a long time in archeological research and radiation dosimetry. In both cases the samples that have been activated by natural and/or man-made radiation sources present a certain number of energy traps that can be emptied by heating. For the datation of ceramics, the heating cycle is extremely fast and brings the sample to more than 300°C in less than a couple of minutes. The magnitude of the light emission under well-standardized conditions is directly linked to the age of the material, which can be determined over several centuries with an accuracy of a few percentage points.

For radiation dosimetry, within or around nuclear plants as well as in open green fields and in the environment at large, the analysts make use, generally, of so-called thermoluminescent solids that are susceptible to "activation" over a large span of doses. Espagnan et al. developed a refined method to that end using lithium fluoride tablets, which are first totally deactivated at 400°C and then exposed to the radiation field at ambient temperature. Reading is done by recording the emission peaks during their progressive "extraction" in the course of a constant-speed (1°C/sec) heating, which generally shows up between 200°C and 400°C.

At the onset of our research work, we tried to operate in the same way with freeze-dried products that had received a strong γ-irradiation (40–60 kGy) at room temperature. The results were disappointing because we were limited to rather low temperatures (below 100°C) and, even so, when the material could stand high temperatures like freeze-dried silica. This is why we attempted to apply to these products the same methodology that we used previously for our investigation on water and solutions.

The dry samples are irradiated at −196°C and deactivated by a gentle rewarming (1.5°C/min) to room temperature. Our preliminary results look promising (Fig. 19).

Figure 20, for instance, shows the behavior of a freeze-dried plug of mannitol (residual moisture 1.5%). We can see that, in comparison with the initial solution (at 10% mannitol), the thermoluminescence of the dry material is some 20 times stronger and does not follow the same pattern. The lower peak (at −152°C) is very narrow whereas the higher one (near −118°C) is substantially depressed. This is not surprising if we assume that the first peak is directly connected to the water molecule itself while the second one seems to be geared to the three-dimensional network of the water molecules within the crystalline lattice.

Additional studies done on freeze-dried silica gels (microbeads gels from Rhodia, Salindres) give some support to this idea. In Figure 21, for instance, we can see that when we shift from a largely hydrated material (like the fresh starting material at around 40% residual moisture) to the low residual moisture

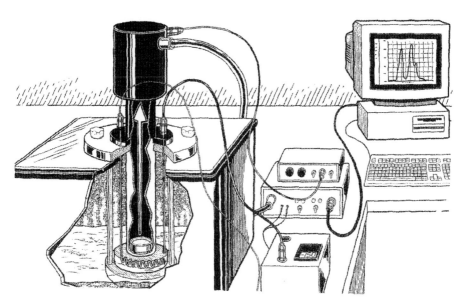

FIGURE 19 Low-temperature thermoluminescence setup. A previously irradiated frozen sample placed in a liquid N_2 cryostat emits light during controlled rewarming. The corresponding glow is recorded by a photomultiplier versus temperature.

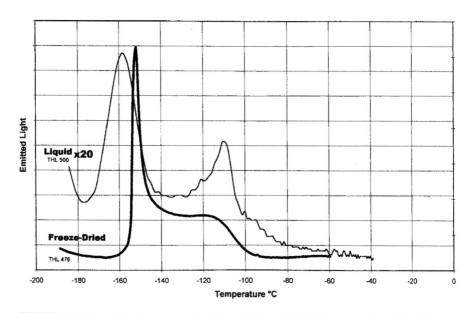

FIGURE 20 Thermoluminescence of a 10 p. 100 mannitol-H_2O system after γ-irradiation at $-196°C$ (30 kGy).

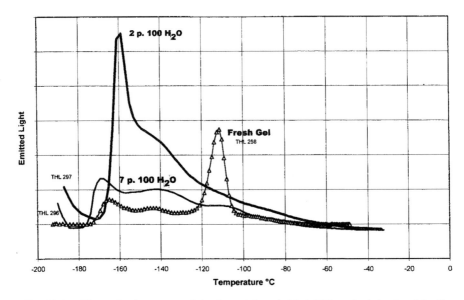

FIGURE 21 Thermoluminescence of fresh and freeze-dried SiO_2 gels (microbeads) after γ-irradiation at $-196°C$ (30 kGy).

freeze-dried gels, the $-120°C/-100°C$ emission is first almost completely erased and then disappears (for 2% residual moisture). At the same time the low-temperature peak increases and sharpens.

We think that low-temperature thermoluminescence applied to freeze-dried solids is susceptible to give us new information on the water "traps" within the solid material, and that with a better knowledge of the temperature, magnitude, and shape of the emission peaks we might be able to beam some light on the role of residual moisture in freeze-dried products and on the mechanisms binding water molecules to their supporting matrix.

UNCONVENTIONAL DEVELOPMENTS: NEW PROSPECTS
Water Is Not the Only Available Solvent

Until now, almost all lyophilization operations have been carried with aqueous solutions, but we know since a long time that water is not the only solvent that can sublimate from the frozen state.

Mineral solvents: We did show, decennia ago, that ammonia and carbon dioxide, which are most interesting solvents, might distill away from the frozen state. The case of carbon dioxide is even a rather remarkable one since it sublimates at high speed, at atmospheric pressure, and at a low temperature of $-78.8°C$. Moreover, it is a nontoxic chemical and leaves no analytically measurable traces in the final dry product. Knowing its exceptional solvent properties in the super-critic state it might become a very valuable extraction and supporting medium for many unstable chemicals and biologicals that need to be stabilized as dry forms.

NH_3 is more difficult to handle but it might offer equally interesting openings for the preparation of transient compounds manufactured in liquid ammonia and unable to be brought to room temperature unless they are freeze-dried from their frozen solution at $-120°C$.

Organic solvents: Many organic solvents are susceptible to be eliminated by sublimation from their frozen solution, for example, dioxane, diethyl amine, chloroform, cyclohexane, dimethylsulfoxide, and benzene. We showed that they could be used for the preparation of delicate biochemicals (like phospholipids). In that case, however, the main problem is the residual amount of solvent that can be found in the final dry product and this, of course, is a redundant issue for the whole pharmaceutical industry. For that reason the use of pure organic solvents in freeze-drying has not really been developed. However, their application as partial constituents in mixed aqueous solutions has given rise to most interesting new processes. This is the so-called cosolvent issue that is dealt with in details further in this book by Dirk Teagarden, who has been pioneering in that field for a long time. It is expected that this approach will find more and more new applications in the future.

Continuous Operations

For more than 65 years freeze-drying has been almost exclusively done within the pharmaceutical and biological industries as a batch-type operation. The starting liquid solutions are distributed in vials or filled in trays, frozen, dried, and then handled as one single batch. To the contrary, the food industry, which had to face large productions, has progressively shifted from batch to semi-continuous and even purely continuous operations. The results are a substantial increase in productivity and a more homogeneous product.

To my feeling, it is ample time for the pharmaceutical and biological industries to take up that challenge and enter the process analytical technology (PAT) constraints with better monitoring and control of their operations.

Indeed, any chemical engineer dreams of a continuous process because it is easy to control and gives manufactured products a standard equal quality. In food areas freeze-drying has not escaped this trend and, as early as in the 60s, semi-continuous to continuous equipments have been designed and built. Leybold was among the very first to do it, and Oetjen developed the continuous quality controlled (CQC) process into an industrial reality for milk products. The operation was, indeed, sequenced into several phases. The frozen products, most often under granular form, were loaded on trays placed on a special carrier, hanged to a monorail that traveled all along the freeze-drying tunnel between heating plates in successive steps through vacuum locks closed by sliding gates. The total cycle time from the entrance lock to the outlet was of the order of several hours.

An alternative to this system was introduced by Atlas who pioneered the so-called "Conrad" system in which the loading of the frozen goods was done tray by tray through a small side lock. Not only coffee and milk, but also vegetables, fish fillets, and meat have been successfully treated in this way.

Quite obviously this technology worked but it was still a semicontinuous process and food industry was eager to develop a fully automated continuous operation for one of its leading products on the international market, that is, instant coffee. I had the opportunity to live this development very closely, and I can tell you that it has been very difficult since most steps of the instant coffee processing had to be revisited.

Thus, it is not surprising that it is only in the mid-1970s that a fully continuous line was introduced. In one of the most advanced designs granulated frozen extract is fed continuously into the dryer through a rotating lock and deposited on a 20 to 30 m long vibrating tray on which it travels as a fluidized bed. It moves, indeed, on its own water vapor cushion all along the heated surface that provides the energy for sublimation. During that long transport, the granules are guided by vertical ribs into parallel channels, and great care is taken to prevent attrition from mechanical shocks between the granules themselves and with the surface and walls of the tray. Though limited, attrition nevertheless does occur and generates a lot of "dust" that has to be stopped by a long semicylindrical screen deployed all over the vibrating tray. In that way, the fines are not carried away with the vapor stream toward the condenser and the pumps.

Under those circumstances, it can be easily understood that a continuous freeze-drying plant is a highly sophisticated piece of equipment that is designed and put together by the food industries themselves, who keep the whole development as strictly confidential and assemble different components purchased from multiple unconnected manufacturers.

The efficiency of a vibrated tray freeze-dryer is enormous and the drying times drop by more than one order of magnitude. We are speaking in terms of minutes instead of hours, and throughput of tens of tons per day is no longer unrealistic. Moreover, the process is "intellectually clean" and the end product is constant in quality.

This is the reason why I have been, for a very long time, dreaming to introduce that technology to the processing of biologicals and pharmaceuticals. Indeed, despite the elaborate design proposed by equipment manufacturers and the care that the drug companies take of their freeze-drying operations, we are still facing the recurrent problem of potential heterogeneity between vials and ampoules within a same single batch. The spatial distribution of 10,000 to 100,000 vials in a multishelf freeze-drying cabinet remains a problem. Some sit close to the door, some right in the middle, others near the condenser, some in the upper shelves, others down, and, let us be serious, they do not dry in the same way. I know that drug manufacturers claim that all the products that they release are within very close given standards and I know, also, that the regulatory agencies keep a close eye on that issue and that they are more and more stringent on the validation tests. It remains, anyhow, that the homogeneity of a single batch is still a matter of deep concern. Can we solve that problem? I do believe so. A very simple idea is then to switch to a semicontinuous or continuous process as this is done in the food industry. Let us forget about the holy paradigm that a vial that has been initially loaded with a given volume of solution has to remain so until it is fully processed, capped with its inside freeze-dried cake, and placed into its commercial retail box. Let us try to consider a process (Fig. 22) in which the initial solution is distributed as individual droplets frozen into spherical granules of given size, continuously fed into a vacuum chamber, and spread on a heated conveyor. On that tray, most probably a vibrated tray, they glide self-suspended on their water vapor cushion as a thin, regular, and fluidized cloud at a well-controlled operating pressure fit for sublimation, say a hundred microbar. At the end of the tray they reach a transfer lock that discharges them on another conveyor placed in a second chamber fit for desorption and secondary drying at pressures of some tens of microbar. Finally, they enter an ultimate lock and are discharged into a receiving bin under dry neutral atmosphere. The efficiency of

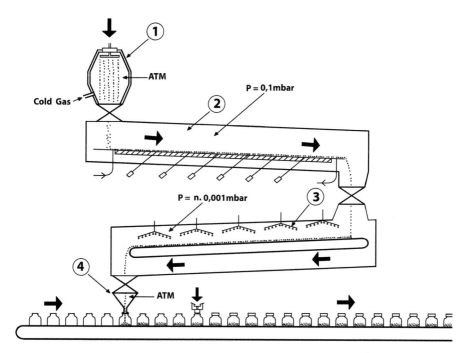

FIGURE 22 Engineering design of a pharmaceutical continuous freeze-drying plant. The system displays four sections.

-In section 1, the liquids are frozen in regular droplets, for instance, as a "rain" falling against a countercurrent of cold air at atmospheric pressure.

-In section 2, the frozen granules of a given standard size are spread throughout a vacuum lock on a heated vibrating tray where they perform their primary drying at a moderate vacuum near 0.1 mbar.

-In section 3, subsequent to a new transfer through a vacuum lock, the already "ice-free" granules are deposited on a conveyor under a controlled infrared or microwave heating at a rather low pressure in the 0.01 to 0.001 mbar range. There they accomplish the secondary drying.

-In section 4, a final lock allows the freeze-dried material to be brought back to atmospheric pressure by injection of a dry, neutral gas such as nitrogen or argon and they are distributed "by the number" into previously sterilized vials that are finally capped.

such a process is tremendous, and if we have granules of 1 to 3 mm size, the whole freeze-drying operation might take less than 30 minutes.

Now, at the end of the road, we have a population of granules, all dried in the same conditions, all equal in quality and size that can be numbered, and fed into dry sterile vials as so many distribution machines can do. The result is a batch of identical vials containing each, say 100 ± 1 freeze-dried granules instead of more or less regular cakes, painfully manufactured over several days. Moreover, the product "elegance" is maximized and, in that dispersed form, its reconstitution is a matter of seconds.

This is not wishful thinking since we do have the basic knowledge, the practical know-how and a multiyear experience of this type of operation. It is just a change of mind, a new approach that deviates from historical practice. Unfortunately, we know that this is sometimes more difficult to pass.

Some people will claim that instant coffee manufacturers do not care about sterility, that individual freezing of small droplets is a difficult undertaking, and that, maybe, mechanically, the granules will not prove resistant to the process. I know quite well that nothing yet has been completely solved but I equally know that we do have industrial solutions to those specific issues (frozen beads generators, isolators, automatic remote handling in sterile environment, etc.) and I do not have the slightest doubt that they can be successfully challenged if, at the same time, some developments are made on the formulation side. This new approach would equally fit better with the new PAT requirements.

More than a hundred years after its discovery, lyophilization (freeze-drying or cryodesiccation) remains a challenging issue that requires an educated mix of classical, basic multidisciplinary disciplines, and new rocketing technologies. This book is the actual demonstration that it is not a *freeze-dried knowledge* but a continuously evolving vivid experience.

FURTHER READINGS

Bordas F, d'Arsonval M. C R Acad Sci Paris 1906; 142:1058 and 1079; 143:567.

Flosdorf EW, Mudd S. Procedure and apparatus for preservation in "lyophile" form of serum and other biological substances. J Immunol 1935; 29:389. (See also Flosdorf EW. Freezing and Drying. New York: Reinhold, 1949:1–280.)

Hauduroy P. Histoire de la technique de lyophilisation. In: Rey LR, ed. Traite de Lyophilisation. Paris: Hermann, 1960:3–16.

Rey LR. Thermal analysis of eutectics in freezing solutions. Ann N Y Acad Sci 1960; 85:513–534.

Rey LR, ed. Fundamental aspects of lyophilization. In: Research and Development in Freeze-Drying. Paris: Hermann, 1964:23–43.

Rey LR. Automatic regulation of the freeze-drying of complex systems. Biodynamica 1961; 8:241–260.

Rey LR. Glimpses into the fundamental aspects of freeze-drying. In: Cabasso VJ, Regamy RD, eds. Freeze-Drying of Biological Products, Biological Standardization Series,. Vol. 36. Basel: S. Karger, 1977:19–27.

Grossweiner LI, Matheson MS. Fluorescence and thermoluminescence of ice. J Chem Phys 1954; 22:1514–1526.

Rey LR. Low temperature thermoluminescence. Nature 1998; 391:418.

Rey LR. Thermoluminescence of ultra-high dilutions of lithium chloride and sodium chloride. Physica A 2003; 323:67–74.

Rey LR. Thermoluminescence of deuterated amorphous and crystalline ices. Radiation Phys Chem 2005; 72:587–594.

Neumann K. Les Problemes de mesure et de reglage en lyophilisation. In: Rey LR, ed. Traite de Lyophiiisation. Paris: Hermann, 1960:1–411.

Oetjen GW, Ehlers W, Hackenberg U, et al. Temperature measurements and control of freeze-drying process. In: Freeze-drying of Foods. Washington, D.C.: National Academy of Sciences and National Research Council, 1962:25–42.

Rieutord L. International patents, 1961.

Espagnan M, Plume P, Marcellin G. Dosimetrie des doses elevees par Thermoluminescence des pics profonds du Fluorure de Lithium: Fli: Mg, Ti. Radioprotection 1991; 26:51–64.

2 Structural and Dynamic Properties of Bulk and Confined Water

Marie-Claire Bellissent-Funel and Jose Teixeira
Laboratoire Léon-Brillouin (CEA-CNRS), CEA Saclay, Gif-sur-Yvette, France

INTRODUCTION

The structural and dynamic properties of bulk water are now mostly well understood in some ranges of temperatures and pressures. In particular, in many investigations using different techniques, such as X-ray diffraction, neutron scattering, nuclear magnetic resonance (NMR), differential scanning calorimetry (DSC), molecular dynamics (MD), and Monte Carlo (MC), simulations have been performed in the deeply supercooled regime (1–10), and in a situation where the effects due to the hydrogen bonding are dominant.

However, in many technologically important situations, water is not in its bulk form, but instead attached to some substrates or filling small cavities. Common examples are water in porous media, such as rock or sand stones, and water in biological material as in the interior of cells or attached to surfaces of biological macromolecules and membranes. This is what we define here as the "confined" or the "interfacial water."

Water in confined space has attracted a considerable interest in the recent years. It is commonly believed that the structure and dynamics of water are modified by the presence of solid surfaces, both by a change of hydrogen bonding and by modification of the molecular motion, which depends on the distance of water molecules from the surface.

Understanding of the modification from bulk liquid water behavior when water is introduced into pores of porous media or confined in the vicinity of metallic surfaces is important to technological problems, such as oil recovery from natural reservoirs, mining, heterogeneous catalysis, corrosion inhibition and numerous other electrochemical processes. Water in porous materials, such as Vycor glass, silica gel, and zeolites, has been actively under investigation because of its relevance in catalytic and separation processes. In particular, the structure of water near layer-like clay minerals (11–12), condensed on hydroxylated oxide surface (13), confined in various types of porous silica (14–22) or in carbon powder (23) has been studied by neutron and/or X-ray diffraction.

In the field of biology, the effects of hydration on equilibrium protein structure and dynamics are fundamental to the relationship between structure and biological function (24–30). In particular, the assessment of perturbation of liquid water structure and dynamics by hydrophilic and hydrophobic molecular surfaces is fundamental to the quantitative understanding of the stability and enzymatic activity of globular proteins and functions of membranes. Examples of structures that impose spatial restriction on water molecules include polymer gels, micelles, vesicles and microemulsions. In the last three cases, since the hydrophobic effect is the primary cause for the self-organization of these structures, obviously the configuration of water molecules near the hydrophilic-hydrophobic interfaces is of considerable relevance.

The microscopic structure of bulk and confined water is currently studied by using X-ray or/and neutron diffraction techniques which are complementary techniques. These diffraction techniques allow to access to the intermolecular pair correlation function $g(r)$ (31) of a system which is the probability density of finding another atom lying in another molecule at a distance r from any atom. In X-ray measurements, $g(r)$ is the pair correlation function of the molecular centers, to a good approximation equal to the oxygen-oxygen correlation function. In neutron measurements, $g(r)$ is the weighed sum of the three partial functions relative, respectively, to the oxygen-oxygen pairs, oxygen-deuterium pairs and deuterium-deuterium pairs. In particular, it is heavily dominated by deuterium-deuterium and oxygen-deuterium partial correlation functions.

The structural and dynamic properties of water may be affected by both purely geometrical confinement and/or interaction forces at the interface. Therefore, a detailed description of these properties must take into account, the nature of the substrate and its affinity to form bonds with water molecules, and the hydration level or number of water layers. To discriminate between these effects, reliable model systems exhibiting hydrophilic or hydrophobic interactions with water are required. This looks the appropriate strategy to be developed to access to some understanding of the behavior of water close to a biological macromolecule, as presented in the following sections.

In the past few years, computer simulations and theoretical treatments of the structure and dynamics of water in different kinds of environments have been undertaken (32–40). Some important results are now available. For instance, molecular dynamics simulations indicate that the water density increases up to 1.5 g/cm^3 in the first few angstroms of the shell around a protein and give information concerning the pair correlation functions and orientations of water molecules (41). Instead, it has been shown experimentally that a thin layer of water vapor is formed between liquid water and a hydrophobic surface (42).

The purpose of this chapter is to account for the more recent developments about the structure and the dynamics of bulk and confined water as a function of temperature. Examples relative to interactions of water molecules with model systems as well as with biological macromolecules will be presented.

THE STRUCTURE AND DYNAMICS OF LIQUID WATER: A SHORT REVIEW

In spite of an enormous amount of experiments performed with liquid water under different external conditions (1–9), many of its properties remain not fully understood. The main reason is the complexity of the intermolecular potential resulting from the formation of intermolecular hydrogen bonds. Such bonds are strongly directional and their study imposes the consideration of quantum effects.

Most of the recent theoretical developments have been achieved by computer simulations of the molecular dynamics using several sophisticated, effective potentials (3,10). Such potentials are written ad hoc to simulate at the best both the microscopic structure and the thermodynamic and transport properties. A general problem is that, either the potential imposes a too much strong and ice-like structure to reproduce the thermodynamic properties, or it reproduces well the pair correlation function $g(r)$ and then, the so-called water anomalies are not well reproduced. Many progresses have been achieved recently and one can classify the different effective potentials in several categories: rigid molecule, flexible molecule (including internal degrees of motion), with enhanced dipole, ab initio, etc.

Heuristic approaches remain, probably for a long time, extremely useful; we will show here that a simple model that focuses on the statistics of the hydrogen bonds is qualitatively sufficient for explaining some of the water anomalous properties.

Every study of liquid water at a molecular level must take into account three features that, together, characterize the microscopic behavior and the thermodynamic properties. These are the tetrahedral symmetry of the molecule, the large number of hydrogen bonds formed between near neighbor molecules and the very short characteristic lifetime of these bonds.

The first feature is a result of the molecular orbital hybridization, which yields an H-O-H angle very close to the ideal tetrahedral angle (109°). As a consequence, in all circumstances, one observes that the coordination number (number of nearest neighbors) is around 4. This is a small number, meaning that there is a large amount of "free space" available for movements, such as the O-O-O bending. This is apparent in the pair correlation function $d(r)$ as defined by:

$$d(r) = 4\pi\rho r[g(r) - 1)] \tag{1}$$

where ρ is the molecular number density.

The function $d(r)$ is obtained by Fourier transform of the scattering function $S(Q)$ measured by X rays or neutron scattering. In the case of X rays, $d(r)$ gives the oxygen positions because the contribution of hydrogen atoms to the scattered intensity is negligible. It shows a well-defined first peak at 2.8 Å, corresponding to the first shell of neighbors, a broad second neighbor peak at 4.5 Å and dies rapidly beyond the third neighbor distance (Fig. 1A) (43). This is in contrast with simple atomic liquids, such as argon, for which a more extended

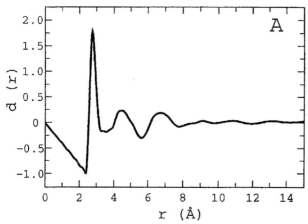

FIGURE 1 **(A)** The function, $d(r)$, related to the pair correlation function, $g(r)$, by $d(r) = 4\pi r\rho[g(r) - 1]$, evaluated from the $S(Q)$ measured by X rays (23,43). **(B)** The pair correlation function, $g(r)$, obtained from neutron-scattering data (44) is shown with the three different partial correlation functions $g_{OO}(r)$, $g_{OD}(r)$ and $g_{DD}(r)$. **(C–E)** Solid line: results of Bellissent-Funel (6); dashed line: results of Soper and Philips (45).

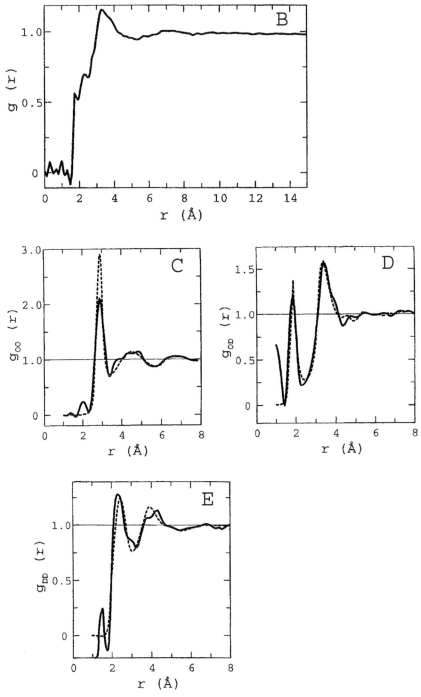

FIGURE 1 (Continued).

range order is clearly observed. Thus, with X rays, the scattering is dominated by $g_{OO}(r)$, while with neutrons all the three partials [$g_{OO}(r)$, $g_{OD}(r)$, and $g_{DD}(r)$] contribute to the total scattered intensity. With neutron scattering combined with H/D isotopic substitution, we can actually measure all three $g(r)$s in water with reasonable precision. Figure 1B depicts the total pair correlation function, $g(r)$ (44), obtained by neutron scattering and shows the three partial correlation functions, $g_{OO}(r)$, $g_{OD}(r)$, and $g_{DD}(r)$ (Fig. 1C–E) (6,45).

The O-D distribution has a well-defined peak at 1.85 Å consisting of about 1.8 hydrogen atoms out to a distance of 2.35 Å from an oxygen atom at the origin. The O-O distribution, in agreement with X-ray diffraction, has a well-defined peak at 2.8 Å consisting of about 4.5 oxygen atoms out to a distance of 2.8 Å from an oxygen atom at the origin. These numbers and distances suggest that the near neighbor coordination of water molecules is well defined and roughly tetrahedral at any instant in time, but that a substantial number of molecules are to be found in other configurations.

The second property is the existence of intermolecular bonds. Spectroscopic studies traditionally classify the bonds in two categories, intact and broken bonds, essentially because of the shape of the intramolecular stretching band of the Raman spectrum of liquid water and its temperature dependence. Such an interpretation is questionable but, whatever is the definition used for the classification of the bonds, a large number of molecules have a strong attractive interaction with their neighbors. In a simple way, one can say that the ensemble of the intact (or with energy beyond some reference energy) bonds constitutes a network well above its percolation threshold. As a consequence, at a given time, essentially all the molecules are part of an "instantaneous gel" (9). It is worth noting that the two preceding properties of water structure explain that, in spite of a very short range local order, the connectivity length of the hydrogen bond network is infinite.

However, and this is the third aspect, the characteristic lifetime of a hydrogen bond is very short: between 10^{-13} and 10^{-12} seconds and this is why viscoelastic properties of a gel structure will never be observed even in short characteristic time experiments. The explanation of such a short time is that hydrogen bond lifetimes are determined by the proton dynamics (3,8). In particular, large amplitude librational movements take the proton from the region easily, between two oxygen atoms, where the energy of the bond is sufficiently large.

Many of the thermodynamic and transport properties of liquid water can be qualitatively understood if one focuses attention on the statistical properties of the hydrogen bond network (9). As an example, let us observe the temperature dependence of density and entropy. As temperature decreases, the number of intact bonds increases and the coordination number is more close to the ideal value 4. Because of the large free volume available, this means that the temperature decrease is associated with an increase of the local molecular volume. Of course, this effect superimposes to the classical anharmonic effects, which dominate at high temperature, when the number of intact bonds is smaller. The consequence of both effects is a maximum on the temperature dependence of the liquid density. This maximum is actually at 4°C for normal water and at 11°C for heavy water. Such a large isotopic effect can also be understood because the larger mass of the deuterium makes the hydrogen bonds more stable.

Entropy decreases with decreasing temperature due to an increase of the local order following the hydrogen bond formation. Similar arguments to those developed above explain the minimum of the temperature dependence of heat capacity.

A more complete discussion of these effects has been done and gives a good explanation of the enhanced anomalies of water observed at low temperature (1,2). More important to notice is a minimum of the isothermal compressibility at 46°C and a sharp increase of the heat capacity at temperatures close and below the melting point, when liquid water is undercooled. The first corresponds microscopically to enhanced density fluctuations corresponding to the existence of short-lifetime low-density regions in liquid water (3). These regions are formed by molecules strongly bound together. Clearly, the percolation mechanism yields a dramatic increase of the size and number of these regions, explaining, incidentally, why homogeneous nucleation takes place at about the same temperature (about 228 K), as the thermodynamic properties seem to diverge. This temperature is also well above the glass transition temperature of liquid water, which is 136 K. The second corresponds to a rapid decrease of the entropy, which approaches the entropy of the crystal, also around 228 K.

The transport properties of liquid water also have a strongly anomalous behavior, in particular at low temperature (1,2). Properties, such as self-diffusion, viscosity, and different relaxation times, show a strong non-Arrhenius temperature dependence, the characteristic activation energy increasing with decreasing temperature. This also corresponds to the fact that, with decreasing temperature, more bonds must be simultaneously broken to allow the movement of a given molecule. At high temperature, when the number of intact bonds is relatively small, the dynamics is similar to that of classical liquid. As the temperature decreases, the activation energy increases. At very low temperatures it is three times larger than at room temperature. Its value corresponds to the energy necessary to break a larger number of bonds at low temperature.

When water is mixed with another liquid, the number of bonds can increase, for instance, in water-ethanol solutions (structure formers) or decrease (structure breakers). However, in all cases, the tetrahedral structure vanishes (except, of course, for isotopic mixtures of light and heavy water). As a consequence, the anomalies of liquid water are strongly reduced upon addition of other components. For instance, 7% of ethanol is sufficient to completely suppress the maximum in the temperature dependence of the density of the mixture (9).

A more detailed study of the properties of water solutions is clearly of major importance in many applications. One can say that it is impossible to speak about a general behavior. Different compounds solubilized in water correspond a priori to different local structures. A very extensive study of salt solutions has been done by Enderby and coworkers (46). Careful neutron-scattering experiments allow the determination of the water shell around several ions. Around a cation, the water molecule is oriented with the oxygen more close to the cation corresponding to a minimization of the energy between the ion and the water dipole. Instead, around an anion, the hydrogen atoms are closer to the ion. The coordination numbers, that is, the number of water molecules that are present inside the first hydration shell and the lifetimes depend very much on the nature of the solute and on its concentration. Around hydrophobic molecules, such as methane, water forms large cages, (clathrates) probably with a geometry close to that of polyhedra, such as icosahedra. Finally, in the presence of a solid substratum, water may form bonds. This is the case, for instance, of glasses where silanol groups, Si-O-H, are present at the interface water silica. It is worth noting that in presence of biological macromolecules, such as peptides, enzymes, proteins, DNA, all these behaviors can be found depending on the nature more or less hydrophilic or hydrophobic of each site or residue. Bonds, in particular, play certainly a major role in the structure of these macromolecules.

STRUCTURE OF CONFINED WATER
Model Systems-Water Interactions
The choice of porous media as model systems depends on two conditions: a well-characterized pore size distribution and surface details. Among the hydrophilic model systems where the structure of confined water has been studied by neutron diffraction, let us mention clay minerals (11–12), and various types of porous silica (14–22). In the latter case, the authors have interpreted their results in terms of a thin layer of surface water with more extensive hydrogen bonding, lower density and mobility, and lower nucleation temperature as compared with bulk water. Recently, the structure of water confined in the cylindrical pores of MCM-41 zeolites with two different pore sizes (21 and 28 Å) has been studied by X-ray diffraction (21) over a temperature range of 223 to 298 K. For the capillary-condensed samples, there is a tendency to form a more tetrahedral-like hydrogen-bonded water structure at subzero temperatures in both the pore sizes.

The more extensive results concern the structure of water confined in a Vycor glass (47), which is a porous silica glass, characterized by a quite sharp distribution of cylindrical interconnected pores and hydrophilic surfaces. Results have been obtained as functions of level of hydration from full hydration (0.25 g water/g dry Vycor) down to 25% hydration and temperature (48). On the basis of the information that the dry density of Vycor is 1.45 g/cm^3, the porosity 28%, and the internal cylindrical pores of cross-sectional diameter 50 Å, the 50% hydrated sample has three layers of water molecules on its internal surface. A 25% hydrated sample corresponds roughly to a monolayer coverage of water molecules.

Results for two levels of hydration of Vycor demonstrate that the fully hydrated case is almost identical to the bulk water and the partially hydrated case is of little difference (Fig. 2). However, the three site-site radial correlation functions are indeed required for a sensible study of the orientational correlations between neighboring molecules and the results of three neutron diffraction experiments on three different isotopic mixtures of light and heavy water have been reported (49).

It is interesting, however, to comment on the level of supercooling possible for heavy water in Vycor. According to Bellissent-Funel et al. (48), for partially hydrated samples, the deepest supercooling is $-27°C$, while for the fully hydrated sample it is $-18°C$. As temperature goes below the limit of supercooling, part of the confined water nucleates into cubic ice. The proportion of cubic ice increases with decreasing temperature. This is in sharp contrast to bulk water, which always nucleates into hexagonal ice.

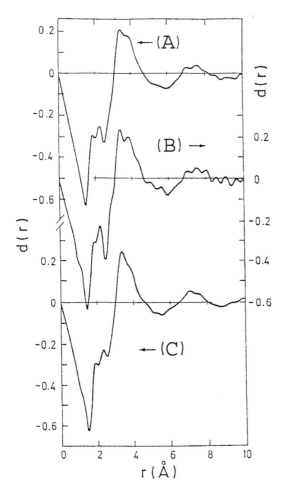

FIGURE 2 Pair correlation function $d(r)$ for (**A**) confined D_2O from fully hydrated Vycor (27°C), (**B**) confined D_2O from partially hydrated Vycor (35°C) as compared with (**C**) bulk water (27°C). *Source*: From Ref. 48.

　　　　Results relative to a 25% hydrated Vycor sample indicate that, at room temperature, interfacial water has a structure similar to that of bulk supercooled water at a temperature of about 0°C that corresponds to a shift of about 30 K (50). The structure of interfacial water is characterized by an increase of the long-range correlations, which corresponds to the building of the H-bond network as it appears in low-density amorphous ice (51). There is no evidence of ice formation when the sample is cooled from room temperature down to −196°C (liquid nitrogen temperature).

　　　　Among hydrophobic model systems, one experimental investigation of particular interest concerns the structure of water contained in a carbon powder (23). The structure of water has been determined both by X-ray and neutron diffraction, as functions of hydration, from room temperature down to 77 K. In agreement with previous work (14–18,48,52,53), this study gave support to the existence of a region near the interface where the properties of water are markedly different from those of the bulk liquid. From X-ray measurements, which yield information about the oxygen-oxygen distribution function, it

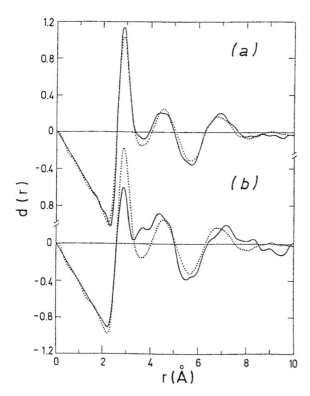

FIGURE 3 X-ray pair correlation function $d(r)$ of water contained in activated carbon, at room temperature, shown by solid lines for 188% (curve a) and 42% hydration (curve b). For comparison, the $d(r)$ of bulk water at the same temperature is also drawn (*dotted line*). *Source*: From Ref. 43.

appears that, at the lowest investigated water content, 42% hydration level, a distortion of the tetrahedral ordering is clearly observed (Fig. 3). Neutron-scattering experiments can be analyzed to describe the intermolecular correlations (Fig. 4). At the same lower level of hydration, the hydrogen bonding is modified and water molecules are more ordered. It is not possible to determine the thickness of the affected layer. However, a crude determination from the specific area indicates that for a hydration equal to 50%, the thickness does not exceed 5 Å. This value must be compared with the computer simulation data (32–37), which indicate that structural modifications do not extend beyond 10 Å from the solid surface. When partially hydrated samples are cooled down to 77 K, no crystallization peak is detected by differential thermal analysis. Both X-ray and neutron diffraction show that an amorphous form is obtained and its structure is different from those of low- and high-density amorphous ice already known (5). This phenomenon looks similar in both hydrophilic and hydrophobic model systems. However, to characterize more precisely the nature of the amorphous phase, the site-site partial correlation functions need to be experimentally obtained and compared with those deduced from molecular dynamics simulations.

Macromolecules-Water Interactions

The structure of water near polymeric membranes (52) has been studied by neutron diffraction. The structure of water confined in a hydrogel has been

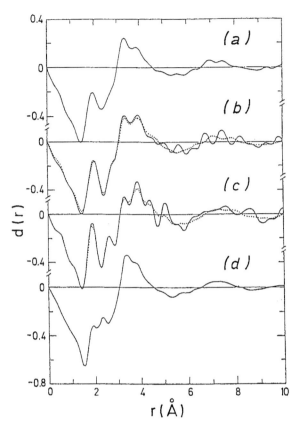

FIGURE 4 Neutron pair correlation function $d(r)$ of water contained in activated carbon, at room temperature, shown by solid lines for 200% (curve a), 42% (curve b) and 25% hydration (curve c). For comparison, $d(r)$ of bulk water at the same temperature is also drawn (curve d) (43). The dotted lines show the result of a smoothing of the experimental data before Fourier transformation. This demonstrates that the additive oscillations, which appear between 3 and 6 Å have a physical meaning in contrast with those seen at higher r values.

investigated by X-ray scattering (54); the distortion seen at the level of the second nearest neighbors has been attributed to the bending of hydrogen bonds.

The amount of information about protein-water correlations is small. For neutron diffraction, deuterated samples are required and difficult to obtain. However, the first results have been obtained in the case of a photosynthetic C-phycocyanin protein for which the X-ray crystallographic structure is known to a resolution of 1.66 Å (55).

C-phycocyanin is abundant in blue-green algae. Nearly 99% deuterated samples of this phycobiliprotein were isolated from the cyanobacteria, which were grown in perdeuterated cultures (56) (99% pure D_2O) at Argonne National Laboratory. This process yielded deuterated C-phycocyanin proteins (d-CPC) that had virtually all the ^1H-C bonds replaced by ^2H-C bonds. One can obtain a lyophilized sample that is similar to amorphous solids as determined by neutron diffraction (53). As it has been defined in previous works (57–59), the level of hydration $h = 0.5$ corresponds to 100% hydration of C-phycocyanin which leads to a coverage of about 1.5 monolayers of water molecules on the surface of the protein (60).

The water (D_2O)-protein correlations at the surface of a fully deuterated amorphous protein C-phycocyanin have been studied by neutron diffraction as a function of both temperature and hydration level (53). The correlation distance of 3.5 Å measured in these diffraction experiments compares well with computer

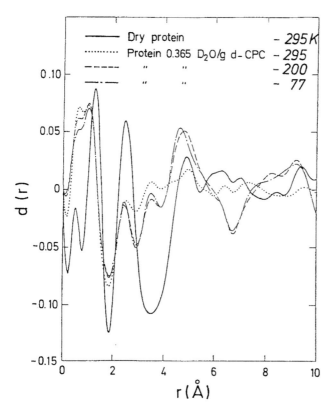

FIGURE 5   Pair correlation function $d(r)$ for a dry d-CPC protein at 295 K and for a D_2O-hydrated ($h = 0.365$) d-CPC protein at different temperatures. *Source*: From Ref. 53.

simulations work on polypeptides and proteins (34,61) and was interpreted as resulting from some increase in the clustering of water molecules (Fig. 5). For the highest hydrated sample ($h = 0.365$), a well-defined peak appears at 3.5 Å. This is the average distance between the center of mass of a water molecule in the first hydration layer and amino-acid residues on the surface of the protein. In the case of the lowest hydrated sample ($h = 0.175$), the perturbation of the structure of protein due to water of hydration is not detectable. It is generally viewed in the literature that at full hydration ($h = 0.5$), there is a complete monolayer of water surrounding the protein (62).

Some similarity between the behavior of water close to C-phycocyanin protein and hydrophilic model systems can be stressed. In fact, for low-hydrated protein samples, no crystallization of water is detectable while, for more than one monolayer coverage, hexagonal ice is formed. Moreover, the peak at 3.5 Å is also detected. However, it should be noted that at the highest hydration level, water nucleates into hexagonal ice at low temperature; this is in contrast with hydrated Vycor where water nucleates into cubic ice.

Some other interesting findings come from the study of glass-liquid transition and crystallization behavior of water trapped in loops of meth-emoglobin chains (63).

DYNAMICS OF CONFINED WATER

Traditionally, the dynamics of interfacial water has been studied by nuclear magnetic relaxation techniques. Halle and coworkers (64–66) have used water ^{2}H and ^{17}O spin relaxation to study water dynamics in the hydration layers of small peptides, globular proteins and in living cells of two microorganisms which is a particularly suitable method for investigating the single-particle dynamics of interfacial water and thus the protein-water interaction.

They have characterized the dynamical heterogeneity of hydration water by performing relaxation measurements over a wide temperature range, extending deeply into the supercooled regime, or by covering a wide frequency range. Protein hydration layers can be described by a power-law distribution of rotational correlation times with an exponent close to 2. This distribution comprises a small fraction of protein-specific hydration sites, where water rotation is strongly retarded, and a dominant fraction of generic hydration sites, where water rotation is as fast as in the hydration shells of small peptides. The generic dynamic perturbation factor is less than 2 at room temperature and exhibits a maximum near 260 K. Water in living cells behaves as expected from studies of simpler model systems, the only difference being a larger fraction of secluded (strongly perturbed) hydration sites associated with the supramolecular organization in the cell. Intracellular water that is not in direct contact with biopolymers has essentially the same dynamics as bulk water. There is no significant difference in cell water dynamics between mesophilic and halophilic organisms, despite the high K^+ and Na^+ concentrations in the latter. This finding is different from a recent report on anomalously slow water diffusion in *Haloarcula marismortui* cells (67).

The free hydration layer studied here differs qualitatively from confined water in solid protein powder samples. With regard to the solvent diffusion constant near protein and silica surfaces, there are reports from other groups showing that it is reduced by a factor of about 5 compared with that of bulk water (68).

An ideally microscopically detailed method for exploring the change in hydrogen bonding patterns as well as the translational and rotational diffusion constants and residence times of water molecules, when they are near surfaces is computer molecular dynamics (CMD). For example, Rossky and coworkers (32,33,35–37) have investigated changes of the structure, hydrogen bonding and dynamics of water molecules when they are adjacent to an atomically detailed hydrophobic surface and to a hydroxylated silica surface. Results of CMD simulations (32,33,35–37) generally indicate that the dynamics of water molecules on protein and silica surfaces where hydrophilic interactions are dominant suffer only a mild slowing down compared with bulk water. More specifically, it has been reported that the retardation is by a factor of ~2 in the protein case and about a factor of 5 in the silica case. Residence times of water in the first hydration layer are typically of the order of 100 ps.

Linse (69) made a similar simulation for water near a charged surface with mobile counterions constituting an electric double layer such as in the interior of a reverse micelle formed by ionic surfactants in oil. He reported that water in the aqueous core of reverse micelles has a reduced rate of translational and rotational motions by a factor of 2 to 4.

These CMD results are still qualitative and somewhat conflicting with the available experimental data (65), largely because of the simplified models used for the surfaces and more certainly because of difficulties in choosing suitable potential functions for the simulations. However, molecular dynamic simulations of the hen egg white lysozyme–Fab D1.3 complex have been reported; both the crystal state and the complex in solution were studied (42). The findings are consistent with the observation by various experimentalists of reduced water mobility in a region extending several angstroms beyond the first hydration layer (64–67), as reported also from CMD simulations (70).

From the above comparison, it seems clear that there are considerable discrepancies in the degrees of slowing down between NMR experiments and CMD. This is especially true for the translational diffusion constant.

A way to resolve these discrepancies has been recently attempted by quasi-elastic and inelastic neutron scattering. Neutron scattering is a powerful and unique tool for studying the self-dynamics of interfacial water; actually the large incoherent scattering cross-section of the protons yields unambiguous results about the individual motions of water molecules (31). In fact, this technique is a method for studying the diffusive motion of atoms in solids and liquids (71). It gives access to the correlation function for the atomic motions which are explored over a space domain of a few angstroms and for times of the order of 10^{-12} seconds. This space and time domain makes the comparison between neutron scattering and CMD better justified. The correlation function can be calculated by various models of the motions of the particles (e.g., Brownian motion, jump diffusion, diffusion in a confined space, rotational motion, etc.). Microscopic properties of the molecules, such as the residence time, the jump length, the self-diffusion coefficients, the hydrogen bond lifetime can be evaluated as well.

Another quantity that can be obtained is the vibrational density of states of mobile protons, in particular the translational and librational modes for water of hydration as compared with that of bulk water.

This method has been used with success for studying the self-dynamics of bulk water as a function of the temperature (72), as previously reported.

Previous studies about dynamics of water near interfaces by quasi-elastic neutron scattering involved the mobility of water on the surface of Nafion membranes (73,74), the diffusive motions and the density of states of water in silica gels (75) and the interfacial melting of ice in graphite and talc powders (76). It is interesting to note that quasi-elastic scattering like effects for extremely small wave vectors can be observed in the pulsed gradient spin-echo NMR experiments. The latter technique has been used for studying diffusion of water in both permeable (77) and connected structures where the effects of confinement can be clearly identified (78).

Model Surface Water Systems

A complete study of the self-dynamics of water close to some well-defined hydrophilic surface as the Vycor surface has been performed by quasielastic and inelastic neutron scattering. It has been done for levels of hydration ranging from full hydration down to the lowest one (25%), corresponding to one monolayer coverage of water molecules. The effect of temperature has also been studied. We report here the main results (79).

The short-time diffusion (few picoseconds) of water molecules close to the Vycor surface has been described in terms of simple models for all the studied samples (80). At short times, the water molecules, close to some hydrophilic surface, perform very local rotational jumps characterized by D_t and τ_1 like in bulk water, but with a longer residence time τ_0 on a given site before diffusing to an adjacent site along the surface with a diffusion coefficient equal to D_{local}. This diffusion is limited to some volume estimated as spherical. For the 25% hydrated sample, the diffusion coefficient measured by NMR appears to be smaller than D_t, which is smaller than D_{local} (81). This is due to the fact that NMR technique measures the long-time and long-range diffusion coefficient.

The effect of the temperature has been followed down to −35°C. The radius of the spherical volume of confinement varies between 5 and 2 Å; it decreases when the temperature is lowered meaning that water molecules are more localized at low temperatures. The observed trend seems reasonable.

The values obtained for D_{local} are low which demonstrates the influence of the hydrophilic groups when one reaches a monolayer coverage of water molecules. Moreover, these values are close to the values of the diffusion coefficient of water molecules at the immediate hydrophilic interface, as determined in a molecular dynamics simulation by Lee and Rossky (35).

Figure 6 shows the vibrational density of states for confined water (52% hydrated Vycor) as compared with that of bulk water at room temperature (3). The density of states of confined water exhibits striking features. The peak associated with the O-O-O intermolecular rotational motions, centered on 6 meV, is much attenuated indicating the reduction of this degree of freedom upon confinement. There is an up-shift of the librational peak at 70 meV, meaning some hindrance of the librational motions because of the presence of the surface. The hindrance of the motions increases when the temperature is lowered.

In hydrophobic environments, such as activated carbon powder, the vibrational density of states for confined water has been determined by inelastic neutron scattering as a function of temperature and compared with bulk water. For the lowest level of hydration, the translational peak around 6 meV and the

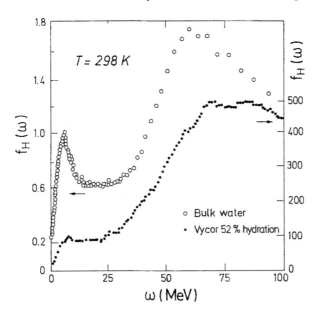

FIGURE 6 Proton vibrational density of states f_H for water contained in 52% hydrated Vycor at 298 K (*solid circles*). For comparison, the corresponding quantity for bulk water (*empty circles*) is also given. *Source*: From Ref. 59.

vibrational peak at 70 meV are less affected than in the Vycor case. However, an up-shift of the librational peak at 70 meV, characteristic of hindered motions, is observed (50).

Biopolymer-Water Systems

The volume of neutron data on biopolymer-water systems has increased during the past years (60,82,83). The situation is more complex because of contributions such as the hydrogen atoms of the protein itself, the possibility of their exchange with water molecules, and the presence of hydrophilic and hydrophobic regions. The studies of single-particle dynamics of hydration water in proteins have been hampered by the fact that about 40% of the constituent atoms in a typical protein molecule are hydrogen atoms, present in the backbone and side chains. The elastic contribution is thus too large for an accurate determination of the dynamical parameters which are characteristic of hydration process. However, by working with a deuterated protein/H_2O system, it has been possible recently to focus on the water dynamics at and near the protein surface (57–59).

At the vicinity of biomolecules, water may adopt very different behaviors depending on the nature of the sites, available free volume, temperature, etc. Because of the difficulty of experiments that can identify local properties among a variety of possibilities, the number of unambiguous results remains scarce. It is plausible that the development of computer simulations of molecular dynamics soon will take into account local environments more precisely (39–41).

Quasi-elastic neutron scattering has been used to describe the motions of water molecules in the vicinity of proteins and other macromolecules. One of the more successful results has been obtained with C-phycocyanin, a protein extracted from blue-green algae, which can be obtained nearly fully deuterated from perdeuterated cultures as mentioned previously (56). In this way, and because of the very large incoherent cross-section of hydrogen atoms, only the motions of water molecules are observed in a quasi-elastic neutron-scattering experiment (57–59). The results presented in Figure 7 show clearly that the total scattered intensity contains two components. The width of the narrow component is imposed by the instrumental resolution. Its area is proportional to the number of water molecules with motions that are too slow to be observed by the technique, that is, typically with a characteristic time longer than several tens of picoseconds. Instead, the wider component depicts the diffusive motions of the other water molecules through a Lorentzian line $L(\omega)$. Its intensity and width can be analyzed as functions of the degree of hydration and of temperature.

The simplest expression that can be written separates the two components in the following way:

$$S_{inc}(Q,\omega) = [P + (1-P)A_0]\delta(\omega) + (1-P)(1-A_0)L(\omega) \qquad (2)$$

where $P = p + q(1 - p)$ contains both the contributions of the p nonlabile protons of the protein and the fraction q of water molecules with a slow dynamics. $A_0(Qa)$ is a mathematical factor, called *elastic incoherent scattering factor*, which takes into account the confinement of the motions within a small volume of size a, assumed spherical for simplicity (80).

Typical results are shown in Figure 8. Figure 8A shows that the number q of "immobile" water molecules increases gradually from 40%, at high temperature, to 100% at around 200 K. This result shows that the number of confined

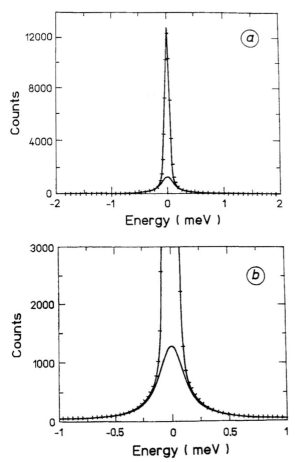

FIGURE 7 A typical quasielastic spectrum for the fully hydrated C-phycocyanin ($h = 0.5$ g water/g protein) at a scattering angle $\theta = 65.4$ K and for $T = 293$ K. Symbols "+" are experimental points and the full line is the fit using equation (2). *Source*: From Ref. 58.

water molecules is small at room temperature but, in contrast, that only at very low temperature (typically at $-50°$C) are their motions totally frozen. The curve shown in Figure 8B depicts the line width of the Lorentzian $L(\omega)$ versus the square of the momentum transfer Q. It shows that, at low values of Q, that is, when one investigates the system at large scales, the molecules appear confined. This is evident from the plateau seen below $Q = 1$ Å^{-1}. The resulting confinement yields a volume of confinement of molecular size ($a \approx 3$ Å). This is in contrast with the behavior of bulk water at the same temperature. The same figure shows that, in the case of bulk water, diffusive motions take place at all scales, as expected for a liquid. Indeed, the line width corresponding to bulk water goes to zero at small values of Q, following the Fick's law, $\Gamma = DQ^2$, where D is the self-diffusion coefficient. At intermediate values of Q, the slope of $\Gamma(Q^2)$ is smaller for confined water than for bulk water. This means that, within the small volume of confinement, the dynamics is hindered.

A numerical analysis of the data obtained at different temperatures shows that the dynamics of confined water molecules is analogous to that of bulk water at temperatures typically 30° lower (84). This can be understood in the following

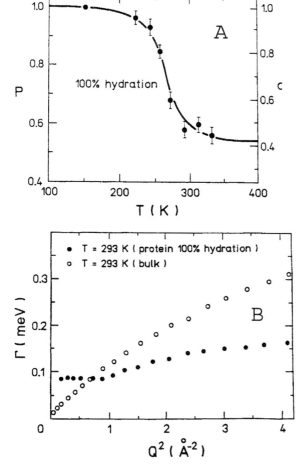

FIGURE 8 **(A)** The variation of the measured P parameter and of the fraction of "immobile water" q versus T for the fully hydrated C-phycocyanin ($h = 0.5$ g water/g protein). **(B)** Half-width at half maximum of the Lorentzian quasi-elastic line (Γ) versus Q^2, for $T = 293$ K for the fully hydrated C-phycocyanin and for the bulk water. *Source:* From Ref. 58.

way: at hydrophilic sites of protein, water molecules form relatively stable hydrogen bonds which keep the molecules confined in a small region of the surface of the protein. Water molecules alternatively form hydrogen bonds with the hydrophilic sites and their characteristic lifetime is longer than in bulk water. At the time scale of the neutron-scattering experiment, the dynamics of hydrogen bond formation concerns only three bonds among the four possible intermolecular bonds, one of them being "blocked" by the hydrophilic site of the protein.

The behavior that we describe appears very general, at least qualitatively. Hydration water from other biomolecules, as well as water confined into small pores, shows a similar slowing down of dynamic properties and a decrease of the temperature at which all the diffusive motions are frozen (59,79,83).

The common features arising from quasi-elastic neutron-scattering studies of water at a Vycor surface or close to a more complex protein surface are presented below. In particular, results of a C-phycocyanin protein at a hydration level, $h = 0.4$, are important since a monolayer coverage of water molecules

TABLE 1 Parameters for Water near the Surface of a C-phycocyanin Protein (Hydrated-Lyophilized Sample, $h = 0.40$)

T (°C)	a (Å)	Confined water (D_{local}, 10^{-5} cm²/sec)	Confined water (D_t, 10^{-5} cm²/sec)	Bulk water (D_t, 10^{-5} cm²/sec)	Confined water (τ_0, ps)	Bulk water (τ_0, ps)
40	4.5	1.28	1.52	3.20	5.9	0.90
25	4.3	0.97	1.20	2.30	6.6	1.10
0	4.0	0.84	0.76	1.10	8.2	3.00

TABLE 2 Parameters for Water Confined in a 25% Hydrated Vycor Sample

T (°C)	a (Å)	Confined water (D_{local}, 10^{-5} cm²/sec)	Confined water (D_t, 10^{-5} cm²/sec)	Bulk water (D_t, 10^{-5} cm²/sec)	Confined water (τ_0, ps)	Bulk water (τ_0, ps)	Confined water (τ_1, ps)	Bulk water (τ_1, ps)
25	4	0.92	2.45	2.30	15	1.10	1.5	1.10
−5	3	0.38	1.36	0.907	20	4.66	1.8	1.57
−15	3	0.26	1.20	0.574	25	8.90	2.0	1.92
−35	2						3.1	

allows the protein to initiate its function (24). Tables 1 and 2 give, respectively, for hydrated protein ($h = 0.4$) and for 25% hydrated Vycor the values of the diffusion coefficients D_{local} and D_t for confined water as compared with the diffusion coefficient D_t of bulk water. The residence time τ_0 (59,79) and the hydrogen bond lifetime τ_1 (59,79) are also given as a function of temperature.

For hydrated protein (Table 1), the values obtained for D_{local} are lower than those of bulk water. They are close to those obtained at the same temperature for 25% H_2O-hydrated Vycor (Table 2) what demonstrates the influence of the hydrophilic groups on the water molecules when one reaches monolayer coverage. This shows that the diffusive motion of water molecules is strongly retarded by interactions with a protein surface.

However, in contrast with the case of water in hydrated Vycor, the values of D_t for hydration water in the protein are smaller than that of bulk water. This is due to some influence of hydrophobic residues of protein, at the vicinity of the protein surface, which, in fact, is not as hydrophilic as that of Vycor. We are, thus, able to detect the effect of the substrate (59).

Figure 9A gives, respectively, the Arrhenius plots of τ_0 for water of hydration at the surface of a protein (59) as compared with those of water in Vycor at different levels of hydration (79) and bulk water (72).

The residence time τ_0 of confined water from 25% hydrated Vycor and hydrated protein are always longer than the residence time of bulk water, at the same temperature. They increase rapidly as either the temperature or the level of hydration decreases. For example, for the 25% hydrated Vycor sample $\tau_0 = 25$ ps at −15°C.

The hydrogen bond lifetimes τ_1 for confined water are close to that of bulk water (79). They have an Arrhenius temperature dependence (Fig. 9B) while the residence time τ_0 does not exhibit such a behavior (Fig. 9A).

Figure 10 gives the evolution of the vibrational density of states for H_2O-hydrated protein C-phycocyanin as a function of temperature for two levels of hydration $h = 0.5$ and 0.25.

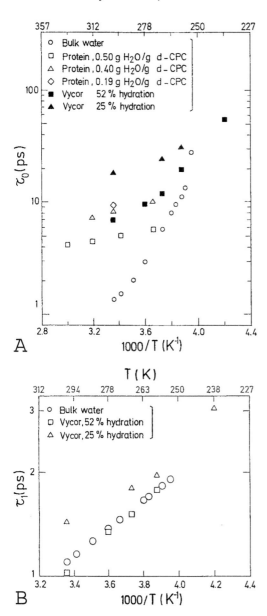

FIGURE 9 (**A**) Arrhenius plot of the residence time τ_0 for different levels of hydration of water: at the surface of H_2O-hydrated d-CPC protein (*empty symbols*), contained in hydrated Vycor (*solid symbols*), as compared with bulk water (*empty circles*). (**B**) Arrhenius plot of the hindered rotations characteristic time, τ_1. This time is a measurement of the hydrogen bond lifetime. *Source:* From Refs. 59 and 79.

One sees that the corresponding peaks for hydration water in protein are also shifted slightly upward compared with the bulk water at the same temperature and as it is observed for water confined in Vycor (Fig. 6). The up-shift of the librational peak increases either as the temperature is lowered or the level of hydration is decreased which reflects the amplified effect of confinement (59). This indicates that both the translational and librational motions of water molecules, near or at protein surface, are slightly more hindered, in agreement with observation from computer simulations.

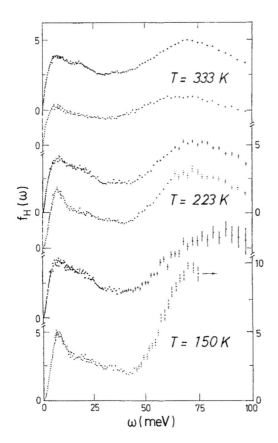

$f_H(\omega)$

$T = 333\ K$

$T = 223\ K$

$T = 150\ K$

$\omega\,(meV)$

FIGURE 10 Proton vibrational density of states f_H for water at surface of H_2O-hydrated d-CPC protein $\theta = 27.5°$, at three temperatures, 333, 223, 150 K and two levels of hydration $h = 0.50$ (*crosses*) and 0.25 (*solid circles*). *Source*: From Ref. 59.

Moreover, in the case of the low level of hydration ($h = 0.25$) the evolution of the density of states of the hydrated protein as a function of the temperature is less pronounced than in the case of $h = 0.5$. This is in agreement with the structural study (53) at the lower hydration ($h = 0.175$), which only detected small changes when the temperature is lowered from room temperature down to 77 K, and with further structural studies of low-hydrated Vycor samples. Low temperatures do not affect significantly the overall structure of the protein or the bound water molecule and no crystallization of water has been observed. This could reflect the fact that at room temperature the interfacial water behaves like a dense, supercooled liquid.

The study of an artificial peptide formed by five molecules of alanine and capped at both extremities, gave information above the very different behavior of water at the vicinity either of the only hydrophilic site of the protein or on its global hydrophobic surface (85). In the first case, the hydrogen bond formed between water and the hydrophilic site is relatively stable and the only detectable motion is the rotation of the water molecule. Instead, the diffusion is important for the water molecules at the vicinity of the hydrophobic surface.

More recent studies focused on the comparison of amino acids that differ only by the existence of a hydrophobic chain, such as glycine (Gly) and leucine (Leu). A detailed comparison of a large number of experiments performed with the trimers Gly-Gly-Gly and Gly-Leu-Gly shows, in particular, that the onset of water dynamics observed at low temperatures (i.e., around 200 K) is sharper for the case of Gly-Leu-Gly because of its hydrophobic chain (86,87). Indeed, in this case, a network of intermolecular hydrogen bonds is formed in a stable way avoiding translational diffusion at low temperatures. In the case of Gly-Gly-Gly, the transition is more gradual, because, prior to translational diffusion, one observes rotational motions at intermediate temperatures. It was also shown that this temperature dependence (sometimes called a *dynamic* or a *glass* transition of the protein) depends drastically on the interaction between water molecules within a first layer on the surface of the protein and on the local hydrophobicity of each site, as well as on its spatial extent.

CONCLUSION

From the more recent findings combining neutron techniques and molecular dynamics simulations, it is now possible to have a more precise picture of confined water. Water, in the vicinity of a hydrophilic surface, is in a state equivalent to bulk water at a lower temperature. As previously demonstrated, this depends on the degree of hydration of the sample. In particular, at room temperature, interfacial water shows a dynamic behavior similar to that of bulk water at a temperature 30 K lower. It behaves like bulk supercooled water.

It appears that the short-time dynamics of water molecules at or near a hydrophilic model surface and at a soluble protein surface is much slower as compared with that of bulk water. It is important to notice that the more significant slow dynamics of interfacial water is reflected in the long residence time for jump diffusion. This suggests that there may be a common underlying mechanism for the slowing down of the single-particle dynamics of interfacial water.

This is the consequence of the confined diffusion theory, which has been used to analyze the quasi-elastic neutron-scattering data. This simple theory gives information on the confinement volume and the slow dynamics of the single-particle motions. To understand the microscopic origin of the confinement and slowing down of motions of water molecules and the exact role played in this context, the theory of kinetic glass transition in dense, supercooled liquids (88,89) has been recently used. This theory leads to some description of the dynamics of confined water in terms of correlated jump diffusion (90) instead of jump diffusion (72). This description looks consistent with molecular dynamics simulations of supercooled water (91) and has been confirmed by high-resolution quasi-elastic neutron-scattering experiments of water from hydrated Vycor (92) and from hydrated C-phycocyanin protein (93).

This more sophisticated way shows a large distribution of residence times for water molecules in the cage formed by the neighboring molecules, which is a more realistic view than the sharp separation of water molecules into two classes, according to their mobility (59). Short-time dynamics results about hydrated myoglobin have been recently interpreted by using this same theory of kinetic glass transition in dense, supercooled liquids (83).

ACKNOWLEDGMENTS

We are grateful to H. L. Crespi for his continuing collaboration in the preparation of the perdeuterated protein C-phycocyanin.

REFERENCES

1. Angell CA. In: Franks F, ed. Water: A Comprehensive Treatise. Vol. 7. New-York, London: Plenum Press, 1981.
2. Lang EW, Ludemann HD. Angew Chem Int Ed Engl 1982; 21:315.
3. Chen SH, Teixeira J. Adv Chem Phys 1985; 64:1.
4. Dore JC. In: Franks F, ed. Water Science Reviews. Vol. 1. Cambridge University Press, 1985.
5. Bellissent-Funel M-C, Teixeira J, Bosio L. J Chem Phys 1987; 87:2231.
6. Bellissent-Funel M-C. In: Dore JC Teixeira J, eds. Hydrogen Bonded Liquids (NATO ASI Series C). Dordrecht: Kluwer Academic Publishers, 1991.
7. Bellissent-Funel M-C, Teixeira J. J Mol Struct 1991; 250:213.
8. Chen SH. In: Dore JC, Teixeira J, eds. Hydrogen-Bonded Liquids (NATO ASI Series C329). Dordrecht: Kluwer Academic Publishers, 1991:289.
9. Teixeira J. J. Phys. IV 3, 1993, C1–C163 and references therein.
10. Poole PH, Sciortino F, Essman U, et al. Nature (London) 1992; 360:324; and references therein.
11. Soper AK. In: Dore JC, Teixeira J, eds. Hydrogen-Bonded Liquids. Dordrecht: Kluwer Academic Publishers, 1991; 329:147.
12. Malikova N, Cadène A, Dubois E, et al. J Phys Chem C 2007; 111:17603.
13. Kuroda Y, Kittaka S, Takahara S, et al. J Phys Chem B 1999; 103:11064.
14. Dore JC, Dunn M, Chieux P. J Physique 1987; 48:C1–C457.
15. Dore JC, Coveney F, Bellissent-Funel M-C. In: Howells WS, Soper K, eds. Recent Developments in the Physics of Fluids. Bristol: Adam Hilger Publishers, 1992:299.
16. Benham MJ, Cook JC, Li JC, et al. Phys Rev B 1989; 39:633.
17. Chen SH, Bellissent-Funel M-C. In: Bellissent-Funel M-C, Dore JC, eds. Hydrogen Bond Networks (NATO ASI C435). Dordrecht: Kluwer Academic Publishers, 1994:307.
18. Zanotti JM, Bellissent-Funel M-C, Chen SHEurophys Lett 2005; 71:91.
19. Liu E, Dore JC, Webber JB, et al. J Phys Condens Matter 2006; 18:10009.
20. Seyed-Yazdi J, Farman H, Dore JC, et al. J Phys Condens Matter 2008; 20:205107.
21. Webber JB, Dore JC, Strange JH, et al. J Phys Condens Matter 2007; 19:415117.
22. Yoshida K, Yamaguchi T, Kittaka S, et al. J Chem Phys 2008; 129:054702.
23. Bellissent-Funel M-C, Sridi-Dorbez R, Bosio L. J Chem Phys 1996; 104:10023.
24. Rupley JA, Careri G. Adv Protein Chem 1991; 41:37.
25. Colombo MF, Rau DC, Parsegian VA. Science 1992; 256:655.
26. Smith J. Q Rev Biophys 1991; 24:227.
27. Steinbach PJ, Brooks BR. PNAS 1993; 90:9135.
28. Lounnas V, Pettitt BM, Findsen L, et al. J Phys Chem 1992; 96:7157.
29. Lounnas V, Pettitt BM. Protein Struct Funct Genet 1994; 18:148.
30. Lounnas V, Pettitt BM, Phillips GN Jr. Biophys J 1994; 66:601.
31. Lovesey SW. Theory of Neutron Scattering From Condensed Matter. 3rd ed. Oxford: Clarendon Press, 1987.
32. Lee CY, McCammon JA, Rossky PJ. J Chem Phys 1984; 80:4448.
33. Levitt M, Sharon R. Proc Natl Acad Sci U S A 1988; 85:7557.
34. Rossky PJ, Lee SH. Chemica Scripta 1989; 29A:93.
35. Lee SH, Rossky PJ. J Chem Phys 1994; 100:3334.
36. Rossky PJ. In: Bellissent-Funel M-C, Dore JC, eds. Hydrogen Bond Networks (NATO ASI C435). Dordrecht: Kluwer Academic Publishers, 1994:337.
37. Netz PA, Dorfmüller T. J Phys Chem B 1998; 102:4875.
38. Marchi M, Sterpone F, Ceccarelli M. J Am Chem Soc 2002; 124:6787.
39. Bizzarri AR, Cannistraro S. J Phys Chem B 2002; 106:6617.
40. Luzar A, Chandler D. Phys Rev Lett 1996; 76:928.
41. Alary F, Durup J, Sanejouand YH. J Phys Chem 1993; 97:13864.
42. Mezger M, Reichert H, Schöder S, et al. PNAS 2006; 103:18401.
43. Kunz W, Bellissent-Funel M.-C, Calmettes P. Structure of water and ionic hydration. In: S. R. Caplan, I. R. Miller, G. Milazzo, eds. Bioelectrochemistry: Principles and Practice. Volume 1: General Introduction. Switzerland: Birkhauser Verlag Basel, 1995, 132–210.
44. Bellissent-Funel M-C, Bosio L, Teixeira J. J Phys Condens Matter 1991; 3:4065.
45. Soper AK, Phillips MG. Chem Phys 1986; 107:47.

46. Enderby JE. In: Bellissent-Funel M-C, Neilson GW, eds. The Physics and Chemistry of Aqueous Ionic Solutions (NATO ASI C205). Dordrecht: Reidel Publishers, 1987:129; and references therein.
47. General information on Vycor Brand Porous thirsty glass, no. 7930, Corning Glass Works, is available from OEM Sales Service, Box 5000, Corning, NY 14830, U.S.A.
48. Bellissent-Funel M-C, Bosio L, Lal JJ. Chem Phys 1993; 98:4246.
49. Bruni F, Ricci MA, Soper AK. J Chem Phys 1998; 109:1478.
50. Zanotti JM. PhD thesis. University of Orsay, France, 1997.
51. Bellissent-Funel M-C, Bosio L, Hallbrucker A, et al. J Chem Phys 1992; 97:1282.
52. Wiggins PM. Prog Polym Sci 1988; 13:1.
53. Bellissent-Funel MC, Lal J, Bradley KF, et al. Biophys J 1993; 64:1542.
54. Bosio L, Johari GP, Oumezzine M, et al. Chem Phys Lett 1992; 188:113.
55. Duerring M, Schmidt GB, Huber R. J Mol Biol 1987; 217:577.
56. Crespi HL. Stable Isotopes in the Life Science. Vienna: IAEA, 1977:111.
57. Bellissent-Funel M-C, Teixeira J, Bradley KF, et al. Physica B 1992; 180,181:740.
58. Bellissent-Funel M-C, Teixeira J, Bradley KF, et al. J Phys I France, 1992; 2:995.
59. Bellissent-Funel M-C, Zanotti JM, Chen SH. Faraday Discuss 1996; 103: 281.
60. Middendorf HD. In: Peyrard M, ed. Nonlinear Excitations in Biomolecules. Les Ulis: Les Editions de Physique, 1995:369.
61. Rossky PJ, Karplus M. J Am Chem Soc 1979; 101:1913.
62. Lee B, Richard FM. J Mol Biol 1971; 55:379.
63. Sartor G, Hallbrucker A, Hofer H, et al. Conference Proceedings of Water Bio-molecule Interactions Bologna: SIF, 1993:143.
64. Qvist J, Persson E, Mattea C, et al. Faraday Discuss 2009; 141:131.
65. Mattea C, Qvist J, Halle B. Biophys J 2008; 95:2951.
66. Persson E, Halle B. PNAS 2008; 105:6266.
67. Tehei M, Franzetti B, Wood K, et al. PNAS 2007; 104:766.
68. Polnaszek CF, Hanggi DA, Carr PW, et al. Analyt Chim Acta 1987; 194:311.
69. Linse P. J Chem Phys 1989; 90:4992.
70. Wong CF, McCammon JA. Isr J Chem Phys 1986; 27:211.
71. Springer T. Quasi-elastic neutron scattering for the investigation of diffusive motions in solids and liquids. Springer Ser Modern Phys 1972; 64.
72. Teixeira J, Bellissent-Funel M-C, Chen SH, et al. Phys Rev A 1985; 31:1913.
73. Volino F, Pineri M, Dianoux AJ, de Geyer A. J Polym Sci 1982; 20:481.
74. Dianoux AJ. In: Bellissent-Funel M-C, Neilson GW, eds. The Physics and Chemistry of Aqueous Ionic Solutions (NATO ASI C205). Dordrecht: Reidel Publishers, 1987:129; and references therein.
75. Ramsay JDF, Poinsignon C. Langmuir 1987; 3:320.
76. Maruyama M, Bienfait M, Dash JG, et al. J Crystal Growth 1992; 118:33.
77. Callaghan PT, Coy A, Halpin TPJ, et al. J Chem Phys 1992; 97:651.
78. Callaghan PT, Coy A, MacGowan D, et al. Nature 1991; 351:467.
79. Bellissent-Funel M-C, Chen SH, Zanotti JM. Phys Rev E 1995; 51:4558.
80. Volino F, Dianoux AJ. Mol Phys 1980; 41:271.
81. Bellissent-Funel M-C, van der Maarel JRC (unpublished results).
82. Middendorf HD, Di Cola D, Cavatorta F, et al. Biophys Chem 1994; 53:145.
83. Settles M, Doster W. Faraday Discuss 1996; 103:269.
84. Teixeira J, Zanotti JM, Bellissent-Funel M-C, et al. Physica B 1997; 234–236:370.
85. Russo D, Baglioni P, Peroni E, et al. Chem Phys 2003; 292:235.
86. Russo D, Ollivier J, Teixeira J. Phys Chem Chem Phys 2008; 10:4968.
87. Russo D, Teixeira J, Ollivier J. J Chem Phys 2009; 130:235101.
88. Götze W, Sjogren L. Rep Prog Phys 1992; 55:241.
89. Cummins HZ, Li G, Du WM, et al. Physica A 1994; 204:169.
90. Chen SH, Gallo P, Bellissent-Funel M-C. Non Equilibrium Phenomena in Super-cooled Fluids, Glasses and Materials. London: World Scientific Publication, 1996.
91. Gallo P, Sciortino F, Tartaglia P, et al. Phys Rev Lett 1996; 76:2730.
92. Zanotti JM, Bellissent-Funel M-C, Chen SH. Phys Rev E 1999; 59:3084.
93. Dellerue S, Bellissent-Funel M-C. Chem Phys 2000; 258:315.

3 | Freezing and Annealing Phenomena in Lyophilization

Jim A. Searles
Aktiv-Dry LLC, Boulder, Colorado, U.S.A.

INTRODUCTION

The freezing step of lyophilization is of paramount importance. It is the principal dehydration step, and it determines the morphology and pore sizes of the ice and product phases. In general, the desired attributes of a lyophilized product are

- consistent high yield of product activity through lyophilization,
- appropriate crystallization (or not) of product and excipient(s),
- glass transition temperature higher than the desired storage temperature (related directly to residual moisture level),
- pharmaceutically elegant, mechanically strong cake,
- rapid reconstitution,
- short, consistent, and robust freeze-drying cycle, and
- stability of all product quality attributes through the intended shelf-life.

The freezing method as well as any intentional or unintentional post-freezing annealing influence many of the above attributes. This chapter will explain the freezing process, the most common freezing methods, and review how freezing and annealing can affect process and product quality parameters including primary and secondary drying rates, surface area, solute crystallization, product aggregation and denaturation, storage stability, reconstitution, and inter- and intrabatch consistency.

PROCESS PHYSICS: THE SUPPLEMENTED PHASE DIAGRAM

A discussion of freezing, annealing, and lyophilization is aided by viewing the process through the "supplemented phase diagram" described by MacKenzie (1). It is an equilibrium freezing point depression diagram supplemented with the glass transition curve [and solute crystallization and/or precipitation curve(s) as appropriate]. Shown in Figure 1 is such a diagram for sucrose using data from Blond et al. (2) and Searles et al. (3) The x-axis is solute concentration in the non-ice phase, and the y-axis is temperature. Sucrose does not crystallize during freeze-concentration, so there is no eutectic point shown in Figure 1. If crystallizing solutes are being used (e.g., mannitol or glycine) then it would be important to add the appropriate eutectic points and phase lines for the crystallizing solute. The diagram will be explained in the context of freezing, annealing, and lyophilizing a 10% (w/w) sucrose solution.

In our example, we will freeze vials containing the sucrose solution on the lyophilizer shelf as the shelf temperature is slowly reduced ("shelf-ramp" freezing). This method is the most prevalent in the pharmaceutical industry, not necessarily because it is the best method but because there are so few choices available. Other freezing methods will be discussed in section "Freezing Methods." Figure 2 shows examples of the shelf and liquid temperatures

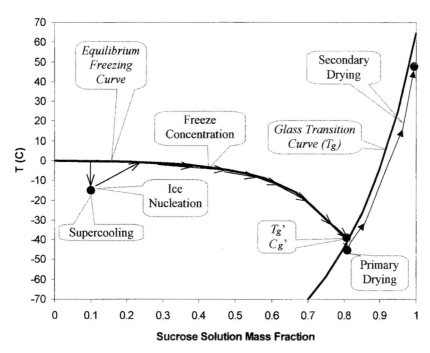

FIGURE 1   Supplemented phase diagram for sucrose. Arrows show freeze-drying process for a 10% sucrose solution.

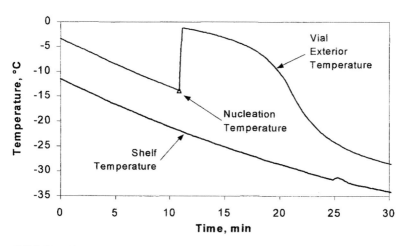

FIGURE 2   Shelf and vial temperature for shelf-ramp cooling. 2 mL of 10% hydroxyethyl starch (HES) in a 5-mL vial instrumented with a 36-gauge externally attached thermocouple. Nucleation occurs at −14°C, after which the supercooling is consumed by the latent heat of ice crystalli-zation. The subsequent solidification of the nucleated volume occurs with a gradual temperature decrease. *Source*: From Ref. 3.

through this process. For the present example we will assume that the liquid height in the vials and the cooling rate are low enough such that the entire sample volume achieves a similar extent of supercooling before nucleation occurs. As discussed further below, this is termed "global supercooling," and the results of this type of freezing are naturally variable because nucleation is spontaneous and not under direct control. When freezing in this manner the nucleation temperature varies considerably from vial to vial (4).

When nucleation occurs in this sample ice crystals grow to encompass the entire liquid volume because the entire liquid volume has nominally the same extent of supercooling. However, only a fraction of freezable water crystallizes during this initial nucleation event because crystallization is exothermic and the supercooling is not sufficient to allow complete solidification. For a sample that nucleates at $-15°C$, about 20% of freezable water crystallizes before the super-cooling is extinguished (4) and the temperature reaches the equilibrium freezing temperature for the newly freeze-concentrated solution. For such a vial the sucrose solution concentration has risen to 24% and the solution temperature is near $-1°C$. However, the shelf temperature continues to decrease, and is now approximately $-22°C$. This provides a strong driving force for the completion of solidification, evidenced by the decreasing sample temperature.

Through the subsequent cooling as the shelf temperature continues to drop to $-50°C$, freeze-concentration continues until the solution reaches the glass transition temperature at maximum freeze-concentration, denoted as T_g'. The corresponding concentration is C_g'. This is shown in Figure 1, and is the intersection of the freezing point depression and glass transition curves. At this point any further thermodynamically favored freeze-concentration is arrested by the high viscosity of the sucrose phase. The mobility of the water in this phase is too low to permit further migration of water to the ice interface for crystallization. The sucrose phase has reached a glassy state characterized by its high viscosity.

It is important to note at this point that the glass transition curve shown is actually one of a family of iso-viscosity curves. The particular one shown corresponds to timescales relevant to our process, which for lyophilization is on the order of days. This means that if we hold a sample under these conditions for several days, we will not observe further freeze-concentration. If we hold it for weeks, however, we would observe further crystallization of ice. The timescale of interest for the period from release to expiry is years. Therefore the T_g' determined using a rapid scanning technique like DSC is just a starting point for understanding the T_g' relevant to lyophilization and the final T_g appropriate for estimations of storage stability. Many have found that the faster the scan rate, the higher the measured glass transition temperature. However, scan rates on the order of days are impractical because the intensity of the thermal transitions is too low to be measurable. Therefore, for determining the sample temperature below which primary drying should be carried out to prevent collapse, T_g' should be a functional description of that temperature below which collapse is not observed. Searles et al. (3) and Ablett et al. (5) present methods to measure T_g' over timescales relevant to lyophilization. In addition, freeze-dry microscopy is commonly used to estimate collapse temperatures and observe other freezing and drying phenomena.

At T_g' the sucrose concentration in the amorphous phase is 81%, show-ing an eightfold concentration increase. A material balance on the ice and amorphous phases reveals that at this point 88% of the water has frozen. The overall concentration of solutes during freeze-concentration increases the potential for degradative processes. As stated by Felix Franks, "chemical reaction rates in part-frozen solutions are substantially higher than in the original dilute solution at room temperature. The rate-enhancing effect of concentration far outweighs the rate-reducing effect of low temperature"(6). Oxidation is a particular case. While the rate constant for an oxidation reaction will decrease with decreasing temperature, the overall rate of such a reaction can increase because the solubility of oxygen in aqueous solutions is inversely proportional to temperature. It increases nearly twofold between 25°C and 0°C. In addition, freeze-concentration can combine with this effect to increase oxygen concentration to 1000 times that at 0°C (7) and increase oxidation rates (8).

Annealing is a hold step at a temperature above the glass transition temperature. Annealing can be carried out as a waypoint during the initial cooling but more commonly it is a postfreezing warming and hold step, followed by recooling. For our present example the shelf temperature would be raised from −50°C to the annealing temperature (e.g., −20°C). The sucrose solution will follow the equilibrium freezing curve and equilibrate at 70% sucrose in the nonice phase. This is the mechanism by which super-T_g' annealing results in *ice melting*: the ice fraction decreases to dilute the sucrose phase. The sample is now well above its glass transition temperature and a number of processes are free to take place. These include ice crystal maturation through Ostwald ripening, the crystallization of solutes, and possibly degradative reactions. These phenomena will be discussed further below.

The freezing step and any post-freezing temperature deviations above T_g will determine the texture or morphology of the product. As discussed in section "Effects on Active Ingredients" below, the morphology of the system has profound impacts on drying rates, protein aggregation, and reconstitution. Figure 3 shows examples of product morphology before and after annealing.

We will now continue our tour of lyophilization using the supplemented phase diagram as our roadmap. Primary drying should take place at a temperature safely below T_g', and although some secondary drying is know to occur during the primary drying phase of lyophilization, by definition primary drying includes only the sublimation of crystalline water from the system, and the solute phase concentration does not change. During secondary drying the shelf temperature is increased (and some decrease the pressure). As the sample dries its T_g increases as shown on Figure 1. During an optimum secondary drying phase, the sample temperature will be raised at such a rate that the product temperature always remains just below the glass transition temperature. As with primary drying, exceeding the glass transition temperature at any time during drying will lead to some extent of product collapse depending on the duration of the event. In rare cases limited collapse is intentionally induced, but in general collapse is cause for rejection of lyophilized pharmaceuticals, so its avoidance is paramount.

The following sections will discuss specific product effects of different freezing methods and annealing steps. We will first review freezing methods, and will then explore how these methods, used in some cases with annealing, have been found to affect product quality and processing attributes.

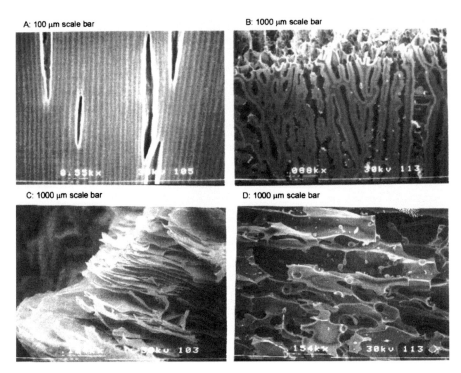

A: 100 µm scale bar

B: 1000 µm scale bar

C: 1000 µm scale bar

D: 1000 µm scale bar

FIGURE 3 Scanning electron micrographs (SEMs) of 10% hydroxyethyl starch (HES) frozen in a vial by liquid nitrogen immersion. (**A**, **B**) Top of the cake (**A** not annealed, **B** annealed); (**C**, **D**) interior of broken cake (**C** not annealed, **D** annealed). *Source*: From Ref. 4.

FREEZING METHODS

Various freezing methods can be used for liquids in glass vials, although some are not appropriate for full-scale GMP production of sterile pharmaceuticals. Principal methods are listed below in order of increasing cooling rate:

1. Slow directional solidification: creating ice nuclei on the bottom of a vial by contact with dry ice, followed by slow freezing on a precooled shelf (9,10)
2. Vacuum-induced freezing (11)
3. Placing vials on a shelf which is then ramped from above freezing temperature to below T_g' ("shelf-ramp" freezing)
4. Placing vials in a freezer or on a lyophilizer shelf which is already below freezing temperature ("precooled shelf" freezing)
5. Immersion in refrigerated heat transfer fluid (e.g., dry ice in alcohol)
6. Blast freezing via forced air and/or sprayed liquid nitrogen (12)
7. Liquid nitrogen immersion freezing

By far the most common method is the third one listed—shelf-ramp freezing. There are no issues with condensation on the shelves during loading, and all lyophilizers can carry out this type of freezing without modification.

Specific ice nucleation techniques have been used for various purposes. Examples include nucleating agents silver iodide and *Pseudomonas syringae* (4), ice fog (13), ultrasound (14–16), vacuum (11,17), and electrical (18,19). The resulting crystallization rate is dependent on the overall solution temperature at the time of ice nucleation, as well as the rate at which heat is removed after nucleation. For example, in freezing method 1 above, the vial and its contents are to be just below 0°C before a bottom corner of the vial is put in contact with dry ice. Then the vial is placed onto a shelf, the temperature of which is then gradually reduced.

While many in the past have interchangeably used the terms "cooling rate" and "freezing rate," it is important to distinguish between them. The cooling rate is the rate at which the vial is cooled. This cooling rate may affect the temperature at which ice nucleates or, more precisely, the regions of the liquid volume over which nucleation occurs. The freezing rate only applies to the postnucleation freezing which in some limited cases is irrelevant for determination of the final ice structure. A true freezing rate is either in terms of a linear front velocity for directional freezing or mass per unit time for bulk freezing operations.

As explained by Searles et al., the following terms are useful when discussing freezing for lyophilization: *primary nucleation* is the initial ice nucleation event (4). *Secondary nucleation* follows primary nucleation, and moves with a velocity on the order of mm/sec to encompass some portion of the liquid volume (20–22). Subsequent to secondary nucleation, *solidification* is completed relatively slowly as the heat of crystallization is transferred from the solidification interface through the already-solidified layer and the vial bottom to the shelf. These terms pertain to freezing by *global supercooling* in which the entire liquid volume achieves a similar level of supercooling, and the secondary nucleation zone encompasses the entire liquid volume (as in the example in the previous section). Given the design of lyophilizer shelf temperature control systems, shelf-ramp freezing is by nature slow and it will, with typical vials and fill volumes, freeze by global supercooling, which yields a low surface area. In contrast, *directional solidification* occurs when a small portion of the volume is supercooled to the point of primary and secondary nucleation. The nucleation and solidification fronts are in close proximity in space and time with the front moving into nonnucleated liquid. Many write about liquid nitrogen immersion freezing inducing less supercooling than slower cooling methods, but more accurately faster cooling results in supercooling over a smaller *volume* before nucleation than slower cooling.

Foam drying has recently gained attention as a promising method of freezing and drying that achieves high retention of activity for vaccines and proteins (23–30). It is a combination of vacuum-induced surface freezing and low-temperature vacuum boiling, which results in a low surface area foamed product.

EFFECTS ON ACTIVE INGREDIENTS

Freezing itself can adversely affect the active ingredient, and the freezing method can also affect the quality of the product through subsequent processing and until expiry. These impacts can result from the low temperature itself, acceleration of degradation reactions, crystallization of the product, product denaturation and aggregation, pH shifts, phase separation, and denaturation at the ice interface. Much of the literature on this subject concerns the stabilization of proteins. The interested reader is referred to comprehensive reviews of protein formulation and lyophilization (31–33). This section will cover how freezing itself can have a major influence on success of the formulation and the lyophilization process. Excipient crystallization will be covered in section "Effects on Solute Crystallization."

Proteins are prone to denaturation at both high and low temperatures. Bhatnagar et al. provide a comprehensive review of the subject (34). Although cold denaturation is frequently mistaken for freezing denaturation, it is known to occur in the absence of freeze-concentration. As early as in 1930, it was reported that the rate of ovalbumin denaturation by urea is higher at 0°C than at 23°C (35). Bhatnagar et al. found ice itself to be responsible for loss of lactate dehydrogenase (LDH) activity during freezing (36). They created concentrated solutions to mimic freeze-concentration and found no degradation. Tang et al. discovered that cryopreservatives sucrose, trehalose, and glycerol can protect proteins from cold denaturation, effectively lowering their cold denaturation temperatures (37).

Several have found faster freezing to result in greater protein aggregation. In freeze-thaw studies, Eckhardt et al. found that the formation of insoluble aggregates of recombinant human growth hormone during freezing increased sharply with increased cooling rates (38). The authors stated possible causes to be surface denaturation or recrystallization during thawing of rapidly frozen samples. Skrabanja et al. report on freezing method and formulation effects on monoclonal antibody aggregation formation (39). Rapid freezing by immersion in CO_2/acetone resulted in greater aggregation after freezing than slow freezing by placing vials into a freezer at −20°C. In another study discussed in the same paper they report on recovery of a recombinant protein after lyophilization. The fast freeze yielded marginally higher yield than freezing by either shelf-ramp or a precooled shelf. Chang et al. showed that denaturation of proteins during freezing is closely related to surface denaturation by quantifying both types of denaturation for a range of proteins (40). A strong correlation ($r = 0.99$) was observed between the tendency of a protein to denature by freezing and its tendency to surface denature. Freezing by liquid nitrogen immersion caused more denaturation than shelf-ramp freezing. Small quantities of surfactant provided protection to the proteins against both types of inactivation (six surfactants were tested). A subsequent study from the same research group showed that surfactants stabilize against surface denaturation by competing with stress-induced soluble aggregates for interfaces, inhibiting subsequent transition to insoluble aggregates (41). Jiang and Nail studied catalase, β-galactosidase, and LDH in phosphate buffers, and found that shelf-frozen samples retained more protein activity than those frozen by liquid nitrogen immersion (42). Freezing by placement in a −40°C freezer yielded even better recovery. In all cases, freeze/thaw as well as lyophilization protein activity retention was found to improve with protein concentration. This is possibly an artifact of constant loss of a given mass of protein. As the concentration is increased, the fraction recovered increases; however, the same mass is lost. Jiang and Nail also found that the activity recovery increased with increasing residual moisture, suggesting that the secondary drying process also contributed to loss.

The trend of greater protein losses via faster freezing continued with the finding by Sarciaux et al. that liquid nitrogen quench freezing of an antibody formulation resulted in more and larger insoluble aggregates than shelf-ramp freezing (43). As discussed in detail above, Hsu et al. showed a clear correlation between surface area and cooling rate by using various cooling methods and vial sizes. They also found that poorer *stability* of their protein correlated inversely with the specific surface area, providing a linkage between the freezing method and rate and product stability through time after lyophilization (44). Stability was even affected by the choice of vial and volume of fill through this linkage to the cooling rate for freezing—smaller fills in smaller vials resulted in greater product specific surface areas. No freeze/thaw protein loss was found. The authors stated that the phenomenon of lower stability for higher surface area samples might arise from possible lack of protection by the arginine excipient for those protein molecules at the surface of the amorphous phase. Citing preliminary results, Patapoff and Overcashier also found that annealing significantly reduced the rate of protein aggregation during storage stability studies (10).

Nema and Avis found contradictory evidence for protein denaturation in that fast freezing by liquid nitrogen resulted (88% recovery) in better retention of LDH activity than shelf-ramp freezing (68%) (45). The authors did not identify a specific mode of action responsible. The protein was formulated neat with no buffers or cryopreservatives. They did test a number of cryoprotectants with shelf-ramp freezing and found three formulations, which afforded protection: 1% w/v bovine serum albumin, 1 M sucrose, and 0.05% w/v Brij 30. It is possible that during shelf-ramp freezing degradative reactions are taking place in the partially freeze-concentrated product phase during the slow solidification after ice nucleation. For this reason one would expect that a postfreezing annealing step would adversely affect any products for which shelf-ramp freezing yields are poorer than those for liquid nitrogen immersion. The finding by Nema that product yields across fast freezing are better is not the only such case—some live viruses respond in the same manner.

Sarciaux et al. found that lyophilization resulted in insoluble aggregates of their bovine IgG formulations, but the damage was not observed after freeze/thaw (43). Freezing by shelf-ramp resulted in less postlyo aggregation than liquid nitrogen immersion freezing, and annealing reduced the damage further. The authors correlated the extent of aggregation with the surface area, and found that aggregation progressed through the secondary drying step of lyophilization. Annealing reduced the percentage of aggregate in the final product from 33% to 12% (46). The reduction was attributed to the lower surface area of the annealed samples. However, a new theory posits that these benefits may have been due to stress relaxation: the publication by Webb et al. reports that liquid nitrogen immersion and lyophilization resulted in greater yields of human interferon-γ than spray-freeze-drying (47). As discussed above, spray-freezing yielded several times greater specific surface area than the immersion method. However, like Sarciaux et al., the authors did not find any freeze-thaw damage caused by spray-freeze-drying, rather the damage was found to occur during the terminal stages of drying. The authors speculated that, instead of surface denaturation, which was ruled out by the ultrahigh cooling rates for spray-freezing, residual stress retained within the solid matrix was responsible for the protein damage. Additional evidence to support their finding is provided by the fact that annealing relieves this stress and reduces the protein loss. Unfortunately, Sonner et al. do not comment on the effects of annealing upon protein recovery (48). However, they did find that while spray-freezing did not damage their protein, the subsequent freeze-drying step did. Shelf-ramp freezing and freeze-drying resulted in a similar level of damage, and polysorbate 80 in the formulation reduced the damage for both types of samples.

Freeze-concentration may bring solutes into concentration ranges where they will phase separate, causing deleterious effects on the product (9,49–55). Heller et al. demonstrated that the increased mobility during annealing facilitated phase separation in a PEG:dextran formulation. The phase separation caused unfolding of recombinant hemoglobin, and they used formulation design strategies (52) and PEGylation of the protein (54) to avoid the damage. The freezing step has also been found to be critical in lyophilization of DNA for gene therapy (56,57).

Recently Cochran and Nail published results showing a correspondence between the ice nucleation temperature and recovery of LDH protein activity (58).

It has long been known that vaccine adjuvants known as "alum" gels (aluminum hydroxide or aluminum phosphate) will be inactivated by coagulation during freezing. Recently, it has been discovered that very fast freezing as in spray-freezing into liquid nitrogen could prevent inactivation (59). Since then, Clausi et al. have found that glass-forming sugars can also prevent coagulation during freeze-drying (60), and Jones-Braun and coworkers developed liquid formulations that are stable when taken to $-20°C$ (61).

Zhai et al. (2004) (62) and Hansen et al. (2005) (63) studied a modified herpes virus intended as a gene therapy vector. Zhai et al. found heavily formulation-dependent effects of different freezing methods on recovery of virus activity through freeze-drying. High concentrations (27%) of sucrose or trehalose achieved the best results, with lower concentrations achieving less in a dose-dependent fashion. The freezing rate studies showed that for 27% sucrose, conventional shelf freezing resulted in titers as high as higher-rate freezing. The other formulation tested in the freezing rate studies contained only 2.5% sucrose. Titer yields were lower regardless of freezing method, but in the absence of high-concentration sucrose conventional shelf freezing did very poorly compared with flash-freezing. Hansen found that with the same freeze-sensitive 2.5% sucrose formulation, flash-freezing as well as rapid thawing were required to retain potency (63).

EFFECTS ON PRODUCT MORPHOLOGY, SURFACE AREA, AND DRYING RATE

The freezing method and cooling rate during freezing have profound impacts on the morphology and surface area of the final product. These parameters can be easily modified by any of the annealing steps (intentional or accidental), and the parameters in turn determine the resistance to vapor flow (affecting the primary drying rate and temperature during drying) as well as the secondary drying rate.

Reviews of early literature on lyophilization freezing and annealing phenomena appear in recent papers by Searles et al. (3,4), parts of which will be recapitulated here. In 1925, Tammann reported that the ice crystal morphology can be strongly influenced by the nucleation temperature (64). Samples frozen at "low supercoolings" yield dendritic structures, whereas "crystal filaments" result from high supercoolings. In 1961, Rey described an annealing step for orange juice that resulted in a twofold increase in the primary drying rate (65).

MacKenzie and Luyet in 1963 showed that annealing 30% gelatin gels resulted in slower primary drying (66). It is likely that the extremely high solute concentration resulted in a low volume fraction of ice (<50%), in which case one would expect annealing to facilitate complete encasement of the ice crystals as the authors reported. Luyet and coworkers published several papers on ice morphology. Luyet and Rapatz identified hexagonal, dendritic, and dispersed spherulitic morphologies in response to different freezing protocols and solutes in aqueous systems (67,68). The morphology was dependent on the freezing temperature, solute, and concentration. In one case the ice morphology in glycerol solutions shifted from hexagonal to "irregular" to spherulitic with increasing supercooling (67).

Quast and Karel in 1968 published results on the effect of freezing method on the subsequent dry layer resistance to gas flow (69). They also provide a thorough review of earlier works on the subject. They studied concentrated coffee (20% and 30% solids) as well as a model food system (10% glucose, 10% microcrystalline cellulose, 2% starch). The samples were frozen in 25 mm diameter glass cylinders 20 mm long, which were frozen by liquid nitrogen immersion, or by incubation in a freezer set at $-5°C$, $-20°C$, or $-40°C$. In addition some samples frozen at $-40°C$ were first seeded with ice crystals. Samples were frozen by the method being tested, freeze-dried, then placed in a gas flow cell for resistance measurement over a range of pressures. Higher solute concentrations led to greater resistance. Liquid nitrogen freezing yielded samples of the lowest resistance, after which, listed in order of increasing resistance, were the $-5°C$, $-20°C$, $-40°C$ seeded, and $-40°C$ methods. Therefore, with the exception of liquid nitrogen immersion, the "faster" the freezing, the higher the resistance. All methods except the $-40°C$ seeded method yielded a layer at the top of the sample of much greater resistance per unit length than the material making up the remainder of the sample. Liquid nitrogen frozen samples cracked extensively during drying.

Thijssen and Rulkens reported in 1969 that the freezing rate is an important determinant of pore size and drying rate in freeze-drying of liquid food products (70). For a 20% dextran solution freeze-dried in slabs, a faster cooling rate during freezing resulted in smaller pores and therefore higher resistance and slower drying rates. In 1969 Blond et al. published results of solidification studies that show the dramatic effects of freezing rate on surface area for polystyrene, starch, and silica gels (71,72). Surface areas after sublimation increased dramatically with higher cooling rates during solidification.

Pikal et al. found that annealing resulted in larger ice crystal sizes for small (5 μL) samples that were frozen rapidly between glass coverslips (73). With sublimation studies in a 13 μL microbalance apparatus, Pikal et al. also showed that annealing resulted in an up to 50% decrease in the normalized dry product resistance during primary drying, but actual drying rates were not presented, and annealing was not tested on products frozen and dried in vials.

Nakamura et al. studied the effect of freezing conditions on the sublimation rate of coffee extract (74). The study was hindered by the cracking of samples frozen with liquid nitrogen, which in their system also affected heat transfer to the sublimation interface. This prevented the authors from making any conclusions about the effect of freezing rate.

Roy and Pikal established an early linkage between the ice nucleation temperature and the subsequent primary drying rate (75). They found that sample vials with internally placed thermocouples nucleated at higher temperatures and completed primary drying sooner than samples without thermocouples. Several have linked the nucleation temperature (extent of global supercooling) to ice crystal morphology (64,67,68).

In 1991 Kochs et al. reported on the effects of the directional solidification process parameters on primary dendritic spacing for samples of 10% hydroxyethyl starch (HES) in a freeze-drying microscope fitted for controlled directional freezing and sublimation (76). The samples were 0.3 mm in thickness, 1.2 mm wide, and 100 mm long. Cooling was applied at one end of the sample, and columnar ice crystals grew in the direction of heat transfer. Faster cooling rates resulted in decreased column spacing, leading to greater resistance and slower drying. The primary dendritic spacing was found to be proportional to the product $v^{-1/4} \cdot G^{-1/2}$ where v is the interface velocity, and G is the temperature gradient at the interface. They also found a linear relationship between the diffusion coefficient for vapor transport through the dried layer, the lamellar spacing, and in turn the primary drying rate.

In a companion paper they examined the effect of freezing conditions on the primary drying rate for 23-mL samples and found higher primary drying rates for samples that had been frozen with slower postnucleation cooling rates during solidification (77). In addition, they analyzed temperature and drying rate data for these samples and found product resistance at the top of the samples to be significantly higher than at the bottom. However, they did not report on the microscopic appearance of the samples.

Dawson and Hockely demonstrated the effect of freezing conditions on morphology for a variety of biological and carbohydrate formulations (78). A 1% (w/v) trehalose formulation frozen by liquid nitrogen exhibited a fine filamentous directional network, whereas freezing by placement on a −50°C shelf yielded a leafy mixed-orientation appearance. The authors state that the rapidly frozen trehalose had less resistance to vapor flow during drying and reconstituted faster than the shelf-frozen samples.

Hsu et al. studied various freezing rates for shelf-ramp freezing, placement on a precooled shelf, and immersion into dry ice/isopropanol (44). For shelf-ramp freezing they obtained lower cooling rates by using greater vial fill volumes in larger vials (keeping the vial:sample volume constant). Vials were instrumented with internal thermocouples near the vial bottom, and the "freezing rate" was calculated as the postnucleation cooling rate. The authors stated that for vials frozen by shelf-ramp, nucleation appeared as a "sudden change in the appearance of the vial contents from a clear liquid to an opaque/translucent slush." This is indicative of global supercooling. On the basis of the author's description of their observations of the freezing of vials in dry ice/isopropanol, freezing by that method yielded directional solidification. For samples frozen by shelf-ramp, the larger samples cooled slower before and after nucleation, achieved less supercooling, had lower surface areas, and better storage stability (discussed further in section "Mechanisms of Morphological Change During Annealing"). The larger samples appear to nucleate at higher temperatures most likely because in a larger sample there would be less temperature homogeneity throughout the liquid volume at the time of nucleation. If when nucleation occurs at the bottom of the vial the temperature of the remaining volume is sufficiently low, the secondary nucleation front will propagate through the entire volume. Samples frozen by a precooled shelf and those frozen by immersion in dry ice/isopropanol exhibited greater cooling rates had larger surface areas, and poorer storage stability (see section "Mechanisms of Morphological Change During Annealing"). When data from all vial sizes and cooling rates were plotted together, the surface area (ranging 0.2–2 m^2/g) was a nonlinear function of the cooling rate, which ranged from 0.3 to 20°C/min.

Chemical composition can have a pronounced effect on morphology. In particular one would think that surfactants have the potential to affect crystal structure and in fact they do. In studies of mannitol, Haikala et al. found that polysorbate 80 concentrations even as low as 0.0001% resulted in altered morphology (79). Increasing concentrations of the surfactant (0.0001–1%) resulted in generally coarser structure in which a fine, "lacy" structure gave way to large plates, longer reconstitution times, and greater mannitol crystallinity (particularly the δ form). Surfactants by definition preferentially populate the interface and thereby modify the thermodynamics of that interface. As discussed in section "Mechanisms of Morphological Change During Annealing" below, surfactants play a role in protecting active ingredients from interfacial damage during freezing.

Milton et al. demonstrated that partial collapse during primary drying could result in progressively decreasing resistance through primary drying itself (80). Their findings were confirmed by direct observation of cake morphology. During drying of lactose near its collapse temperature small holes developed in the plate-like structures of the solid phase, lowering the resistance to vapor flow. Overcashier et al. studied the time course of product resistance to water vapor flow during the primary drying of recombinant humanized antibody and two placebo formulations over a range of pressures and shelf temperatures (81). All samples were shelf-ramp frozen, product temperatures were measured via internal thermocouples, and drying rates were determined gravimetrically by removing and weighing samples at intervals during the runs. Product resistances were calculated from the drying rate and temperature data, and the authors reported that in all cases resistance increased as the depth of dried sample increased. Although drying rates were constant (weight loss was linear with time), the calculated resistance increased because of the increase in measured product temperature through time. The resistance per unit thickness decreased from the top of the dried layer to the bottom. In addition they found, as had Milton et al. (80), that samples dried at higher temperatures had lower product resistance than their counterparts dried at lower product temperatures. This they attribute to localized collapse during primary drying, facilitated by higher temperatures.

Franks in 1998 had suggested that the stochastic nature of nucleation results in heterogeneity among samples (82). In a study published in 2001, Searles et al. confirmed this in their examination of the effect of nucleation temperature during shelf freezing on the freezing mechanism, morphology, and primary drying rate (4). Using varying sample particulate content, vial scoring, and ice nucleating agents, they found the primary drying rate to correlate inversely with the extent of supercooling. To determine the nucleation temperature externally mounted thermocouples were used so as to not interfere with nucleation itself. Both global supercooling and directional solidification mechanisms were found to be possible for their system when frozen via shelf-ramp freezing, but that the latter only occurred with the aid of ice nucleating agents. Global supercooling freezing resulted in dispersed spherulitic morphology, and directional solidification was characterized by lamellar plate morphology. Ice crystal size is inversely correlated to the extent of supercooling; thereby, the number of nuclei correlates directly with the extent of supercooling (83). In cases where the supercooling exceeded 5°C, freezing took place via global supercooling. Sample cooling rates of 0.05 to 1°C/min had no effect on nucleation temperatures and drying rate. Therefore, within the "global supercooling" freezing regime the stochastic nucleation process is in control. Its stochastic nature is the cause of significant drying rate (and therefore temperature) heterogeneity for samples frozen by global supercooling (4). Nucleation temperature heterogeneity may also result in variation in other morphology-related parameters such as surface area and secondary drying rate. Factors such as particulate content and vial condition, which influence ice nucleation temperature, must be carefully controlled to avoid, for example, lot-to-lot variability during cGMP production. The presence of operators has been found to be a significant source of particulates (84), so the proximity of an operator during filling could result in atypically high particulate loading for a number of samples within a lot, and may lead to a subset of samples within a lot nucleating earlier and drying faster. If the factors influencing nucleation temperature are not controlled and/or inadvertently changed during through process development and scale-up a lyophilization cycle that was successful on the research scale may fail during large-scale production.

In a follow-on study, Searles et al. reported that postfreezing annealing can reduce freezing-induced heterogeneity in sublimation rates (3). In addition, they found annealing to result in severalfold drying rate increases. Aqueous solutions of HES, sucrose, and HES:sucrose were frozen either by shelf-ramp or by liquid nitrogen immersion. Samples were then annealed for various durations over a range of temperatures and partially lyophilized to determine the primary drying rate. The drying rate results are shown in Figure 4. In some cases, annealing for only 30 minutes gave a substantial rate increase. Higher annealing temperatures and longer durations of annealing correlated with increased drying rates, but all drying rates appeared to be constrained by a maximum theoretical value. The cause was later identified to be a shift in control of the sublimation rate from mass transfer (water vapor transit through the dried layer of product) to energy transfer (energy from the shelf to the ice in the vial) due to the decreased mass transfer resistance of annealed samples (85).

The morphologies of fully dried liquid nitrogen frozen samples were examined using scanning electron microscopy (3). Figure 3 shows the morphologies of annealed and not-annealed liquid nitrogen frozen samples. Annealing resulted in the merging of the fine lamellar plate structures, reduction in surface area, larger pore sizes, and larger and more numerous holes on the cake surface of annealed samples. The mechanisms behind these changes are discussed in section "Effects on Solute Crystallization." A wide range of postannealing resolidification cooling rates did not affect the primary drying rate, and annealed HES samples dissolved slightly faster than their not-annealed counterparts because of better wetting characteristics.

Annealing below T_g' did not result in increased drying rates. On the basis of that finding, the authors proposed a new annealing-lyophilization method of T_g' determination which can be carried out with only a balance and a freeze-dryer, measures T_g' over any timescale desired, and has the additional

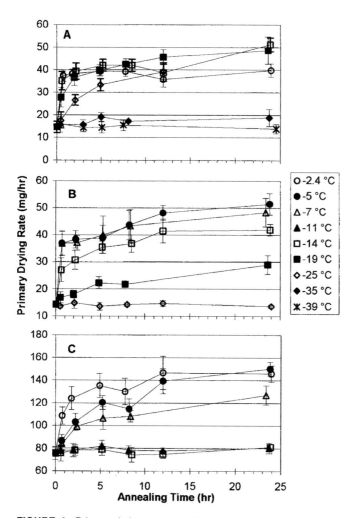

FIGURE 4 Primary drying rate as a function of annealing time and temperature for (**A**) 10% sucrose; (**B**) 5% sucrose, 5% hydroxyethyl starch (HES); (**C**) 10% HES. Groups A, B, and C were dried under different conditions. *Source*: From Ref. 4.

advantage that a large number of candidate formulations can be evaluated simultaneously (3).

Roth, Winter, and Lee used an in situ microbalance to measure the primary drying rate throughout primary drying for product in vials frozen by a variety of methods (86). "Slow" freezing was by shelf-ramp from 10°C to −40°C at −0.14°C/min, "moderate" freezing was carried out in the same manner but cooling was at −1.7°C/min, and liquid nitrogen immersion was used as the "fast" freezing method. The microbalance enabled Roth et al. to observe changes in the drying rate through the entire period of primary drying. Drying rates through time for the slow and moderate freezing methods were virtually indistinguishable, but the fast freezing method resulted in samples that dried faster than the other methods.

Patapoff and Overcashier tested freezing methods and annealing on primary drying rates, sample temperature during primary drying, dried product resistance, appearance of the freeze-dried cakes, and protein aggregation (10). The latter will be discussed in section "Mechanisms of Morphological Change During Annealing" below. They tested a recombinant human antibody formulation. Cooling via immersion into a dry ice/ethanol bath resulted in the greatest product resistance and therefore slower sublimation rates and higher product temperatures. Shelf-ramp freezing and ice crystal seeded shelf-ramp freezing each resulted in successively less resistance. Samples that were frozen using the standard shelf-ramp method and annealed had the lowest resistance. The dry ice/ethanol and standard methods also formed a more resistive layer at the top of the sample. They also found that a placebo version of their formulation exhibited completely different freezing and drying characteristics than the active-containing formulation: the placebo dried significantly faster.

Vacuum-flask freeze-dryers are used in many laboratories for bulk lyophilization. They consist of a vacuum and condenser system to which glass flasks are attached externally. The liquid within the flasks is frozen by evaporative cooling. Kramer et al. used this principle to freeze solutions in vials within a freeze-dryer (11). With vials loaded on a $10°C$ shelf, they reduced the pressure to 760 mT to induce the formation of a 1 to 3 mm thick layer of ice on the top of the solution within five minutes. The shelf temperature was then ramped down to $-40°C$ to complete solidification and lyophilization was continued normally. The drying rates and morphologies of these samples were compared with those frozen by conventional shelf-ramp. In addition, annealing was tested on samples frozen by both means. The evaporative nucleation method resulted in large "chimney-like" ice crystals, and the drying rates of some formulations were up to 20% faster. Annealing also resulted in increased drying rates. Shelf-ramp frozen sample morphology was spherulitic, and annealing of these samples increased the size of those spheroidal ice crystals.

Spray-freeze-drying is a process in which fine droplets are sprayed through an ultrasonic nozzle directly into liquid nitrogen. The frozen droplets are then placed into vials and freeze-dried in a standard lyophilizer. They are of particular interest for pulmonary and epidermal delivery. Sonner et al. recently reported on the use of this technology using trypsinogen as a model protein (48). They produced spherical particles with diameters from 20 to 90 µm that contain high internal porosity. Sonner et al. compared the properties of 20% trehalose samples that were spray-freeze-dried and annealing during primary drying to those that were not annealed. Annealing caused the particles to shrivel. The authors tested the postdrying hygroscopicity of the particles by measuring moisture uptake rates. They found that the longer the particles were annealing during drying, the lower the subsequent moisture uptake rates. They theorized that annealing reduced the specific surface area of the particles, slowing water adsorption. Protein recovery results will be discussed in section "Mechanisms of Morphological Change During Annealing" below.

Webb et al. studied several formulations of human interferon-γ, comparing liquid nitrogen immersion vial freezing with spray-freezing (47). Vial immersion resulted in a directional lamellar morphology, and the freeze-dried cakes were severely cracked. Spray-freezing yielded spherical particles similar to Sonner et al. (48). Specific surface areas for the spray-freeze-dried samples were 6.8 to 15 m^2/g, three to seven times those of their counterparts that were filled into vials and frozen by liquid nitrogen immersion. The liquid nitrogen immersion formulations contained more water at the end of freeze-drying than their spray-freeze-dried counterparts, and reconstituted much more rapidly. The rapid reconstitution was a visually violent process with extensive bubble formation and some splattering of liquid onto the sides of the vials. Dissolution rates were not affected by the presence of 0.03% polysorbate 20. Annealing of the liquid nitrogen immersion frozen samples alleviated the cracking which occurred during drying, resulting in a more pharmaceutically elegant appearance. The internal microstructure of these samples was greatly simplified by annealing, leading to severalfold surface area reductions, faster primary drying, and slightly higher moisture levels after lyophilization. The annealed samples did not exhibit the bubbling and splattering during reconstitution, and reconstitution times were longer. Annealing also had profound impacts on the spray-freeze-dried samples. Surface areas of these samples were reduced by up to 30-fold. Regardless of the freezing method all samples tended toward a surface area of 0.5 to 1.0 m^2/g after annealing. After annealing, the spray-frozen spheres appeared as agglomerated hollow shells, indicating that the spheres had collapsed outward toward their outer shells, and that some merging of adjacent shells had occurred. Annealing of these samples also resulted in faster primary drying and higher final moisture contents (47). In 1990 Pikal et al. showed that for solutions of moxalactam disodium the *secondary* drying rate is inversely proportional to the specific surface area (m^2/g) of the sample (87). The fact that annealing decreases surface areas provides the linkage necessary to state that those of Pikal et al. corroborate the observation by Webb et al.

The stress relaxation and morphological changes that result from annealing can affect reconstitution. Those changes that allow more efficient water vapor transport during drying may also improve wetability of the porous matrix. For example, the holes formed in the "skin" layer on the surface of the cake may allow easier liquid penetration. However, simplification of the matrix morphology will increase the thickness of the matrix structures and reduce the surface area, which could slow dissolution. Webb et al. also found that annealed spray-freeze-dried samples dissolved slightly faster than their not-annealed counterparts during reconstitution (47). Conversely, however, the annealed lyophilized samples dissolved more slowly. This is in contrast to results of Searles et al. who found that the same formulation dissolved slightly faster after annealing; however, the drying protocols for the two studies were not the same (3).

In 2004, Abdelwahed et al. reported that aqueous solutions of nanocapsules exhibit responses to annealing that are very similar to those observed for other materials (88). Annealing above the formulation's T_g' increased ice crystal size, opened holes on the cake's surface film, decreased the sublimation rate, and decreased the temperature during primary drying. Secondary drying kinetics were slowed for a sucrose formulation, but not for one with PVP.

Liu et al. studied effects of freezing and annealing conditions on cake structure and drying rate for a product with a high fill height of 15% cyclodextrin (17). They found that slow shelf-ramp freezing at $-0.3°C/min$ resulted in a slow-drying product for which the cake was denser in the core than at the top or bottom. This is because, while initial ice nucleation was from the vial bottom, ice broke off and floated to the top, causing ice growth to start from there as well. The converging ice fronts from bottom and top caused solute concentration in the middle of the vial. A two-step freezing procedure (from

+5°C to −10°C at −1°C/min, held for one hour, then cooled to −45°C at −1°C/min) alleviated the concentrated core by shifting the mechanism to "global super-cooling" as described in this chapter. Annealing yielded significant reductions in primary drying time with no adverse effect on product quality.

An interesting case of inadvertent annealing is described by Wallen et al. (12) They examined a case of higher numbers of rejected vials originating from the higher shelves of a production lyophilizer. The product in question is frozen in a liquid nitrogen-freezing tunnel, in which liquid nitrogen is sprayed onto filled vials. The frozen vials are then loaded onto precooled shelves of the lyophilizer. Use of product temperature monitors found that, owing to the shelf indexing scheme and natural convection of room air, the vials loaded onto the topmost shelves were warming. This "inadvertent annealing" appears to have caused a higher collapse rate.

Blue and Yoder describe a 60% reduction in reconstitution time resulting from annealing a concentrated monoclonal antibody formulation (89).

MECHANISMS OF MORPHOLOGICAL CHANGE DURING ANNEALING

For amorphous solutes, annealing above T_g' will, by the definition of T_g', result in melting of ice into neighboring nonice regions (Fig. 1). The increased water content and higher temperature increase both the bulk mobility of and the amorphous phase and diffusional mobility of all species in that phase. The increased bulk mobility of the amorphous phase during annealing allows it to relax into physical configurations of lower free energy. Surface free energy (also known as surface tension) (γ) is defined as free energy per unit area, and consequently has units of energy per unit area (85). Therefore there will always be a driving force for a contiguous volume to reduce its surface area so long as it is not already a perfect sphere (the shape which possesses the minimum possible surface-to-volume ratio).

The structures in freeze-concentrated systems inevitably possess junctions, edges, and other departures from the perfect sphere. Each of these three-dimensional features has two principal radii of curvature. The *pressure* inside such a structure with positive radii of curvature can be shown to be

$$P = 2\gamma \left(\frac{1}{r_1} + \frac{1}{r_2} \right) + P_o$$

where γ is the surface free energy (energy/length^2), r_1 and r_2 are the radii of curvature, and P_o is the surrounding pressure. Consider a needle-shaped structure. The tip will have very small radii of curvature, and the radii of curvature for material in the body of the needle will be much greater. Using the equation above we can deduce that the pressure in the tip will be higher than the pressure in the body. Given sufficiently low viscosity, this pressure difference will (during our timescale of observation) cause *bulk flow* of material from the tip into the body, and the tip will actually retract: the needle will become more sphere-like in shape. Structures with negative internal radii, such as the concave surface at the junction of two spheres in contact, will have a *lower* pressure than the surroundings, leading to a flow of material to these areas. Sintering and accretion of high surface area ice particles into a single lower surface area structure are manifestations of this principle (90,91).

Another fundamental relationship of surface chemistry is the Kelvin equation, which states that smaller radii of curvature induce a higher vapor pressure.

$$\Delta G = RT \ln \frac{P_v}{P_v^0} = \gamma V \left(\frac{1}{r_1} + \frac{1}{r_2} \right)$$

In this equation ΔG is the increase in molar free energy, R is the universal gas constant, T is temperature, P_v^0 is the normal vapor pressure, P_v is the vapor pressure over a curved interface, V is the specific volume, and r is the radius of curvature. Ice crystal regions with smaller radii will melt preferentially because of their higher free energy because smaller ice crystals have higher vapor pressures than larger ones. Searles et al. found that in some cases only a very short annealing period was required to achieve a maximum sublimation rate (3). The authors believed this to be due to the preferential melting of the smallest ice crystals upon initial heating for annealing. Recooling did not result in new crystal growth because the volume was already nucleated. Rather, upon recooling, the fewer larger remaining crystals simply grew. In that sense annealing can be thought of as a partial "refreezing" step.

Ostwald ripening (recrystallization) can also occur in these systems. It is a phenomenon by which dispersed crystals smaller than a critical size decrease in size as those larger than the critical size grow, and it can be either diffusion or surface-attachment limited. The food science literature contains several examples of diffusion-limited Ostwald ripening in aqueous carbohydrate solutions (model ice creams) (92–96). A common aspect of these reports is that no increase in ice crystal size is observed below T_g'.

Searles et al. observed that annealing opened up holes on the surface of the lyophilized cake (3). This would occur via drainage of the amorphous film into adjacent junctions. Drainage of the surface film ("skin") into the consolidated junctions is apparent in Figure 5. The thinnest area of the film between the junctions is most susceptible to instabilities, and rupture would be followed by film retraction from the edges with small positive curvature to the junctions because of their negative radii of curvature. On examining Figure 3, the lamellar plates can be seen as they join the surface film from below. From Figure 4B (note the difference in scale) it can be seen that some of the plates fused together, since

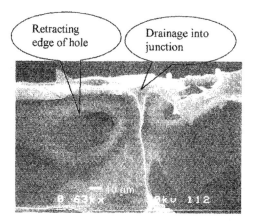

FIGURE 5 Evidence of drainage into a junction as a result of annealing: 10% hydroxyethyl starch, which had been frozen by immersion into liquid nitrogen, annealed, then lyophilized. Retracting hole edges as well as a filling junction are evident.

the ~5-μm not-annealed spacing has been consolidated into up to 100 μm spacing, with the implication that up to 20 lamellae were fused. Shalaev and Franks show evidence of similar mechanisms in the formation of globular structures due to dendrite retraction due to partial collapse (97).

NONAQUEOUS COSOLVENT SYSTEMS

As discussed above, the chemical makeup of a sample will play a large role in its freezing behavior. Nonaqueous cosolvent systems are no exception. Readers are referred to Teagarden and Baker for a comprehensive review of these systems in lyophilization (98). Advantages of such systems include "increased drug wetting or solubility, increased sublimation rates, increased predried bulk solution or dried product stability, decreased reconstitution time, and enhancement of sterility assurance of the predried bulk solution" (98). Disadvantages are requirements for additional manufacturing safety precautions, potential unique equipment modifications for solvent condensation, qualification of appropriate GMP purity, and increased regulatory scrutiny. This section will be confined to changes that cosolvents impart to freezing behavior.

Kasraian and DeLuca studied sucrose formulations with and without tertiary butyl alcohol (TBA) (99). They found that inclusion of TBA resulted in the formation of large, columnar ice crystals that appear characteristic of directional solidification. Sucrose with TBA was found to dry at 1/10th the rate, and the cake appearance indicated the presence of spheroidal ice crystals. Product resistance was dominated by a layer at the top of the cake for the sucrose-only formulation, whereas for the TBA-containing samples resistance an order of magnitude lower and increase appeared constant throughout the cake depth. Surface area measurements confirmed that the specific surface area of the TBA formulation was over 10 times that of the sucrose-only formulation (8.6 vs. 0.67 m^2/g) consistent with the high surface-to-volume ratio of columnar compared with spheroidal shapes. Others have found the addition of TBA to improve sublimation rates for their formulations (98).

In studies of sucrose lyophilization in a water/TBA cosolvent system Wittaya-Areekul and Nail found that the method of freezing affected the level of residual TBA after drying (100). Freezing by shelf-ramp resulted in lower residual TBA levels than liquid nitrogen immersion freezing. The authors theorized that the slower cooling rate facilitated more complete crystallization of the TBA during freezing. It may be that these crystalline regions sublimate more rapidly than noncrystalline TBA can diffuse and desorb from the solid phase.

Wittaya-Areekul et al. found that annealing facilitated more complete crystallization of TBA in an aqueous solution of tobramycin, which resulted in lower levels of residual TBA after lyophilization (101). While residual TBA was correlated with extent of TBA crystallization, residual isopropyl alcohol (IPA), which is an impurity in TBA, was highest with the annealed case (describe the other conditions). IPA does not crystallize, thus the lower surface area and increased diffusion path length resulted in the annealed samples having the highest levels of residual IPA.

In two cited cases annealing was found to facilitate greater removal of *tert*-butanol, which the authors ascribed to crystallization of the cosolvent during annealing. Annealing also improved the uniformity of moisture levels among the samples.

EFFECTS ON SOLUTE CRYSTALLIZATION

Many have found solute crystallinity to be strongly dependent on the freezing method and annealing. This is understandable given the extreme freeze-concentration that occurs through the freezing process. Supersaturation of some components can result if freezing is too rapid to allow complete crystallization. Dire consequences can result. If, for example, mannitol is not completely crystallized by the completion of lyophilization it may crystallize during subsequent storage, releasing waters of hydration and causing the lyophilized cake to seemingly spontaneously collapse. Mannitol is of great interest because it is widely used as a bulking agent for lyophilization, but it is only effective as a bulking agent when in the crystalline form. In addition to the dangers of incomplete crystallization, one must be aware of the fact that mannitol crystallization can cause vial breakage if the freezing step is not carried out appropriately (102,103). Glycine is also used as a crystalline bulking agent (104). These agents provide mechanical strength to the cake during drying and storage, and, if fully crystalline, provide excellent protection against collapse. A recent report cites a significance of crystalline mannitol—it can prevent collapse of a sucrose cake during drying at $-10°C$ even though sucrose alone will collapse at $-40°C$ (105).

Pikal-Cleland showed that the presence of glycine reduced the freeze concentration–induced pH drop in phosphate buffers depending on the buffer and phosphate concentrations (106). The greatest benefit was with 10 mM (rather than 100 mM) buffer, and glycine concentrations ≤ 50 mM. The authors found that the reduction in pH change was the result of decreased crystallization of disodium phosphate, and theorized that glycine hindered nucleation of the disodium phosphate by forming a glycine sodium salt. Interestingly for 10 mM phosphate buffer, the presence of glycine concentrations ≥ 100 mM resulted in greater pH drops than for no glycine at all. This was linked to more complete crystallization of disodium phosphate. Recovery of LDH activity after freeze-thaw was improved from $\sim 80\%$ to 100% by the use of intermediate glycine concentrations (75–250 mM) regardless of the pH shift during freezing.

Cannon and Trappler found that the freezing method affected the crystalline polymorph of 70 mM mannitol solutions (107). Crystalline content was evaluated after lyophilization. Shelf-ramp freezing produced mostly the δ form with a minor presence of the α polymorph, and freezing on a precooled shelf led to mostly α with minor β content. Addition of a four-hour hold step at $-20°C$ during the shelf-ramp protocol yielded a sample that contained only the δ polymorph. Annealing a shelf-ramp frozen sample at $-20°C$ for one hour resulted in only the β form.

A very important topic related to solute crystallization is the extreme pH drops that can occur when freezing solutions containing phosphate buffer systems. A recent publication by Gomez et al. provides a thorough characterization of the phenomenon (108). The pH drop is caused by precipitation of the disodium salt during freeze-concentration. They present dramatic time course pH data through the freezing process, illustrating that for solutions with an initial pH of 7.4, the pH after freezing can be as low as 4. Sodium chloride also crystallizes during freeze-concentration (109,110).

In some cases product crystallization can occur. Oguchi studied the freezing of methyl *p*-hydroxybenzoate (MPHB) and found that the freezing rate affected crystallization of the MPHB (111). Rapid freezing resulted in amorphous MPHB while slow freezing caused crystalline inclusion complexes of the product. Slow freezing was by incubation at −13°C, and rapid freezing was by immersion in liquid nitrogen. Williams and Schwinke reported on their efforts to produce crystalline pentamidine isethionate (PI) after lyophilization (112). Normal shelf-ramp freezing and lyophilization of their 100 mg/mL formulation resulted in product crystallization within only about 20% of the samples. This was improved to 80% using a 15-hour −5°C prefreezing hold that exploited the fact that at 5°C the solubility of their product was only 39 mg/mL. Macroscopic appearance was markedly different as well. Those which were frozen subsequent to product crystallization were "fluffy and the particles clung together like a cotton ball," while the conventionally frozen sample cakes appeared hard and compact. Photostability of fully crystalline material was better as well. Chongprasert et al. used a similar method for PI (113). The product was held at 2°C prior to freezing to induce precipitation of a PI trihydrate that converted to an anhydrous crystal form during lyophilization. A different anhydrous crystal form was observed to result from liquid nitrogen immersion freezing. The authors found that when shelf-ramp frozen, the PI sometimes crystallized from solution before ice nucleation, and in the other cases crystallized as the other form subsequent to ice nucleation. In some cases both types were found within different regions of the same product cake. Therefore in this case the stochastic nature has been found to have a profound affect on a product quality attribute.

Skrabanja et al. presented an interesting consequence of phosphate precipitation (39). They studied a solution of sucrose in a buffer of citric acid and dibasic sodium phosphate. Nonenzymatic browning was observed after one month of storage at 60°C, caused by a pH drop resulting from the crystallization of sodium phosphate. The pH drop had caused hydrolization of sucrose into fructose and glucose. Glucose is a reducing sugar that could then bind to a protein product in the formulation via the Maillard "browning" reaction.

Izutsu et al. reported in 1993 that amorphous mannitol provides some protection to β-galactosidase during freeze-thaw, but that crystallization during annealing reduced the protein-stabilizing effects of the solute (114,115). Greater activity of annealed samples remained after freeze-thaw than after lyophilization-reconstitution, confirming that damage also occurs throughout the process. In 1994 the same research group published results of a follow-on study in which they used annealing to crystallize mannitol in three different formulations, each containing a different protein (115). In each case, annealing led to mannitol crystallization, which caused protein inactivation to an extent proportional to the degree of mannitol crystallinity.

Pyne and Suryanarayanan found that glycine forms an amorphous freeze-concentrate when quench-cooled in liquid nitrogen, but crystallizes as the β-form when the solution is frozen by either rapid (20°C/min) or slow (2°C/min) cooling in a DSC (116). When these pure glycine solutions were freeze-dried, the quench-cooled solutions formed γ-glycine during primary drying, and at the end of secondary drying both β- and γ-glycine were found. These measurements were made using novel in situ x-ray diffraction during lyophilization. Samples frozen by either the rapid or slow methods contained only the β form after freeze-drying, a result that did not depend on the choice of primary drying temperature used (from −35°C to −10°C). However for quench-cooled solutions the β-glycine content did not change, but the γ-glycine crystalline content of the samples increased in samples dried at a higher primary drying temperature. It was found that the glycine concentration also affected the crystallization behavior for quench-cooled samples in that those dried at the lowest temperature of −35°C contained only β-glycine.

Pyne and Suryanarayanan also found the effect of annealing to depend on freezing rate (116). For solutions that had been frozen at 20°C/min or 2°C/min, annealing at −10°C led to an increase in the existing β-glycine content. As noted above, quench-cooled glycine did not crystallize, but upon heating to −65°C an unidentified crystalline form appeared, which gave way to β-glycine at −55°C. Annealing at −10°C caused a transition from β-glycine to γ-glycine. However, annealing at −20°C only caused increases in the β-glycine content, and −35°C annealing showed miniscule increases in β content with no detectable γ-glycine present.

The crystals grown during annealing can serve to impede the flow of water vapor through the dry layer (117). Highly crystallized mannitol was found to block the pathways for water vapor escape, contributing to the increase in the dry layer resistance and thus the longer times for primary drying. Freeze-dried cakes without annealing had lower dry layer resistances because partial collapse created larger channels for water vapor escape (117).

PRIMARY DRYING OPTIMIZATION THROUGH ANNEALING

Searles found that annealing decreased the resistance to vapor flow to such an extent that control of the process shifts from mass transfer to energy transfer (85). Figure 6 shows predicted drying rate (panel A) and product temperature (panel B) as a function of product resistance for products lyophilized in vials. The heat transfer characteristics are particular for a specific vial and the chamber pressure (which in this case is 26.6 Pa). As product mass transfer resistance is increased, the drying rate decreases and the product temperature increases. However, for products with a low product resistance (in this case less than approximately $1E + 4$ Pa·sec·m^2/kg), very slight drying rate decreases will be seen as product resistance is increased. This is because at low product resistance, the product resistance is not *controlling*. Rather, the process is controlled by heat transfer. For high product resistance, the drying rate is very sensitive to product resistance and mass transfer is controlling.

The product temperature trends shown in panel B are very interesting. Note that at high product resistance the product temperature tends toward that of the shelf. At low product resistance the product temperature tends toward the temperature at which the vapor pressure of ice equals the chamber pressure. In this case that temperature is −33°C. As shelf temperature is increased, drying rate increases and product temperature increases *if* the product resistance is high enough. Note that for low product resistance, shelf temperature increases yield higher drying rates but barely-measurable product temperature increases. Similar to the boiling of water in a pot, additional heat input does not increase the temperature of the ice but merely increases the sublimation rate. Therefore low product resistance to mass transfer renders the product temperature insensitive to additional energy input.

The lyophilization of samples in vials on the shelf of a freeze-dryer presents a unique challenge in that the perimeter vials receive additional energy during sublimation and dry at up to twice the rate as vials on the interior of the shelf (85). Therefore, the process development scientist/engineer must design the lyophilization cycle such that perimeter vials do not collapse because of the

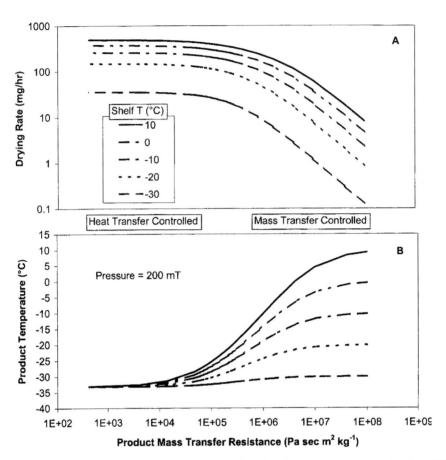

FIGURE 6 Simulations of (**A**) drying rate and (**B**) ice interface temperature as a function of the product mass transfer resistance and shelf temperature: 2-mL fill in a 5-mL vial, 26.6 Pa (200 mT) chamber pressure. Note the change from heat to mass transfer control over the drying rate and temperature.

excessive energy input, but primary drying must be long enough to accommodate the vials in the center of the shelf (which can be drying at half the rate). However if the vials were to have been sufficiently annealed, all would dry at the nearly same temperature (close to that dictated by the chamber pressure). Not only can the overall drying rate be increased to the maximum that the equipment can support, heterogeneity in product temperatures can be greatly reduced. Such a strategy will allow use of the maximum sublimation rate which the equipment can support and will improve product consistency.

THE FUTURE
While this chapter has summarized many examples of freezing and annealing effects on the product and the lyophilization process (some of which are quite surprising), few of the citations in this chapter are for work devoted to better

understanding the freezing process itself. Pharmaceutical active ingredients generally fall into one of the two categories: characterized or uncharacterized. Can it be said that *lyophilized* dosage forms are fully characterized, even if the active ingredient is fully characterized? When a fully characterized active ingredient is lyophilized we must be mindful of the limited extent to which we can characterize attributes of the final lyophilized product. For example, aspects of product cake structure and surface area are not well understood or easy to measure and are part of what is accepted as the "natural" heterogeneity within a lyophilized batch. Do we understand how a shift in the freezing protocol will affect stability of the product? Without a better understanding of how freezing actually takes place, we cannot hope to fully characterize our lyophilized dosage forms. For example, how and why will cake morphology be affected by a change in vial? Why do cakes sometimes appear with a directional morphology in the bottom half of the cake and spherulitic in the top half? How do we correlate cake appearance (e.g., extent of collapse) with product quality for deviation investigations? What cooling rates during freezing will result in the "same" product? When freezing on a precooled shelf, what conditions are necessary to achieve directional solidification? How will the cake morphology change with a slight change in the freezing protocol? Which forms of crystalline mannitol will arise from a slight change in the protocol? These questions are worth answering—failing to pursue them will lead to increased regulatory scrutiny, occasional product development and manufacturing deviation failures, and inefficient use of our existing lyophilization infrastructure.

CONCLUSIONS

Freezing is the major dehydration step of lyophilization. It determines the texture of the final product. Changes to the freezing protocol, thermal history during freezing, vial, volume of fill, particulate levels, ice nucleation properties, or any postfreezing temperature excursions above T_g' (intentional or not) can result in morphological changes, phase separation, product degradation, changes in the crystallization behavior of the solutes or the product, product stability, changes in drying rates (which can result in altered moisture levels or product collapse), and altered levels of residual cosolvent. Beware that many of the process factors affecting freezing can change through the scale-up/technical transfer process, leading to risk of failure at full-scale if not adequately controlled.

Two fundamental types of freezing behavior have been identified: directional solidification and global supercooling. Directional solidification results in a directional lamellar morphology with connected pores, and global supercooling creates spherulitic ice crystals. The directional morphology has a higher specific surface area, and results in higher primary drying rates due to the connected, continuous nature of the pores.

Within a given type of freezing one can say that "faster freezing" leads to higher surface areas, smaller more numerous ice crystals, and slower primary drying rates. For directional solidification, freezing rate control equates to cooling rate control, but below a critical cooling rate one would expect a loss of directional solidification at least in some portion of the volume being frozen. For freezing by global supercooling Hsu et al. have altered the "freezing rate" by changing the vial sizes (44). Searles et al. found that *within the global supercooling regime* the extent of supercooling and therefore "freezing rate" were independent of shelf cooling rate (4).

However one *cannot* say that in general "faster freezing" means higher surface areas, smaller more numerous ice crystals, and slower primary drying rates because liquid nitrogen immersion provides much faster cooling than shelf-ramp freezing. Liquid nitrogen immersion results in higher surface areas but faster primary drying rates due to the completely different type of morphology that results.

Different freezing methods and annealing steps are powerful tools to manipulate numerous product quality and productivity attributes. By understanding how freezing and annealing affect our products and processes, we can devise better freezing and annealing protocols to increase lyophilization plant capacity utilization and improve product quality and consistency. Do not blindly accept the results of your freezing method: investigate alternative freezing methods and test the effects of annealing steps.

REFERENCES

1. MacKenzie AP. The physico-chemical environment during the freezing and thawing of biological materials. In: Duckworth R, ed.Water Relations of Foods, London: Academic Press, 1975.
2. Blond G, Simatos D, Catte M, et al. Modeling of the water-sucrose state diagram below 0°C. Carbohydr Res 1997; 298:139–145.
3. Searles J, Carpenter J, Randolph T. Annealing to optimize the primary drying rate, reduce freezing-induced drying rate heterogeneity, and determine Tg′ in pharmaceutical lyophilization. J Pharm Sci 2001; 90(7):872–887.
4. Searles J, Carpenter J, Randolph T. The ice nucleation temperature determines the primary drying rate of lyophilization for samples frozen on a temperature-controlled shelf. J Pharm Sci 2001; 90(7):860–871.
5. Ablett S, Izzard MJ, Lillford PJ. Differential scanning calorimetric study of frozen sucrose and glycerol solutions. J Chem Soc Faraday Trans 1992; 88(6):789–794.
6. Franks F. Freeze-drying: a combination of physics, chemistry, engineering, and economics. Jpn J Freezing Drying 1992; 38(5):5–16.
7. Li S, Schoneich C, Borchardt RT. Chemical instability of protein pharmaceuticals: mechanisms of oxidation and strategies for stabilization. Biotechnol Bioeng 1995; 48(5):490–500.
8. Franks F. Freeze drying: from empiricism to predictability. Cryo Letters 1990; 11:93.
9. Heller MC. Causes and consequences of polymeric phase separations in protein formulations during lyophilization. University of Colorado Chemical and Biological Engineering PhD thesis, 1998.
10. Patapoff T, Overcashier D. The importance of freezing on lyophilization cycle development. BioPharm 2002; 15(3):16–21, 72.
11. Kramer M, Sennhenn B, Lee G. Freeze drying using vacuum induced surface freezing. J Pharm Sci 2002; 91:433–443.
12. Wallen AJ, Van Ocker SH, Sinacola JR, et al. The effect of loading process on product collapse during large-scale lyophilization. J Pharm Sci 2009; 98(3):997–1004.
13. Rambhatla S, Ramot R, Bhugra C, et al. Heat and mass transfer scale-up issues during freeze drying: II. Control and characterization of the degree of supercooling. AAPS PharmSciTech 2004; 5(4):e58.
14. Nakagawa K, Hottot A, Vessot S, et al. Influence of controlled nucleation by ultrasounds on ice morphology of frozen formulations for pharmaceutical proteins freeze-drying. Chem Eng Process 2006; 45:783–791.
15. Hottot A, Nakagawa K, Andrieu J. Effect of ultrasound-controlled nucleation on structural and morphological properties of freeze-dried mannitol solutions. Chem Eng Res Des 2008; 86:193–200.

16. Passot S, Trelea IC, Marin M, et al. Effect of controlled ice nucleation on primary drying stage and protein recovery in vials cooled in a modified freeze-dryer. J Biomech Eng 2009; 131(7):074511.
17. Liu J, Viverette T, Virgin M, et al. A study of the impact of freezing on the lyophilization of a concentrated formulation with a high fill depth. Pharm Dev Technol 2005; 10(2):261–272.
18. Petersen A, Schneider H, Rau G, et al. A new approach for freezing of aqueous solutions under active control of the nucleation temperature. Cryobiology 2006; 53(2):248–257.
19. Petersen A, Rau G, Glasmacher B. Reduction of primary freeze-drying time by electric field induced ice nucleus formation. Heat Mass Transf 2006; 42:929–938.
20. Muhr A, Blanshard J. Effect of polysaccharide stabilizers on the rate of growth of ice. J Food Technol 1986; 21:683–710.
21. Budiaman E, Fennema O. Linear rate of water crystallization as influenced by temperature of hydrocolloid suspensions. J Dairy Sci 1987; 70:534–546.
22. Blond G. Velocity of linear crystallization of ice in macromolecular systems. Cryobiology 1988; 25:61–66.
23. Abdul-Fattah AM, Truong-Le V, Yee L, et al. Drying-induced variations in physicochemical properties of amorphous pharmaceuticals and their impact on stability (I): stability of a monoclonal antibody. J Pharm Sci 2007; 96(8):1983–2008.
24. Abdul-Fattah AM, Truong-Le V, Yee L, et al. Drying-induced variations in physicochemical properties of amorphous pharmaceuticals and their impact on Stability II: stability of a vaccine. Pharm Res 2007; 24(4):715–727.
25. Pisal S, Wawde G, Salvankar S, et al. Vacuum foam drying for preservation of LaSota virus: effect of additives. AAPS PharmSciTech 2006; 7(3):60.
26. Truong-Le V. Preservation of bioactive materials by freeze dried foam. U.S.A. Patent 7135180 B2, 2006.
27. Truong-Le V. Preservation of bioactive materials by freeze dried foam. U.S.A. Patent 7,381,425 B1, 2006.
28. Bronshtein V. Preservation by foam formation. U.S.A. Patent 5,766,520, 1998.
29. Bronshtein V, Bracken KR, Cambell JG. Bulk drying and the effects of inducing bubble nucleation U.S.A. Patent 6,884,866, 2005.
30. Bronshtein V, Bracken KR, Livers RK, et al. Industrial scale barrier technology for preservation of sensitive biological materials at ambient temperatures. U.S.A. Patent 6,306,345, 2001.
31. Carpenter JF, Chang BS, Garzon-Rodriguez W, et al. Rational design of stable lyophilized protein formulations: theory and practice. Pharm Biotechnol 2002; 13:109–133.
32. Carpenter J, Prestrelski S, Anchordoguy T, et al. Interations of stabilizers with proteins during freezing and drying. ACS Symp Ser 1994; 567:134.
33. Wang W. Lyophilization and development of solid protein pharmaceuticals. Int J Pharm 2000; 203(1–2):1–60.
34. Bhatnagar BS, Bogner RH, Pikal MJ. Protein stability during freezing: separation of stresses and mechanisms of protein stabilization. Pharm Dev Technol 2007; 12(5): 505–523.
35. Hopkins F. Denaturation of proteins by urea and related substances. Nature 1930; 126:383.
36. Bhatnagar BS, Pikal MJ, Bogner RH. Study of the individual contributions of ice formation and freeze-concentration on isothermal stability of lactate dehydrogenase during freezing. J Pharm Sci 2007; 97(2):798–814.
37. Tang XC, Pikal MJ. The effect of stabilizers and denaturants on the cold denaturation temperatures of proteins and implications for freeze-drying. Pharm Res 2005; 22(7):1167–1175.
38. Eckhardt B, Oeswein J, Bewley T. Effect of freezing on aggregation of human growth hormone. Pharm Res 1991; 8(11):1360.
39. Skrabanja A, De Meere L, De Ruiter R, et al. Lyophilization of biotechnology products. PDA J Pharm Sci Technol 1994; 48(6):311.

40. Chang BS, Kendrick BS, Carpenter JF. Surface-induced denaturation of proteins during freezing and its inhibition by surfactants. J Pharm Sci 1996; 85(12):1325–1330.
41. Kreilgaard L, Jones LS, Randolph TW, et al. Effect of tween 20 on freeze-thawing- and agitation-induced aggregation of recombinant, human factor XIII. J Pharm Sci 1998; 87(12):1597–1603.
42. Jiang S, Nail S. Effect of process conditions on recovery of protein activity after freezing and freeze-drying. Eur J Pharm Biopharm 1998; 45:249.
43. Sarciaux JM, Mansour S, Hageman MJ, et al. Effects of buffer composition and processing conditions on aggregation of bovine IgG during freeze-drying. J Pharm Sci 1999; 88(12):1354–1361.
44. Hsu C, Nguyen H, Yeung D, et al. Surface denaturation at solid-void interface-a possible pathway by which opalescent particulates form during the storage of lyophilized tissue-type plasminogen-activator at high-temperatures. Pharm Res 1995; 12(1):69–77.
45. Nema S, Avis K. Freeze-thaw studies of a model protein, lactate dehydrogenase, in the presence of cryoprotectants. J Parenter Sci Technol 1992; 47(2):76–83.
46. Sarciaux JM, Hageman MJ. Effects of bovine somatotropin (rbSt) concentration at different moisture levels on the physical stability of sucrose in freeze-dried rbSt/ sucrose mixtures. J Pharm Sci 1997; 86(3):365.
47. Webb SD, Cleland JL, Carpenter JF, et al. Effects of annealing lyophilized and spray-lyophilized formulations of recombinant human interferon-gamma. J Pharm Sci 2003; 92(4):715–729.
48. Sonner C, Maa YF, Lee G. Spray-freeze-drying for protein powder preparation: particle characterization and a case study with trypsinogen stability. J Pharm Sci 2002; 91(10):2122–2139.
49. Izutsu K, Shigeo K. Phase separation of polyelectrolytes and non-ionic polymers in frozen solutions. Phys Chem Chem Phys 2000; 2:123.
50. Izutsu K, Kojima S. Freeze-concentration separates proteins and polymer excipients into different amorphous phases. Pharm Res 2000; 17(10):1316–1322.
51. Heller MC, Carpenter JF, Randolph TW. Effects of phase-separating systems on lyophilized hemoglobin. J Pharm Sci 1996; 85(12):1358–1362.
52. Heller MC, Carpenter JF, Randolph TW. Manipulation of lyophilization-induced phase separation: implications for pharmaceutical proteins. Biotechnol Prog 13(5): 590–596.
53. Heller MC, Carpenter JF, Randolph TW. Conformational stability of lyophilized PEGylated proteins in a phase-separating system. J Pharm Sci 1999; 88(1):58–64.
54. Heller MC, Carpenter JF, Randolph TW. Protein formulation and lyophilization cycle design: prevention of damage due to freeze-concentration induced phase separation. Biotechnol Bioeng 1999; 63(2):166–174.
55. Her LM, Deras M, Nail SL. Electrolyte-induced changes in glass transition temperatures of freeze-concentrated solutes. Pharm Res 1995; 12(5):768.
56. Allison SD, Molina MC, Anchordoquy TJ. Stabilization of lipid/DNA complexes during the freezing step of the lyophilization process: the particle isolation hypothesis. Biochim Biophys Acta 2000; 1468(1–2):127–138.
57. Molina MC, Allison SD, Anchordoquy TJ. Maintenance of nonviral vector particle size during the freezing step of the lyophilization process is insufficient for preservation of activity: insight from other structural indicators. J Pharm Sci 2001; 90(10): 1445–1455.
58. Cochran T, Nail SL. Ice nucleation temperature influences recovery of activity of a model protein after freeze drying. J Pharm Sci 2009; 98(9):3495–3498.
59. Maa YF, Zhao L, Payne LG, et al. Stabilization of alum-adjuvanted vaccine dry powder formulations: mechanism and application. J Pharm Sci 2003; 92(2):319–332.
60. Clausi AL, Merkley SA, Carpenter JF, et al. Inhibition of aggregation of aluminum hydroxide adjuvant during freezing and drying. J Pharm Sci 2008; 97(6):2049–2061.
61. Braun LJ, Tyagi A, Perkins S, et al. Development of a freeze-stable formulation for vaccines containing aluminum salt adjuvants. Vaccine 2009; 27(1):72–79.

62. Zhai S, Hansen RK, Taylor R, et al. Effect of freezing rates and excipients on the infectivity of a live viral vaccine during lyophilization. Biotechnol Prog 2004; 20(4): 1113–1120.
63. Hansen RK, Zhai S, Skepper JN, et al. Mechanisms of inactivation of HSV-2 during storage in frozen and lyophilized forms. Biotechnol Prog 2005; 21(3):911–917.
64. Knight C. The Freezing of Supercooled Liquids. Princeton: Van Nostrand, 1967.
65. Rey LR, Bastien MC. Biophysical aspects of freeze-drying. In: Fisher F, ed. Freeze-Drying of Foods. Washington DC: National Academy of Sciences, 1961:25.
66. MacKenzie AP, Luyet BJ. Effect of recrystallization upon the velocity of freeze-drying. Prog Refrigeration Sci Technol 1963; 6:1573.
67. Rapatz G, Luyet B. Patterns of ice formation in aqueous solutions of glycerol. Biodynamica 1966; 10(204):69–80.
68. Luyet B, Rapatz G. Patterns of ice formation in some aqueous solutions. Biodynamica 1958; 8(156):1–68.
69. Quast D, Karel M. Dry layer permeability and freeze-drying rates in concentrated fluid systems. J Food Sci 1968; 33:170.
70. Thijssen H, Rulkens W. Effect of freezing rate on rate of sublimation and flavour retention in freeze-drying. Int Inst Refrigeration Commun 1969; 10:99–113.
71. Blond G, Medas M, Merle R, et al. Study of the porous texture resulting from the freeze-drying of certain aqueous and non-aqueous systems. Int Inst Refrigeration Commun 1969; 10:59–69.
72. Simatos D, Blond G. The porous texture of freeze dried products. In Goldblith S, Rey L, Rothmayr W, eds. Freeze Drying and Advanced Food Technology. London: Academic Press, 1975, 401.
73. Pikal MJ, Shah S, Senior D, et al. Physical chemistry of freeze-drying: measurement of sublimation rates for frozen aqueous solutions by a microbalance technique. J Pharm Sci 1983; 72(6):635–650.
74. Nakamura K, Kumagai H, Yano T. Effect of freezing conditions on freeze drying rate of concentrated liquid foods. In: Food Engineering and Process Applications; Proceedings of the Fourth International Congress on Engineering and Food, Maguer ML, Jelen P. University of Alberta, Faculty of Agriculture and Forestry. London: Elsevier, 1986:445.
75. Roy M, Pikal M. Process control in freeze-drying: determination of the end point of sublimation drying by an electronic moisture sensor. J Parenter Sci Technol 1989; 43(2):60–66.
76. Kochs M, Korber C, Nunner B, et al. The influence of the freezing process on vapor transport during sublimation in vacuum-freeze-drying. Int J Heat Mass Transf 1991; 34(9):2395–2408.
77. Kochs M, Korber C, Heschel I, et al. The influence of the freezing process on vapor transport during sublimation in vacuum freeze-drying of macroscopic samples. Int J Heat Mass Transf 1993; 36(7):1727–1738.
78. Dawson P, Hockley D. Scanning electron microscopy of freeze-dried preparations: relationship of morphology to freeze-drying parameters. Dev Biol Stand 1991; 75:185.
79. Haikala R, Eerola R, Tanninen VP, et al. Polymorphic changes of mannitol during freeze-drying: effect of surface-active agents. PDA J Pharm Sci Technol 1997; 51(2): 96–101.
80. Milton N, Pikal M, Roy M, et al. Evaluation of manometric temperature measurement as a method of monitoring product temperature during lyophilization. PDA J Pharm Sci Technol 1997; 51(1):7.
81. Overcashier D, Patapoff T, Hsu C. Lyophilization of protein formulation in vials: investigation of the relationship between resistance to vapor flow during primary drying and small-scale product collapse. J Pharm Sci 1999; 88(7):688.
82. Franks F. Freeze-drying of bioproducts: putting principles into practice. Eur J Pharm Biopharm 1998; 45:221.
83. Fennema O, Powrie W, Marth E. Low-Temperature Preservation of Foods and Living Matter. New York: Marcel Dekker, 1973.

84. Chandler S, Trissel L, Wamsley L, et al. Evaluation of air quality in a sterile-drug preparation area with an electronic particle counter. Am J Hosp Pharm 1993; 50:2330–2334.
85. Searles JA. Heterogeneity phenomena in pharmaceutical lyophilization. University of Colorado Chemical and Biological Engineering PhD Thesis, 2000.
86. Roth C, Winter G, Lee G. Continuous measurement of drying rate of crystalline and amorphous systems during freeze-drying using an in situ microbalance technique. J Pharm Sci 2001; 90(9):1345–1355.
87. Pikal MJ, Shah S, Roy ML, et al. The secondary drying stage of freeze drying: drying kinetics as a function of temperature and chamber pressure. Int J Pharm 1990; 60:203–217.
88. Abdelwahed W, Degobert G, Fessi H. Freeze-drying of nanocapsules: impact of annealing on the drying process. Int J Pharm 2006; 324(1):74–82.
89. Blue J, Yoder H. Successful lyophilization development of protein therapeutics. Am Pharm Rev 2009:90–96.
90. Itagaki K. Some surface phenomena of ice. J Colloid Interface Sci 1967; 25:218–227.
91. Jellinek HHG, Ibrahim SH. Sintering of powdered ice. J Colloid Interface Sci 1967; 25:245–254.
92. Hagiwara T, Hartel RW. Effect of sweetener, stabilizer, and storage temperature on ice recrystallization in ice cream. J Dairy Sci 1967; 79:735.
93. Sutton RL, Cooke D, Russel A. Recrystallization in sugar/stabilizer solutions as affected by molecular structure. J Food Sci 1997; 62(6):1145.
94. Sutton RL, Evans RL, Crilly JF. Modeling ice crystal coarsening in concentrated disperse food systems. J Food Sci 1994; 59(3):1227.
95. Sutton RL, Lips A, Piccirillo G. Recrystallization in aqueous fructose solutions as affected by locust bean gum. J Food Sci 1994; 61(4):746–748.
96. Sutton RL, Lips A, Piccirillo G, et al. Kinetics of ice recrystallization in aqueous fructose solutions. J Food Sci 1996; 61(4):741.
97. Shalaev E, Franks F. Changes in the physical state of model mixtures during freezing and drying: inpact on product quality. Cryobiology 1996; 33:14–26.
98. Teagarden DL, Baker DS. Practical aspects of lyophilization using non-aqueous co-solvent systems. Eur J Pharm Sci 2002; 15(2):115–133.
99. Kasraian K, DeLuca P. Thermal analysis of the tertiary butyl alcohol-water system and its implications on freeze-drying. Pharm Res 1995; 12(4):484.
100. Wittaya-Areekul S, Nail S. Freeze-drying of tert-butyl alcohol/water cosolvent systems: effects of formulation and process variables on residual solvents. J Pharm Sci 1998; 87(4):491.
101. Wittaya-Areekul S, Needham GF, Milton N, et al. Freeze-drying of tert-butanol/water cosolvent systems: a case report on formation of a friable freeze-dried powder of tobramycin sulfate. J Pharm Sci 2002; 91(4):1147–1155.
102. Williams NA, Lee Y, Polli GP, et al. The effects of cooling rate on solid phase transitions and associated vial breakage occurring in frozen mannitol solutions. J Parenter Sci Technol 1986; 40(4):135–141.
103. Williams N, Dean T. Vial breakage by frozen mannitol solutions: correlation with thermal characteristics and effect of stereoisomerism, additives, and vial configuration. J Parenter Sci Technol 1991; 45(2):94.
104. Milton N, Nail SL. The crystallization kinetics of glycine hydrochloride from "frozen" solution. Cryo Letters 1997; 18:335–342.
105. Johnson RE, Kirchhoff CF, Gaud HT. Mannitol-sucrose mixtures—versatile formulations for protein lyophilization. J Pharm Sci 2002; 91(4):914–922.
106. Pikal-Cleland KA, Cleland JL, Anchordoquy TJ, et al. Effect of glycine on pH changes and protein stability during freeze-thawing in phosphate buffer systems. J Pharm Sci 2002; 91(9):1969–1979.
107. Cannon A, Trappler E. The influence of lyophilization on the polymorphic behavior of mannitol. PDA J Pharm Sci Technol 2000; 54(1):13–22.

108. Gomez G, Pikal M, Rodriguez-Hornedo N. Effect of initial buffer composition on pH changes during far-from-equilibrium freezing of sodium phosphate buffer solutions. Pharm Res 2001; 18(1):90–97.
109. Pikal MJ. Freeze-Drying of Proteins—Process, Formulation, and Stability. ACS Symp Series 1994; 567:120.
110. Pikal MJ. Freeze-drying of proteins part I: process design. BioPharm 1990; 3(8):18.
111. Oguchi T, Terada K, Yamamoto K, et al. Freeze-drying of drug-additive binary systems. I. Effects of freezing condition on the crystallinity. Chem Pharm Bull (Tokyo)1989; 37(7):1881–1885.
112. Williams N, Schwinke D. Low temperature properties of lyophilized solutions and their influence on lyophilization cycle design: pentamidine isethionate. PDA J Pharm Sci Technol 1994; 48:135–139.
113. Chongprasert S, Griesser U, Bottorff A, et al. Effects of freeze-dry processing conditions on the crystallization of pentamidine isethionate. J Pharm Sci 1998; 87(9):1155.
114. Izutsu K, Yoshioka S, Terao T. Decreased protein-stabilizing effects of cryoprotectants due to crystallization. Pharm Res 1993; 10(8):1232–1237.
115. Izutsu K, Yoshioka S, Terao T. Effect of mannitol crystallinity on the stabilization of enzymes during freeze-drying. Chem Pharm Bull (Tokyo) 1994; 42(1):5–8.
116. Pyne A, Suryanarayanan R. Phase transitions of glycine in frozen aqueous solutions and during freeze-drying. Pharm Res 2001; 18(10):1448–1454.
117. Lu X, Pikal MJ. Freeze-drying of mannitol-trehalose-sodium chloride-based formulations: the impact of annealing on dry layer resistance to mass transfer and cake structure. Pharm Dev Technol 2004; 9(1):85–95.

Phase Separation of Freeze-Dried Amorphous Solids: The Occurrence and Detection of Multiple Amorphous Phases in Pharmaceutical Systems

Adora M. Padilla
KBI Biopharma, Inc., Durham, North Carolina, U.S.A.

Michael J. Pikal
School of Pharmacy, University of Connecticut, Storrs, Connecticut, U.S.A.

INTRODUCTION

In the past 25 years the biotechnology industry has boomed with more than 100 protein-based drugs making their way into clinical trials and ultimately to market (1). The task of stabilizing these proteins for an extended shelf life is not trivial and many times requires the skillful design of a freeze-dried solid dosage form (2). Because of the susceptibility of proteins to denature and undergo a variety of chemical reactions (3,4) (i.e., deamidation, oxidation, hydrolysis, and β elimination) in the liquid state, the solid state is often chosen as the most practical approach, although the drying process involved produces many stresses for a protein (5,6). Such stresses are induced by low temperature, formation of ice, and a subsequent concentration of the solute (7). The real benefit of drying is seen long-term, during storage, due to the removal of water, which facilitates many chemical degradation processes, as well as immobilization of the molecule. Degradations require molecular mobility, and drying inhibits molecular mobility and therefore slows physical and chemical degradation (8).

The process of freeze-drying or lyophilization is a costly and time-consuming unit operation, but is often the technology of choice for protein stabilization. Lyophilization involves three main steps: freezing, primary drying or sublimation of ice, and secondary drying or desorption of unfrozen water (9). During each stage of the process the protein is subjected to a variety of stresses, or perturbations to the physical and chemical stability. During the freezing phase of the process the product is often cooled to $-40°C$ or lower and held for a period of time until the system is completely solid, with all "freezable water" converted to ice. During this time, the protein is exposed to the ice-water interface, low temperatures, and freeze-concentrated solutes, all of which may be detrimental for protein stability (10). The addition of protective excipients or stabilizers is often essential in protecting proteins from such stresses. These stabilizers include, but are not limited to, buffer salts, surfactants, sugars, and polymers (2). Stabilizers are selectively chosen to protect the protein of interest during the process of freeze-drying and storage, with storage stability often presenting an even larger problem than the in-process stability. It is often speculated that phase behavior, particularly phase separation, during processing of the dosage form may play a significant role in stability. For example, it is well known that high molecular weight of one or more components facilitates phase separation, particularly in the case of polymers, and this effect is due to the relatively small entropic gain upon mixing of polymer components, which then may allow domination of the free energy change by a larger positive enthalpic contribution to the free energy of mixing (11). Additionally, various process-related factors may contribute to phase changes. Rapid freezing could bring the sample well below the T_g' very quickly, resulting in almost immediate immobilization in a glassy state, and therefore phase separation would be kinetically hindered (12). However, annealing during the freezing stage of the cycle is often used as a means of crystallizing a bulking agent, assuring uniform and reproducible ice crystal size and morphology to facilitate scale-up, and/or increasing the size of ice crystals to obtain lower resistance and increase mass transfer, that is, faster drying (9). Annealing at a temperature above the glass transition of the maximally freeze-concentrated solute not only allows mobility for changes in the ice structure but also allows mobility that could result in phase separation of incompatible formulation components when the thermo-dynamically stable system is phase separated (12). The cooling rate and degree of supercooling could play a role in phase behavior, as well.

The stability of proteins during freeze-drying has been an area of great interest for researchers in biopharmaceutical formulation design (4,13,14), and discussions are often focused on the mechanisms of stabilization associated with drying or removal of water from the system. Two main mechanisms are often proposed to rationalize stability trends in the stabilization of proteins: (*i*) the "water substitution" mechanism (15) and (*ii*) the "vitrification" mechanism. Briefly, the first mechanism hypothesizes that stabilizers such as sucrose and trehalose in a system replace water's H-bonding with the protein as water is removed during the drying process, thereby maintaining the native structure and stability of the protein. (15) The latter states that mobility of the protein is limited by the formation of a glass during the freezing and drying process, hence stabilizing the protein arises by limiting the attainment of equilibrium in a glassy solid (14,16). That is, even when thermodynamically the protein should unfold and degrade, kinetic limitations prevail and pharmaceutical stability is assured because limited mobility in the system prevents significant change on the timescale of relevance. Both these mechanisms require a molecular dispersion of the protein in the stabilizer. That is, a single amorphous phase is necessary for stabilization by the intended stabilizer. It is quite likely that current methods for detection of phase separation have failed, at least on some occasions, to detect the presence of phase separation in formulations (17). Thus, formulation candidates move forward in the development process and into long-term storage testing. The stability implications of phase separation may not be observed until well into long-term storage.

During freezing the solute is concentrated in the interstitial spaces of the ice. It is well known that concentration of solutes and temperature both play a role in phase separation in many relevant systems including polymer-polymer (11,18,19), polymer-protein (20), protein-polysaccharide (21–23), protein-salt (24), protein-polysaccharide-salt (23,25), and saccharide-polymer systems (26). These systems are typically of most interest to the polymer and food science industries. However, all of these systems have direct relevance to the formulation of proteins in the pharmaceutical industry. The phase separation of proteins is also of particular interest in medicine where the separation of crystallins in the eye plays an important role in cataracts development (27). Very little is known regarding the likelihood for phase separation in freeze-dried formulations, particularly those, which contain protein as the active molecule and low–molecular weight saccharides as stabilizers. Because, in large part, of limited detection methodology, few studies are aimed at understanding and preventing phase separation in amorphous freeze-dried samples. However, it is widely appreciated that phase separation may exist and when it does occur may well be responsible for poor stability behavior of trial formulations.

The implications of undetected phase separation on protein stability and overall product quality are a serious concern. Proteins that are inadequately protected during the freeze-drying process are subject to structural changes not only upon processing but also during storage (5,28,29). In fact, it is likely that the protein may not be significantly degraded by lack of a suitable amount of stabilizer during processing but still, because of alterations in structure that present a "reactive conformation" during storage, would present stability problems during storage. Several studies report the propensity for proteins to degrade upon storage in systems of low stabilizer content (5,28,29). Phase separation results in most of the protein being present in a protein-rich phase, which of course contains low levels of stabilizer. With the appropriate means of detection it is possible that the presence of phase separation could be detected during the screening of potential formulation candidates. Thus, formulations that may lead to stability failure are rejected early and development can proceed more efficiently.

Methodology in the pharmaceutical industry that is aimed at detecting amorphous phase separation has primarily been limited to differential scanning calorimetry (DSC) and scanning electron microscopy (SEM). DSC is used to detect multiple T_g's, which are typically indicative of multiple phases. The T_g' is the glass transition of the maximally freeze-concentrated solute and represents the transition of the solute phase from a viscous liquid to a brittle glass during freezing. Although it is recognized that some systems contain what is called a T_g'', this event can be distinguished from T_g' by annealing the sample, which results in an upward shift in T_g'' and an eventual deletion of the event, while the higher transition, T_g', remains unaffected (30). SEM is used to visually observe a change in topography within a dried formulation, which might indicate phase separation. Clearly, both techniques have their limitations with DSC lacking sensitivity for detection of a T_g' and T_g in protein-rich phases and some polymer-rich phases (20,31). Additionally, SEM relies on a visual distinction between a topographical feature and a neighboring continuous phase (32). Thus, the technique is often quite subjective and may only suggest phase separation if such a feature is observed. Moreover, the absence of topographical differences does not always equate to the absence of phase separation.

Several studies have been reported regarding detection of phase separation in polymer systems and biopolymer food systems, and studies that highlight parameters affecting phase separation in such systems have also appeared. Very little research has been aimed at the study of amorphous phase separation as a result of freeze-drying in pharmaceutical systems. The vast majority of this research has come from two groups, that is, Izutsu et al. (11,20,24,33–35) and Heller et al. (12,17,32,36–38). These research groups have demonstrated the potential for multiple amorphous phases to form in a freeze-dried solid. It is the purpose of this chapter to summarize the current state of knowledge of amorphous phase separation in freeze-dried systems and to identify promising methods of detection of phase separation using technology developed in other fields of science. We do not focus on systems which result in crystalline phases. Detection of crystalline phases is relatively straightforward and general formulation knowledge, if used correctly, can lead the formulator away from formulations that will likely give crystallization problems. Additionally, we highlight possible solutions to the problem of phase separation in pharmaceutically relevant freeze-dried systems through formulation and process design related changes.

THERMODYNAMICS OF PHASE SEPARATION DURING FREEZING

The classical representation of phase separation can be best approached from the basic thermodynamics of mixing of a simple two-component ideal solution (39). Consider a system of component A and component B. The free energy of mixing, ΔG_{mix}, is given by equation (1), whereas the entropy of mixing, ΔS_{mix}, is represented by equation (2). The thermodynamics of mixing in an ideal solution is controlled by the entropy contribution. That is, since in an ideal solution, $\Delta H_{mix} = 0$, the free energy change is completely determined by the entropy change.

$$\Delta G_{mix} = \Delta H_{mix} - T\Delta S_{mix} \tag{1}$$

$$\Delta S_{mix} = -nR(x_A \ln x_a + x_B \ln x_B) \tag{2}$$

However, we rarely encounter ideal solutions in real practice. Thus, interactions are typically present and the solution deviates, at least moderately, from ideal solution behavior. A regular solution is the next level of complexity and is a useful model for many real solutions, even if only for qualitative understanding. In many cases ΔH_{mix} is positive and opposes mixing to at least some extent. Therefore, it is the competition between enthalpy and entropy that ultimately dictates the phase separation process. In a regular solution a dimensionless quantity referred to as the exchange (or interaction) parameter (χ_{AB}) is defined (40). The exchange parameter describes the energetic cost of starting with pure A and B and transferring one molecule of B into a medium of pure A and one molecule of A into a medium of pure B. The equation defining χ_{AB} is shown in equation (3) (40), where w_{AB} is a measure of the energy of interaction between A and B molecules, w_{AA} and w_{BB} represent the energy of interaction between like molecules A or B, z describes the number of nearest neighbors (or coordination number), k is Boltzman's constant, and T is temperature. Therefore, if A-B interactions dominate over A-A and B-B interactions there will be a preference for A molecules to be next to B molecules, that is, mixing will tend to be spontaneous. On the other hand, if like molecule interactions dominate there will be a preference for phase separation, but whether phase separation actually occurs will depend on the magnitude of the entropy of mixing. An expression for the ΔG_{mix} in a regular solution can be described by equation (4) (40), where X_A and X_B are mole fractions of A and B, respectively. The first two terms of equation (4) describe the entropic term of the equation and the last term describes the enthalpic contribution, ΔH_{mix}. We note that, in a regular solution, the entropy of mixing is the same as for an ideal solution, which for any real solution is only an approximation. If the interactions between like molecules, that is, A-A and B-B, dominate over interactions between unlike molecules, A-B, then ΔH_{mix} is very positive and may dominate the free energy change, and if so, from equation (4), $\Delta G_{mix} > 0$, and thus phase separation is thermodynamically favored. The model described is referred to as "the regular solution model" (40,41).

$$x_{AB} = \frac{z}{kT}\left(w_{AB} - \frac{w_{AA} + w_{BB}}{2}\right) \tag{3}$$

$$\Delta G_{mix} = RTX_A \ln X_A + RTX_B \ln X_B + RT\chi_{AB}X_AX_B \tag{4}$$

It should be obvious that entropy plays a significant role in phase separation. If mixing results in a very small entropic gain, as is the case with many polymer-polymer mixtures, a given positive change in enthalpy on mixing can more easily dominate, resulting in a positive free energy change on mixing and giving phase separation, at least for a system at equilibrium. Polymers behave differently than small molecules, and their size and structure are factors that impact the thermodynamics of mixing. Equation (5) describes ΔG_{mix} for a polymer mixture (40,42), where ϕ_A and ϕ_B represent volume fractions of A and B, respectively. n_A and n_B are degrees of polymerization for each polymer, A and B, respectively. The degree of polymerization is determined by calculating the number of repeat monomer units in the average polymer chain from the molecular weight of the polymer chain divided by the molecular weight of the monomer. χ_{AB} is the exchange parameter previously described, R is the ideal gas constant, and T is the temperature.

$$\frac{\Delta G_{mix}}{RT} = \frac{\phi_A}{n_A} \ln \phi_A + \frac{\phi_B}{n_B} \ln \phi_B + \chi_{AB}\phi_A\phi_B \tag{5}$$

Upon mixing of two polymers, A and B, the entropy and enthalpy are dependent on the volume fraction occupied by the polymers. Polymers are not able to fully utilize the available volume increase upon mixing because of their size and connectivity (43). Thus, large–molecular weight polymers experience a smaller entropic gain per unit weight than a "small molecule," and the thermodynamics of mixing are more easily dominated by a larger enthalpic contribution, resulting in phase separation. The above model is known as the Flory-Huggins theory.

Typically, a phase-separating system exists as a single phase at temperatures above or below a critical temperature and at specific concentrations. However, some solutions do have both an upper and lower critical solution temperature, although this is not typically the case for systems of pharmaceutical freeze-drying significance. Rather, formulations for freeze-drying typically have an upper critical solution temperature only, meaning that below that temperature phase separation is thermodynamically spontaneous. In Figure 1A, at temperature T_1, there exists two phases at two different compositions of components A and B, that is, compositions a and b. However, there exists only one phase above the upper critical temperature, T_{co}. A plot of the free energy of a system, which results in phase separation is given in Figure 1B. When ΔG_{mix} is greater than zero the system spontaneously phase separates into two phases. Thus, the presence of two free energy minima in Figure 1B means two stable compositions and therefore demands the formation of two phases to provide a thermodynamically stable system.

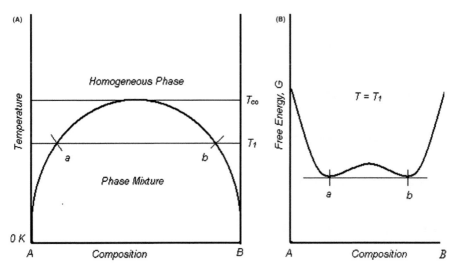

FIGURE 1 (**A**) Phase diagram of a simple two-component system. (**B**) Free energy diagram of a phase-separated two-component system. The presence of two free energy minima results in phase separation. *Source*: Adapted from Ref. 67.

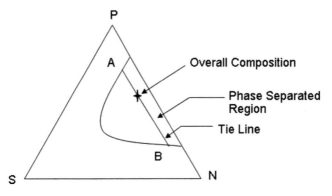

FIGURE 2 Schematic of ternary phase diagram of polymer-solvent-nonsolvent system. The overall composition is noted and falls on the tie line in the phase-separated region of the diagram. The tie line connects the two resulting phases, phase A, a concentrated polymer phase, and phase B, a dilute polymer phase. *Source*: From Ref. 42.

The understanding of the thermodynamics driving phase separation during the freezing process itself is not as straightforward as the simpler systems outlined above might suggest. First, even a simple freeze-dried system is typically at least a three-component system (Fig. 2) in which there is a stabilizer or polymer (*P*), an active molecule (*N*) and a solvent (*S*, typically water) (42). The composition that results in phase separation lies on the tie line in the thermodynamically unstable region of the phase diagram (i.e., the binodal, which divides the phase diagram into regions of miscible and immiscible mixtures). The resulting phases that occur are labeled by A, a polymer-rich phase, and B, a polymer-deficient phase. When control of pH is necessary a small amount of buffer salt is also used, as is normally the case with proteins. The small amount of salt may not significantly impact the overall thermodynamics of mixing. However, a large amount of buffer salt, which is not generally recommended for a variety of reasons (44), could have a significant impact on phase behavior and is discussed later in this chapter. Additionally, there are other factors to consider during the process, such as the freeze concentration effect that occurs as a result of removal of water from the solute phase via crystallization. This concentration effect can be very dramatic, resulting in concentrations as high as 50 times the original solution concentration (45). Several models exist for the understanding of phase separation in dilute polymer solutions. The most widely used and accepted model is the Flory-Huggins theory [eq. (5)], which uses a lattice model to predict the compatibility of polymers. However, the strong influence of orientation-dependent interaction forces (i.e., hydrogen bonding) makes the Flory-Huggins theory inadequate for predicting phase separation in systems of direct interest to freeze-drying (17,19). Many studies have focused on modifying the Flory-Huggins theory to incorporate the effects of such interactions (19,46). The results of such work may be useful for the prediction of phase separation during freezing, but only one such thermodynamic model has been developed thus far.

In the late 1960s, Edmond and Ogston developed a model to study phase separation in dilute polymer solutions, which uses the osmotic virial equation truncated after the second term (47). Although this model was developed for dilute systems and does not directly apply to concentrated, frozen systems, it has been extrapolated to such systems to assess the likelihood of phase separation during the freezing step. In the late 1990s, Heller et al. directly applied the model to freeze-dried systems (17). Although there were several approximations of dubious validity involved, they were able to demonstrate the potential for phase separation to occur during the freezing process (17). Additionally, they addressed the presence of what they refer to as a "phase separation envelope" (Fig. 3), which introduces the kinetic factor that arises because of an increase in viscosity and eventual glass transition at subzero temperatures. Essentially, phase separation may be thermodynamically favored because of concentration and temperature effects, but kinetically hindered because of immobilization of molecules in the glassy state. Thus, there is a period of time in the early freezing process, prior to the onset of the glass transition, which is critical for the occurrence of phase separation.

The most widely studied systems that show phase separation are polymer systems, which can include protein-polysaccharide (23,48), sugar-protein-polysaccharide (49), protein-protein, protein-surfactant (50), polymer-polymer (18), and polymer-polyelectrolyte systems (24,33). Phase separation in such systems is generally the rule, rather than the exception, and is dependent on temperature, concentration, molecular weight, and the presence of other components such as salts. In a freeze-dried system this would directly translate to low temperatures and high concentration, the most unstable condition for many of the systems listed above. As stated previously, the main contribution to phase separation in polymer-containing systems is the large molecular weights of the polymers that limits the entropic gain upon mixing and large orientation-dependent positive enthalpies of interaction upon mixing, resulting in phase separation (19). It is widely speculated that such phase separation occurs during the freezing process with proteins and their intended stabilizers, as a protein can be viewed as a polymer. However, proteins are structurally different than synthetic polymers, and little proof exists for such occurrences in pharmaceutically relevant formulations.

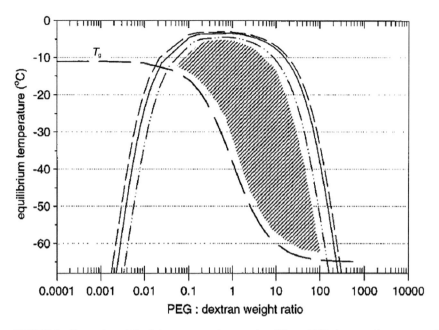

FIGURE 3 Illustration of the "phase separation envelope" for a PEG:dextran phase-separating system. The shaded region below the parabolic lines represents the temperature range at which phase separation is likely to occur. The nonshaded region lies below the glass transition, meaning phase separation is thermodynamically favored but kinetically hindered. *Source*: From Ref. 17.

PROCESS AND FORMULATION FACTORS
The Role of the Lyophilization Cycle in Phase Separation
The process of lyophilization and its impact on product quality and stability are topics of interest for many pharmaceutical scientists. It is well known that many of the common problems, (i.e., protein stability, product appearance, crystallization of buffer salts, etc.) can be circumvented by the addition of stabilizers or by manipulation of the lyophilization process itself (2,6,9,51). However, with little information available on amorphous phase separation in freeze-dried systems, changes in the process or formulation to avoid phase separation would be extremely empirical. In fact, there are few methods for determining whether a given product is, in fact, phase separated. At this point, there is much speculation surrounding phase separation in common protein formulations, but with little direct evidence. However, the study of model polymer systems, which are known to phase separate, has provided some understanding of phase separation during the freezing process. These model systems have allowed for identification of some potential process and formulation factors that may contribute to or inhibit phase separation in pharmaceutically relevant systems.

It is perhaps only the freezing stage of the freeze-drying process, which can impact phase separation on a timescale of practical interest. In this stage of the process, there is still significant mobility due to the presence of water, whether it is prior to ice crystallization or after ice crystallization. Additionally, during freezing the solution temperature is significantly decreased, and because of limitations in heat transfer, cooling is often slow, thereby giving increased residence time in the concentrated, aqueous state (9) and kinetically allowing for the formation of new phases.

There are several process parameters directly related to the potential for phase separation (1) cooling rate, (2) degree of supercooling, and (3) annealing (9,12). Cooling rate and degree of supercooling are somewhat interdependent and their roles may be quite significant. The degree of supercooling refers to the difference between the equilibrium freezing point and the temperature at which ice crystals first form in the sample (52). The initial concentration post freezing depends on the degree of supercooling since the amount of ice that can first form depends on the product of the heat capacity and the degree of supercooling. The cooling rate will govern how quickly a given degree of supercooling is reached, but additionally could allow for a greater degree of supercooling since formation of nuclei is a kinetic process. Annealing is typically used to crystallize an excipient that may remain amorphous upon initial freezing or to control ice morphology and/or increase the size of the ice crystals via Ostwald ripening (9). This step involves increasing the product temperature to a temperature above the glass transition temperature of the freeze concentrate, T_g', to allow increased mobility. The duration of the annealing process and the increased mobility in this liquid state may provide an opportunity for amorphous phase separation if a two-phase system is thermodynamically stable.

There are a few studies in the pharmaceutical literature that have evaluated the effects of some of these lyophilization process parameters on phase separation during freezing. In particular, the effect of cooling rates on the secondary structure of hemoglobin in a PEG 3350 and dextran T500 system using FTIR were studied (12). The loss in structure was used to indicate the presence of phase separation, known to occur in the model polymer system chosen, which induced a destabilizing effect on the structure of the protein, as a large fraction of the stabilizer was removed from the protein phase. With an increase in cooling rate, an increase in the α-helical band depth was observed, indicating a more nativelike structure and at least moderated phase separation. Therefore, they concluded that the fast cooling quickly bypasses this phase separation envelope and moves into a glassy state where phase separation does not occur on the timescale of the process. However, their fast cooling procedure involved submersion of a small volume of solution into liquid nitrogen or dry ice/acetone, which is not representative of the cooling rates achievable in a manufacturing process. Unfortunately, the rate of cooling in a freeze-dryer is limited by the freeze-dryer itself to several degrees per minute or less (53). Additionally, it must be noted that with practical solution volumes in vials, the product cools more slowly than the shelf because of heat transfer limitations (54). Thus, the product will see a relatively slow cooling. Further, once nucleation and crystallization of ice occurs the latent heat warms the temperature of the product back up to near the equilibrium freezing point. This increase in temperature could easily allow for significant phase separation before the system cools back to a more viscous state in which the molecules are essentially immobilized and phase separation is kinetically hindered. The authors observed a loss in secondary structure in a hemoglobin/PEG/dextran system nucleated at −1°C and cooled slowly. The −1°C represents the equilibrium freezing point of the system, which is essentially the temperature to which the sample would reach immediately upon the initial formation of ice. This process more closely mimics the thermal history of a product being frozen in a vial.

The effect of nucleation temperature itself, or degree of supercooling, on phase separation in susceptible formulations has not been systematically studied to date. However, one would expect that at lower nucleation temperatures, that is, closer to the T_g' of the system, the extent of phase separation would be less, since exposure to a mobile environment above the T_g' would be at least of more limited duration. With a very small volume where latent heat effects are small and the total heat capacity is small, T_g' would be reached quickly. However, with most practical examples, the latent heat effect is large, driving the product temperature nearly up to the equilibrium freezing point, and the heat capacity is large, but even here, the time to T_g' would decrease as the nucleation temperature decreases. That is, a lower nucleation temperature would usually mean a low shelf temperature and therefore a large temperature difference between shelf and vial contents, which would shorten the time from nucleation to T_g'. A lesser extent of phase separation is in fact observed in the case of freezing with liquid nitrogen, which would typically give the greatest supercooling and perhaps also somewhat limited latent heat effects with the small volumes used (12). However, in some studies larger supercooling effects have been observed at slow cooling rates (0.5°C/min) (53). A moderate cooling rate of about 1°C/min is often desirable for uniformity in freezing, which will result in moderate degrees of supercooling, a relatively fast freezing rate (53) and likely limit phase separation. Control of nucleation temperature is possible at least on a lab scale and perhaps on a manufacturing scale using an ice fog technique (52,55), which introduces a cold stream of dry nitrogen into the chamber with the product equilibrated at the intended nucleation temperature. The cold nitrogen creates a cloud of ice in the humid chamber and uniformly seeds crystallization in the vials. Control of the nucleation process may allow for control of phase separation, provided the latent heat effect does not completely control the thermal history. Clearly, there is a need for systematic studies in this area.

One would expect that the annealing process would facilitate phase separation when it is thermodynamically favored, and such effects were demonstrated by Heller et al. (12) and Izutsu et al. (24) using model systems. It is, however, not clear how long an annealing process need be to impact a formulation's tendency to phase separate. Clearly this would be dependent on the formulation components as well as the temperature history of the product and the proximity of sample temperature and T_g'. While it will likely be necessary to evaluate each formulation, it may be possible to set some general guidelines for typical stabilizer-drug/protein conditions if reliable methods for detecting phase separation were readily available.

Choice of Excipients

Though the design of the process is important and could dictate whether phase separation will occur, to some extent, the underlying problem should also be addressed from a formulation standpoint. The compatibility of formulation components during the process is critical for avoiding phase separation. While limited information is available for pharmaceutically relevant components, there are some studies that have provided a better understanding of the roles of various types of formulation components on phase behavior during freezing.

The effect of salts, sugars, and molecular weight on phase separation in polymer-containing systems, which are known to phase separate during the freezing process, have been studied (11). The studies have shown that immiscibility in polymer systems increases with an increase in molecular weight of the polymers. This trend may be important to consider when formulating high–molecular weight APIs, such as antibodies. Phase diagrams for small proteins and antibodies are frequently developed in the field of protein crystallization, and a necessary requirement for protein crystallization is the presence of a metastable liquid-liquid phase-separated system (56,57). Though proteins are structurally different than polymers, their propensity to phase separate may be correlated, in part, to molecular weight. However, it also seems likely that electrostatic properties will be important and perhaps even dominate any effect of size. Studies involving the addition of salts show that salts can prevent the occurrence of phase separation in some systems by shielding the electrostatic effects (24). While some salts have little to no effect on phase separation, such as Na_2SO_4 and Na_2HPO_4 (a salting-out salt), other salts, such as NaSCN (a salting-in salt) inhibit phase separation in some polymer systems (24). Aside from shielding effects, salts may play another critical role by formation of a glass and immobilization of the formulation matrix. The addition of NaCl versus KCl and its impact on phase separation in a model PEG/dextran phase-separating system has been studied, and it was determined that when KCl was substituted for NaCl in a formulation, phase separation was prevented (32). The authors speculated that KCl forms a glass with rapid cooling and kinetically hinders phase separation, and thereby also prevents structural damage to the protein. In our view, while this observation is interesting and perhaps useful, the proposed mechanism seems unlikely. However, the important message is that it may be possible to prevent phase separation during processing by adding the appropriate salt to a formulation, although from a freeze-drying formulation standpoint one should generally strive to minimize salt content. Salts have low glass transitions and may lower the T_g' significantly, making primary drying a longer, more costly process (58).

Sugars or disaccharides are perhaps the most important class of stabilizers that can be added to a freeze-dried formulation, particularly when the API is a protein (59). Disaccharides do generally stabilize well, and it may be argued that they play more than one role in stabilizing the protein (15,29). As previously discussed, sugars are believed to substitute for water H-bonding to the surface of a protein but also form an amorphous solid or glass, which limits molecular mobility thereby slowing the rate of any degradation unfolding or degradation process in the protein (15,60). Unfortunately, little is known about the potential for phase separation in protein-sugar systems. One study reported phase separation in freeze-dried mixtures of trehalose and lysozyme, using near infrared spectroscopy (NIR) as a means of detection (61). Nonoverlapping absorption bands were identified for lysozyme and trehalose and used to document regions of different composition in dried samples. The authors demonstrated homogeneity in a supercritical fluid (SCF) dried 1:1 trehalose:lysozyme sample and heterogeneity in a 1:1 trehalose:lysozyme freeze-dried sample. A weakness of the NIR study was the penetration depth, which was reported to be 30 to 180 μm. Such a large penetration depth would likely result in sampling from more than one domain, assuming the domains are on the order of about 10 μm. However, the presence of heterogeneity in the freeze-dried trehalose:lysozyme sample does suggest that phase separation might be present in protein:disaccharide systems. This conclusion seems inconsistent with the concept of the sugar protecting the protein from unfolding during the drying process by substituting for the hydrogen bonding of water. Strong interaction with the protein should mean there is little chance of the protein and sugar separating into different phases.

A bulking agent is often added to a freeze-dried formulation for mechanical strength of the cake and for aesthetic reasons to produce an elegant cake (2). Mannitol is the most common bulking agent, largely because it crystallizes easily to form a rigid and "elegant" cake, and the high eutectic temperature means primary drying is simple and fast. The addition of a bulking agent to a phase-separating hemoglobin/PEG/dextran model system prevented the structural damage to hemoglobin (12). That is, a system with mannitol remaining amorphous [2% (w/w)] had a greater loss in secondary structure than a system with crystalline mannitol [5% (w/w)]. Both formulations were annealed at $-7°C$, allowing for crystallization of the mannitol in the 5% (w/w) sample. The authors speculated that crystallization of mannitol inhibited the formation of multiple amorphous phases, assumed to be damaging to protein structure, by dividing the amorphous matrix into smaller volumes, thereby kinetically preventing nucleation and growth of two amorphous phases. The remaining amorphous phase, once mannitol crystallizes out, consists of PEG and dextran, a well-studied phase-separating system. Previous studies reported on the phase separation of protein out of the stabilizer phase (PEG) and into the dextran phase resulted in loss in protein secondary structure (36). Thus, the formation of small volumes prevented the separation of protein into a non-PEG phase.

Nonionic surfactants are an important excipient class in freeze-dried formulations and are typically added to protect the protein from denaturation at interfaces, such as the ice-water interface. At least one study has demonstrated the potential for phase separation to occur between surfactants and proteins (50). The authors demonstrate the propensity for bovine serum albumin (BSA) and sodium dodecyl sulfate (SDS) to phase separate under specific conditions of high ionic strength and low pH. Although this system is not a typical freeze-dried formulation, since SDS is not an acceptable surfactant for parenteral use, the observation does suggest that under certain conditions phase separation may occur in protein-surfactant systems. It should be noted that since high surfactant concentrations can have a negative effect on protein stability (59), a phase-separated protein-surfactant system may be destabilizing. That is, if the protein were to separate into a phase rich in surfactant that protein would likely be destabilized.

The active molecule and its properties may play a role in phase separation. For example, the isoelectric point of the protein is important when choosing the formulation pH and when choosing excipients. Phase separation has been shown to occur in protein-polysaccharide mixtures at pHs above the protein's isoelectric point because of repulsive electrostatic interactions and altered affinities to water (21). Additionally, the molecular weight or size may be a significant factor in formulation compatibility. As previously stated, molecular weight is a critical parameter in the phase separation of polymer systems (11) and could also be a significant variable in formulating proteins, particularly very high–molecular weight proteins such as antibodies. Again, although structurally different than polymers, proteins can phase separate at low temperatures and high concentrations, which is a desirable attribute when attempting to crystallize proteins (56,57).

A systematic study focused on evaluating the effects of protein molecular weight and other relevant properties, that is, pI, is necessary to critically assess what relationship these properties have in the role of phase separation.

IMPACT OF AMORPHOUS PHASE SEPARATION ON PRODUCT QUALITY

The formation of two phases, one rich in stabilizer and one rich in protein has obvious negative implications for stability. However, because of the limited ability to detect multiple amorphous phases in freeze-dried products, particularly protein products, little quantitative information is available regarding the impact of this type of phase separation on stability and product quality, in general. What is obvious is that if a protein were to separate from its intended stabilizer the protein would be highly susceptible to degradation during processing and storage (4,37,62). Additionally, the occurrence of such phase behavior could result in phases rich in stabilizer, which may convert to a more energetically favorable crystalline state upon storage, at least for stabilizers that easily crystallize (63), thereby releasing moisture into the remaining amorphous phase, lowering the T_g, and causing catastrophic degradation. The presence of the unwanted crystalline phase could also result in longer reconstitution times. Thus, it is clear that phase separation could negatively impact product quality.

In-Process Stability Issues

Since phase separation occurs during the freeze-drying process, product stability may be negatively impacted during processing. For example, if a protein requires a sugar for hydrogen bonding during the drying process, but is present in a phase deficient in sugar, that protein may partially unfold during the process, causing in-process degradation and/or decreased storage stability due to the presence of the more reactive unfolded form during storage. The negative impact of phase separation on protein stability has been demonstrated using a model phase-separating PEG-dextran system and a protein, hemoglobin (37). The protein's secondary structure was monitored with use of FTIR as a measure of stability. Process-related phase separation that occurred as a result of freeze concentration resulted in a significant loss in α-helical band depth, as measured by FTIR. It was evident from the results that secondary structure was severely perturbed by the protein separating into a protein-rich (and stabilizer-poor) phase. As previously discussed, the effect of various process parameters on the secondary structure of hemoglobin has been reported. The effects of various process parameters were demonstrated not only to impact the ability of a system to phase separate but also to impact the detrimental effects that this phase separation has on the protein when it does occur during the process (12).

Storage Stability Issues

Certainly the goal in developing a freeze-dried product is production of a shelf-stable product, which can be stored for typically two years at its intended storage temperature, ideally controlled room temperature, but in practice, mostly refrigerated storage. If phase separation is present but does not affect the product negatively during processing it is possible that the formulation may be tested and considered a viable formulation for storage stability, only to fail stability specifications during storage. Undetected, the phase separation may have led to structural changes and/or perhaps crystallization of excipients during storage. The stability of proteins in the absence of stabilizer has been well studied and documented (15,62,64). The stabilizer is normally a necessary requirement for maintaining protein stability, and formation of a protein-rich phase will generally seriously destabilize the protein in that phase, which may well be most of the protein in the formulation. Further, as noted earlier, formation of a stabilizer-rich phase could result in crystallization during storage, subsequent release of moisture into the surrounding protein-rich phases and accelerating degradation. Results of the effect of sucrose/glucose and trehalose/glucose freeze-dried matrices on protein activity of G6PDH (glucose-6-phosphate dehydrogenase) upon accelerated storage have been reported (26). Upon accelerate storage, two amorphous phases, one rich in glucose and the other rich in sucrose or trehalose, appeared in the DSC thermograms, and some of the amorphous disaccharide phase converted to the crystalline form. Sucrose underwent amorphous phase separation more readily, followed by the formation of crystalline sucrose. The result was dramatic inactivation of the protein in the sucrose system. The trehalose system resisted total conversion to the crystalline form, resulting in a smaller loss in protein activity. It should be noted that both systems studied had relatively high water contents of approximately 6% (w/w) and 8% (w/w) for sucrose/glucose and trehalose/glucose, respectively. These water contents are much higher than conventional, and whether or not the results would have been the same for systems of more typical water content is an open question. However, the results obtained do demonstrate that the ultimate result of amorphous phase separation, crystallization, does in fact damage protein stability.

If the protein requires a specific molar ratio of stabilizer to protein (64,65) for acceptable stability, and phase separation produces a nonoptimal stabilizer to protein ratio, product stability will not be acceptable if the system does phase separate. Several studies concluded that a lyoprotectant to protein mole ratio of at least 360:1 was necessary for adequate stabilization of a monoclonal antibody (64,65). This ratio was sufficient to stabilize the protein at accelerated storage conditions (40°C) for longer than two years (65). The optimal sugar:protein weight ratio necessary for structural stabilization of several proteins has been reported to be about 1:1(66). In the absence of stabilizer there was a clear loss in native structure, as detected by FTIR. A recent study on accelerated storage stability of human growth hormone (hGH) in the absence of stabilizer reported much larger chemical degradation rate constant and aggregation rate constant as compared with that in samples containing stabilizer (28). With an increase in stabilizer from zero stabilizer to a 6:1 weight ratio of stabilizer to protein a decreasing trend in rate constants, from 8.2 to as low as 0.38 for chemical degradation and from 3.4 to as low 0.07 for aggregation ($\sqrt{t}$), was observed. Clearly, creation of a protein-rich phase-by-phase separation would greatly compromise stability.

METHODS OF DETECTION

Most of the phase behavior studies in the pharmaceutical literature use two main methods of detection (1) DSC and (2) SEM. This section will discuss some of the studies on phase behavior using these two techniques. We also discuss some alternate techniques that have been used to study phase behavior in other science and engineering fields.

Differential Scanning Calorimetry

The amorphous state offers some unique characteristics in comparison with its crystalline counterpart. Amorphous solids offer the opportunity to molecularly disperse a labile drug into a matrix of stabilizer, allowing for direct contact between the stabilizer and drug, which can significantly increase storage stability. DSC is commonly used to determine the onset temperature of global molecular mobility, which is termed the glass transition temperature, T_g. Glass transition temperatures depend on the composition of the phase, so DSC has the ability to monitor phase separation in amorphous systems that show glass transitions by DSC, at least when the phase compositions are sufficiently different to exhibit separate T_g values (about 10°C apart). DSC may be used to identify the presence of multiple amorphous phases in either a frozen sample or a dried sample by detection of multiple glass transitions (Fig. 4) (18,21). However, there are limitations to this technique for application in the detection of phase separation. One such limitation is that detection of a T_g by DSC requires a significant change in

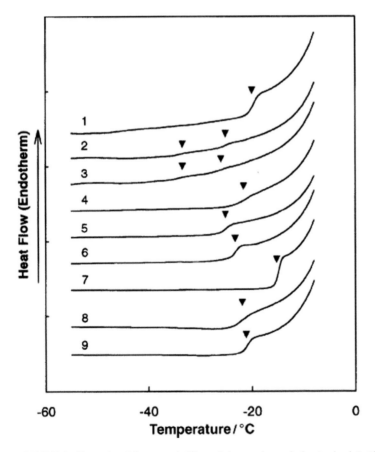

FIGURE 4 Example of the use of differential scanning calorimetry in detecting multiple glass transitions in freeze-concentrated solutions. One to nine represent mixtures of different polyelectrolytes and nonionic polymers. The presence of two T_g's indicates phase separation in some of the mixtures. *Source*: From Ref. 33.

heat capacity over a relatively modest temperature range. However, sometimes the ΔC_p is small and occurs over a large temperature range, as is the case for proteins and some polymers in the dried solid state, that is, strong glass formers (67). Normally it is easier to detect a $T_g{}'$, glass transition of the maximally freeze-concentrated solute, because of the presence of water in the amorphous solute phase. However, even here, it is not always easy to detect the $T_g{}'$ of a protein phase (20) or, indeed, to detect any phase present in small amounts. It has been reported that if a system is less than about 20% phase-separated, DSC will be unable to detect the glass transition of the phase present in the smaller amount (68).

Despite the limitations of DSC, it has been successfully used to identify the presence of amorphous phase separation during the freezing process in some systems, thereby documenting the possibility of phase separation during freeze-drying. Model systems that have been studied are mainly limited to polymer-containing systems, which are well known to undergo phase separation. The identification of multiple T_g's using DSC has been extensively used to study the miscibility of protein-polymer, polyelectrolyte-polymer, and polymer-polymer systems during freezing and freeze-drying (11,20,24,33,35).

More recently, a new version of the DSC known as hyper-DSC (PerkinElmer) or high–ramp rate DSC (HRR-DSC) has received attention for its potential in detecting glass transitions of proteins and high–molecular weight polymers, which have previously been difficult or impossible to detect using either conventional or modulated DSC. Results for the application of the technique to dried proteins suggests the potential for the technique to be used in detection of the glass transition of a protein-rich phase, and the technique may therefore be useful in the detection of multiple phases in a freeze-dried system (69). The technique increases the sensitivity of standard DSC by increasing the heating rate and allowing for the same observable heat flow over a shorter time period, allowing for a very small ΔC_p to be observable at the glass transition. The results (69), obtained from second scan data (i.e., scan a fresh sample past the transition, cool, and scan a second time), suggest the T_g of proteins occurs at very high temperatures, much higher than reported values of T_g determined by extrapolation of T_g of protein:disaccharide mixtures (i.e., systems that do show a T_g by conventional DSC) to zero excipient concentration. In fact, the authors performed their own extrapolation to zero sucrose content for lysozyme:sucrose systems and report a 30°C lower glass transition temperature, also determined from glass transition data of second heat scans, than that directly measured for the pure protein using HRR-DSC (69). The authors speculate that the difference is a result of deviation from ideal mixing in the lysozyme:sucrose mixtures. In this study, however, all reported T_g values of proteins determined by HRR-DSC are analyzed from a second heating scan, which follows a first scan above 200°C. Most proteins undergo thermal denaturation before or close to 200°C in the dried state (70–72), so it seems likely the authors were evaluating the T_g of the unfolded protein. The authors do report good agreement between their measured T_g values and loss in mechanical rigidity using thermal mechanical analysis (TMA), but it is not clear whether the authors applied a first heat treatment to the TMA sample, as well, similar to the DSC studies. Thus, in our view, it is not fully established that the HRR-DSC technique can determine the T_g of a phase rich in protein that has not been thermally denatured. The technique of HRR-DSC needs further investigation before its application to the detection of glass transitions in strong glass formers is "validated."

Scanning Electron Microscopy

SEM is a useful tool in the study of freeze-dried solids when using it to assess cake morphology and pore structure. In addition, SEM has been used to observe what appears to be phase separation in freeze-dried solids (12,17,32). A significant limitation of SEM in the study of phase separation is the ambiguous nature of the observation, which is a visual detection of a topographical feature assumed to represent phase separation.

Historically, the use of SEM in the study of phase-separating freeze-dried solids has been limited to systems which are expected to phase separate. Therefore, it is perhaps easier to identify a feature within a cake as a "different" phase. SEM has been extensively used in the study of PEG/dextran systems to observe phase separation (12,17,32). Figure 5 illustrates an example of the use of SEM in the detection of phase separation in freeze-dried systems. The onset of a new phase is observable by the appearance of a budlike droplet. It should be pointed out that SEM is used extensively in the study of phase separation in the polymer industry (73,74). In these cases, the phases are typically very large, morphologically different, and easily identifiable using SEM. So, the application of this technique to pharmaceutical systems is certainly attractive. However, the use of SEM as a primary technique in identifying phase separation in freeze-dried systems, where the morphology or structure may not visibly differ from the neighboring phase, is difficult at best.

Dielectric Techniques

Thermally stimulated current spectroscopy (TSC) and Dielectric relaxation spectroscopy (DRS) are both dielectic techniques, which rely on molecular mobility responses in the presence of an electric field. Specifically, TSC is a temperature domain dielectric technique, which takes advantage of the ability of a molecule's dipole to reorient in the presence and absence of an electric field as a function of temperature as a transition temperature is passed (75,76). The low frequency of TSC, 10^{-4} to 10^{-2} Hz range, allows for improved sensitivity and resolution (77). It presents a unique opportunity to study mobility in the glassy state, particularly when a large ΔC_P at T_g is not observable, as is the case with strong glass formers. This offers an alternative to DSC in the study of phase

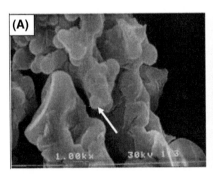 

FIGURE 5 Example of SEM in the detection of phase separation. SEM images of a PEG/dextran phase-separating system at ratios of (**A**) 1.0 and (**B**) 500. Phase separation is observed in (**A**) as budding (*white arrow*) appears with the emergence of a new phase. *Abbreviation*: SEM, scanning electron microscopy. *Source*: From Ref. 17.

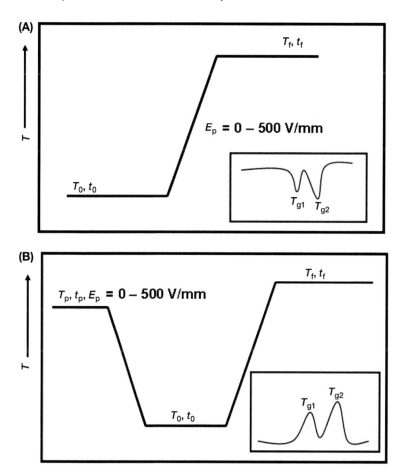

FIGURE 6 Experimental modes of thermally stimulated current spectroscopy. (**A**) Depolarization mode and (**B**) polarization mode. Insets illustrate the recorded mobility response for a two-phase system, that is, the presence of two glass transition temperatures.

separation when looking for different glass transitions associated with multiple amorphous phases.

The principle of TSC is based on the same principles as a parallel plate capacitor. The instrument consists of an electrode, which is placed in contact with the sample, that is, the dielectric, which rests on the surface of another electrode. An electric field is applied to the sample. The technique can be run in two different modes, thermally stimulated depolarization mode (TSDC) and thermally stimulated polarization mode (TSPC), both shown in Figure 6. The first relies on some knowledge of the glass transition or relaxation process temperature of interest prior to the experiment. Briefly, the temperature is held at a temperature slightly above or at the T_g of the sample and held for a period of time, t_p, allowing for polarization equilibration in the presence of an electric field, E_p, at which time the mobile molecules will align with the electric field. The sample is then cooled quickly to a temperature, T_0, for a time, t_0, where the molecular motions are quenched on the timescale of the experiment. The electric field is then removed, and the temperature is linearly ramped to a temperature, T_f, above the glass transition for a time, t_f. The depolarization current is then measured as temperature increases and a pulse of current indicates cooperative molecular motion, which is interpreted as T_g. The TSPC mode works by cooling the sample to a temperature, T_0, for a time, t_0, with the molecular dipoles randomly oriented. The electric field, E_p, is then applied, and the temperature is ramped linearly, resulting in polarization of the sample at the mobility temperature, that is, T_g. The TSPC mode might be the most useful for the applications to phase separation because it does not involve heating the sample to temperatures at which phase separation may be kinetically allowed to occur in samples where none existed previously.

Similar to TSC, DRS, a time domain technique, also relies on a capacitor plate configuration. However, the DRS technique is capable of probing relaxation processes in a wider range of frequencies, 10^{-4} and 10^{10} Hz (75), and is therefore capable, at least in principle, of measuring processes in the glassy state (slow processes) as well as in the supercooled liquid state (relatively fast processes) (78). A step voltage is applied to the sample held isothermally, and the response current is measured as a function of time, which can then be converted into the frequency domain via Laplace transform. Dividing the response current by the applied voltage results in information pertaining to the physical properties of the sample (1) the permittivity (ε') and, (2) conductivity (σ), which results from DC conductivity and the dielectric loss (79).

Dielectric techniques have not been well established in the study of freeze-dried systems, but are seeing increased applications in this area (75,79). The use of TSC in the pharmaceutical industry has mainly been limited to the study of molecular mobility in small nonpolar organic molecules, typically prepared by quenching from a melt (80–83). Freeze-dried amorphous solids have not been extensively investigated using this technique. Lyophilized β-cyclodextrin and quinapril mixtures were analyzed with the use of TSC. Phase separation was identified in lyophilized mixtures, including some mixtures that did not demonstrate two T_g values by DSC, demonstrating the better sensitivity of TSC relative to conventional DSC (84). Another recently published study looked at T_g and sub-T_g relaxation in dried powders and dried proteins (85). The study focused on dried PVP powders and freeze-dried porcine somatropin (pST), BSA, and IgG. Sub-T_g relaxation events were believed to be different from the global relaxation events responsible for the α-relaxation at the T_g of a protein. The authors suggested that the peaks observed are characteristic of the protein's internal molecular mobility. The differences observed between the "T_g" determined by TSC and the T_g determined by extrapolation to zero sugar content from calorimetric T_g of protein:sugar mixtures were suggested to be due to a decoupling of translational and whole molecule rotational motion (85). In at least one protein, pST, the T_g determined by TSC agreed well with the calorimetric T_g. Therefore, it may be possible to use TSC to investigate phase-separated mixtures of protein and stabilizer by identification of multiple T_g values. However, there does appear to be a substantial difference between some of the protein T_g values obtained by extrapolation of DSC mixture data and those directly determined by TSC. Perhaps the origin of these relaxations is not related to the α-relaxation of the molecule. Melt-quenched amorphous trehalose was also investigated using TSDC (86). These results indicate an astonishing absence of a detectable glass transition using TSDC. Rather, this study only showed, consistently, a secondary relaxation peak at approximately 60°C, well above the reported β-relaxations and well below the T_g. The speculation was that this secondary relaxation is unique to trehalose and molecules structurally similar to trehalose. We suggest that this observation may well be an artifact of the technique. This technique certainly warrants further investigation in the application to freeze-dried materials and particularly in studies of phase behavior in these systems, but investigations should be conducted with caution. It was recently stated in an article reviewing the application of TSC to pharmaceutical science that "dielectric techniques are in a development phase similar to that of DSC a few decades ago"(76). Accepting this assessment, thermal analysis using dielectric techniques may well be the future of materials science evaluations in the pharmaceutical industry, but these techniques clearly require further research before applications become routine.

Contrary to use in the pharmaceutical industry, dielectric techniques such as TSC and DRS have been extensively used for the study of phase separation in the polymer industry. Such techniques are useful for studying miscibility behavior in systems that do not allow detection of the glass transition using standard DSC techniques (87). Poly butanediol terapthalic acid (PBT) blends and tetramethylene ether glycol (PTMEG) were shown, using TSC, to exhibit two distinct glass transitions as a result of complete phase separation of the PBT and PTMEG (87). Several polymer phase-separating systems have been studied using both TSC and DRS (87,88). Moreover, the ability of the two techniques to probe different and overlapping modes of mobility within a sample allows for better understanding of the nature of the processes responsible. Both techniques are known to be subject to artifacts, so it is essential that the user is aware of the potential for these artifacts to confound the interpretation of results (79).

Raman Microscopy

Raman spectroscopy is a technique that has seen an increase in use and applications through the last few decades in the pharmaceutical industry. Like its spectroscopic counterpart, infrared spectroscopy, Raman probes the vibrational modes of a molecule and thus molecular and structural information can be obtained (89). Also, similar to IR, Raman can obtain information regarding inter- and intramolecular interactions (89). The Raman effect involves the inelastic scattering of photons. In short, a transfer of energy occurs between an incoming photon and a molecule with a subsequent emission of the photon at a virtual energy state, either resulting in Stokes (absorption of energy by the molecule) or anti-Stokes (loss in energy by the molecule) (90,91). Typically, the Stokes shift is measured and referred to in Raman scattering since the anti-Stokes process requires the molecule to be in an excited state (90). Thus, the signal for anti-Stokes is typically much weaker than the Stokes signal, as few molecules are in an excited state under ambient temperature conditions. Raman has been used to investigate polymorphs (92,93), tablet heterogeneity (94), protein structure (64,95), excipient interactions (96,97), formulation screening (98), and in-line process monitoring (99,100). More specifically, the use of Raman microscopy, which couples a Raman spectrometer to an optical microscope equipped with a motorized x-y-z stage, has drawn increasing interest (89,94,101,102). With the ability to focus the laser onto a sample and obtain chemical and spatial information from characteristic bands corresponding to components within a sample, the technique offers many possibilities for sample investigation. With modern instrumentation and optics, spatial resolution of less than 1 μm is achieved, and with confocal capabilities penetration depth can be minimized allowing for surface analysis on the micron level (101). The technique of Raman mapping collects data at points across a sample allowing for probing of exact areas of a sample of micron size. Raman offers many advantages over IR, and even more so, IR microscopy. Perhaps the most important advantage of Raman is due to the weak Raman scattering of water. Interference from moisture is typically a serious problem in IR data collection, as water must be subtracted out from the spectra. Raman microscopy is also capable of achieving better spatial resolution (≈ 1 μm), than IR microscopy (≈ 10 μm) (101).

Although Raman microscopy has become an important tool in the pharmaceutical industry, it has yet to be directly applied to the study of amorphous phase separation in freeze-dried solids. Interestingly, it has been used to look at mannitol phase behavior during freezing, using a cold stage for temperature control for the purpose of understanding mannitol crystallization during the freeze-drying process (103). The authors noted that the technique is capable of identifying the location and form of crystalline mannitol within a sample. Additionally, they were capable of studying polymorphism issues associated with annealing, and found that the β-form of mannitol dominates when annealing. Although this example does not outline Raman microscopy's application to amorphous phase behavior, it does suggest the potential for this technique to be used in studying phase behavior in freeze-dried systems, as well as in frozen systems, which would be particularly useful in the study of phase separation in pharmaceutically relevant freeze-dried systems.

Raman microscopy actually has an extensive history in other fields of science, particular the materials science field (104–106), food sciences (107,108), and biophysical sciences (109). Many groups have improved sampling and analysis techniques to provide the most accurate and useful data pertaining to phase behavior in polymer and biopolymer systems. The ability of Raman mapping to characterize phase separation between two biopolymers, κ-carrageenan and gellan, has been reported (108). The phase behavior of PET/HDPE polymer blends using confocal Raman microscopy has been studied (104). The authors mapped a 60×60 μm region at 2 μm step sizes of a blended sample and were able to identify the distribution of the polymers within the sample area. Their data are shown in Figure 7 as an example of a Raman mapping image, where Figure 7A is an image for the 1062 cm^{-1} band intensity, representative of HDPE in the sample and Figure 7B is an image for the 1725 cm^{-1} band intensity, representative of the PET in the sample. This example clearly demonstrates the power of Raman microscopy when nonoverlapping characteristic bands are identifiable for individual components within a sample. This technique holds significant promise for the study of amorphous phase behavior in freeze-dried systems.

X-Ray Diffraction/Scattering

X-ray techniques are useful for probing structural characteristics of a solid or crystalline sample. X rays are electromagnetic radiation (10^{-2}–10^2 Å), which are elastically (no transfer of energy) scattered by matter (110). Specifically, they are scattered from the electrons within an atom or molecule, and therefore structural and packing information within a solid can be gained. Small-angle X-ray scattering (SAXS) and wide-angle X-ray diffraction (WAXD) are two X-ray

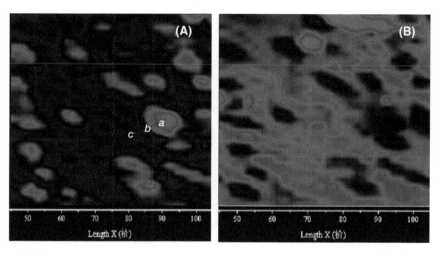

FIGURE 7 Raman mapping images for PET/HDPE blends; **(A)** Raman image at band 1062 cm^{-1}, where *a*, *b*, and *c* represent different phase compositions and **(B)** Raman image at band 1725 cm^{-1}. *Source*: From Ref. 104.

techniques commonly used (110). SAXS, as the name suggests, detects scattered X rays from a sample at very low angles (0.1°–10°) and is used to look at structural inhomogeneities on the nanometer scale. WAXD detects scattered X rays, Bragg peaks, at wide angles, providing information of atomic spacing on the pm scale; thus, this is an excellent tool for looking at crystal structure. Typical radiation sources are X-ray tubes or a synchrotron source, the latter involving accelerating electrons in a closed circular vacuum storage ring at speeds near the speed of light (110). This technique results in X rays covering a broader range of the spectrum and a much greater flux of X rays, which is not achievable in a normal laboratory. Thus the use of synchrotron radiation allows for increased sensitivity in studies of structure in solids.

The use of X-ray techniques in the study of phase behavior in freeze-dried systems has mainly been limited to crystalline phases. This is due to the ease of detecting crystalline phases in comparison with identification of amorphous phases, and additionally, if crystallization is present in a sample, the intensity of the crystalline components typically make it difficult to detect the weaker bands associated with the amorphous phase. Detection of crystalline sodium phosphate at levels as low as 1 mM concentrations using SAXS from a synchrotron source have been reported, which clearly demonstrates the sensitivity of this technique (111). Recent work on using pair distribution function (PDF) analysis of powder X-ray diffraction (PXRD) data to detect phase separation in a model amorphous freeze-dried polymer dispersion has been presented (112). A typical diffractogram of an amorphous solid presents a broad halo characteristic of the amorphous matrix. Information can be extracted from the amorphous halo via the use of PDFs, which represents a weighted probability of finding two atoms separated by a particular distance (112). Thus, PDFs describe the short-range order within a sample and can be obtained via Fourier transform of PXRD data (110). At least, if phase separation is essentially complete in a mixture, the PDF (or PXRD) data for the mixture may be modeled as a linear combination of pure component reference patterns, weighted according to the known composition of that component within a sample. This calculated pattern is compared with the mixture experimental patterns, and if the calculated pattern accurately describes the experimental mixture pattern the sample is judged to be phase separated (113). The results presented by the authors on PDF analysis of lyophilized mixtures confirm the presence of phase separation in a model PVP-dextran system, and these results are confirmed by the presence of two glass transitions in the DSC thermogram. They also noted phase separation in a trehalose-PVP system where the domain size was estimated to be in the nanometer range, which demonstrates the ability of the technique to detect phase separation in systems with small domain sizes. Since the technique does not rely on a change in heat capacity at the glass transition, the application of PDF analysis on XRPD of protein formulations may yield useful information regarding phase separation in such formulations. A disadvantage of the technique is the inability to distinguish between completely miscible systems and phase separation into phases that contain significant levels of both components.

SAXS methods are commonly employed for detecting phase separation in polymer-polymer phase-separating systems (114–118). The identification of polyisobutylene (PIB) nanophases within samples of PHEMA-1-PIB (poly(2-hydroxyethyl methacrylate)-1-polyisobutylene) using SAXS has been reported (116). This is interesting, given the size of the domains, which were reported on the nanometer scale, and demonstrates the ability of the technique to detect the presence of such small domains. Microphases of disordered amorphous polymers in a block copolymer blend using SAXS from a synchrotron source have been detected (117). Examples such as these illustrate the capability of X-ray techniques in detecting phase separation, which would be particularly valuable when the size of the phase domains lies below the limit of detection of many of the other techniques presented. However, a more extensive evaluation of the application of X-ray techniques to pharmaceutical freeze-dried systems is necessary to properly assess the application to the study of amorphous phase separation. Of course, it is not practical to expect that every formulation might be screened using a synchrotron source, but the use of X-ray techniques should facilitate development of a better understanding of phase behavior in freeze-dried systems, thus allowing for more general guidelines to be established for processing and formulation development.

Other Techniques

The review of potential techniques for investigating amorphous phase separation would not be complete without mentioning two spectroscopic studies focused on detection of phase separation in dried solids. One such study used solid-state NMR to identify T_1 relaxation times in sugar:lysozyme spray-dried samples (119). The authors identified phase separation in 1:1 and 9:1 sugar:protein formulations only when stored under high moisture conditions. Not surprisingly, the sugars crystallized under these conditions. Strong interactions between sugar and protein were indicated with a decreased mobility correlating to increased sugar content (at 0% RH). This would suggest that phase separation is unlikely under these conditions. However, another group studied heterogeneity in lysozyme:trehalose freeze-dried systems using NIR as a method of detection (61). They report heterogeneity in freeze-dried trehalose:lysozyme mixtures when compared with pure components. By identification of non-overlapping spectral absorption bands, the authors determined changes in the ratio of the nonoverlapping bands and found a significant difference when comparing the freeze-dried sample (heterogeneous) with a SCF-dried sample (homogeneous) and the pure components. A physical mixture of trehalose and lysozyme was used as a positive control for the technique and demonstrated an expected even larger degree of heterogeneity. The spatial resolution of the NIR technique is reported to be about 10 μm, with a penetration depth ranging from 30 to 180 μm. These resolutions constitute a significant handicap of the method for application in freeze-dried systems in which phase domains are expected to be much less than the reported penetration depth. Thus, one would expect that penetration into multiple phases may mask phase separation in some samples. Better control of penetration depth and spatial resolution would improve the technique for the purposes discussed here. Fundamentally, both techniques should be able to identify phase separation on some level. Further evaluation of these techniques for the study of phase separation might confirm the potential application to relevant freeze-dried systems.

CONCLUSIONS

Phase separation in freeze-dried amorphous solids is an area that has not been sufficiently studied to date. Perhaps this lack of attention is due to the lack of suitable methodology for detecting phase separation in systems of practical interest. Additionally, the nature of these dried solids (i.e., dry, fluffy, porous, hygroscopic, etc.) presents challenges in adapting some techniques to the study of freeze-dried samples. With the proper tool and appropriate sample preparation, it is likely that many of the techniques presented in this chapter would be useful in the study of phase behavior in freeze-dried systems. Of particular promise is Raman microscopy. With ever-evolving and improving instrumentation, this technique has significant promise for the study of phase separation in samples of practical importance to freeze-dried protein formulations. Additionally, computational methodologies applied to XRPD data are continually being developed for the study of phase separation in molecular dispersions, and have the potential of being successfully applied to protein formulations (112,113,120). The current state of dielectric techniques is hampered by a poor understanding of the origin of the many artifacts that can occur. Thus, these techniques, TSC and DRS, are not easily applied, and interpretation of results involves a considerable amount of ambiguity. A better understanding of the nature of the processes that give rise to relaxation events detectable by these methods may improve the application of these techniques to the study of phase separation in pharmaceutical freeze-dried systems.

Detection of phase separation could allow for modifications in formulation and process parameters as a means of circumventing the problem and resulting product instability. We expect that the application of new techniques to this field of study will continually progress. There is no doubt that improved methodology is necessary, as detection of detrimental phase separation upon processing and prior to storage stability testing would certainly lead to reduced product development time lines and reduction in the cost of product development.

REFERENCES

1. Dudzinski DM, Kesselheim AS. Scientific and legal viability of follow-on protein drugs. N Engl J Med 2008; 358(8):843–849.
2. Pikal MJ. Freeze-drying of proteins part II: formulation selection. BioPharm 1990; 3(9):26–30.
3. Manning MC, Patel K, Borchardt RT. Stability of protein pharmaceuticals. Pharm Res 1989; 6(11):903–918.
4. Lai MC, Topp EM. Solid-state chemical stability of proteins and peptides. J Pharm Sci 1999; 88(5):489–500.
5. Wang W. Lyophilization and development of solid protein pharmaceuticals. Int J Pharm 2000; 203:1–60.
6. Pikal MJ. Freeze-drying of proteins: process, formulation, and stability. ACS Symp Ser 1994; 567(Formulation and Delivery of Proteins and Peptides):120–133.
7. Franks F. Cryobiochemistry—responses of proteins to suboptimal temperatures. In: Biophysics and Biochemistry at Low Temperatures. Cambridge: Cambridge Universtiy Press, 1985.
8. Pikal MJ. Mechanisms of protein stabilization during freeze-drying and storage: the relative importance of thermodynamic stabilization and glassy state relaxations dynamics. In: Rey L, May JC, eds. Freeze-Drying/Lyophilization of Pharmaceutical and Biological Products. New York: Marcel Dekker; 2000:161–198.
9. Pikal MJ. Freeze-drying of proteins. Part I: process design. BioPharm (Duluth, MN, United States) 1990; 3(8):18–20, 2–4, 6–8.
10. Bhatnagar BS, Bogner RH, Pikal MJ. Protein stability during freezing: separation of stresses and mechanisms of protein stabilization. Pharm Dev Technol 2007; 12:505–523.
11. Izutsu K, Aoyagi N, Kojima S. Effect of polymer size and cosolutes on phase separation of poly(vinylpyrrolidone) (PVP) and dextran in frozen solutions. J Pharm Sci 2005; 94(4):709–717.
12. Heller MC, Carpenter JF, Randolph TW. Protein formulation and lyophilization cycle design: prevention of damage due to freeze-concentration induced phase separation. Biotechnol Bioeng 1999; 63(2):166–174.
13. Liao YH, Brown MB, Martin GP. Investigation of the stabilisation of freeze-dried lysozyme and the physical properties of the formulations. Eur J Pharm Biopharm 2004; 58:15–24.
14. Franks F, Hatley RHM, Mathias SF. Materials science and the production of shelf-stable biologicals. BioPharm 1991.
15. Carpenter JF, Crowe JH. An infrared spectroscopic study of the interactions of carbohydrates with dried proteins. Biochemistry 1989; 28(9):3916–3922.
16. Franks F. Solid aqueous solutions. Pure and Appl Chem 1993; 65(12):2527–2537.
17. Heller MC, Carpenter JF, Randolph TW. Application of a thermodynamic model to the prediction of phase separations in freeze-concentrated formulations for protein lyophilization. Arch Biochem Biophys 1999; 363(2):191–201.
18. Krause S. Polymer-polymer miscibility. Pure Appl Chem 1986; 58(12):1553–1560.
19. Yu M, Nishiumi H, de Swan Arons J. Thermodynamics of phase separation in aqueous solutions of polymers. Fluid Phase Equilib 1993; 83:357–364.
20. Izutsu KI, Kojima S. Freeze-concentration separates proteins and polymer excipients into different amorphous phases. Pharm Res 2000; 17(10):1316–1322.
21. Jara FL, Pilosof AMR. Glass transition temperature of protein/polysaccharide co-dried mixtures as affected by the extent and morphology of phase separation. Thermochimica Acta 2009; 487:65–73.
22. Tolstoguzov V. Foods as dispersed systems. Thermodynamic aspects of composition-property relationships in formulated food. J Therm Anal Calorim 2000; 61(2): 397–409.
23. de Kruif CG, Tuinier R. Polysaccharide protein interactions. Food Hydrocolloids 2001; 15(4–6):555–563.
24. Izutsu K, Heller MC, Randolph TW, et al. Effect of salts and sugars on phase separation of polyvinylpyrrolidone-dextran solutions induced by freeze-concentration. J Chem Soc Faraday Trans 1998; 94(3):411–417.

25. Tolstoguzov V. Compositions and phase diagrams for aqueous systems based on proteins and polysaccharides. Int Rev Cytol 2000; 192:3–31.
26. Sun WQ, Davidson P. Protein inactivation in amorphous sucrose and trehalose matrixes: effects of phase separation and crystallization. Biochim Biophys Acta 1998; 1425(1):235–244.
27. Annunziata O, Pande A, Pande J, et al. Oligomerization and phase transitions in aqueous solutions of native and truncated human âB1-Crystallin. Biochemistry 2005; 44:1316–1328.
28. Pikal MJ, Rigsbee D, Roy ML, et al. Solid state chemistry of proteins: II. The correlation of storage stability of freeze-dried human growth hormone (hGH) with structure and dynamics in the glassy solid. J Pharm Sci 2008; 97(12):5106–5121.
29. Chang L, Shepherd D, Sun J, et al. Mechanism of protein stabilization by sugars during freeze-drying and storage: native structure preservation, specific interaction, and/or immobilization in a glassy matrix? J Pharm Sci 2005; 94(7):1427–1444.
30. Chang L, Milton N, Rigsbee D, et al. Using modulated DSC to investigate the origin of multiple thermal transitions in frozen 10% sucrose solutions. Thermochim Acta 2006; 444(2):141–147.
31. Pikal MJ, Rigsbee DR, Roy ML. Solid state chemistry of proteins: I. Glass transition behavior in freeze dried disaccharide formulations of human growth hormone (hGH). J Pharm Sci 2007; 96:2765–2776.
32. Heller MC, Carpenter JF, Randolph TW. Manipulation of lyophilization-induced phase separation: implications for pharmaceutical proteins. Biotechnol Prog 1997; 13(5):590–596.
33. Izutsu K, Shigeo K. Phase separation of polyelectrolytes and non-ionic polymers in frozen solutions. Phys Chem Chem Phys 2000; 2(1):123–127.
34. Izutsu K, Yoshioka S, Kojima S. Phase separation and crystallization of components in frozen solutions: effect of molecular compatibility between solutes. ACS Symp Ser 1997; 675 (Therapeutic Protein and Peptide Formulation and Delivery):109–118.
35. Izutsu K, Yoshioka S, Kojima S, et al. Effects of sugars and polymers on crystallization of poly(ethylene glycol) in frozen solutions: phase separation between incompatible polymers. Pharm Res 1996; 13(9):1393–1400.
36. Heller MC, Barbieri DM, Randolph TW, et al. The effects of polymer liquid-liquid phase separation on hemoglobin during freezing and drying. Book of Abstracts, 211th ACS National Meeting, New Orleans, LA, March 24-28 1996:BIOT-140.
37. Heller MC, Carpenter JF, Randolph TW. Effects of phase separating systems on lyophilized hemoglobin. J Pharm Sci 1996; 85(12):1358–1362.
38. Heller MC, Carpenter JF, Randolph TW. Conformational stability of lyophilized PEGylated proteins in a phase-separating system. J Pharm Sci 1999; 88(1):58–64.
39. Atkins P. Physical Chemistry. 6th ed. New York: W.H. Freeman and Company, 1998.
40. Dill KA, Bromberg S. Molecular Driving Forces: Statistical Thermodynamics in Chemistry and Biology. New York: Garland Science, Taylor and Francis Group, 2003.
41. Hildebrand JH, Scott RL. Regular Solutions. Englewood Cliffs: Prentice-Hall, Inc., 1962.
42. Sperling LH. Introduction to Physical Polymer Science. New York: John Wiley and Sons, Inc., 1992.
43. Jonsson B, Lindman B, Holmberg K, et al. Surfactants and Polymers in Aqueous Solution. Chichester, UK: John Wiley and Sons, 1998.
44. Pikal MJ. Freeze-drying. In: Swarbrick J, Boylan JC, eds. Encyclopedia of Pharmaceutical Technology. 2nd ed. Marcel Dekker, 2002:1299–1326.
45. Randolph TW. Phase separation of excipients during lyophilization: effects on protein stability. J Pharm Sci 1997; 86(11):1198–1203.
46. Pessoa Filho PA, Mohname RS. Thermodynamic modeling of the partitioning of biomolecules in aqueous two-phase systems using a modified Flory-Huggins equation. Process Biochemistry 2004; 39:2075–2083.
47. Edmond E, Ogston AG. An approach to the study of phase separation in ternary aqueous systems. Biochem J 1968; 109(569–576).

48. Turgeon SL, Beaulieu M, Schmitt C, et al. Protein-polysaccharide interactions: phase-ordering kinetics, thermodynamic and structural aspects. Curr Opin Colloid Interface Sci 2003; 8(4,5):401–414.
49. Schorsch C, Clark AH, Jones MG, et al. Behavior of milk protein/polysaccharide systems in high sucrose. Colloids Surf B Biointerfaces 1999; 12(3–6):317–329.
50. Guo XH, Chen SH. Observation of polymerlike phase separation of protein-surfactant complexes in solution. Phys Rev Lett 1990; 64(16):1979–1982.
51. Pikal MJ. Formulation of proteins for freeze drying: theoretical concepts and practical guidelines. Book of Abstracts, 211th ACS National Meeting, New Orleans, LA, March 24–28, 1996:BIOT-135.
52. Rambhatla S, Ramot R, Bhugra C, et al. Heat and mass transfer scale-up issues during freeze drying: II. Control and characterization of the degree of supercooling. AAPS PharmSciTech 2004; 5(4):e58.
53. Tang X, Pikal Michael J. Design of freeze-drying processes for pharmaceuticals: practical advice. Pharm Res 2004; 21(2):191–200.
54. Pikal MJ. Use of laboratory data in freeze drying process design: heat and mass transfer coefficients and the computer simulation of freeze-drying. J Parenter Sci Technol 1985; 39(3):115–138.
55. Patel SM, Bhugra C, Pikal MJ. Reduced pressure ice fog technique for controlled ice nucleation during freeze-drying. AAPS PharmSciTech 2009; [Epub ahead of print].
56. Ahamed T, Esteban BNA, Ottens M, et al. Phase behavior of an intact monoclonal antibody. Biophysical J 2007; 93:610–619.
57. Haas C, Drenth J. Understanding protein crystallization on the basis of the phase diagram. J Cryst Growth 1999; 196(388–394).
58. Her LM, Deras M, Nail SL. Electrolyte induced changes in glass transition temperatures of freeze-concentrated solutes. Pharm Res 1995; 12(5):768–772.
59. Carpenter JF, Pikal MJ, Chang BS, et al. Rational design of stable lyophilized protein formulations: some practical advice. Pharm Res 1997; 14(8):969–975.
60. Franks F. Biophysics and Biochemistry at Low Temperatures. Cambridge: Cambridge University Press, 1985.
61. Jovanovic N, Gerich A, Bouchard A, et al. Near-infrared imaging for studying homogeneity of protein-sugar mixtures. Pharm Res 2006; 23(9):2002–2013.
62. Arakawa T, Prestrelski SJ, Kenney WC, et al. Factors affecting short-term and long-term stabilities of proteins. Adv Drug Deliv Rev 2001; 46:307–326.
63. Izutsu K, Yoshioka S, Kojima S. Physical stability and protein stability of freeze-dried cakes during storage at elevated temperatures. Pharm Res 1994; 11(7):995–999.
64. Sane SU, Wong R, Hsu CC. Raman spectroscopic characterization of drying-induced structural changes in a therapeutic antibody: correlating structural changes with long-term stability. J Pharm Sci 2004; 93(4):1005–1018.
65. Cleland JL, Xanthe L, Kendrick B, et al. A specific molar ratio of stabilizer to protein is required for storage stability of a lyophilized monoclonal antibody. J Pharm Sci 2001; 90(3):310–321.
66. Wang B, Tchessalov S, Cicerone M, et al. Impact of sucrose level on storage stability of proteins in freeze-dried solids: II. Correlation of aggregation rate with protein structure and molecular mobility. J Pharm Sci 2009; 98(9):3145–3166.
67. Hill JJ, Shalaev EY, Zografi G. Thermodynamic and dynamic factors involved in the stability of native protein structure in amorphous solids in relation to levels of hydration. J Pharm Sci 2005; 94(8):1636–1667.
68. Elliott SR. Physics of Amorphous Materials. 2nd ed. Essex: Longman Scientific and Technical, 1990.
69. Katayama DS, Carpenter JF, Manning MC, et al. Characterization of amorphous solids with weak glass transitions using high ramp rate differential scanning calorimetry. J Pharm Sci 2008; 97(2):1013–1024.
70. Sochava IV. Heat capacity and thermodynamic characteristics of denaturation and glass transition of hydrated and anhydrous proteins. Biophys Chem 1997; 69(1):31–41.

71. Sochava IV, Smirnova OI. Heat capacity of hydrated and dehydrated globular proteins. Denaturation increment of heat capacity. Food Hydrocolloids 1993; 6(6):513–524.
72. Pikal MJ, Rigsbee D, Roy ML. Solid state stability of proteins III: calorimetric (DSC) and spectroscopic (FTIR) characterization of thermal denaturation in freeze dried human growth hormone (hGH). J Pharm Sci 2008; 97(12):5122–5131.
73. Pezzin APT, Duek EAR. Miscibility and hydrolytic degradation of bioreabsorbable blends of poly(p-dioxanone) and poly(L-lactic acid) prepared by fusion. J Appl Polym Sci 2006; 101(3):1899–1912.
74. Cabanelas JC, Serrano B, Baselga J. Development of cocontinuous morphologies in initially heterogeneous thermosets blended with poly(methyl methacrylate). Macromolecules 2005; 38(3):961–970.
75. Craig DQM. Dielectric Analysis of Pharmaceutical Systems. London: Taylor and Francis Publishers, 1995.
76. Lacabanne C, Bauer M. Thermally stimulated currents: a tool for pharmaceutical science. Am Pharm Rev 2007; 10(1):66–72.
77. Gun'ko VM, Zarko V, Goncharuk EV, et al. TSDC spectroscopy of relaxational and interfacial phenomena. Adv Colloid Interface Sci 2007; 131:1–89.
78. Bhugra C, Shmeis RA, Krill SL, et al. Predictions of onset of crystallization from experimental relaxation times I-correlation of molecular mobility from temperatures above the glass transition to temperatures below the glass transition. Pharm Res 2006; 23(10):2277–2290.
79. Pearson DS, Smith G. Dielectric analysis as a tool for investigating the lyophilization of proteins. Pharm Sci Technol Today 1998; 1(3):108–117.
80. Boutonnet-Fagegaltier N, Menegotto J, Lamure A, et al. Molecular mobility study of amorphous and crystalline phases of a pharmaceutical product by thermally stimulated current spectroscopy. J Pharm Sci 2002; 91(6):1548–1560.
81. Bhugra C, Shmeis RA, Krill SL, et al. Different measures of molecular mobility: comparison between calorimetric and thermally stimulated current relaxation times below T_g and correlation with dielectric relaxation times above Tg. J Pharm Sci 2008; 97(10):4498–4515.
82. Shmeis RA, Wang Z, Krill SL. Amorphous pharmaceutical compound molecular mobility assessment and excipient influence. Proceedings of the 31st NATAS Annual Conference on Thermal Analysis and Applications 2003; 055/1–/8.
83. Ramos JJM, Correia NIT, Taveira-Marques R, et al. The activation energy at Tg and the fragility index of indomethacin, predicted from the influence of the heating rate on the temperature position and on the intensity of thermally stimulated depolarization current peak. Pharm Res 2002; 19(12):1879–1884.
84. Li J, Guo Y, Zografi G. The solid-state stability of amorphous quinapril in the presence of b-cyclodextrins. J Pharm Sci 2002; 91(1):229–243.
85. Reddy R, Chang LL, Luthra S, et al. The glass transition and subT$_g$-relaxation in pharmaceutical powders and dried proteins by thermally stimulated current. J Pharm Sci 2008.
86. Ramos JJM, Pinto SS, Diogo HP. The slow molecular mobility in amorphous trehalose. ChemPhysChem 2007; 8:2391–2396.
87. Sauer BB, Avakian P, Cohen GM. Studies of phase separated copolymer blends using thermally stimulated currents and dielectric spectroscopy. Polymer 1992; 33(13):2666–2671.
88. Tsonos C, Apekis L, Zois C, et al. Microphase separation in ion-containing polyurethanes studied by dielectric measurements. Acta Mater 2004; 52:1319–1326.
89. Whitley A, Barnett S. Advances and useful applications of Raman spectroscopy, imaging, and remote sensing. SPIE 1998; 3261:250–259.
90. Turrell G. The Raman effect. In: Turrell G, Corset J, eds. Raman Microscopy: Developments and Applications. San Diego: Academic Press, 1996:1–25.
91. Skoog DA, Holler JF, Nieman TA. Principles of Instrumental Analysis. 5th ed. Philadelphia: Hartcourt Brace College Publishers, 1998.
92. Taylor LS. Raman spectroscopy as a tool to probe solid state form of tablets. Am Pharm Rev 2001; 4(4):60–67.

93. Pratiwi D, Fawcett JP, Gordon KC, et al. Quantitative analysis of polymorphic mixtures of ranitidine hydrochloride by Raman spectroscopy and principal component analysis. Eur J Pharm Biopharm 2002; 54:337–341.

94. Zhang L, Henson MJ, Sekulic SS. Multivariate data analysis for Raman imaging of a model pharmaceutical tablet. Anal Chim Acta 2005; 545:262–278.

95. Sane SU, Cramer SM, Przybycien TM. A Holistic approach to protein secondary structure characterization using amide I band Raman spectroscopy. Anal Biochem 1999; 269:255–272.

96. Taylor LS, Zografi G. Sugar-polymer hydrogen bond interactions in lyophilized amorphous mixtures. J Pharm Sci 1998; 87(12):1615–1621.

97. Taylor LS, Langkilde FW, Zografi G. Fourier transform raman spectroscopic study of the interaction of water vapor with amorphous polymers. J Pharm Sci 2001; 90(7): 888–901.

98. Sasic S, Clark DA, Mitchell JC, ET AL. Raman line mapping as a fast method for analyzing pharmaceutical bead formulations. Analyst 2005; 130:1530–1536.

99. Wikstrom H, Lewis IR, Taylor LS. Comparison of sampling techniques for in-line monitoring using raman spectroscopy. Appl Spectrosc 2005; 59(7):934–941.

100. Romero-Torres S, Wikstrom H, Grant ER, et al. Monitoring of mannitol phase behavior during freeze-drying using non-invasive raman spectroscopy. PDA J Pharm Sci Technol 2007; 61(2):131–145.

101. Turrell G, Delhaye M, Dhamelincourt P. Characteristics of raman microscopy. In: Turrell G, Corset J, eds. Raman Microscopy: Developments and Applications. San Diego: Academic Press, 1996:27–49.

102. McGeorge G. Combining raman spectroscopy and microscopy to support pharmaceutical development. Am Pharm Rev 2003; 6(3):94–99.

103. Beattie JR, Barrett LJ, Malone JF, et al. Investigation into the subambient behavior of aqueous mannitol solutions using temperature-controlled Raman microscopy. Eur J Pharm Biopharm 2007; 67:569–578.

104. Huan S, Lin W, Sato H, et al. Direct characterization of phase behavior and compatibility in PET/HDPE polymer blends by confocal raman mapping. J Raman Spectrosc 2007; 38:260–270.

105. Janik H, Palys B, Petrovic ZS. Multiphase-separated Polyurethanes studied by micro-raman spectroscopy. Macromol Rapid Commun 2003; 24:265–268.

106. Sammon C, Mura C, Eaton P, et al. Raman microscopic studies of polymer surfaces and interfaces. Analusis 2000; 28(1):30–34.

107. Pudney PDA, Hancewicz TM, Cunningham DG, et al. A novel method for measuring concentrations of phase separated biopolymers: the use of confocal Raman spectroscopy with self-modelling curve resolution. Food Hydrocolloids 2003; 17:345–353.

108. Pudney PDA, Hancewicz TM, Cunningham DG. The use of confocal Raman spectroscopy to characterise the microstructure of complex biomaterials: foods. Spectroscopy 2002; 16:217–225.

109. Percot A, Lafleur M. Direct observation of domains in model stratum corneum lipid mixtures by raman microspectroscopy. Biophys J 2001; 81:2144–2153.

110. Roe RJ. Methods of X-ray and Neutron Scattering in Polymer Science. New York: Oxford University Press, 2000.

111. Varshney DB, Kumar S, Shalaev EY, et al. Solute crystallization in frozen systems–use of synchrotron radiation to improve sensitivity. Pharm Res 2006; 23(10): 2368–2374.

112. Newman A, Engers D, Bates S, et al. Characterization of amorphous API:polymer mixtures using X-ray powder diffraction. J Pharm Sci 2008; 97(11):4840–4856.

113. Ivanisevic I, Bates S, Chen P. Novel Methods for the Assessment of Miscibility of Amorphous Drug-Polymer Dispersions. J Pharm Sci 2009; 98(9):3373–3386.

114. Thiyargarajan P. Characterization of materials of industrial importance using small-angle scattering techniques. J Appl Crystallogr 2003; 36:373–380.

115. Fairclough JPA, Hamley IW, Terrill NJ. X-ray scattering in polymers and micelles. Radiat Phys Chem 1999; 56:159–173.

116. Domjan A, Erdoedi G, Wilhelm M, et al. Structural studies of nanophase-separated poly(2-hydroxyethyl methacrylate)-l-polyisobutylene amphiphilic co-networks by solid-state NMR and small-angle X-ray scattering. Macromolecules 2003; 36(24):9107–9114.
117. Inomata K, Liu L-Z, Nose T, et al. Sychrotron small-angle X-ray scattering studies on phase separation and crystallization of associated polymer blends. Macromolecules 1999; 32:1554–1548.
118. Runt JP, Zhang X, Miley DM, et al. Phase behavior of poly(butylene terephthalate)/polyarylate blends. Macromolecules 1992; 25:3902–3905.
119. Suihko E, Forbes RT, Apperley D. A solid-state NMR study of molecular mobility and phase separation in co-spray-dried protein-sugar particles. Eur J Pharm Sci 2005; 25:105–112.
120. Bates S, Zografi G, Engers D, et al. Analysis of amorphous and nanocrystalline solids from their X-ray diffraction patterns. Pharm Res 2006; 23(10):2333–2349.

5 The Use of Microscopy, Thermal Analysis, and Impedance Measurements to Establish Critical Formulation Parameters for Freeze-Drying Cycle Development

Kevin R. Ward
Biopharma Technology Ltd., Hampshire, U.K.

Paul Matejtschuk
National Institute for Biological Standards and Control, Health Protection Agency, Potters Bar, U.K.

INTRODUCTION

Historically, many freeze-drying cycles have been arrived at on an empirical basis, and while some of these cycles may produce consistently acceptable products, in recent years it has become more widely accepted that cycles should be developed on a more rational and scientific basis with an understanding of the formulation in question (1,2). There is therefore a need to provide documentary evidence to support the rationale of how a formulation has been selected and the corresponding freeze-drying cycle has been developed on the basis of that specific formulation, particularly with regard to its critical temperature(s) and its response to freezing, sublimation, and desorption. This in turn has led to the advent of more commercially available equipment and methods, which together can provide a reasonably complete characterization of a material prior to freeze-drying, so that processing conditions may be selected on a case-by-case basis. In this volume, Wang describes a logical stepwise approach to formulation characterization for freeze-drying. In this chapter, we expand on some of the formulation parameters that are pertinent to achieving a freeze-dried product with acceptable appearance on the macroscopic and microscopic scales, aspects of which may be considered qualitative parameters that are difficult to assess. Secondly, we describe some of the more widely used, well-established methods that enable characterization of a formulation with a view to achieving structural integrity in the dried product and discuss some of the issues surrounding data interpretation and their relevance to freeze-drying. Thirdly, we outline some recent advances in this field with regard to new methodologies, and present and discuss some data from a novel method of electrical impedance analysis that may help understand how a formulation may behave at the microscopic level as it undergoes freeze-drying. Finally, we discuss some of the issues surrounding data interpretation from the various formulation characterization methods and their relevance to the development of a freeze-drying cycle for a pharmaceutical or biological product.

ACHIEVING ACCEPTABLE FREEZE-DRIED PRODUCT APPEARANCE

Together with activity, water content, and stability, the macroscopic integrity of a lyophile is one of the fundamental parameters typically assessed for any freeze-dried pharmaceutical and biological materials, with the aim being to produce a material that occupies the same volume as the initial fill, with a high surface area to volume ratio. Indeed, it has been well established that appearance often goes hand in hand with activity and stability, since a product that displays a macroscopic loss of structure often contains higher levels of water than its "intact" counterpart, has a lower specific surface area that can lead to increased reconstitution time, and will mean that the various components are in more intimate contact with each other during and after freeze-drying.

Product appearance itself may be assessed in a number of ways, but usually this is restricted to the use of qualitative, relatively subjective methods such as judging color, degree of shrinkage, and level of uniformity by eye, which provides information solely on the macroscopic structure. These parameters are often used as part of an assessment to indicate the success of technology transfer or scale-up. In terms of quantifying various aspects of the microscopic structure, estimates of total surface area may be made using the Brunauer-Emmett-Teller (BET) method, which calculates surface area from experimental measurements of multilayer gas adsorption onto a material (3). Hibbert et al. (4) explain how this has been applied to freeze-dried powders, while Rey (5) further discusses the practicalities and limitations of using this method and provides data allowing calculation of the specific surface area of a real-life pharmaceutical material. Porosity can be assessed using scanning electron microscopy (SEM), although SEM relies on representative sampling of the lyophile and that the preparation of the sample for imaging does not itself cause changes in the sample. Some of the methods currently under investigation for quantifying aspects of lyophile substructure and various physical phenomena brought about by the freeze-drying process itself (some of which have even been applied in process during freezing and/or drying) include Fourier transform infrared spectroscopy (FTIR) (6), temperature scanning FTIR (7), near infrared (NIR) spectroscopy (8), dielectric relaxation spectroscopy (9), Raman spectroscopy (10,11), and temperature-controlled Raman microscopy (12).

At the most fundamental level, to achieve a lyophile with acceptable macroscopic appearance, the primary objective for crystalline materials is not to exceed their eutectic temperature and that for amorphous materials is not to exceed their glass transition temperature or collapse temperature between the initial cooling phase and the end of the sublimation process. However, often a formulation will contain a mixture of crystallizing and amorphous components, in which case, the critical temperature may not be so obvious. One approach is to assume complete phase separation will occur, assess the individual components separately, then aim to cool the starting material to below the lowest of the critical temperatures and maintain it below this temperature until the end of the sublimation process. For example, an aqueous solution containing lactose and sodium chloride may have a T_g' of $-32°C$ (lactose) and a T_{eu} of $-21°C$ (NaCl) in the frozen state, and therefore if complete separation occurred, the critical temperature may logically be assumed to be $-32°C$. However, what this approach does not take into account is the fact that the presence of lactose may inhibit crystallization of the sodium chloride and that in a real freeze-drying situation amorphous sodium chloride may be present. We have observed in our laboratory that while a solution of 1% (w/v) lactose and 1% (w/v) NaCl tends to collapse around $-30°C$—implying that most or all of the NaCl is crystalline—a solution containing 1% lactose and 0.3% NaCl collapses around $-46°$, suggesting the presence of a phase within the frozen structure consisting of amorphous NaCl and/or a (combined) amorphous mixture of NaCl and lactose. A similar pattern of behavior has also been observed in our laboratories when the concentration of sucrose included as an excipient in an influenza antigen, preparation was varied between 1% and 2% (w/v).

A further consideration here is the possibility of microcollapse, as discussed by Wang later in this volume and elsewhere (13,14). A simple example of microcollapse, and one that we have frequently observed in our laboratories, is the one that occurs in formulations containing mannitol together with amorphous components (which may be excipients and/or the active ingredient). A solution containing 2% mannitol and 1% glucose where the mannitol crystallizes during freeze-drying (either as a controlled event during a deliberate annealing step or perhaps in a less-controlled manner during sublimation) would be expected to comprise separate phases in the frozen (and drying) structure where mannitol exhibits a T_{eu} of $-1.4°C$ but glucose undergoes viscous flow around $-41°C$, thus microcollapsing onto the scaffold of crystalline mannitol. In this case, a conservative estimate of the critical temperature might be considered to be $-41°C$, although we have observed that in such a mixture total loss of structure at the macroscopic level often does not occur until above $-20°C$. This is still significantly lower than the eutectic temperature of crystalline mannitol alone; whether this is attributable to significant levels of viscous flow in the amorphous glucose phase at this temperature that was able to counter the mechanical strength afforded by crystalline mannitol or whether indeed some of the mannitol persisted in the amorphous phase because of the presence of amorphous glucose and/or the thermal history of the sample, was not clear. This example illustrates the point that combining crystalline and amorphous components does not always give a single critical temperature that can be predicted from the behavior of the individual components, and that exceeding a particular ratio could lead to step changes in critical temperature with respect to the macroscopic structure due to phase/state changes in one or more of the components, which may also be a function of thermal history. Similar observations were made by Adams and Irons for mixtures of sodium chloride and lactose (15).

Conceivably, in mixtures where the eutectic temperature (T_{eu}) of the crystalline phase is lower than the collapse temperature (T_c) of the amorphous phase, a phenomenon may occur where the crystalline solute phase melts onto the "backbone" of the rigid amorphous phase when freeze-dried above T_{eu}. We have notionally termed this phenomenon "micromelting", since it is analogous to microcollapse. An example of this that has been observed in our laboratories is that of a biological product containing a small amount of calcium chloride as part of a buffer system; in this case, while macroscopic collapse was not observed until temperatures above $-30°C$ by FDM, the use of other methods suggested significant mobility changes around $-53°C$, the eutectic temperature of calcium chloride. When the formulation was freeze-dried above $-53°C$ no shrinkage was observed at the macroscopic level, although there was evidence of heterogeneity in the form of small visible "spots" on the base of the cake.

It has been noted in studies focusing on specific formulations that there are instances where it yields no measurable benefit to keep below the lower (or lowest) critical temperature, since lyophilization will, by definition, take longer at lower temperatures and may not confer any greater stability on the final product (13). However, from a wider point of view, it is likely that this is a factor that should be assessed on a case-by-case basis, since the extent of microcollapse or micromelting in a formulation will depend on factors such as the proportion of material undergoing the microcollapse or micromelting event and its effect on the structure and of the dried material, together with any changes in the dynamics of the residual water. For example, if a microcollapsed amorphous saccharide component harbors more residual water than its noncollapsed counterpart this water could be "released" over time (perhaps during crystallization of that component over time in the lyophile) and affect the remaining structure and the kinetics of any unwanted reactions in the dried state. While an understanding of the macroscopic behavior of a formulation during lyophilization may not represent a very full characterization, often this information has been sufficient and has been shown to relate directly with the performance of a product when dried at a series of temperatures. However, in other cases, an understanding of the macroscopic structure alone is not sufficient to fully appreciate how that formulation will behave during freezing and drying or to predict the resulting stability of the dried product. For instance, it may be key to know whether the active substance (and may be also other components) is in the amorphous or crystalline state, which may not affect the structure of the cake but could be critical to stability and/or activity and have regulatory compliance implications. We discuss this issue further in the section on electrical impedance analysis below.

There is evidence that the apparent T_g' of a formulation can vary with concentration, but this may be due to the formation of more constrained glasses during analysis (which typically employs faster cooling rates than would operationally be the case in a freeze-dryer). Thus, a glass may be formed that does not truly reflect the maximally concentrated amorphous phase, which is the generally accepted definition of T_g', but instead a real amorphous sample may contain higher levels of unfrozen water (16). While it may be possible to predict the T_g' for mixtures of amorphous components that form a single phase within the frozen material, this is not always straightforward in more complex mixtures, as highlighted recently by Izutsu et al. (17).

TECHNIQUES FOR ASSESSING CRITICAL FORMULATION PARAMETERS

Over the past several decades, a number of methods have been developed that allow the characterization of a material prior to freeze-drying to assess those parameters that are significant in enabling it to be lyophilized to yield a product with acceptable macroscopic appearance. Wang provides a historical overview of the development of freeze-drying microscopy (FDM) and shows how information obtained using FDM can be taken with data from differential scanning calorimetry (DSC) to build up a picture of a material prior to drying (13). There are a number of texts that list and describe in great detail the large number of thermal methods that can be used to characterize pharmaceutical materials in general (18–20), yet freeze-drying is rarely mentioned specifically, apart from in a small number of cases. The review by Kett et al. (16) provides a comprehensive overview of some of the thermal methods that are widely used in the characterization of materials specifically for freeze-drying purposes. In this section, we expand on some of the information that can be gained from the use of FDM in addition to simply evaluating the collapse temperature of a material, provide a brief description of some of the more widely used thermal methods, and discuss some of the issues with data interpretation in the light of the relevance of the information to formulation and lyo-cycle development.

Freeze-Drying Microscopy

Cryomicroscopy and FDM have been employed as methods of evaluating the freezing and lyophilization behavior of simple solutions and formulations since the middle of the 20th century, ever since eminent researchers such as Rey and Mackenzie showed how this information could be useful in-process development (21,22). There are now commercially available systems that enable temperatures to be controlled to within a fraction of a degree and pressure to be controlled and monitored throughout the analytical process. Digital cameras have further allowed images to be captured and saved in galleries that enable retrospective analysis of information to allow the collapse event to be pinpointed to within a fraction of a degree, as well as the ability to compile images into video format. We define the onset of the collapse event (the lower temperature limit giving first visible signs of viscous flow) as $T_{c\ (onset)}$ and the endpoint of the event (upper temperature limit where no observable drying structure remains) as $T_{c\ (endpoint)}$. Others have defined these events as T_{oc} and T_{fc}, respectively (23).

Operationally, with modern FDM designs, pressure control is rendered insignificant by virtue of the fact that the sample is in good thermal contact with the drying block and is in the format of a thin layer sandwiched between two cover slips, as depicted in Figure 1. Employing such a format means that heat input to the sample by convection is minimized; however, the pressure in the sample chamber should be lower than the vapor pressure of ice at the corresponding sample temperature. Additionally, the use of a relatively thin sample minimizes the effect of sublimation cooling on the sample, which would otherwise render it at a lower temperature than the temperature-controlled block.

FDM provides the ability to determine a number of events occurring in a material as it is frozen and subsequently dehydrated under vacuum, including nucleation temperature (although this is somewhat of a misnomer, given the random nature of ice crystal formation), eutectic melting, collapse, microcollapse (in certain instances), the effect of annealing a sample on ice crystal structure

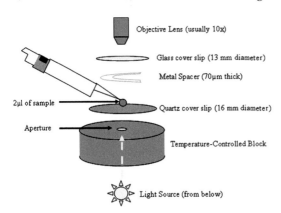

FIGURE 1 Sample format employed for freeze-drying microscopy analysis using Lyostat2 (Biopharma Technology Limited, Winchester, U.K.).

and solute crystallization, and the formation of a surface skin, where a band of concentrated solute is forced to the edge of the sample during ice formation. Indeed, FDM is the only method that allows the collapse temperature and skin formation potential to be determined. While the formation of a surface skin may be partly a function of the cooling rate used (and the temperature gradients thereby introduced within a sample because of the main mechanism of heat removal being from direct contact with the shelf), we have noted in our laboratories that certain formulations display a higher propensity to form a surface skin than others. Examples of the above-mentioned events are given in Figures 2 to 8 below. In all cases, images were taken using a Lyostat2 freeze-drying

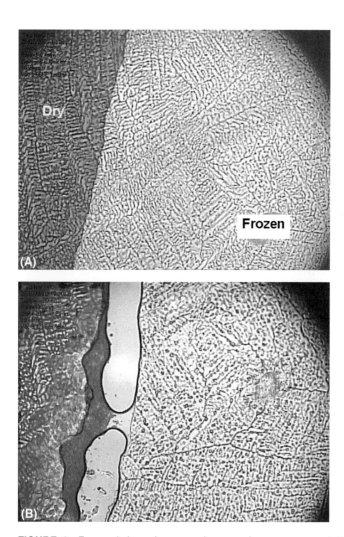

FIGURE 2 Freeze-drying microscopy images of an aqueous solution of 4% (w/v) sodium chloride: (**A**) at $-24°C$ where drying takes place below the eutectic melt temperature; (**B**) at $-20.7°C$, which is above the eutectic temperature, showing the visible effects of eutectic melting.

microscope (Biopharma Technology Ltd., Winchester, U.K.), using the format depicted in Figure 1 with sample volume 2µl and sample thickness 70µm.

Figure 2 shows a sample of sodium chloride solution drying below and above its eutectic temperature, illustrating the fact that the eutectic temperature can be identified very clearly using FDM because of the marked visible change in sample appearance that occurs at T_{eu}. Figure 3 shows a sample of sucrose solution drying below, at the onset of, and above, its collapse temperature, demonstrating the effectiveness of FDM as a method of identifying the collapse event. Figure 4 illustrates the visual indication of skin formation that a sample may exhibit, even when cooled rapidly from the liquid state. Figure 5 depicts an FDM image taken around the collapse temperature of a minor component (glucose) in the presence of a more concentrated crystalline component (mannitol), highlighting separate areas of dense structure between apparent voids in the dried material. Figure 6 shows the affect of annealing a frozen sample on ice crystal growth, which can be assessed on a semiquantitative basis if a calibrated microscope lens is employed. Figure 7 shows how FDM can be set up to employ a polarized light source, while Figure 8 illustrates how this technique may be employed to qualitatively assess whether solute crystallization occurs during freezing and/or sublimation and whether this can be induced by warming of the sample. Zhai et al. also demonstrated that FDM can also be used to compare primary drying rates for different formulations or for a single formulation following the use of different freezing and drying conditions (24).

Differential Thermal Analysis

Differential thermal analysis (DTA) involves cooling or warming a sample and a reference (usually the same standard of water used to prepare the sample) under identical conditions (25–28), observing the differences in their temperatures (ΔT) and plotting this as a function of temperature or time. This enables detection of any measurable endothermic or exothermic transitions in the sample that do not occur in the reference material. A historical perspective of DTA is provided by Ozawa (29). DTA was the first method to be employed for the systematic investigation of the thermal characteristics of frozen solutions (30), where a series of transitions were reported prior to the melting event upon warming. In particular, an apparent heat capacity increase was noted prior to the main melting endotherm; this event was termed "antemelting." Limitations of DTA as a method arise from some of the practical aspects such as the placement of the temperature measuring devices (usually thermocouples) within the sample and reference holders, the thermal resistance that can restrict the flow of heat to and from the temperature measuring devices, and the relative lack of sensitivity to low energy or adiabatic events when compared with that in more advanced calorimetric methods (16). However, DTA can prove a simple yet effective technique when dealing with relatively simple solutions that undergo changes that are accompanied by a significant heat flow.

Differential Scanning Calorimetry and Modulated Differential Scanning Calorimetry

While DTA provides a measure of the difference in temperatures between a sample and a reference material, DSC provides an additional level of detail in

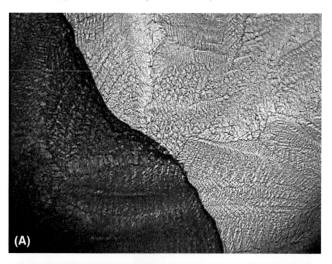

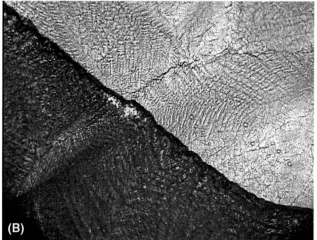

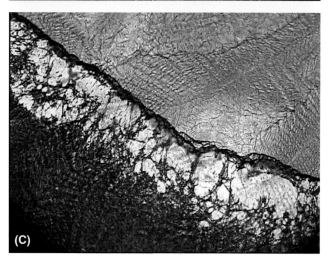

FIGURE 3 Freeze-drying microscopy images of an aqueous solution of 2% (w/v) sucrose: (**A**) at −31.8°C, with good visible structure obtained when dried below the collapse temperature; (**B**) the onset of collapse at −30.8°C, and (**C**) drying at −28.3°C, above the end-point of collapse event.

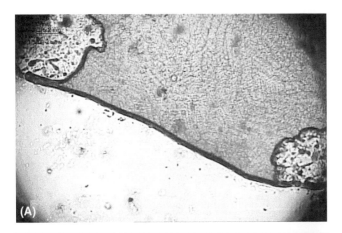

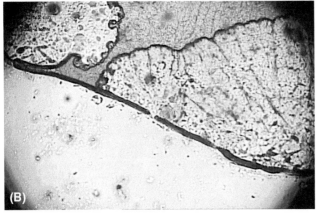

FIGURE 4 Freeze-drying microscopy images showing the formation of a concentrated band of solute at the perimeter of the sample which impedes drying and gives a sublimation front that is not parallel with the edge of the sample. (**A**) The sample immediately after drying commences; (**B**) the same sample following further progression of the drying front.

that it also enables the determination of the heat flow associated with the observed events. This can be useful when studying crystallization events, since it can provide an indication of the proportion of a particular material crystallizing during analysis, if the specific heat of crystallization is known. However, as discussed in the section below regarding the issues arising with the interpretation of analytical data, glass transitions in amorphous materials often remain elusive with conventional DSC because of the dilute nature of many frozen solutions. Additionally, a peak known as a relaxation endotherm is often reported to appear in the same temperature region as the glass transition itself, but which can precede the glass transition by several degrees and may be a function of sample ageing or a mismatch in cooling and warming rates (16). Where solutions are sufficiently concentrated and where no relaxation

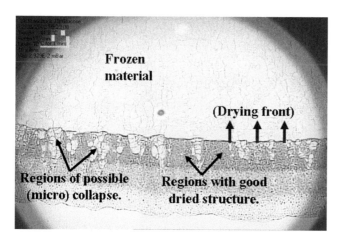

FIGURE 5 Freeze-drying microscopy image of an aqueous solution of 2% (w/v) mannitol + 1% (w/v) glucose at −41.8°C, showing distinct areas of good structure (attributed to crystalline mannitol) and observable voids in the drying structure (attributed to the microcollapse of glucose within the structure).

endotherm appears, data from conventional DSC can give a clear result. Indeed, in this volume, Wang gives a brief overview of DSC and presents some of his own data to show how it can be applied to formulations to be freeze-dried, and also to the freeze-dried product itself. However, where solutions are dilute or where the true glass transition is obscured by the accompanying relaxation endotherm, further analysis may be necessary to ascertain the "true" glass transition. A further development of DSC technology that can provide clarification in such cases is modulated (temperature) DSC (abbreviated to MDSC or MTDSC). This technique allows a modulated heating rate to be superimposed over the linear heating rate that is traditionally applied to a sample, enabling the deconvolution (by software) of the conventional thermogram data to obtain the underlying linear and modulated responses (31). From the deconvoluted data, it may be observed which events—or components of an apparently single event—respond to the modulated (sinusoidal) heating rate and which do not. The result of this calculation allows reversing events, such as melts or glass transitions, to be displayed on one output graph, while nonreversing events, such as crystallizations or endothermic relaxations, appear on a separate output graph. Thus, a glass transition may be observed separately from its accompanying (but often wider in temperature range) relaxation endotherm. To minimise artefacts, it is recommended that at least six modulations take place between the onset and the endpoint of each thermal event; therefore, the frequency of the modulations should be set accordingly. From a practical standpoint, this often means that low heating rates are necessary—since glass transitions are usually relatively narrow in temperature range and six modulations are required during this range—which makes analysis longer than for conventional DSC. An example of the

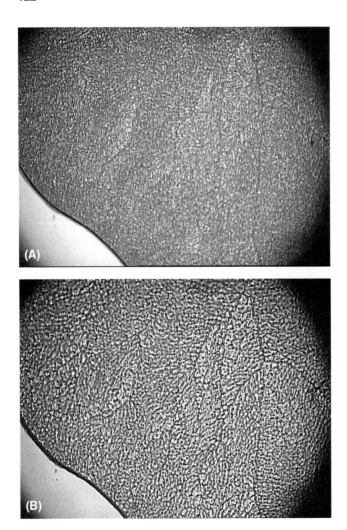

FIGURE 6 Freeze-drying microscopy images showing the effect of annealing on ice crystal changes in an aqueous solution initially cooled to $-40°C$ then warmed to $-10°C$ and held at this temperature: **(A)** 0 minutes and **(B)** 15 minutes.

MDSC warming profile for an aqueous solution of lactose is shown in Figure 9, on which a clear glass transition can be seen in the reversing signal ($-30.3°C$) and weakly in the total heat flow signal ($-32.7°C$).

Thermomechanical Analysis and Dynamic Mechanical Analysis

Thermomechanical analysis (TMA) may be used to examine the effect of temperature on the length or volume of a sample while subjected to a constant

Camera ⟶

Analyzer ⟶

Sample ⟶

Polarizer ⟶

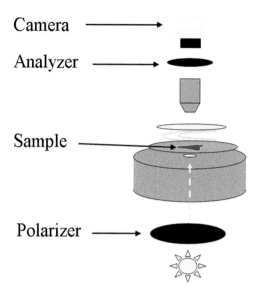

FIGURE 7 Sample format employed for polarized light freeze-drying microscopy analysis using Lyostat2 (Biopharma Technology Limited, Winchester, U.K.).

mechanical stress (32). For frozen solutions, the stress is applied though a flat-ended probe in a temperature-controlled cell and the deformity of the frozen sample is measured as it is heated through the melt or glass transition. This method has been employed to investigate the T_g' values of disaccharide solutions for food application (33–35) and the effect of polymers on the viscoelastic properties of sucrose (36).

Dynamic mechanical analysis (DMA) is a technique that measures the stiffness of a substance in response to an applied force (37). While it is widely used in the study of solids such as polymers, plastics, and metals, it can also be used to measure the glass transition of both the frozen glassy state (38) and the dry state in amorphous powders (39). The technique has been used in the food industry to examine the T_g' of sugars (40). Either the modulus, a measure of how stiff the sample is, or the "tan delta," a measure of the damping factor, can be followed. At a glass transition there is a marked fall in modulus and a maximum in the tan delta. Although some form of sample holder must be used to hold a liquid or powdered solid sample to freeze it, a material can be selected that does not show a change in the stress modulus or tan delta at the temperature range of interest. Although not a commonly used technique in laboratories that employ freeze-drying, the method may have application where other thermal analytical techniques are unresolving. In particular, it is easy to separate the T_g' from the ice melt peak, which may not be the case, for instance, with DSC. The temperature of the tan delta maximum should vary with frequency of the force applied and this can be a useful diagnostic of the type of thermal event that is occurring. An example of T_g' determination by DMA is illustrated in Figure 10 for a sample of porcine heparin, which shows a T_g' at $-18°C$.

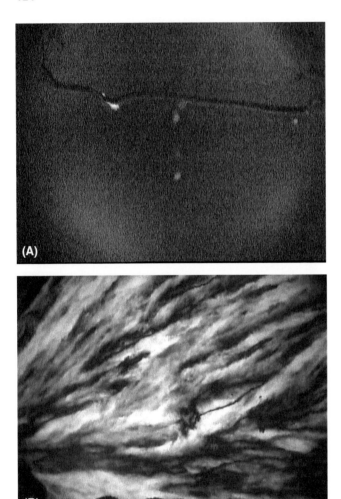

FIGURE 8 Freeze-drying microscopy images of a simple aqueous solution of an API taken under plane polarized light, showing the effect of temperature on solute crystallization: (**A**) sample solution "quench cooled" to −40°C at 50°C/minute, to produce amorphous solute (absence of transmitted light through sample due to lack of source light rotation); (**B**) same sample when temperature raised to −18°C during sublimation—areas of transmitted light signify rotation of source due to angle of rotation in API crystals. *Abbreviation*: API, active pharmaceutical ingredient.

Other Methods

There are a number of other thermal methods that, while not widely employed as yet, have been described that provide additional and complementary information about the behavior of a material prior to, and following, lyophilization. Atomic force microscopy (AFM) is now well established as a form of microthermal

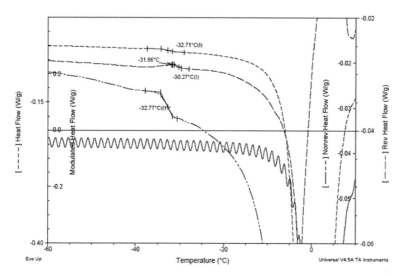

FIGURE 9 Modulated differential scanning calorimetry thermogram, showing warming profile for a frozen aqueous solution of 2% (w/v) lactose. A clear glass transition can be seen as a baseline step in the reversing signal.

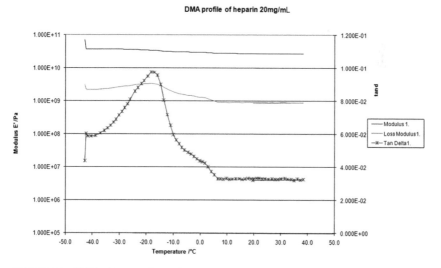

FIGURE 10 DMA of heparin (20mg/mL in water) 20μL loaded in steel pocket with heating rate of 5°C/min and frequency of 1 Hz. Glass transition at −18°C with secondary event around 0°C, probably due to ice melt. Run on a Tritec 2000 DMA (Triton Technology Ltd., Nottinghamshire, U.K.) by John Gearing (Gearing Scientific Ltd, Ashwell, U.K.). *Abbreviation*: DMA, dynamic mechanical analysis.

analysis that can be applied to the study of pharmaceutical materials (41) and may also be adapted for use at controlled temperatures, including subambient temperatures ("Cryo-AFM") for the purposes of biological imaging (42). Wu et al. recently reported a new methodology for Cryo-AFM that did not cause damage to samples of frozen aqueous trehalose solutions (38). Dielectric spectroscopy (DS) can provide an indication of structural and morphological changes in a sample in response to an applied alternating current. Application of this electric field subjects the sample to a harmonically varying stress where the dipolar molecules align with the field and relax when the field is removed. The rate of relaxation in the material is termed the "decay," which provides a measurement of the mobility within the material. A detailed description of the technique, together with its principles and the resulting sample parameters that can be calculated by applying various mathematical equations to the data obtained from DS, is given by Craig (43,44). For the purposes of freeze-drying, it has been shown that the glass transition in the frozen state may be identified by carrying out a series of experiments where either the temperature or the frequency is varied while keeping the other parameter constant, and plotting the resulting sample response against the varied parameter. Dielectric measurements have been used to obtain information about the molecular mobility of amorphous materials above the glass transition in the solid state (45), where the results correlated well with T_g data obtained using DSC analysis. DS equipment with different frequency ranges has been developed to allow for the analysis of liquids, frozen solutions, and solid materials.

Issues with Data Interpretation and Application in the Freeze-Drying Process

Knowledge of the and critical temperature(s) of a formulation is useful in freeze-drying in that they enable the freeze-drying scientist to define the maximum allowable temperature to which that formulation should be subjected during freezing and sublimation, while knowledge of the fundamental behavior can also help define suitable cooling rates and whether annealing would be beneficial in terms of controlling solute behavior. As with all forms of analysis, the data obtained from the techniques described in this chapter require some degree of interpretation, firstly with regard to what events or qualities they represent, and secondly with regard to their relevance and application to the formulation and cycle development process.

With FDM the level of subjectivity is perhaps more apparent than with other methods because of the obvious need for the analyst to interpret visual images. Additionally, it is well recognized that the temperature (and range) of a collapse event does have a dependence on the methodology employed in preparing and analyzing a sample (46). Some workers have provided anecdotal evidence of having dried a product apparently above its T_c and having obtained a product of acceptable appearance. In such cases it may be possible that the analytical methodology and subjectivity of the collapse temperature measurement, when taken together with practical issues such as product temperature probe placement and the significant temperature gradients that can exist between the drying front and the probed portion of the sample during primary drying, could explain the apparent willingness of a sample to dry with good macroscopic structure, even above its collapse temperature.

As mentioned above, glass transitions can be difficult to determine because of the low energy of the heat capacity change associated with glass transitions of frozen solutions, which may often be further obscured by the presence of a relaxation endotherm in the region of interest on the thermogram (16). Additionally, a number of instances have been reported where a frozen solution was observed to exhibit several endothermic discontinuities in the heat capacity signal, which gives the appearance of there being more than one glass transition. This was first reported in the late 1960s (30,47). To a large extent the assignment of each of the transitions remains disputed (38,48). This could be due to several genuine glass transitions arising from a mixture that displays phase separation, as discussed by Padilla and Pikal in chapter 4 of this volume. In a recent study of freeze-dried protein-polysaccharide mixtures by Jara and Pilosof (49), a sample with a degree of phase separation of more than 50% yielded three separate glass transitions, which were attributed to a protein-rich phase, a polysaccharide-rich phase, and a mixed interface. However, two apparent glass transition events have also been observed for frozen aqueous solutions containing only one solute. Shalaev and Franks (50) and Ablett et al. (51) hypothesized that the lower transition (Tr_1) reflected the true glass transition, while the higher transition (Tr_2) was attributed to the onset of dissolution of ice crystals. A recent study by Wu et al. (38) where MTDSC was employed in combination with cryoatomic force microscopy (cryo-AFM) and a novel DMA technique to analyze the behavior of a frozen trehalose solution supported this hypothesis, with the two transitions observed in the reversing heat flow signal at $-35.4°C$ (denoted "Tr_1") and $-27.9°C$ (denoted "Tr_2") being attributed to softening and the loss of ice crystals, respectively. Contrastingly, data from a study reported by Sacha and Nail (48), where a range of saccharide solutions were studied using MTDSC and FDM, supported the hypothesis that both events were glass transitions, which was confirmed by the construction of Lissajous plots for samples at different temperatures. Interestingly, from a freeze-drying standpoint, it was observed in this case that the temperatures obtained for Tr_2 in each case corresponded better with T_c than did Tr_1.

Almost all but the least sophisticated freeze-dryer will be designed with product temperature probes—usually type T thermocouples or Pt100 resistance thermometers—so that the freeze-drying scientist can gain some understanding of what temperatures a product experiences in the freeze-dryer. Product temperature is customarily measured at the base of a product, for three reasons: firstly, it should be the warmest part of the frozen structure during sublimation; secondly, it is the last part of the product to dry, thus providing the possibility of sublimation endpoint prediction; and thirdly, it reduces probe placement variability from container to container and batch to batch, and allows somewhat for different operator judgements. Some freeze-dryers also offer an indirect way of measuring the temperature of the drying front, for example, using manometric temperature measurement (52), which can provide information as to the "mean batch behavior" during most of the sublimation stage of the process.

To exercise prudence in freeze-drying cycle development, it is standard practice to allow a "safety margin" of 2°C to 5°C during cycle development (53); in other words, the probed product temperature is maintained 2°C to 5°C lower than the critical temperature following the initial freezing step (apart from during annealing, if employed) and throughout sublimation. This safety margin is designed to allow for a number of variables, some of which may be difficult to measure, predict, or model mathematically, including variations in temperature, pressure, and radiative heating effects within the freeze-dryer, the effect of the temperature probe on the sample with regard to the nonprobed samples within a batch, and a whole plethora of scale-up issues. These issues are discussed in later chapters in this volume.

Critical temperatures can still be employed during freeze-drying, even if no possibility exists to probe product temperature directly. This is the case for products that are frozen in pellets prior to sublimation or where a simplistic freeze-dryer is used in the early stages of product development. In such instances a "safe"—yet perhaps somewhat inefficient—cycle may be devised by employing shelf (or chamber) temperatures that are several degrees below the critical temperature of the frozen material under sublimation and relying on the sublimation cooling effect to maintain the product temperature below its critical temperature; this approach should work, provided the source of heat into the material does not exceed the cooling effect of the sublimation process on the product.

For an amorphous material, collapse occurs at the drying front while the remainder of the frozen material remains unaffected (unless a different transition occurs within the frozen structure in the same temperature range). Therefore, if the base of a product is probed, the drying front will, by definition, be colder because of the sublimation cooling effect at the front itself. Therefore, during the sublimation process, if the base of the material is kept below the collapse temperature of the interface a safety margin already exists, but which will reduce in magnitude as the rate of drying decreases over time (because of the increased resistance to vapor flow from the dried layer). Contrastingly, for crystalline materials, if the temperature anywhere in the frozen material exceeds the eutectic temperature part of the eutectic solid will melt and may evaporate under vacuum, leading to eruption of the material and resulting in a product with poor aesthetic appearance. Therefore, although there will be a temperature gradient between the base of the material and the drying front, this does not constitute a safety margin as it does with amorphous materials, since the base of the material is at risk of melting. Thus the use of a greater safety margin may be advisable when freeze-drying crystalline materials.

ELECTRICAL IMPEDANCE ANALYSIS: A NOVEL METHOD THAT MAY ALLOW GREATER INSIGHT INTO MICROCOLLAPSE/ MICROMELTING

The use of electrical resistance (ER) analysis or freezing resistance analysis (FRA) has been widely reported for the study of frozen materials that are subsequently to be freeze-dried (21,54–58). Here, the ER of a material to a low-voltage alternating current is measured, as it is cooled and rewarmed. Traditional devices that measure ER (often termed "eutectic analyzers") have often been observed to miss subtle changes that may turn out to be significant and do not tend to give clear indications of changes in nonionic materials such as proteins, polymers, and saccharides. Furthermore, the term eutectic analyzer may often be considered a misnomer, since a high proportion of materials that are lyophilized do not give a true eutectic but instead persist in a glassy state. To this end, on the basis of observations made by Rey (59), who established that analysis using a function of impedance ($Z\sin\varphi$) can provide more detailed information about the behavior of the frozen solute, a novel instrument has been

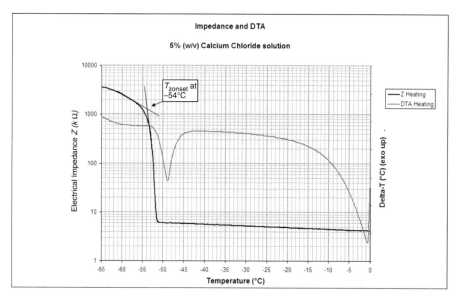

FIGURE 11 Example of a $Z\sin\varphi$ and differential thermal analysis warming profile for an aqueous solution of 5% (w/v) calcium chloride.

developed to measure a function of electrical impedance ($Z\sin\varphi$) of frozen solutions to indicate changes in ionic/molecular mobility that may be relevant to freeze-drying. Using this instrument (Lyotherm2, Biopharma Technology Limited, Winchester, U.K.), $Z\sin\varphi$ was measured from below $-100°C$ to $>0°C$ for a wide range of simple solutions and more complex formulations at a fixed frequency of 1000 Hz and plotted as a function of temperature. DTA was carried out simultaneously using the same instrument to indicate thermal events.

In many cases, for the wide range of solutions tested, $Z\sin\varphi$ analysis provided data on changes in mobility within the frozen solute phase upon warming that could be related to freeze-drying behavior. Specifically, the temperature of the event denoted as T_{Zonset} (the onset of a sharp decrease in $Z\sin\varphi$) was observed to correlate well with known individual softening or melting events.

Firstly, for simple crystallizing solutions, T_{Zonset} and an endotherm in the DTA curve were seen to correlate well with the eutectic temperatures. Figure 11 shows a thermogram for an aqueous solution of 5% (w/v) calcium chloride, where T_{Zonset} was calculated to occur at $-54°C$, just prior to the eutectic melt observed in our laboratories to occur at $-53°C$ (using FDM).

Secondly, for simple amorphous solutions the temperature of T_{Zonset} observed with the Lyotherm2 instrument was often observed to correlate with the expected T_g'. For an aqueous solution of 10% (w/v) sucrose, the observed T_{Zonset} closely matched the published T_g' of $-32°C$ and T_c of $-31°C$ (60).

Thirdly, in aqueous solutions containing amorphous components together with solutes that would normally tend toward crystallization in the absence of other solutes, a clear T_{Zonset} was often obtained. For a solution containing mannitol (2% w/v) and glucose (1% w/v), T_{Zonset} was observed to be $-41°C$. Interestingly, freeze-drying this solution either below T_{Zonset} or just below the

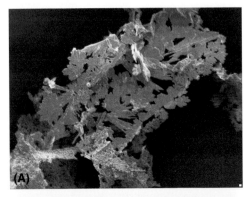

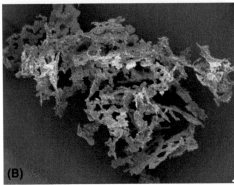

FIGURE 12 Scanning electron microscopy images showing different lyophile
structures obtained for an aqueous solution of mannitol (2% w/v) and glucose (1%
w/v) freeze-dried (**A**) at −50°C and (**B**) at
−18°C, following an initial annealing step
to encourage mannitol crystallization.

observed T_c (−16°C) yielded no macroscopic collapse, but further analysis of the
resulting lyophiles by SEM revealed that the samples freeze-dried below T_{Zonset}
appeared to exhibit different microscopic surface structure compared with those
dried between T_{Zonset} and T_c, as shown in Figure 12.

Fourthly, for more complex solutions, T_{Zonset} often correlated well with T_c,
even when no thermal event was observed in the temperature range of interest
using DTA or MDSC. A multicomponent cell culture medium was seen to
exhibit mobility changes as low as −71°C during warming, with several events
occurring prior to the T_{Zonset} of −45°C, as shown in Figure 13. It can be seen
from this thermogram that Zsinφ shows a decrease around −71°C (upper left of
the graph) before almost immediately increasing at −67°C; this correlates well
with the crystallization exotherm observed in the same temperature range on the
DTA curve, suggesting that Zsinφ, when measured at 1000 Hz, is sensitive
enough to detect softening in the structure of the frozen material immediately
prior to crystallization (decrease in Zsinφ), as well as the crystallization event
itself (leveling or increase in Zsinφ). Additionally, it is interesting to note that the
T_{Zonset} value of −45°C correlated well with the T_c of −43°C observed for the
same solution using FDM.

Finally, for a series of real biological formulations, it was noted that Zsinφ
analysis often yielded a clear indication of changes in frozen state mobility even
when MDSC analysis did not yield a clear glass transition, and that in many
cases, T_{Zonset} was significantly lower than the T_c observed using FDM; some

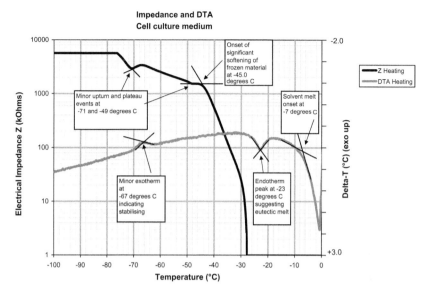

FIGURE 13 $Z\sin\varphi$ and differential thermal analysis warming profile for a multicomponent solution, showing a range of transitions observed in the frozen state.

TABLE 1 T_{Zonset} Values Observed for Simple and Multicomponent Frozen Solutions and Comparison with Corresponding T_c Onset Temperature (by Freeze-Drying Microscopy; No Annealing)

Formulation (% values in w/v)	T_{Zonset} (°C), Lyotherm2	Onset T_c (°C), Lyostat2
NaCl (0.9) + HSA (0.5)	−46	−18.4
NaCl (0.9) + HSA (1.0)	−64	−20.4
NaCl (0.9) + HSA (5.0)	−60	−23.0
Egg allantoic fluid (undiluted)	−60	−50
Potassium phosphate buffered HSA (0.2) + trehalose (0.1)	−58	−50
HSA (0.1) + casein (0.3) in PBS	−59	−50

Abbreviation: HSA, human serum albumin.

examples of such cases are shown in Table 1. While some of the discrepancies may arise as a consequence of the different methodologies employed in carrying out analyses using these two very different techniques, the fact that low temperature mobility changes were observed at all is one that arguably warrants further investigation into the freezing behavior of these formulations.

In the mixtures of sodium chloride and human serum albumin (HSA), the low T_{Zonset} values may be attributable to the inhibition of salt crystallization by the presence of HSA, which may well have implications on product behavior in a real freeze-drying situation. In these cases, the situation is likely to be more complex than simply the micromelting or microcollapsing of one component onto the other; rather, it might well involve the microcollapsing of a "mixture" of amorphous salt and HSA onto the backbone of either amorphous HSA alone or amorphous HSA and crystalline NaCl. In more complex formulations it is difficult to assign the difference between T_{Zonset} and T_c to the behavior of a particular solute or group of solutes without further analysis; however, in such cases, and as illustrated by some of the above examples, $Zsin\varphi$ analysis could prove useful in simply alerting the freeze-drying scientist to potential issues that may warrant further investigation as part of the formulation and cycle development process.

Therefore, as demonstrated in the five examples described above, impedance ($Zsin\varphi$) analysis may not only be able to provide data to corroborate T_{eu}, T_g', or T_c, as measured using conventional methods, but may also offer additional evidence of changes within the structure of a range of frozen solutions at temperatures below that of gross changes observed using FDM and/or where no significant event such as T_g' may be detected using MDSC. Such changes may be important in understanding the potential for microscopic product defects occurring during the lyophilization process and the stability of the final lyophilized product itself, especially when viewed in light of the fact that in amorphous materials some degree of mobility still exists below the glass transition temperature (61). It is well established that the efficiency of the sublimation process can be markedly increased by employing higher temperatures (62), yet the need exists to balance processing efficiency against the requirement to avoid, or at least minimize, product defects. Traditionally, gaining an understanding of the macroscopic behavior of a formulation has often been considered sufficient to allow minimization of such defects; however, $Zsin\varphi$ analysis may also provide an additional level of detail that could inform the freeze-drying scientist of possible microscopic defects that may result from changes in frozen state mobility.

CONCLUSIONS

It is becoming essential to demonstrate increasing levels of knowledge of the freezing and drying properties of a pharmaceutical formulation when contemplating its freeze-drying as part of the production process. There is no single analytical method that will provide a complete picture of how a formulation will respond to the freeze-drying process, but a number of methods are available that can provide complementary information that allow a formulation to be better understood, and it is recommended that more than one method is employed so that aspects of formulation behavior that are relevant to the freeze-drying process, are not overlooked. From a practical standpoint, FDM can often provide most of the necessary information to establish how a formulation can be freeze-dried to yield a product with acceptable macroscopic appearance. However, certainly for more complex formulations, other techniques can add to the information gained from FDM and provide a deeper level of understanding of any underlying changes that may occur in the formulation in the frozen state or during the sublimation process, which may either help explain the events observed using FDM or alert the freeze-drying scientist to other changes that may not be observed using FDM alone. Traditional thermal methods may enable mechanical or energetic changes to be evaluated, with some of these methods—even though they may be costly—also allowing analysis of the lyophilized product itself. However, even the more sensitive methods such as MDSC can struggle to detect subtle events such as glass transitions in dilute frozen solutions. DMA may provide an alternative to DSC for analysis of frozen solutions in that it can enable T_g' (and also T_g for the lyophilized material) to be established, and while the glass transition tends to be broader, it may be less affected by the ice melting endotherm. Electrical impedance analysis is a relatively new method and has been demonstrated to detect events that other methods appear to miss, some of which may be relevant to freeze-drying, including mobility changes in regions of the frozen matrix that could alert the freeze-drying scientist to the possibility of microcollapse or micromelting occurring in a formulation. Such information is useful not only in formulation development but also in enabling the development of a lyophilisation cycle that is tailored to the requirements of that product.

ACKNOWLEDGMENTS

The authors would like to thank Professor Louis Rey for his invaluable input to the electrical impedance technology, Dr Vicky Kett for useful discussions on thermal analysis, and Kiran Malik and Thomas Peacock for their assistance with the practical laboratory analyses.

REFERENCES

1. Franks F. Freeze-drying: from empiricism to predictability. Cryo Letters 1990; 11:93–110.
2. Franks F. Freeze Drying of Pharmaceuticals & Biopharmaceuticals, Principles & Practice. Cambridge: RSC Publishing, 2007.
3. Brunauer S, Emmett PH, Teller E. Adsorption of gases in multi-molecular layers. J Am Chem Soc 1938; 60(2):309–319.
4. Hibbert BD, Lovegrove J, Tseung ACC. A critical examination of a cryochemical method for the preparation of high surface area semiconducting powders. Part 3: factors which determine surface area. J Mater Sci 1987; 22:3755–3761.
5. Rey L. Glimpses into the realm of freeze-drying: fundamental issues. In: Rey L, May J, eds. Freeze-Drying/Lyophilization of Pharmaceutical and Biological Products. 2nd ed. New York: Marcel Dekker, 2004:1–32.
6. Schiffter H, Vonhoff S. The determination of structural changes of biopharmaceuticals in freeze-drying using Fourier transform infrared spectroscopy. Eur Pharm Rev 2009; 14(2):57–64.
7. Imamura K, Ohyama K, Yokoyama T, et al. Temperature scanning FTIR analysis of interactions between sugar and polymer additive in amorphous sugar-polymer mixtures. J Pharm Sci 2008; 97(1):519–528.
8. Brülls M, Folestad S, Sparén A, et al. Applying spectral peak area analysis in near-infrared spectroscopy moisture assays. J Pharm Biomed Anal 2007; 44:127–136.
9. Derbyshire HM, Feldman Y, Bland CR, et al. A study of the molecular properties of water in hydrated mannitol. J Pharm Sci 2002; 91(4):1080–1088.
10. De Beer TRM, Alleso M, Goethals F, et al. Implementation of a process analytical technology system in a freeze-drying process using Raman Spectroscopy for in-line process monitoring. Anal Chem 2007; 79(21):7992–8003.
11. Xie Y, Cao WJ, Krishnan S, et al. Characterization of mannitol polymorphic forms in lyophilized protein formulations using a multivariate curve resolution (MCR)-based Raman spectroscopic method. Pharm Res 2008; 25(10):2292–2301.
12. Beattie JR, Barrett LJ, Malone JF, et al. Investigation into the subambient behaviour of aqueous mannitol solutions using temperature-controlled Raman microscopy. Eur J Pharm Biopharm 2007; 67(2):569–578.
13. Wang DQ. Formulation characterization. In: Rey L, May J, eds. Freeze-Drying/Lyophilization of Pharmaceutical and Biological Products. 2nd ed. New York: Marcel Dekker, 2004:213–238.
14. Liu W, Wang DQ, Nail SL. Freeze-drying of proteins from a sucrose-glycine excipient system: effect of formulation composition on the initial recovery of protein activity. AAPS PharmSciTech 2005; 6(2):E150–E157.

15. Adams GDJ, Irons LI. Some implications of structural collapse during freeze-drying using Erwinia caratovora L-asparaginase as a model. J Chem Technol Biotechnol 1993; 58:71–76.
16. Kett V, McMahon D, Ward K. Freeze-drying of protein pharmaceuticals—the application of thermal analysis. Cryo Letters 2004; 25(6):389–404.
17. Izutsu K, Kadoya S, Yomota C, et al. Freeze-drying of proteins in glass solids formed by basic amino Acids and dicarboxylic acids. Chem Pharm Bull (Tokyo) 2009; 57(1):43–48.
18. Gabbott P, ed. Principles and Applications of Thermal Analysis. Oxford: Blackwell, 2008.
19. Ford JL, Timmins P. Pharmaceutical Thermal Analysis: Techniques and Applications. Chichester: Ellis Horwood, 1989.
20. Haines PJ, ed. Principles of Thermal Analysis and Calorimetry. Cambridge: RSC Publishing, 2002.
21. Rey LR. Thermal analysis of eutectics in freezing solutions. Ann N Y Acad Sci 1960; 85:513–534.
22. Mackenzie AP. Apparatus for microscopic observations during freeze-drying. Biodynamica 1964; 9(186):213–222.
23. Meister E, Gieseler H. Freeze-dry microscopy of protein/sugar mixtures: drying behavior, interpretation of collapse temperature and a comparison to corresponding glass transition data. J Pharm Sci 2009; 98(9):3072–3087.
24. Zhai S, Taylor R, Sanches R, et al. Measurement of lyophilisation primary drying rates by freeze-drying microscopy. Chem Eng Sci 2003; 58:2313–2323.
25. MacInnes WM. Dynamic mechanical thermal analysis of sucrose solutions. In: Blanshard JMV, Lillford PJ, eds. The Glassy State in Foods. Loughborough: Nottingham University Press, 1993:223–248.
26. Ozawa T. Thermal analysis—review and prospect. Thermochim Acta 2000; 355 (1):35–42.
27. Shalaev EY, Valaksin NA, Rjabicheva TG, et al. Solid-liquid state diagrams of the pseudo-ternary systems hydrolysed protein-sucrose-water: implication for freeze-drying. Cryo Lett 1996; 17:183–194.
28. Wesolowski M. Thermoanalytical methods in pharmaceutical technology. J Therm Anal 1992; 38:2239–2245.
29. Ozawa T. Retrospection of DTA automation, J Jap Soc Therm Anal 2000; 27:3–16.
30. Luyet B, Rasmussen D. Study by differential thermal analysis of the temperatures of instability of rapidly cooled solutions of glycerol, ethylene glycol, sucrose and glucose. Biodynamica 1968; 10:167–191.
31. Reading M. Modulated differential scanning calorimetry—A new way forward in materials characterization. Trends Polym Sci 1993; 1(8):248–253.
32. Price DM. Thermomechanical and thermoelectrical methods. In: Haines PJ, ed. Principles of Thermal Analysis and Calorimetry. Cambridge: Royal Society of Chemistry Press, 2002:94–128.
33. Le Meste M, Huang V. Thermomechanical properties of frozen sucrose solutions J Food Sci 1992; 57:1230–33.
34. Levine H, Slade L. Thermomechanical properties of small carbohydrate-water glasses and 'rubbers.' Kinetically metastable systems at sub-zero temperatures. J Chem Soc Faraday Trans 1 1988; 84:2619–2633.
35. Sahagian ME, Goff HD. Thermomechanical effect of freezing rate on the thermal, mechanical and physical aging properties of the glassy state in frozen sucrose solutions, Thermochimica Acta 1994; 246:271–283.
36. Blond G. Mechanical properties of frozen model solutions, J. Food Eng. 1994; 22:253–269.
37. Duncan J. Principles and applications of mechanical thermal analysis In: Gabbott P, ed. Principles and Applications of Thermal Analysis. Oxford: Blackwell, 2008:119–163.
38. Wu J, Reading M, Craig DQM. Application of calorimetry, sub-ambient atomic force microscopy and dynamic mechanical analysis to the study of frozen aqueous trehalose solutions. Pharm Res 2008; 25(6):1396–1404.
39. Royall PG, Huang CY, Tang SWJ, et al. The development of DMA for the detection of amorphous content in pharmaceutical powdered materials. Int J Pharm 2005; 301:181–191.

40. Cruz IB, Oliveira JC, MacInnes WM. Dynamic mechanical thermal analysis of aqueous sugar solutions containing fructose, glucose, sucrose, maltose and lactose. Int J Food Sci Technol 2001; 36:539–550.
41. Pollock HM, Hammiche A. Micro-thermal analysis: techniques and applications. J Phys D Appl Phys 2001; 34:R23–R53.
42. Han W, Mou J, Sheng J, et al. Cryo atomic force microscopy: a new approach for biological imaging at high resolution. Biochemistry 1995; 34:8215–8220.
43. Craig DQM, ed. Dielectric Analysis of Pharmaceutical Systems. London: Taylor & Francis, 1995.
44. Craig DQM. Dielectric spectroscopy as a novel analytical technique within the pharmaceutical sciences. STP Pharma Sciences 1995; 5:421–428.
45. Andronis V, Zografi G. The molecular mobility of supercooled amorphous Indomethacin as a function of temperature and relative humidity. Pharm Res 1998; 15:835–842.
46. Pikal MJ, Shah S. The collapse temperature in freeze-drying: dependence on measurement methodology and rate of water removal from the glassy phase. Int J Pharm 1990; 62:165–186.
47. Rasmussen D, Luyet B. Complementary study of some non-equilibrium phase transitions in frozen solutions of glycerol, ethylene glycol, glucose and sucrose. Biodynamica 1969; 10:319–331.
48. Sacha G, Nail S. Thermal analysis of frozen solutions: multiple glass transitions in amorphous systems. J Pharm Sci 2009; 98(9):3397–3405.
49. Jara F, Pilosof A. Glass transition temperature of protein/polysaccharide co-dried mixtures as affected by the extent and morphology of phase separation. Thermochimica Acta 2009; 487(1–2):65–73.
50. Shalaev EY, Franks F. Structural glass transitions and thermophysical processes in amorphous carbohydrates and their supersaturated solutions. J Chem Soc Faraday Trans 1995; 91:1511–1517.
51. Ablett S, Clark AH, Izzard M, et al. Modelling of heat capacity-temperature data for sucrose-water. J Chem Soc Faraday Trans 1992; 88:795–802.
52. Milton N, Pikal MJ, Roy ML, et al. Evaluation of manometric temperature measurement as a method of monitoring product temperature during lyophilization. PDA J Pharm Sci Technol 1997; 51:7–16.
53. Trappler E. Validation of lyophilization: equipment and process. In: Costantino H, Pikal M, eds. Biotechnology: Pharmaceutical Aspects. Lyophilization of Biopharmaceuticals. Arlington: AAPS Press, 2004:43–74.
54. De Luca P, Lachman L. Lyophilization of pharmaceuticals. I. effect of certain physical-chemical properties. J Pharm Sci 1965; 54(4):617–624.
55. Ma X, Wang DQ, Bouffard R, et al. Characterization of Murine monoclonal antibody to tumor necrosis factor (TNF-MAb) formulation for freeze-drying cycle development Pharm Res 2001; 18(2):196–202.
56. Jennings T. Thermal-analysis instrumentation for lyophilisation. Medical Device and Diagnostic Industry. 1980; 11:49–57.
57. Chen T, Oakley DM. Thermal analysis of proteins of pharmaceutical interest. Thermochimica Acta 1995:229–244.
58. Mackenzie AP. Changes in electrical resistance during freezing and their application to the control of the freeze-drying process. Refrigeration Science and Technology, IIF-IIR Commission, Tokyo, Japan, 1985:155–163.
59. Rey L. Glimpses into the realm of freeze-drying: classical issues and new ventures. In: Rey L, May J, eds. Freeze-Drying/Lyophilization of Pharmaceutical and Biological Products. 1st ed. New York: Marcel Dekker, 1999:1–30.
60. Adams GDJ. Freeze-drying of biological materials, drying technology. 1991; 9(4):891–923.
61. Yoshioka S, Aso Y. Glass transition-related changes in molecular mobility below glass transition temperature of freeze-dried formulations, as measured by dielectric spectroscopy and solid state nuclear magnetic resonance. J Pharm Sci 2005; 94(2):1445–1455.
62. Pikal MJ. Freeze-drying of proteins. Part I: process design. BioPharm 1990; 3(8):18–27.

6 The Relevance of Thermal Properties for Improving Formulation and Cycle Development: Application to Freeze-Drying of Proteins

Stéphanie Passot, Ioan Cristian Tréléa, Michèle Marin, and Fernanda Fonseca
UMR 782 Génie et Microbiologie des Procédés Alimentaires, AgroParisTech, INRA, Thiverval-Grignon, France

INTRODUCTION

Although freeze-drying is the process of choice for improving the shelf life of biological products such as proteins, the process itself and the subsequent long-term storage generate several stresses responsible for protein denaturation (1–4). An appropriate choice of protective excipients is thus needed to inhibit freezing- and drying-induced damage and to preserve biological activity during processing, storage, and rehydration (1,5,6). Moreover, the active ingredient is usually present in very low concentrations (μg/mL or ng/mL) in the formulation, and excipients also act as bulking agents to increase the dry matter of the formulation and to obtain an elegant freeze-dried cake. Different molecules alone or in combination and at various concentrations can be used, making the formulation step complex, hazardous, and time-consuming and not always resulting in the highest quality product attainable. To obtain successful results when developing freeze-dried protein, the formulation has to fulfill the following five criteria (5):

- Inhibition of protein unfolding during freezing and drying
- A glass transition temperature of the freeze-dried product that is higher than the storage temperature
- Low residual water content (<3%)
- A strong and elegant cake structure
- The minimization of protein chemical degradation during storage (addition of antioxidant, conditioning in nitrogen atmosphere, etc.)

Obtaining an elegant freeze-dried cake structure requires performing the drying stages at a product temperature that does not exceed a critical value—the maximum allowable product temperature (T_{max})—usually defined as being a few degrees below the collapse temperature (T_{coll}). The (T_{coll}) is defined as the temperature above which the freeze-dried product loses macroscopic structure and collapses during freeze-drying (7–9). The thermal properties of the formulation including eutectic crystallization temperature (T_{cr}), eutectic melting temperature (T_{eut}), glass transition temperature of the maximally freeze-concentrated phase (T_g'), and ice melting temperature are used to determine the value of T_{max} and to thus design and optimize the process. Many approaches have been explored to determine the thermal properties of formulations for freeze-drying, including differential thermal analysis (DTA), differential scanning calorimetry (DSC) (6,10,11), electrical resistance analysis (ERA) (12), and freeze-drying microscopy

(FDM) (9,13,14). The measurement of the thermal properties of a formulation and their utilization for freeze-drying cycle design have been reviewed and discussed in the previous chapter. In this chapter, we will investigate deeper into the use of thermal properties for improving the freeze-drying process. For example, we will examine how knowledge about these can help in solving the following two process-related problems:

1. Intra- and interbatch heterogeneity: By applying the same process conditions (shelf temperature and chamber pressure) to the same product configuration (formulation, vial) in the same freeze-dryer, diverse product qualities can be obtained between different freeze-drying cycles and within the same batch as well.
2. Process scale-up: By changing from the laboratory to the manufacturing scale, collapsed freeze-dried products can be observed.

Solving these problems and obtaining freeze-dried products of high and reproducible quality require identifying the critical formulation attributes and the critical process variables. The composition of the formulation is essential for limiting these problems and for designing robust and short cycles. The thermal properties obtained by DSC and FDM can be used for (*i*) predicting the formulation's physical behavior under processing conditions of freezing and drying and (*ii*) understanding the impact of excipients on this behavior. Excipients that exhibit a negative effect on the thermal properties by lowering the T_{max} value, for example, can thus be replaced by excipients that increase the value of T_{max} or at least lower their concentration. Furthermore, by relating thermal properties to protein stability, critical parameters can be identified and easily integrated into a quality management system and a quality-by-design approach.

PREDICTION OF THE FORMULATION'S PHYSICAL BEHAVIOR DURING FREEZING AND DRYING

The freeze-drying process involves freezing of the sample followed by ice removal by sublimation (primary drying) and, finally, removal of unfrozen or sorbed water by desorption (secondary drying). The freezing step is the first and the shortest step of the process but determines the ice crystal morphology and the physical state of the formulation's components that, in turn, influences both the sublimation and desorption rates and the quality of the freeze-dried product (elegance of the cake and protein stability) (15–18). Moreover, the physical state of the excipients in the frozen temperature range determines the value of the maximal allowable product temperature (T_{max}) not to be exceeded during the drying stages and especially during the sublimation step. Solutions to be freeze-dried are generally dilute and multicomponent systems, with water contents exceeding 80%. Ice crystallizes first on cooling. Then, depending on the properties of the residual cryoconcentrated solution, the system can display one of the following three behaviors:

1. The residual cryoconcentrated solution forms a kinetically stable amorphous phase with a solute composition identical to that of the initial solution. For these systems, the value of T_{max} is associated with the glass transition temperature of the maximally freeze-concentrated phase (T_g') (14,19,20).

2. One of the solute crystallizes during cooling; the remaining solution forms a kinetically stable glass with a solute composition that is completely different from that of the initial solution. The value of T_{max} is associated with the eutectic temperature of the crystalline solute.
3. The residual solution forms a "doubly unstable" glass, and one of the solute can crystallize during heating. The majority of the formulation involving a crystalline bulking agent such as mannitol or glycine belongs to this category. The value of T_{max} depends on the extent of the bulking agent crystallisation allowed by the thermal history of the sample. The collapse of the product will thus occur when the product temperature during the sublimation step will exceed either the glass transition temperature of the maximally freeze-concentrated phase (T_g') when no excipient crystallization is observed, or the eutectic melting temperature (T_{eut}) when complete solute crystallization occurs, or an intermediate temperature comprised between T_g' and T_{eut} when incomplete crystallization occurs.

Moreover, the physical state of the formulation's excipients also determines protein activity preservation. Protein stability depends on interactions between excipients, especially on the interaction between proteins and stabilizers such as sugars or polyols by hydrogen bonding (21,22). Crystallization of the stabilizer components or the crystalline bulking agent (mannitol) during drying or storage can compromise the stability of the active ingredient (23–25).

Systems involving a crystalline bulking agent display considerable complexities in their phase and state behavior during freezing and drying and require a deep and relevant characterization of their physical behavior. For example, when developing a new formulation, it is essential to know the extent of the bulking agent crystallization and the minimal weight fraction of crystalline bulking agent required to reach extensive crystallization during freezing. DSC and FDM are the analytical tools of choice to obtain this information. By performing cooling and heating runs at different scanning rates, annealing samples at different temperatures, and subjecting samples to a variety of heat/cool/heat cycles, the physical behavior of a formulation can be predicted during freezing and drying. The effect of formulation composition and thermal treatment (cooling rate, annealing treatment) on the thermal properties (T_g', T_{cr}, T_{eut}, T_{coll}) of formulations that include mannitol or glycine as crystalline bulking agents is reported in Tables 1 and 2. These tables are not exhaustive but aim at illustrating the complex physical behavior of those excipients.

Ma et al. (12) (Table 1) have investigated the effect of cooling rate and annealing treatment (during or after the freezing step) on the physical behavior of a formulation involving 2% glycine, 1% maltose, and 2% of a protein active ingredient. Applying a cooling rate of 3°C/min or 1°C/min did not permit the complete crystallization of glycine during the freezing step, and a moderate exothermic event was observed during warming, suggesting the completion of glycine crystallization. This incomplete crystallization of glycine during the freezing step resulted in a low value of T_g' ($T_g' = -43$°C for a cooling rate of 1°C/min) because of the presence of amorphous glycine, a glycine solution with a very low T_g' value of -70°C. From a practical point of view, this value of T_g' obtained when applying a regular freezing procedure is too low for the formulation to be freeze-dried. The addition of an annealing treatment (during or after the freezing step) made it possible to reach the maximum degree of glycine

(text continues on page 147)

TABLE 1 Summary of Thermal Properties of Formulations Including Glycine as the Crystalline Bulking Agent Determined by DSC and FDM: Impact of Solute and Freezing Conditions on the Physical State of Glycine

Composition of the formulation				Differential scanning calorimetry				Freeze-drying microscopy			References
Other solutes	Total solid content (%)	Weight fraction of glycine(%)	Cycle	T_g' (°C)	T_{cr} (°C)	T_{eut} (°C)	State of glycine after freezing	T_{coll} (°C)	State of glycine after freezing	Freeze-dried cake	
Sucrose	7	0	CR = 1.5°C/min HR = 3.5°C/min	−32	—		A	−33	A	Collapsed	Kasraian, 1998 (26)
		14		−36	—		A	−33	A	Collapsed	
		29		−40	—		A	−33	A	Collapsed	
		43		−44	—		A	−33	PC after annealing (30 min at −17°C)	Collapsed	
		57		−48	−19	−5	PC	−20/−15	C after annealing (15 min at −17°C)	Elegant	
		71		−52	−31	−5	PC	−10	C after annealing (5 min at −26°C)	Elegant	
		86		—	—	−5	C	−10	C	Elegant	
		100		—	—	−5	C	−7	C	Elegant	
Sucrose	5.1	33	CR = 10°C/min HR = 10°C/min	−40							Lueckel, 1998 (27)
	4	50		−44							
	4.5	67		−49							
	4.5	48		−44							
	4.5	45		−42.5							
	4.5	42		−41.5							

(continued)

TABLE 1 Summary of Thermal Properties of Formulations Including Glycine as the Crystalline Bulking Agent Determined by DSC and FDM: Impact of Solute and Freezing Conditions on the Physical State of Glycine (*Continued*)

Composition of the formulation			Differential scanning calorimetry					Freeze-drying microscopy			References
Other solutes	Total solid content (%)	Weight fraction of glycine(%)	Cycle	T_g' (°C)	T_{cr} (°C)	T_{eut} (°C)	State of glycine after freezing	T_{coll} (°C)	State of glycine after freezing	Freeze-dried cake	
Sucrose	5.1	33	CR = 10°C/min	−40							Lueckel, 1998 (27)
	4	50	HR = 10°C/min	−33.5							
	4.5	67	Annealing at −30°C for 2 hr	−32							
	4.5	48		−43.1							
	4.5	45		−41.9							
	4.5	42		−40.7							
Sucrose	5.1	67	CR = 10°C/min	−39							
Lysine-HCl	4	50	HR = 10°C/min	−44							
Maltose (1%) API (2%)	5	40	CR = 1°C/min HR = 1°C/min	−43	−31		PC	−20/−10 Complete at −7°C	C after annealing at −20°C		Ma, 2001 (12)
			CR = 1.5°C/min Sub-annealing treatment at −4°C for 60 min HR = 1.5°C/min	−26			C				
			CR = 3°C/min HR at 3°C/min	−71; −48	−28		A or PC				
			CR = 3°C/min HR = 3°C/min Annealing at −15°C	−61 ; −26			C				
PEG	5	80	CR = 10°C/min HR = 10°C/min	–	–	−25; −5	C				Passot, 2005 (6)
Sucrose	5	80		−69	−38; −21		A or PC				
Sucrose (1%) + PEG (1%)	6	67		−69	−36; −21		A or PC				

Formulation			Conditions								Reference
Sucrose (1%) + Tween 80 (0.02%)	5	80		−69	−57; −36		A or PC				
Maltose	5	80		−69	−37; −20		A or PC				
PVP (25)	5	80	CR = 10°C/min HR = 10°C/min	−68; −34			A or PC				
PVP (25) (1%) + PEG (1%)	6	67		−72; −24		−16	A or PC				
PVP (25) (1%) + Tween 80 (0.02%)	5	80		−67; −32			A or PC				
Maltodextrin (DE 6)	5	80		−72	−45; −29		A or PC				
Sucrose	5	80	CR = 1.5°C/min Sub-annealing treatment at −4°C for 60 min HR = 20°C/min	−38		−3	C	−16	C	Elegant	Passot, 2005 (6)
Sucrose (1%) + PEG (1%)	6	67		−38		−19; −4	C	−15	C	Elegant	
Sucrose (1%) + Tween 80 (0.02%)	5	80		−39		−4	C	−18	C	Elegant	
Maltose	5	80		−37		−4	C	−15	C	Elegant	
PVP (25)	5	80		−28		−4	C	−15	C	Elegant	
PVP (25) (1%) + PEG (1%)	6	67		−26		−18; −4	C	−18	C	Elegant	
PVP (25) (1%) + Tween 80 (0.02%)	5	80		−27		−4	C	−15	C	Elegant	
Maltodextrin (DE 6)	5	80		−20		−4	C	−10	C	Elegant	

(continued)

TABLE 1 Summary of Thermal Properties of Formulations Including Glycine as the Crystalline Bulking Agent Determined by DSC and FDM: Impact of Solute and Freezing Conditions on the Physical State of Glycine (*Continued*)

Composition of the formulation			Differential scanning calorimetry					Freeze-drying microscopy			References
Other solutes	Total solid content (%)	Weight fraction of glycine(%)	Cycle	T_g' (°C)	T_{cr} (°C)	T_{eut} (°C)	State of glycine after freezing	T_{coll} (°C)	State of glycine after freezing	Freeze-dried cake	
	1	0		−28.4							
	1	10	CR = 10°C/min	−34			A				
	1	20	HR = 5°C/min	−37.5			A				
	1	30		−42			A				
	1	40		−46			A				
Raffinose	1.5	67	CR = 10°C/min	−31			C			Elegant	Chatterjee, 2005 (28)
	1.6	63	HR = 5°C/min	−31			C			Elegant	
	1.8	56	Annealing treatment (5 hr)	−31			C			Elegant	
	1	0		−34			A				
	1	10	CR = 10°C/min	−38			A				
	1	20	HR = 5°C/min	−42			A				
	1	30		−45			A				
	1	40		−50			A				
Trehalose	1	50		−54			A				
	1.5	67	CR = 10°C/min	−36.1			C			Elegant	Chatterjee, 2005 (28)
	1.6	63	HR = 5°C/min	−36.1			C			Elegant	
	1.8	56	Annealing treatment (5 hr)	−36.1			C			Elegant	

Abbreviations: DSC, differential scanning calorimetry; FDM, freeze-drying microscopy; T_g', glass transition temperature of the maximally freeze-concentrated phase; T_{cr}, crystallization temperature of solute; T_{eut}, eutectic melting temperature of solute; T_{coll}, collapse temperature determined by freeze-drying microscopy; A, amorphous; PC, partially crystalline; C, crystalline; CR, cooling rate; HR, heating rate; API, active pharmaceutical ingredient; PVP, polyvinylpyrrolidone; PEG, polyethylene glycol.

TABLE 2 Summary of Thermal Properties of Formulations Including Mannitol as the Crystalline Bulking Agent Determined by DSC and FDM: Impact of Solute and Freezing Conditions on the Physical State of Mannitol

Composition of the formulation			Differential scanning calorimetry				Freeze-drying microscopy			References
Other solutes	Total solid content (%)	Weight fraction of mannitol (%)	Cycle	T_g' (°C)	T_{cr} (°C)	State of mannitol after freezing	T_{coll} (°C)	State of mannitol after freezing	Freeze-dried cake	
	7.5	67	CR = 10°C/min HR = 10°C/min	−38		A				
	10	50		−37		A				
	7.5	33		−34		A				
	7.5	67	CR = 0.5°C/min HR = 10°C/min	−32	−18	A or PC				
Sucrose	10	50		−37	−14	A or PC				Lueckel, 1998 (27)
	7.5	33		−35		A				
	7.5	67	CR = 10°C/min HR = 10°C/min Annealing at −24°C for 2 hr	−31.8		C				
	10	50		−31		C				
	7.5	33		−34		A				
Sodium tetraborate (20 mM)	5.4	93		−36	Not observed	A				
Sodium tetraborate (50 mM)	6	83		−34	Not observed	A				
Sodium tetraborate (100 mM)	7	71	CR = 20°C/min HR = 5°C/min	−31	Not observed	A				Izutsu, 2003 (29)
Sodium tetraborate (200 mM)	9	56		−25	Not observed	A				
Sodium tetraborate (300 mM)	11	45		−22	Not observed	A				
None	4	100		−55; −32	−40; −26	A or PC				
Tween 80 (0.02%)	4	100	CR = 10°C/min HR = 10°C/min	−56; −32	−39; −25	A or PC				Passot, 2005 (6)
Sucrose	5	80		−44; −31	−27	A or PC				
PVP (25)	5	80		−58; −30; −25	−37; −20	A or PC				

(continued)

TABLE 2 Summary of Thermal Properties of Formulations Including Mannitol as the Crystalline Bulking Agent Determined by DSC and FDM: Impact of Solute and Freezing Conditions on the Physical State of Mannitol (*Continued*)

Composition of the formulation			Differential scanning calorimetry				Freeze-drying microscopy			References
Other solutes	Total solid content (%)	Weight fraction of mannitol (%)	Cycle	T_g' (°C)	T_{cr} (°C)	State of mannitol after freezing	T_{coll} (°C)	State of mannitol after freezing	Freeze-dried cake	
PVP (25) (1%) + Tween 80 (0.02%)	5	80	CR = 10°C/min HR = 10°C/min	−48; −32; −27	−38; −21	A or PC				Passot, 2005 (6)
None	4	100			−52	C	−7	C	Elegant	
Tween 80 (0.02%)	4	100	CR = 10°C/min HR = 10°C/min	−30		C	−9	C	Elegant	
Sucrose	5	80		−38		C	−20	C	Elegant	
PVP (25)	5	80	Annealing at −20°C for 60 min	−30		C	−14	C	Elegant	
PVP (25) (1%) + Tween 80 (0.02%)	5	80		−31		C	−15	C	Elegant	
Sucrose	7.26	31		−40	Not observed	A				Liao, 2005 (30)
	6.75	52		−42	Not observed	A				
	6.74	60	CR = 20°C/min HR = 5°C/min	−43	Not observed	A				
	6.26	66		−43.5	Not observed	A				
	7.5	67		−43.5	Observed	C				
	6.67	75		−43.5	Observed	C				
Sucrose (5%) + protein (20 mg/mL)	9.26	24		−38	Not observed	A				
Sucrose (3.2%) + protein (20 mg/mL)	8.75	40		−38	Not observed	A				
Sucrose (2.7%) + protein (20 mg/mL)	8.74	46	CR = 20°C/min HR = 5°C/min	−42	Not observed	A				
Sucrose (2.1%) + protein (20 mg/mL)	8.29	50		−42	Not observed	A				
Sucrose (2.5%) + protein (20 mg/mL)	9.5	53		−44	Not observed	A				
Sucrose (1.7%) + protein (20 mg/mL)	8.67	58		−44	−21	A or PC				

Formulation							Reference
Sucrose (5%) + protein (10 mg/mL)	8.26	27	CR = 20°C/min HR = 5°C/min	−39.1	Not observed	A	Liao, 2005 (30)
Sucrose (2.7%) + protein (10 mg/mL)	7.74	52		−43.5	Not observed	A	
Sucrose (1.7%) + protein (10 mg/mL)	7.67	65		−44	−21	A or PC	
Sucrose (5%) + protein (50 mg/mL)	12.26	18		−34.7	Not observed	A	
Sucrose (2.7%) + protein (50 mg/mL)	11.74	34	CR = 20°C/min HR = 5°C/min	−37.5	Not observed	A	
Sucrose (1.67%) + protein (50 mg/mL)	11.67	43		−38.5	−15	A or PC	
Trehalose			CR = 20°C/min HR = 5°C/min	−41; −32	−26	A or PC	Liao, 2006 (31)
	5.69	64	CR = 20°C/min HR = 5°C/min Annealing at −28°C 60 min or at −18°C for 15 min		−32	C	
Trehalose + protein	5.86	62	CR = 20°C/min HR = 5°C/min	−41; −32	−23	A or PC	
Sucrose	5	80	CR = 10°C/min HR = 10°C/min	−42; −32; −28	Observed	A or PC	Hawe, 2006 (32)
		60		−41	Observed	A or PC	
		50		−40	Observed	A or PC	
		40		−39	Not observed	A	
		20		−36.5	Not observed	A	
Sucrose (1%) + NaCl (0.05%)	5.05	79	CR = 10°C/min HR = 10°C/min	−42			
Sucrose (1%) + NaCl (0.1%)	5.1	78		−42.5			
Sucrose (1%) + NaCl (0.2%)	5.2	77		−43.5			

(continued)

TABLE 2 Summary of Thermal Properties of Formulations Including Mannitol as the Crystalline Bulking Agent Determined by DSC and FDM: Impact of Solute and Freezing Conditions on the Physical State of Mannitol (*Continued*)

Composition of the formulation			Differential scanning calorimetry				Freeze-drying microscopy			References
Other solutes	Total solid content (%)	Weight fraction of mannitol (%)	Cycle	T_g' (°C)	T_{cr} (°C)	State of mannitol after freezing	T_{coll} (°C)	State of mannitol after freezing	Freeze-dried cake	
Sucrose (2%) + NaCl (0.05%)	5.05	59		−41						
Sucrose (2%) + NaCl (0.1%)	5.1	59	CR = 10°C/min HR = 10°C/min	−42						
Sucrose (2%) + NaCl (0.2%)	5.2	58		−43						
Sucrose (2.5%) + NaCl (0.05%)	5.05	50		−40.5						
Sucrose (2.5%) + NaCl (0.1%)	5.1	49	CR = 10°C/min HR = 10°C/min	−41						
Sucrose (2.5%) + NaCl (0.2%)	5.2	48		−42						Hawe, 2006 (32)
Sucrose (3%) + NaCl (0.05%)	5.05	40		39.5						
Sucrose (3%) + NaCl (0.1%)	5.1	39	CR = 10°C/min HR = 10°C/min	−41						
Sucrose (3%) + NaCl (0.2%)	5.2	38		−41						
Sucrose (4%) + NaCl (0.05%)	5.05	20		−37.5						
Sucrose (4%) + NaCl (0.1%)	5.1	20	CR = 10°C/min HR = 10°C/min	−38						
Sucrose (4%) + NaCl (0.2%)	5.2	19		−39						

Abbreviations: DSC, differential scanning calorimetry; FDM, freeze-drying microscopy; T_g', glass transition temperature of the maximally freeze-concentrated phase; T_{cr}, crystallization temperature of solute; T_{coll}, collapse temperature determined by freeze-drying microscopy; A, amorphous; PC, partially crystalline C, crystalline; CR, cooling rate; HR, heating rate; PVP, polyvinylpyrrolidone; PEG, polyethylene glycol.

crystallization and resulted in a sharp increase of the T_g' value ($T_g' = -26°C$) of the remaining phase mainly composed of the protein active ingredient, maltose, and a small amount of amorphous glycine.

The DSC thermograms obtained with this type of formulation that includes a crystalline bulking agent are usually complex, exhibiting several events and event overlappings that are difficult to interpret. To illustrate this complex behavior, warming thermograms obtained when a solution including 4% mannitol and 1% PVP (polyvinylpyrrolidone) was submitted to different thermal treatments are presented in Figure 1: a cooling rate of 10°C/min (curve a) or 1°C/min (curve b), and an annealing treatment of 1 hour at −20°C (curve c) (6). When applying a cooling rate of 10°C/min (curve a), several thermal events were observed in the heating scan: a weak glass transition event at −58°C (arrow 1); several exothermic events in the range of −40°C to 0°C, which were attributed to the crystallization of both mannitol and unfrozen water (34); and an endothermic event at −1°C because of the melting of ice and mannitol. The interpretation of the DSC heating curves is complicated by the overlap of the glass transition event with the exothermic event of crystallization of both mannitol hydrate and the unfrozen water. When applying a cooling rate of 1°C/min to the sample (curve b), the profile of the heating scan changed with the decrease and removal of some exothermic events, and two endothermic events clearly appeared at around −25°C (arrow 2), which were ascribed to glass transition events by Cavatur et al. (34). The weak glass transition event at around −55°C remained (arrow 3). Applying a

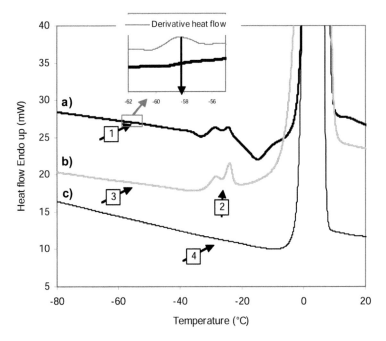

FIGURE 1 DSC heating profile of the frozen formulation MnP (4% mannitol, 1% PVP, and 10 mM Tris-HCl) submitted to different thermal treatments. Curve (a) depicts cooling and heating at 10°C/min; curve (b), cooling at 1°C/min and heating at 20°C/min; and curve (c), cooling at 10°C/min, annealing at −20°C for 60 minutes, and heating at 20°C/min. *Source*: From Ref. 33.

cooling rate of 1°C/min resulted in a decrease of the area of the crystallization peak of mannitol but did not allow its complete crystallization during freezing. When applying an annealing step at −20°C for 60 minutes (curve c), the DSC heating curve did not exhibit any exothermic event, suggesting the completion of mannitol crystallization during annealing. Furthermore, only a weak glass transition event at −30°C was observed (arrow 4). Freezing the MnP formulation resulted in a complex structure with a large crystalline phase composed of mannitol and a minor amorphous phase embedding the proteins and composed of PVP and low mannitol content. The resulting T_g' value is highly dependent on the thermal treatment applied to the sample.

The studies of Ma et al. (12) and Passot et al. (6) are based on a unique value of weight fraction of the crystalline agent: 40% glycine and 80% mannitol, respectively (Table 1). Kasraian et al. (26) investigated the physical behavior of solutions of various ratios of glycine/sucrose during freezing by DSC and FDM and reported the complete glycine crystallization following a cooling rate of 1.5°C/min for a glycine weight fraction higher than 71% (Table 1). The crystallization of glycine, even partial, makes it possible to perform primary drying at high product temperatures, with a value of T_{max} close to the eutectic melting temperature of glycine (−4°C). Elegant freeze-dried cakes were obtained for weight ratios of glycine higher than 57%. To develop a formulation that makes it possible to design a short freeze-drying process, it is therefore crucial to know the minimal weight fraction of crystalline bulking agent required to obtain its complete crystallization during freezing for the range of freezing rates used in the freeze-drying process.

Chatterjee et al. (28) have quantified the extent of glycine crystallization in frozen solution in a ternary model system containing water and crystallizable (glycine) and noncrystallizable (raffinose or trehalose) components by applying a systematic approach (Table 1). Aqueous solutions including different

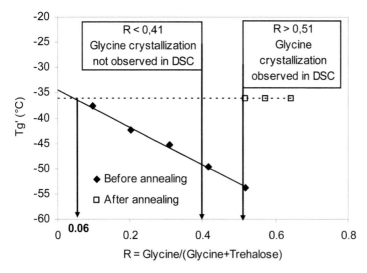

FIGURE 2 Glass transition temperature of freeze-concentrated phase (T_g') of aqueous solution including glycine and trehalose. T_g' values from the first warming DSC curve are compared with T_g' values after annealing. *Source*: Adapted from Ref. 28.

glycine-to-carbohydrate weight ratios were investigated by DSC and variable X-ray diffractometry and by applying two thermal treatments (cooling rate of 10°C/min and prolonged annealing). Figure 2 displays the evolution of the T_g' values of frozen solutions of glycine and trehalose with increasing glycine weight fractions. At glycine weight fractions higher than 0.51, glycine crystallization was observed in the DSC warming curve. These systems were subjected to a prolonged annealing treatment. The T_g' values of the annealed systems appear to be independent of the initial solution composition and close to the T_g' value of a frozen aqueous trehalose solution. For glycine weight fractions lower than 0.41, glycine crystallization was not detected and the T_g' value of the solution decreased linearly with the glycine weight fraction. The intersection of both curves makes it possible to quantify the amorphous glycine fraction in the freeze-concentrated phase after the maximal glycine crystallization during cooling and/or annealing treatment. The maximally freeze-concentrated phase is thus composed of 94% trehalose and 6% glycine. On the basis of these results, a simplified state diagram can be built for the ternary water-glycine-trehalose system (Fig. 3). This triangle can be divided into three regions:

- Region 1 (Water-W'-Trehalose triangle): contains all compositions that form a kinetically stable glass, irrespective of thermal annealing steps.
- Region 3 (Water-Glycine-G triangle): contains all compositions that readily crystallize during cooling or subsequent warming.
- Region 2 (Water-G'-W' triangle): this intermediate region contains all compositions that do not crystallize during the experimental timescale.

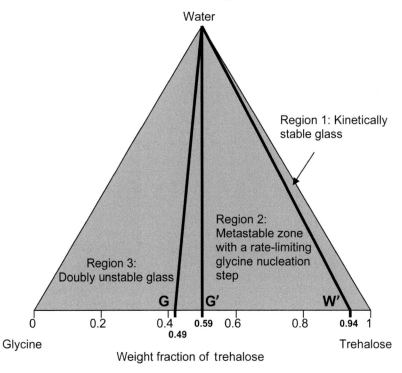

FIGURE 3 Water-glycine-trehalose state diagram. *Source*: From Ref. 28.

The weight fraction of glycine should be higher than 0.5 in a glycine-carbohydrate formulation (glycine-to-carbohydrate weight ratio > 1) to ensure excessive glycine crystallization in the final freeze-dried product. The possibility of further glycine crystallization during product storage is therefore sharply reduced and protein integrity will be preserved.

Considering a water/glycine/sucrose system, Shalaev and Kanev (35) reported that 22% glycine remains amorphous in the amorphous sucrose freeze concentrate, despite annealing treatments. From the point of view of the composition triangle (Fig. 3), this would imply that the area of an analogous Water-W'-Sucrose triangle would be much larger than the one observed in the threhalose (or raffinose) system by Chatterjee et al. (28) These results indicate a lower crystallization potential of glycine in sucrose than in threhalose or raffinose systems.

DSC provides useful information on the physical behavior of solutions during freezing, but many difficulties are usually encountered for interpreting the thermograms. The DSC thermogram may depict several transitions, especially in the case of formulations including crystalline bulking agents. Extensive experience in DSC data analysis is needed to identify the maximum allowable temperature (T_{max}) or temperature range among the observed transitions that, in turn, will make it possible to predict how readily a formulation can be freeze-dried.

By allowing the direct microscopic observation of a thin section of material as it freezes and dries, FDM makes it possible to assess the structure of the drying material over a range of conditions. The T_{coll} is defined as the lowest temperature of overall loss of the initial frozen structure during sublimation. For example, when examining the DSC warming scan of a sample involving 4% glycine, 1% sucrose, and 1% PEG (polyethylene glycol) obtained after applying an annealing treatment to the sample (Table 1), the following events were observed: a glass transition event at $-38°C$, two endothermic events at $-19°C$ and $-4°C$ ascribed to the melting of PEG and the melting of glycine and ice, respectively (6). When analyzing this solution by FDM, the collapse of the dried product during sublimation was observed at $-18°C$ with the melting of the PEG crystals. In this case, a temperature difference of $20°C$ was observed between the glass transition temperature of the maximally freeze-concentrated phase (T_g') and the T_{coll}. From the data reported in Tables 1 and 2 concerning the samples submitted to annealing treatment to promote the crystallization of the bulking agent, the values of T_{coll} are much higher than the values of glass transition temperature of the maximally freeze-concentrated phase. If the T_{coll} is used to fix the value of the product temperature (T_{max}) to be applied during sublimation instead of T_g', the sublimation time will be significantly reduced. Pikal (36) reported that a product temperature increase of $1°C$ would result in a 13% reduction of primary drying time. Therefore, the T_{coll} determined by FDM appeared to be a more relevant parameter than T_g' for process optimization, considering the retention of the freeze-dried structure.

Until now, the discussion has been focused on the physical behavior of formulation involving a crystalline component. When considering formulations involving only amorphous components, their behavior during freezing and drying is easier to describe and predict. During freezing, these systems form a kinetically stable glass and the T_{coll} is close to the glass transition temperature of the maximally freeze-concentrated phase, differing by less than $3°C$ (6,14). A state diagram can be used to describe the phase change of the binary system during freezing and drying (37,38). However, some pseudo-binary systems can exhibit phase separation during freezing with the formation of two amorphous phases. Single-phase solutions containing combinations of two nonionic polymers (e.g., PVP and dextran), or polyelectrolytes and nonionic polymers (e.g., DEAE-dextran and dextran), or nonionic polymers and buffer salts (e.g., PVP and phosphate buffer) have been observed to separate into two amorphous phases in a frozen solution (39–43). This phase separation can result in protein degradation (44,45) and should thus be avoided during the freezing step. By reproducing the freezing conditions in DSC experiments, the phenomenon of phase separation in amorphous formulation can be detected with the appearance of multiple glass transition events in the thermogram, with events corresponding to the glass transition temperature of the freeze-concentrated phase of the binary aqueous solution of each component of the formulation. Furthermore, the DSC analysis of the freeze-dried powder can confirm the phase separation phenomenon.

DSC and FDM provide more information than an estimation of the maximal value of product temperature not to be exceeded during sublimation (T_{max}). By varying the experimental conditions, the physical behavior of the formulation during freezing and drying can be predicted. This knowledge is essential for process design and scale-up and for identifying the potential risk of protein degradation. In the following section, the impact of the formulation components on the physical behavior will be examined to improve formulation composition that, in turn, will make it possible to develop a robust freeze-drying cycle.

EFFECT OF THE SOLUTES ON THE FORMULATION'S PHYSICAL BEHAVIOR: SELECTION OF EXCIPIENTS

A model formulation for protein freeze-drying usually includes five components: a buffer that does not acidify during freezing (Tris, histidine, citrate), specific pH/ligands that optimize the thermodynamic stability of protein, a stabilizer (generally a disaccharide such as sucrose or trehalose) to inhibit protein unfolding and provide a glassy matrix, a bulking agent (mannitol, glycine, hydroxyethyl starch, or dextran) to ensure physical stability, and a nonionic surfactant to reduce protein aggregation. A slight increase or decrease of one of the excipients can lead to significant changes in the formulation's physical behavior during freezing and drying. Excipients can have a positive or a negative impact on the thermal properties and especially on the value of T_{max}. The composition of the formulation is a key factor for developing a robust and short freeze-drying process, and optimization of the freeze-drying process should be integrated into the formulation development step.

When considering formulations involving only amorphous excipients, the glass transition temperature of the maximally freeze-concentrated phase (T_g') is closely related to the T_g' values of binary aqueous solutions of each excipient and can be easily predicted with the following simple linear relationship (19,46):

$$T_{g'_M} = \frac{\sum_{i=1}^{n} C_i T_{g i}'}{\sum_{i=1}^{n} C_i}$$

TABLE 3 Predicted Values of Glass Transition Temperature of Freeze-Concentrated Phase (T_g') of Solutions Including Sucrose (S) and Glucose (G) or Dextran (D) or Tris-HCl Buffer (T)

		Predicted T_g' (°C)		
		Sucrose + (glucose or dextran or Tris-HCl buffer) solution (5%)		
	5% S	1% (G or D) or 0.1% T	2.5% (G or D) or 0.5% T	4% (G or D) or 1% T
Glucose (G) ($T_g' = -43$°C)	−32	−34.2	−37.5	−40.8
Dextran (D) ($T_g' = -10$°C)		−27.5	−21	−14.1
Tris-HCl (T) ($T_g' = -65$°C)		−32.7	−35.3	−38.6

where $T_{g'i}$ and C_i are the T_g' (in °C) and the solute concentration in water (in solids, % wet basis), respectively, of each component i, and $T_{g'M}$ is the glass transition temperature of the formulation.

This simple relationship could be used to evaluate the impact of excipients on the T_g' and, consequently, on the T_{max} value of the formulation. Table 3 illustrates the impact of adding different concentrations of solutes [glucose (G), dextran (D), and Tris-HCl buffer (T)] on the T_g' value of a sucrose solution. Increasing the concentration of solutes with high T_g' values such as dextran (−10°C) enhances the T_g' value of the sucrose solution (−27.5°C to −14.1°C while increasing dextran concentration from 1 to 4%). Conversely, the addition of glucose with a lower T_g' value (−43°C) than sucrose depresses the T_g' value of the solution (−34.2°C to −40.8°C while increasing glucose concentration from 1 to 4%). Furthermore, the addition of only 1% Tris-HCl in a sucrose solution results in the same T_g' depression as the addition of 4% glucose. The addition of glucose or Tris-HCl buffer results in a decrease of T_g' of the formulation, which can make the freeze-drying process difficult or impossible to carry out. Components with high T_g' values and therefore a positive impact on the $T_{g'M}$ should thus be given priority when developing formulations. The addition of polymers or polysaccharides such as dextran or PVP in the formulation makes it possible to obtain T_g' (and T_{max}) values comparable to those obtained when using a crystalline bulking agent (glycine or mannitol). As mentioned in the previous paragraph, formulations involving crystalline bulking during freezing exhibit complex physical behavior during freezing that are highly dependent on the thermal history of the sample. Using amorphous bulking agent can thus be an interesting alternative strategy for improving process productivity without compromising protein stability. This strategy has been applied with success by various authors (6,47).

Salts such as NaCl and buffer are important addition to the formulation to control pH and to ensure optimal chemical and physical stability of proteins. These salts are widely used in the purification and concentration steps of active ingredients. The concentrated solution of protein used for preparing the formulation to be freeze-dried contains various amounts of salts. Since salts exhibit low T_g' values, slight variations in salt concentrations may lead to large modifications of T_{max}. Figure 4 illustrates the effect of NaCl content on the T_{coll} of sucrose and dextran solutions determined by FDM (9). For weight ratios of NaCl lower than 3%, the addition of NaCl sharply depresses the T_{coll} value of

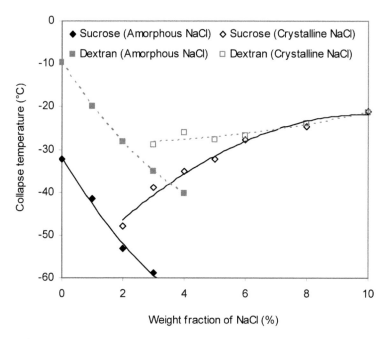

FIGURE 4 Effect of weight fraction of NaCl on the collapse temperature of solutions of sucrose and dextran determined by freeze-drying microscopy. *Source*: From Ref. 9.

solutions of sucrose or dextran. A 2% increase of NaCl content results in the decrease of T_{coll} by 20°C. NaCl acts as a plasticizer by increasing the amount of unfrozen water in the amorphous phase leading to T_{coll} depression. For higher weight ratios of NaCl, the salt crystallizes during freezing and the T_{coll} becomes closer to the T_{eut} of the NaCl (−26°C).

The crystallization of NaCl has a positive effect on the T_{coll} value of a sucrose solution (T_{coll} increases from −32°C to −26°C) but a negative effect in the case of dextran solution (T_{coll} decreases from −10°C to −26°C). As another example, Hawe and Freiss (32) investigated the effect of NaCl and mannitol addition on the T_g' of a sucrose solution (Fig. 5). The glass transition temperature of the maximally freeze-concentrated phase is lowered by both mannitol and NaCl. Even low NaCl concentration (<0.2%) can significantly depress the T_g' values. Since the salt concentration may vary from one batch to another after the protein purification stages, it is therefore of paramount importance to determine the thermal properties of each lot of purified and concentrated protein.

Buffering salts also act as plasticizers due to their low T_g' values and induce depression of the T_g' of the formulation. Moreover, a buffer component may crystallize during freezing, producing significant pH changes that are usually undesirable for protein stability. Crystallization and pH changes of phosphate buffer at sub-zero temperatures were studied in detail in the presence of different metal ions and for a wide range of pH and concentrations (48,49). Shalaev et al. (50) evaluated the crystallization behavior and the glass transition temperature of the maximally freeze-concentrated phase (T_g') of different

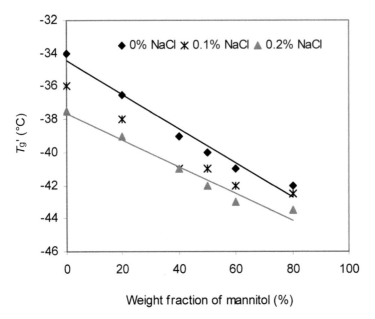

FIGURE 5 Effect of NaCl on the glass transition temperature of the maximally freeze-concentrated phase (T_g') of sucrose formulations with increasing mannitol concentrations. *Source*: From Ref. 32.

buffers (sodium tartrate, sodium malate, potassium citrate, and sodium citrate buffers) in the frozen state for a wide range of pH using DSC. Unexpected T_g' evolution with pH was observed for several buffers, with T_g' reaching a maximum at pH 3 to 4. Such complex T_g' behavior has been explained by the balance between an increase in viscosity due to the increase in the electrostatic interactions with increasing pH, and a decrease in viscosity due to the increase in the amount of unfrozen water in the freeze-concentrate solution. The authors observed crystallization of buffer salts in the DSC heating scan only for the succinate and tartrate buffers. The lack of crystallization observed for citrate and malate buffers during the DSC experiment does not necessarily mean that these salts will not crystallize during an actual freeze-drying process and, conversely, the observed crystallization of some buffer salts does not mean that these salts will systematically crystallize. Solute crystallization is highly dependent on numerous factors: cooling rate, presence of other solutes, sample size, etc. It is therefore tricky to analyze buffer crystallization using DSC, and the transposition to the real behavior of a formulation during freeze-drying can be difficult. However, by varying the cooling rate and the formulation composition, it is possible to obtain information about the buffer behavior during freezing and to compare different buffer behaviors.

When considering formulations involving a crystalline bulking agent (mannitol or glycine), excipients can decrease the value of the maximum allowable product temperature not to be exceeding during the sublimation step (T_{max}) of a formulation by acting as inhibitors of the crystallization of bulking agents (Tables 1 and 2) (51,52). This inhibitory effect of an excipient on the

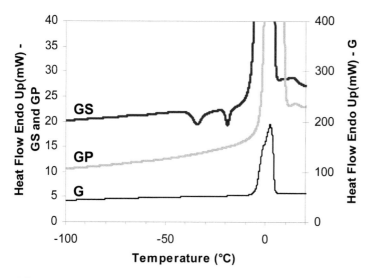

FIGURE 6 DSC heating profile of solutions of 4% glycine (G), 4% glycine and 1% sucrose (GS), and 4% glycine and 1% PVP (GP). The solutions were cooled to −120°C at 10°C/min and then heated to 25°C at 10°C/min.

crystallization of mannitol or glycine can be investigated by DSC. Figure 6 shows the DSC heating scans of solutions of pure glycine (G), glycine and sucrose (GS), and glycine and PVP (GP) when cooled at 10°C/min. The DSC heating scan of glycine only exhibits an endothermic event at −4°C, which is attributed to the melting of ice and glycine. When 1% sucrose was added to the glycine solution, the profile of the DSC heating scan was modified and several thermal events appeared: a weak glass transition event at around −69°C and two exothermic events ascribed to the crystallization of glycine (52–54). These events were also observed when replacing sucrose by maltose or maltodextrin (Table 1). The crystallization of glycine was totally or partially inhibited during the cooling step and occurred during warming. When sucrose was replaced by PVP, the two exothermic events were not observed and the DSC heating scan exhibited only two weak glass transition events at −67°C and −32°C, corresponding to the T_g' values of the individual components, glycine and PVP, respectively. When a cooling rate of 1.5°C/min was applied and an annealing treatment was added, only a weak glass transition event at −28°C was observed during warming. These results suggest that PVP is a stronger inhibitor of glycine crystallization than sucrose.

The inhibitory effect of excipient on mannitol crystallization was also studied by various authors (Table 2). In contrast to glycine application where glycine must be in a crystalline form after the freezing step [the low T_g' value of glycine (−70°C) is not compatible with the process], it may be worthwhile to maintain mannitol in the amorphous state. Mannitol is only able to preserve protein via molecular interactions when it is in an amorphous state (24,25). Several approaches have been described for producing amorphous mannitol by adding, for example, NaCl, boric acid, or sodium tetraborate (55,56).

TABLE 4 Summary of Thermal Properties of Solutions Including Various Contents of an Active Ingredient (bacteria or protein) Determined by DSC (T_g') and FDM (T_{coll} and $T_{\mu coll}$)

	T_g' (°C)	$T_{\mu coll}$ (°C)	T_{coll} (°C)	$T_{coll} - T_g'$ (°C)	References
Supernatant (S)	−42	−43	−41	1	
Lb + S (1:10)	−42	−42	−39	3	
Lb + S (1:2)	−36	−28	−22	14	
Lb + S (1:1)	−37	−29	−23	14	Fonseca,
Lb + S (1.5:1)	−30	−20	−12	18	2004 (19,57)
Lb + S (2:1)	−27	−17	−8	19	
Lb (washed)	Nd	Nd	Nd		
St + S (1:1)	−47	−37	−27	20	
Formulation (F)	−33	−33	−31	2	
F + 10 mg/mL protein	−31	−32	−30	1	
F + 20 mg/mL protein	−31	−29	−24	7	
F + 40 mg/mL protein	−29	−27	−23	6	Colandene,
F + 60 mg/mL protein	−27	−27	−17	10	2007 (58)
F + 80 mg/mL protein	−26	−19	−12	14	
F + 100 mg/mL protein	−26	−21	−13	13	

Abbreviations: Lb, *Lactobacillus bulgaricus*; St, *Streptococcus thermophilus*; S, supernatant from centrifugation of bacteria fermentation medium; Nd, not detected; T_g', glass transition temperature of freeze-concentrated phase; T_{coll}, temperature corresponding to the beginning of collapse phenomenon observed by freeze-drying microscopy; T_{coll}, temperature corresponding to complete collapse.

The active ingredient is usually present in very low concentrations in the formulation and can be considered as having no impact on the thermal properties. However, sometimes its concentration cannot be neglected. Few studies have investigated the impact of active ingredients on the formulation's physical behavior during freezing and drying (Table 4). Fonseca et al. (19,57) studied the physical behavior of lactic acid bacterial suspension by DSC and FDM. For simple binary and ternary aqueous solutions with either none or a low concentration of lactic acid bacteria, the T_g' and T_{coll} values are close differing by less than 3°C, and the collapse phenomenon occurs within a small range of temperatures (1–3°C). When the concentration of bacteria increases in the solution, the T_{coll} increases and the values are significantly higher than the corresponding T_g' value, with $T_{coll} - T_g'$ difference greater than 10°C. Lactic acid bacterial cells add a degree of robustness to the freeze-dried product, thus allowing the use of higher sublimation temperatures during primary drying than expected from the protective supernatant alone. The same effect was observed by Colandene et al. (58) on a formulation that included an increasing concentration of a protein active ingredient. Figure 7 illustrates the effect of the protein weight fraction on the T_g' and T_{coll} values. Linear relationships are observed between the protein weight fraction in the formulation and the T_g' and T_{coll} values. At low protein concentrations, the glass transition temperature of the maximally freeze-concentrated solution (T_g') determined by DSC is similar to the T_{coll} determined by FDM. However, at higher protein concentrations, the difference between T_{coll} and T_g' becomes progressively larger.

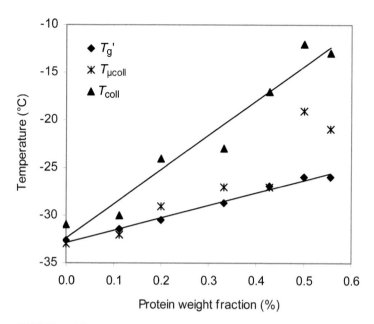

FIGURE 7 Effect of protein (monoclonal antibody) weight fraction on the values of glass transition temperature of the maximally freeze-concentrated phase (T_g') and collapse temperatures ($T_{\mu coll}$: beginning of the collapse phenomenon; T_{coll}: complete collapse) determined by freeze-drying microscopy. *Source*: From Ref. 58.

Liao et al. (30) studied the inhibitory effect of a protein active ingredient, a human monoclonal antibody, on mannitol crystallization. The presence of the antibody, even at moderate concentrations (20 mg/mL), dramatically inhibited mannitol crystallization even under aggressive annealing conditions. This inhibitory effect was concentration-dependent and would be even more pronounced if the protein concentration were increased (>20 mg/mL).

The understanding of the impact of each excipient on the formulation's physical behavior during freezing and drying is crucial when developing a new freeze-dried protein formulation. An adequate choice of excipients and excipients concentration is thus needed not only to preserve protein but also to obtain a formulation's physical behavior during freezing that is relatively independent of the operating conditions and that results in a high value of T_{max}. By fulfilling these needs, it will make it possible to design short and robust freeze-drying processes that are easily transferable from one piece of equipment to another. A fundamental question must now be addressed: Can these thermal properties be related to protein stability or does collapse during freeze-drying cause a significant decrease in protein stability?

RELATIONSHIP BETWEEN THERMAL PROPERTIES, PROTEIN STABILITY, AND CYCLE DESIGN

There has been much speculation over the years regarding the need of freeze-drying below T_g' to maintain protein stability. However, few data are available. The formulations including both stabilizer and crystalline bulking agent have a

T_{coll} close to the eutectic temperature of the bulking agent added (eutectic temperature for glycine or mannitol $\approx -3°C$), which is typically much higher than the T_g' of most stabilizers (T_g' for sucrose and trehalose $\approx -34°C$ and $-28°C$, respectively). If drying is carried out above T_g', the amorphous phase will undergo viscous flow that is undetectable (i.e., micro-collapse) due to the rigid matrix provided by the crystalline component. If the physical stability of the freeze-dried cake is preserved, protein degradation could exist due to the increase of molecular mobility linked to the viscous flow of the amorphous phase. The knowledge of the effect of the thermal history of the product during freeze-drying on long-term protein stability is thus essential. Table 5 summarizes the result of the main studies concerning the effect of collapse or microcollapse on protein activity. The study of this parameter is sensitive since factors other than product temperature can influence protein degradation. For example, the protein stability is strongly related to the residual moisture content of the freeze-dried product.

Some authors including Jiang and Nail (59), Wang et al. (61), Chatterjee et al. (62), and Colandene et al. (58) observed no significant effect of freeze-dried cake structure and collapse, in particular, on protein activity recovery after freeze-drying and even after storage (Table 5). Moreover, Jiang and Nail (59) studied three model proteins and observed a large decrease in the recovery of activity only for the lactate dehydrogenase (LDH). The authors explained this result by the sensitivity of LDH to freeze-thaw damage since some melting of frozen solution during primary drying could occur. These results may also be partially confused with the effect of residual moisture in the freeze-dried material. It was concluded in this study that the most important factor for protein activity recovery after freeze-drying was the amount of residual moisture in the freeze-dried matrix and that the collapsed material, having a lower specific surface area, tends to have a high residual moisture level.

Other authors such as Lueckel et al. (60) and Passot et al. (33) observed a significant effect of the thermal history applied to the product during the freeze-drying process on protein activity recovery. Lueckel et al. (60) demonstrated that collapse of the freeze-dried cake results in an increased level of aggregation of IL-6 in a glycine/sucrose formulation. They also reported that the use of an annealing step increased the level of aggregation of IL-6 in a glycine/sucrose or mannitol/sucrose formulation after both freeze-drying and storage, compared to a control without an annealing step. The authors speculated that this destabilization arises from increased time above T_g' in the frozen system during the freeze-drying cycle. Passot et al. (33) investigated the effect of performing sublimation at product temperatures below and above T_g' on the long-term stability of freeze-dried proteins by applying a systematic approach. Figure 8 illustrates the loss of antigenic activity of both toxins that were freeze-dried in a formulation including 4% mannitol and 1% PVP after six-months of storage at 4°C as a function of the temperature difference between the product temperature obtained during sublimation and the glass transition temperature of the freeze-concentrated phase ($T_{product} - Tg'$). The antigenic activity measured after freeze-drying was used as the reference state to calculate the loss of antigenic activity during storage. The losses of antigenic activity of both toxins increased from 0% when sublimation was performed at a product temperature lower than T_g' to 25% when the product temperature was higher than T_g'. When carrying out sublimation at product temperatures higher than T_g', the amorphous phase

TABLE 5 Summary of the Main Results of the Different Studies Dealing with the Effect of the Thermal History of the Product During the Primary Drying Step on Protein Stability

	Formulation	Storage time	Structure retention Water content (%)	Activity recovery/Aggregates[a] (%) After FD	5°C	40°	Microcollapse Water content (%)	Activity recovery/Aggregates (%) After FD	5°C	40°C	Collapse Water content (%)	Activity recovery/Aggregates (%) After FD	5°C	40°C	References
Catalase				92								88			Jiang, 1998 (59)
β-GA				95								85			
LDH				95								50			
Interleukin-6	Glycine (2%) Sucrose (2%)	9 mo	1.9	1[a]		2[a]	0.6	1.9[a]		3[a]	3.8	1[a]		39[a]	Lueckel, 1998 (60)
	Glycine (1.7%) sucrose (3.4%)	9 mo	0.9	0[a]		0[a]					5.6	1[a]		>50[a]	
	Mannitol (4%) Sucrose (2%)	6 mo	0.6	2[a]		4[a]	1.4	2[a]		17[a]					
rFVIII	Glycine (2.2%)	78 wks	0.79	90	85	38	0.95	89	95	38	1.58	92	84	64	Wang, 2004 (61)
α-amylase	Sucrose (1%)		0.74	87	95	84	0.95	96	96	89	1.24	131	98	87	
LDH	Glycine Raffinose			95				90				95			Chatterjee, 2005 (62)
	Glycine Trehalose			100				95				92			
Toxin A	PVP (4%) Sucrose (1%)	6 mo	1.5	70	0%[b]						1.5	80	5–20%[b]		Passot, 2007 (33)
	Mannitol (4%) PVP (4%)	6 mo	1.5	60	0%[b]		1.5	65	10–25%[b]						
Toxin B	PVP (4%) Sucrose (1%)	6 mo	1.5	65	0%[b]						1.5	70	2–20%[b]		
	Mannitol (4%) PVP (4%)	6 mo	1.5	78	0%[b]		1.5	85	5–30%[b]						
Monoclonal antibody	Sucrose (8%)	6 mo	0.3–0.5	100	100	98					0.3–0.5	100	99	98	Colandene, 2007 (58)

[a]% of aggregates
[b]Losses of protein activity during storage
Abbreviations: FD, freeze-drying; PVP, polyvinylpyrrolidone; β-GA, β-galactosidase; LDH, lactate dehydrogenase; rFVIII, recombinant factor VIII.

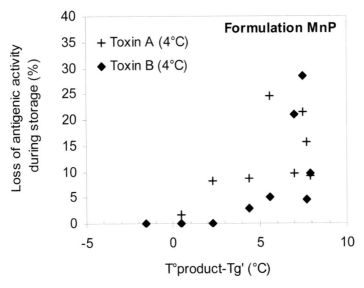

FIGURE 8 Effect of the product temperature reached during primary drying on the loss of antigenic activity of both toxins that were freeze-dried in a formulation including a crystalline bulking agent after six months of storage at 4°C. *Source*: From Ref. 33. *Abbreviations*: T_g', glass transition temperature of the freeze-concentrated phase determined by DSC ($T_g' = -30°C$ and Tcoll $= -14°C$ for the formulation); T°product, product temperature during primary drying; MnP, 4% mannitol, 1% PVP.

undergoes viscous flow (visually undetected) and could result in increased protein instability, and thus, affect the protein shelf life.

The shelf life of freeze-dried proteins can thus be affected by the thermal history of the product during the freeze-drying process. Performing primary drying below the T_{coll} will make it possible to maintain the physical stability of the freeze-dried product (i.e., the structure of the cake). When the T_{coll} is close to T_g', it can also be a good indication for protein activity preservation even after storage. However, when the T_{coll} is higher than T_g', which is the case when the formulation includes a crystalline bulking agent, maintaining the product temperature below the glass transition temperature of the freeze-concentrated phase (T_g') could be necessary to enhance the activity recovery of the protein after storage.

Optimization of the freeze-drying cycle requires the knowledge of the T_{max} value, the maximum allowable product temperature not to be exceeded during the process and especially during the primary and secondary drying steps. When relating T_{max} to the freeze-dried protein stability, T_{max} can be considered as a product quality indicator. When the glass transition temperature is defined as T_{max}, the evolution of the glass transition temperature of the formulation during the freeze-drying process can be modeled by using the Gordon–Taylor equation and integrated into a simple model of the process. Tréléa et al. (63) developed software based on a simple one-dimensional dynamic model of heat and mass transfer that includes the glass transition temperature as product quality criteria. Figure 9 illustrates the evolution of the product temperature and

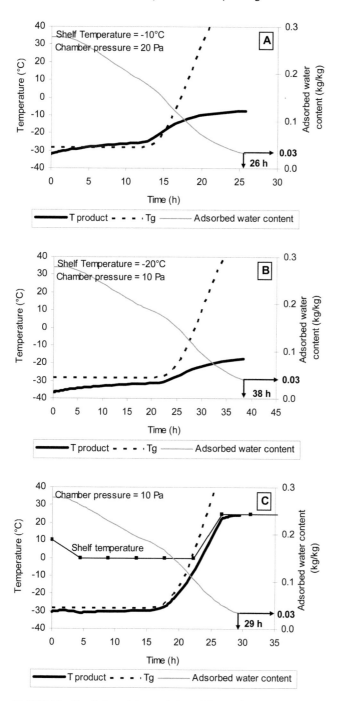

FIGURE 9 Simulation of the evolution of the product temperature ($T_{product}$), the glass transition temperature (T_g), and the adsorbed water content of a formulation including 4% PVP and 1% sucrose during the drying steps of the freeze-drying process for various profiles of shelf temperature and chamber pressure. The simulation was obtained by using the model developed by Tréléa et al. (63).

the glass transition temperature of a formulation including 4% PVP and 1% sucrose for different profiles of shelf temperature and chamber pressure. When applying a chamber pressure of 20 Pa and a shelf temperature of $-10°C$ during primary drying (Fig. 9A), the product temperature exceeds the glass transition temperature after five hours of sublimation, resulting in the collapse of the product. When decreasing the shelf temperature and the chamber pressure (Fig. 9B), the product temperature remains below the glass transition temperature throughout the process and a drying time of 38 hours is obtained. The drying time can be further reduced by manipulating the shelf temperature profile (Fig. 9C). The shelf temperature must be decreased over time because the self-cooling effect due to ice sublimation decreases when the mass transfer resistance through the dry layer increases with time. During the secondary drying, the moisture content of the whole product decreases, the glass transition temperature increases throughout, and the shelf temperature can be significantly increased up to a limit imposed by the thermal sensitivity of the biological product.

CONCLUSION

In this chapter we have demonstrated that DSC and FDM are powerful analytical tools that must be used right from the beginning of the development phase of a new freeze-dried protein formulation and not be limited to process development. The composition of the formulation is a key element for optimizing both the product quality and the process. Among the problems of product quality heterogeneity, the nonconformity of the visual aspect of the freeze-dried cake is a frequent trouble observed not only within the same batch but also among different batches. These problems could be avoided by a deep and relevant physical characterization of the formulation and an adequate choice of excipients.

Formulations involving crystalline bulking agent such as mannitol or glycine are widely used to improve the visual appearance of the freeze-dried cake. Their phase and state behavior during freezing and drying are complex and highly dependent on the extent of the solute crystallization that, in turn, is influenced by the composition and the thermal history of the formulation. The freezing step and the stochastic nature of ice nucleation, in particular, generate a lot of heterogeneity in the freezing and drying behavior of the vials. These heterogeneities will result in heterogeneous extents of solute crystallization, heterogeneous values of T_{coll}, and, therefore, in large variability of the product quality (involving the visual aspect of the freeze-dried product and eventually the protein stability). When a crystalline bulking agent is used, the development of a robust formulation, that means a formulation displaying a physical behavior independent of the freezing configuration, needs the quantification of the impact on the extent of solute crystallization of different factors: (*i*) each component of the formulation, (*ii*) the freezing conditions, and (*iii*) small changes of formulation composition.

Using an amorphous bulking agent can be considered as an interesting alternative when developing formulation. The physical behavior of a formulation involving only amorphous components is easy to describe and appears independent of the thermal history. In this case, the challenge is to combine disaccharides with low value of T_{coll} with polymers or polysaccharides to increase the value of the maximum allowable product temperature (T_{max}) not to be exceeding during the process without compromising the protein stability. Small changes in the salt content in the purified and concentrated proteins can lead to significant increase or decrease of the T_{coll}. Combining the determination of the T_{coll} of the formulation before freeze-drying and the use of a simple simulation tool for predicting the drying kinetics, the process variables of the primary and secondary drying steps can be adapted to optimize both the process and the product quality.

Even if few data concerning the effect of the product thermal history during the process on protein stability are today available, this parameter should not be ignored and, furthermore, its effect should be quantified during the development step when the quality management system is moved toward the quality-by-design approach.

ABBREVIATIONS

DSC: differential scanning calorimetry
FDM: freeze-drying calorimetry
T_g': glass transition temperature of the maximally freeze-concentrated phase
T_{eut}: eutectic melting temperature of a solute
T_{cr}: crystallization temperature of a solute
T_{coll}: collapse temperature (determined by FDM)
T_{max}: maximum allowable product temperature

REFERENCES

1. Carpenter JF, Prestrelski SJ, Arakawa T. Separation of freezing- and drying-induced denaturation of lyophilized proteins using stress-specific stabilization. Arch Biochem Biophys 1993; 303(2):456–464.
2. Lai MC, Topp EM. Solid-state chemical stability of proteins and peptides. J Pharm Sci 1999; 88(5):489–500.
3. Prestrelski SJ, Arakawa T, Carpenter JF. Separation of freezing- and drying-induced denaturation of lyophilized proteins using stress-specific stabilization: II. Structural studies using infrared spectroscopy. Arch Biochem Biophys 1993; 303(2):465–473.
4. Prestrelski SJ, Tedeschi N, Arakawa T, et al. Dehydration-induced conformational transitions in proteins and their inhibition by stabilizers. Biophys J 1993; 65(2):661–671.
5. Carpenter JF, Chang BS, Garzon-Rodriguez W, et al. Rational design of stable lyophilized protein formulations: theory and practice. Pharm Biotechnol 2002; 13: 109–133.
6. Passot S, Fonseca F, Alarcon-Lorca M, et al. Physical characterisation of formulations for the development of two stable freeze-dried proteins during both dried and liquid storage. Eur J Pharm Biopharm 2005; 60:335–348.
7. Bellows RJ, King CJ. Freeze-drying of aqueous solutions maximum allowable. Cryobiology 1972; 9:559–561.
8. Bellows RJ, King CJ. Product collapse during freeze-drying of liquid foods. AICHE Symp Ser 1973; 69(132):33–41.
9. MacKenzie AP. Collapse during freeze-drying—qualitative and quantitative aspects. In: Goldblith SA, Rey L, Rothmayr WW, eds. Freeze-Drying and Advanced Food Technology. New York: Academic Press, 1974:277–307.
10. Chang B, Randall C. Use a subambient thermal analysis to optimize protein-lyophilization. Cryobiology 1992; 29:632–656.
11. Her LM, Nail SL. Measurement of glass transition temperatures of freeze-concentrated solutes by differential scanning calorimetry. Pharm Res 1994; 11(1): 54–59.

12. Ma X, Wang W, Bouffard R, et al. Characterization of murine monoclonal antibody to tumor necrosis factor (TNF-MAb) formulation for freeze-drying cycle development. Pharm Res 2001; 18(2):196–202.

13. Nail SL, Her LM, Proffitt CPB, et al. An improved microscope stage for direct observation of freezing and freeze drying. Pharm Res 1994; 11(8):1098–1100.

14. Pikal MJ, Shah S. The collapse temperature in freeze drying: dependence on measurement methodology and rate of water removal from the glassy state. Int J Pharm 1990; 62:165–186.

15. Kochs M, Korber C, Nunner B, et al. The influence of the freezing process on vapour transport during sublimation in vacuum-freeze-drying. Int J Heat Mass Transf 1991; 34(9):2395–2408.

16. Pikal MJ, Shah S, Senior D, et al. Physical chemistry of freeze-drying: measurement of sublimation rates for frozen aqueous solutions by a microbalance technique. J Pharm Sci 1983; 72(6):635–650.

17. Sarciaux J, Mansour S, Hageman MJ, et al. Effects of buffer composition and processing conditions on aggregation of bovine IgG during freeze-drying. J Pharm Sci 1999; 88(12):1354–1361.

18. Searles JA, Carpenter JF, Randolph TW. The ice nucleation temperature determines the primary drying rate of lyophilization for samples frozen on a temperature-controlled shelf. J Pharm Sci 2001; 90(7):860–871.

19. Fonseca F, Passot S, Cunin O, et al. Collapse temperature of freeze-dried lactobacillus bulgaricus suspensions and protective media. Biotechnol Prog 2004; 20:229–238.

20. Sun WQ. Temperature and viscosity for structural collapse and crystallization of amorphous carbohydrate solutions. Cryo Letters 1997; 18:99–106.

21. Allison SD, Chang B, Randolph TW, et al. Hydrogen bonding between sugar and protein is responsible for inhibition of dehydration-induced protein unfolding. Arch Biochem Biophys 1999; 365(2):289–298.

22. Carpenter JF, Crowe JH. An infrared spectroscopic study of the interactions of carbohydrates with dried proteins. Biochemistry 1989; 28(9):3916–3922.

23. Izutsu K, Kojima S. Excipient crystallinity and its protein-structure-stabilizing effect during freeze-drying. J Pharm Pharmacol 2002; 54:1033–1039.

24. Izutsu K, Yoshioka S, Terao T. Decreased protein-stabilizing effects of cryoprotectants due to crystallization. Pharm Res 1993; 10(8):1232–1237.

25. Izutsu K, Yoshioka S, Terao T. Effect of mannitol crystallinity on the stabilization of enzymes during freeze-drying. Chem Pharm Bull (Tokyo) 1994; 42(1):5–8.

26. Kasraian K, Spitznagel TM, Juneau JA, et al. Characterization of the sucrose/glycine/water system by differential scanning calorimetry and freeze-drying microscopy. Pharm Dev Technol 1998; 3(2):233–239.

27. Lueckel B, Bodmer D, Helk B, et al. Formulations of sugars with amino acids or mannitol—influence of concentration ratio on the properties of the freeze-concentrate and the lyophilizate. Pharm Dev Technol 1998; 3(3):325–336.

28. Chatterjee K, Shalaev EY, Suryanarayanan R. Partially crystalline systems in lyophilization. I. Use of ternary state diagrams to determine extent of crystallization of bulking agent. J Pharm Sci 2005; 94(4):798–808.

29. Izutsu K, Rimando A, Aoyagi N, et al. Effect of sodium tetraborate (borax) on the thermal properties of frozen aqueous sugar and polyol solutions. Chem Pharm Bull (Tokyo) 2003; 51(6):663–666.

30. Liao X, Krishnamurthy R, Suryanarayanan R. Influence of the active pharmaceutical ingredient concentration on the physical state of mannitol—implication in freeze-drying. Pharm Res 2005; 22(11):1978–1985.

31. Liao X, Krishnamurthy R, Suryanarayanan R. Influence of processing conditions on the physical state of mannitol—implications in freeze-drying. Pharm Res 2006; 24(2):370–376.

32. Hawe A, Friess W. Impact of freezing procedure and annealing on the physicochemical properties and the formation of mannitol hydrate in mannitol-sucrose-NaCl formulations. Eur J Pharm Biopharm 2006; 64(3):316–325.

33. Passot S, Fonseca F, Barbouche N, et al. Effect of product temperature during primary drying on the long-term stability of lyophilized proteins. Pharm Dev Technol 2007; 12(6):543–553.
34. Cavatur RK, Vermuri NM, Pyne A, et al. Crystallization behavior of mannitol in frozen aqueous solutions. Pharm Res 2002; 19(6):894–900.
35. Shalaev EY, Kanev AN. Study of the solid-liquid state diagram of the water-glycine-sucrose system. Cryobiology 1994; 31:374–382.
36. Pikal MJ. Use of a laboratory data in freeze drying process design: heat and mass transfer coefficients and the computer simulation of freeze drying. J Parenter Sci Technol 1985; 39(3):115–138.
37. Blond G, Simatos D, Catté M, et al. Modeling of the water-sucrose state diagram below 0°C. Carbohyd Res 1997; 298:139–145.
38. MacKenzie AP. Collapse during freeze-drying—qualitative and quantitative aspects. In: Goldblith SA, Rey L, Rothmayr WW, eds. Freeze-drying and Advanced Food Technology. New York: Academic Press, 1974, 277–307.
39. Heller MC, Carpenter JF, Randolph TW. Manipulation of lyophilization-induced phase separation: implications for pharmaceutical proteins. Biotechnol Prog 1997; 13(5):590–596.
40. Her L, Deras M, Nail SL. Electrolyte-induced changes in glass transition temperatures of freeze-concentrated solutes. Pharm Res 1995; 12(5):768–772.
41. Izutsu K, Heller MC, Randolph TW, et al. Effect of salts and sugars on phase separation of polyvinylpyrrolidone-dextran solutions induced by freeze-concentration. J Chem Soc Faraday Trans 1998; 94(3):411–417.
42. Izutsu K, Kojima S. Freeze-concentration separates proteins and polymer excipients into different amorphous phases. Pharm Res 2000; 17(10):1316–1322.
43. Izutsu K, Yoshioka S, Kojima S, et al. Effects of sugars and polymers on crystallization of poly(ethylene glycol) in frozen solutions: phase separation between incompatible polymers. Pharm Res 1996; 13(9):1393–1400.
44. Heller MC, Carpenter JF, Randolph TW. Effects of phase separating systems on lyophilized hemoglobin. J Pharm Sci 1996; 85(12):1358–1362.
45. Randolph TW. Phase separation of excipients during lyophilization: effects on protein stability. J Pharm Sci 1997; 86(11):1198–1203.
46. Fonseca F, Obert JP, Béal C, et al. State diagrams and sorption isotherms of bacterial suspensions and fermented medium. Thermochim Acta 2001; 366:167–182.
47. Allison SD, Manning MC, Randolph TW, et al. Optimization of storage stability of lyophilized actin using combinations of disaccharides and dextran. J Pharm Sci 2000; 89(2):199–214.
48. Larsen SS. Studies on stability of drugs in frozen systems VI: the effect of freezing upon pH for buffered aqueous solutions. Arch Pharm Chem Sci Ed 1973; 1:41–53.
49. Murase N, Franks F. Salt precipitation during the freeze-concentration of phosphate buffer solutions. Biophys Chem 1989; 34:289–300.
50. Shalaev EY, Johnson-Elton TD, Chang B, et al. Thermophysical properties of pharmaceutically compatible buffers at sub-zero temperatures: implications for freeze-drying. Pharm Res 2002; 19(2):195–201.
51. Kim AI, Akers MJ, Nail SL. The physical state of mannitol after freeze-drying: effects of mannitol concentration, freezing rate and a noncrystallizing cosolute. J Pharm Sci 1998; 87(8):931–935.
52. Pyne A, Chatterjee K, Suryanarayanan R. Solute crystallization in mannitol-glycine systems—implications on protein stabilization in freeze-dried formulations. J Pharm Sci 2003; 92(11):2272–2283.
53. Chang BS, Fischer NL. Development of an efficient single-step freeze-drying cycle for protein formulations. Pharm Res 1995; 12(6):831–837.
54. Chongprasert S, Knopp SA, Nail SL. Characterization of frozen solutions of glycine. J Pharm Sci 2001; 90(11):1720–1728.
55. Yoshinari T, Forbes RT, York P, et al. Crystallisation of amorphous mannitol is retarded using boric acid. Int J Pharm 2003; 258(1–2):109–120.

56. Izutsu K, Ocheda SO, Aoyagi N, et al. Effects of sodium tetraborate and boric acid on nonisothermal mannitol crystallization in frozen solutions and freeze-dried solids. Int J Pharm 2004; 273(1-2):85–93.
57. Fonseca F, Passot S, Lieben P, et al. Collapse temperature of bacterial suspensions: the effect of cell type and concentration. Cryo Letters 2004; 25(6):425–434.
58. Colandene JD, Maldonado LM, Creagh AT, et al. Lyophilization cycle development for a high-concentration monoclonal antibody formulation lacking a crystalline bulking agent. J Pharm Sci 2007; 96(6):1598–1608.
59. Jiang S, Nail SL. Effect of process conditions on recovery of protein activity after freezing and freeze-drying. Eur J Pharm Biopharm 1998; 45(3):249–257.
60. Lueckel B, Helk B, Bodmer D, et al. Effects of formulation and process variables on the aggregation of freeze-dried interleukin-6 (IL-6) after lyophilization and on storage. Pharm Dev Technol 1998; 3(3):337–346.
61. Wang DQ, Hey JM, Nail SL. Effect of collapse on the stability of freeze-dried recombinant factor VIII and alpha-amylase. J Pharm Sci 2004; 93(5):1253–1263.
62. Chatterjee K, Shalaev EY, Suryanarayanan R. Partially crystalline systems in lyophilization. II. Withstanding collapse at high primary drying temperatures and impact on protein activity recovery. J Pharm Sci 2005; 94(4):809–820.
63. Tréléa IC, Passot S, Fonseca F, et al. An interactive tool for the optimization of freeze-drying cycles based on quality criteria. Dry Technol 2007; 25(4–6):741–751.

7 Freezing- and Drying-Induced Perturbations of Protein Structure and Mechanisms of Protein Protection by Stabilizing Additives

John F. Carpenter and Ken-ichi Izutsu
University of Colorado Health Sciences Center, Denver, Colorado, U.S.A.

Theodore W. Randolph
University of Colorado, Boulder, Colorado, U.S.A.

INTRODUCTION

There are numerous unique, critical applications for proteins in human healthcare (1–3). However, even the most promising and effective protein therapeutic will not be of benefit if its stability cannot be maintained during packaging, shipping, long-term storage, and administration. For ease of preparation and cost containment by the manufacturer and ease of handling by the end user, an aqueous protein solution often is the preferred formulation. However, proteins are readily denatured (often irreversibly) by the numerous stresses arising in solution, for example, heating, agitation, freezing, pH changes, and exposure to interfaces or denaturants (4). The result is usually inactive protein molecules and aggregates, which compromise clinical efficacy and increase the risk of adverse side effects (5). Even if its physical stability is maintained, a protein can be degraded by chemical reactions (e.g., hydrolysis and deamidation), many of which are mediated by water. Thus, inherent protein instability and/or the logistics of product handling often precludes development of aqueous, liquid formulations (6,7). Also, simply preparing stable frozen products, which is relatively straightforward, is not a practical alternative because the requisite shipping and storage conditions are not technically and/or economically feasible in many markets.

The practical solution to the protein stability dilemma is to remove the water. Lyophilization (freeze-drying) is most commonly used to prepare dehydrated proteins, which, theoretically, should have the desired long-term stability at ambient temperatures. However, as will be described in this review, recent infrared spectroscopic studies have documented that the acute freezing and dehydration stresses of lyophilization can induce protein unfolding (8–11). Unfolding not only can lead to irreversible protein denaturation, even if the sample is rehydrated immediately, but can also reduce storage stability in the dried solid (12,13).

Moreover, simply obtaining a native protein in samples rehydrated immediately after lyophilization is not necessarily indicative of adequate stabilization during freeze-drying or predictive of storage stability. Many proteins unfold during lyophilization but readily refold if rehydrated immediately (8,11,14). Without directly examining the structure in the dried solid, it is not possible to know whether an unfolded protein with poor storage stability is present or not.

"This chapter is a direct repeat of the text that appeared in *Freeze-Drying/Lyophilization of Pharmaceutical and Biological Products, Second Edition, Revised and Expanded* (Rey L and May J, eds.) 2004, Marcel Dekker, Inc., New York."

To develop a protein formulation that minimizes protein unfolding during freezing and drying, it is crucial that the specific conditions (e.g., pH and specific stabilizing ligands) for optimum protein stability be established *and* the appropriate nonspecific stabilizing additives (i.e., those excipients that generally stabilize any protein) be incorporated into the formulation. Other physical factors—the glass transition temperature and the residual moisture of the dried solid—must also be optimized to assure storage stability in the dried solid (reviewed in Ref. 15). These aspects of developing a lyophilized protein formulation will not be considered here because they are addressed in other chapters in this volume as are the interplay between formulation, lyophilization cycle design, cake structure, and long-term stability of proteins (15). Here we will describe how to design formulations that protect proteins during both freezing and drying and the mechanisms by which additives stabilize proteins and, also importantly, fail to do so. In addition, we will give an overview of the use of infrared spectroscopy to directly monitor protein conformation in frozen and dried samples. This structural information is crucial for the rational development of stable, lyophilized protein formulations.

We wish to emphasize that the principles and mechanisms to be discussed should be generally applicable to any protein. However, each protein has unique physicochemical characteristics, which often manifest themselves as specific routes of chemical and physical degradation during storage. Although we will not address chemical degradation directly in this chapter (see earlier works in Refs. 15–17), it is important to realize that minimizing unfolding during freezing and drying can reduce such degradation during lyophilization and subsequent storage (13). Currently, it is not possible to predict if degradation of a given protein will be inhibited by simply designing a formulation to maintain native structure, nor is it clear as to why the efficacy of "general" protein stabilizers often varies depending on the protein being studied. Thus, there is a great need to increase the fundamental understanding of the mechanisms by which protein stabilizers act and to document, by case studies, the applicability of the general rules to individual proteins. With sufficient effort by academic and industrial researchers, this can be an iterative process in which progress can be made toward developing a general strategy for protein formulation that can be rationally modified for the successful lyophilization of each new protein product.

PROTEIN STABILIZATION DURING LYOPHILIZATION/REHYDRATION

Much of the early research on protein stabilization during lyophilization was with labile enzymes, which were found to be irreversibly inactivated, presumably due to aggregation of nonnative molecules, to varying degrees after rehydration. As such, attempts at improving the recovery of activity were focused on the entire process of lyophilization and rehydration. It was not known at what point(s) during the process the damage arose and the stabilizers were operative. Also, usually these studies tested the capacity of nonspecific stabilizers (i.e., those that will generally protect any protein) to prevent irreversible protein denaturation (i.e., aggregation) and inactivation. However, for practical purposes, the first step in increasing the resistance of a given protein to lyophilization-induced damage is to choose the specific conditions that provide the greatest stability to that protein. In general, any factor that alters the free energy of unfolding in solution will tend to have the same qualitative effect during lyophilization. For example, the stability of many enzymes during freeze-thawing is altered by the presence of substrates, cofactors, and/or allosteric modifiers (18). Even for nonenzyme proteins, specific ligands can be important components of the formulation. For example, the stability of fibroblast growth factors is greatly increased in the presence of heparin or other polyanionic ligands (reviewed in Ref. 19). The *pH* and specific ligands that confer optimum stability often are known from purification protocols, preformulation studies, and/or earlier efforts at designing a liquid formulation.

However, most proteins are not adequately stabilized by specific solution conditions. Of the nonspecific stabilizers that have been tested, sugars have been shown to stabilize the most proteins during lyophilization and have been known to have this property for the longest time. To our knowledge, the first published report is the 1935 paper by Brosteaux and Eriksson-Quensel (20) in which they described the protection during dehydration/rehydration of several proteins by sucrose, glucose, and lactose. Subsequent detailed comparisons of sugars documented that usually disaccharides provide the greatest stabilization (4,8,21,22). For protection during the lyophilization cycle itself, both reducing and nonreducing disaccharides are effective. However, reducing sugars (e.g., lactose and maltose) can degrade proteins during storage via the Maillard reaction (protein browning), a process that can be accelerated at intermediate residual moisture contents (22,23). Therefore, the choice of disaccharides is essentially limited to the nonreducing sugars, sucrose and trehalose. Since, as of early 1998, trehalose has not been used in any Food and Drug Administration (FDA) approved parenteral product, sucrose is usually the first choice for commercial protein drug formulations.

Although the data are much more limited, polyvinylpyrrolidone (PVP) and bovine serum albumin (BSA) have also been shown to protect a few tetrameric enzymes, that is, asparaginase, lactate dehydrogenase (LDH), and phosphofructokinase (PFK), during lyophilization and rehydration (24,25). Another class of compounds that has been found to be useful in freeze-dried formulations are nonionic surfactants. For example, sucrose fatty acid monoester, 3-[(3-cholamidopropyl)-dimethylammonio]-1-propa-nesulfonate (CHAPS), and Tweens have been found to increase recovery of β-galactosidase activity after freeze-drying and rehydration (26). Various surfactants have been shown to protect LDH during freeze-drying and rehydration (27). Hydroxypropyl-β-cyclodextrin, which is surface active (28,29), inhibited the inactivation of recombinant tumor necrosis factor (30), interleukin-2 (31,32), and LDH (27) during freeze-drying/rehydration.

MECHANISMS OF STABILIZATION OF PROTEINS BY SUGARS DURING DEHYDRATION

Most protein pharmaceuticals are multicomponent systems that contain protein, buffer salts, bulking agents, and stabilizers. Each component has its intended role in the formulation. For example, often a crystallizing excipient (e.g., mannitol or glycine) is chosen as a bulking agent (15). In contrast, numerous studies have documented that stabilization of a protein during dehydration requires the presence of a compound that remains at least partially amorphous. When a protein formulation is frozen, the protein partitions into the non-ice phase with other amorphous components. The interaction between the protein and these amorphous components must be maintained during the entire freeze-drying process to assure recovery of a native protein in the dried solid and after rehydration (8,9,33–36).

Although most carbohydrates used for protein formulations remain amorphous in frozen solutions and during drying (e.g., sucrose and trehalose), some exhibit eutectic phase separation from frozen solutions (34–39). For example, mannitol readily crystallizes during freeze-drying, but the degree of crystallization can be manipulated by altering processing conditions and formulation components (34–39). In the concentration range where it remains mostly amorphous, mannitol has been shown to protect enzymes during freeze-drying in a concentration-dependent manner (35,36). A relatively high mass ratio of protein to mannitol will serve to inhibit mannitol crystallization, whereas with excess mannitol, crystallization and loss of stabilization arise. Similarly, substantial stabilization has been achieved with solutes (including buffer salts) that alone crystallize but in combination interfere with each other's crystallization. For example, Izutsu et al. (35) found that with a sufficiently high ratio of potassium phosphate to mannitol, mannitol remained amorphous and protected LDH during freeze-drying. However, when there was excess mannitol, its crystallization obviated protein protection. Similarly, Pikal et al. (40) found that appropriate ratio of mannitol and glycine resulted in a sufficiently large amorphous fraction to protect freeze-dried human growth hormone.

Although it is well established that an amorphous excipient is needed to protect proteins during dehydration, the nature of the protective interaction of amorphous solutes with the protein in the dried solid has been a subject of controversy in the literature. There are at least two nonexclusive mechanisms proposed. Before describing these mechanisms, we wish to emphasize that neither mechanism alone is sufficient to fully explain stabilization during lyophilization. Both mechanisms focus only on the effect of stabilizers during the terminal stress of dehydration and essentially ignore the freezing step. As documented below, no matter what the nature of the interaction of the additive with the dried protein, the most important factor is that the additive(s) prevent unfolding during *both* freezing and dehydration.

Proponents of one mechanism state that proteins are simply mechanically immobilized in the glassy, solid matrix during dehydration (41). The restriction of translational and relaxation processes is thought to prevent protein unfolding, and spatial separation between protein molecules (i.e., "dilution" of protein molecules within the glassy matrix) is proposed to prevent aggregation. Although it is clear that protective additives must partition with the protein into the amorphous phase of the dried sample, simply forming a glassy solid does not assure protein stabilization. First, if all that were needed to prevent damage to a protein is the formation of a glass, then the protein by itself should be stable. Clearly, this is not the case because proteins themselves should form an amorphous phase in the dried solid (42); however, most unprotected proteins are denatured during lyophilization (8–14). In some cases adding another protein (e.g., BSA), which should simply add to the mass of the final protein glass, confers protection (25).

One might further qualify the mechanism by proposing that the requisite mechanical restriction to unfolding and aggregation can only be achieved if another amorphous compound is present to provide immobilization and spatial separation of the protein drug molecules. However, then the question becomes what amount of additive is sufficient to provide the desired physical properties of the dried solid, which are not achieved with the protein alone? This question has not been answered or addressed in the literature. However, it is expected that, in general, the capacity of an additive to protect a protein specifically during dehydration should depend on the final additive protein mass ratio. Increasing this ratio will favor spatial separation and immobilization of the protein within the glassy matrix. Also, the mass ratios between all compounds in the dried solid affect the influence of the compounds on each other's crystallization (e.g., with glycine and mannitol).

Several studies have shown that formation of a glassy phase by an additive, even when it is used in large excess relative to the protein, is not a sufficient condition for acute stabilization of proteins during lyophilization. For example, formulations of 100 mg/mL interleukin-l receptor antagonist, prepared with sucrose concentrations ranging from 0% to 10% (wt/vol), formed a glass during lyophilization and all had glass transition temperatures of $66 \pm 2°C$ (13). Yet only in formulations containing $\geq 5\%$ sucrose was lyophilization-induced unfolding prevented. Tanaka et al. (43) have found that the capacity of carbohydrates to protect freeze-dried catalase decreased with increased carbohydrate molecular weight. Dextrans were the largest and least effective of all of the carbohydrates tested, and the larger the dextran molecule the less it stabilized catalase. Although they did not determine whether their dried samples were amorphous, it is well known that as the molecular weight of the carbohydrate is increased, the glassy state is formed more readily (44–46). In addition, more recent studies have shown (T. Randolph, M. Zhang, S. Prestrelski, T. Arakawa, and J. Carpenter, unpublished data) that PFK was not protected, and LDH was inactivated further, by dextran during freeze-drying and rehydration. Differential scanning calorimetry documented that the dried samples were amorphous. The potential mechanistic bases for these observations will be described below. For now, it is important to stress the conclusion that it is necessary for stabilizing additives to remain amorphous to protect proteins during lyophilization, but glass formation alone appears not to be sufficient for stabilization of proteins against the severe stress of dehydration.

There are several studies supporting the other mechanism, which is often referred to as the water replacement hypothesis. According to this hypothesis, sugars protect labile proteins during drying by hydrogen bonding to polar and charged groups as water is removed, thus preventing drying-induced denaturation of the protein. For example, in early studies, using solid-state Fourier transform infrared (FTIR) spectroscopy, it was found that the band at 1583 cm^{-1} in the spectrum for lysozyme, which is due to hydrogen bonding of water to carboxylate groups, is not present in the spectrum for the dried protein (33). When lysozyme is dried in the presence of trehalose or lactose, the carboxylate band is retained in the dried sample, indicating that the sugar is hydrogen-bonding in the place of water. Similar results have been obtained with α-lactalbumin and sucrose (8). More recently, it has been documented that the carboxylate band can be titrated back by freeze-drying lysozyme in the presence of increasing concentrations of either trehalose or sucrose (S. Allison and J. Carpenter, unpublished observations). This effect correlates directly with an increased inhibition of protein unfolding in the presence of increasing amounts of sugar.

Three other recent studies on enzyme preservation provide further support for the water replacement mechanism. Tanaka et al. (43) have found that the capacity of a saccharide to protect catalase during freeze-drying is inversely related to the size of the saccharide molecule. They suggest that as the size of the saccharide increases, steric hindrance interferes with hydrogen bonding between the saccharide and the dried protein. In support of this contention, recent experiments have shown that the carboxylate band is only minimally detectable in the infrared spectrum of lysozyme freeze-dried in the presence of dextran (D. Barberi, T. Randolph, and J. Carpenter, unpublished observation). In addition, Tanaka et al. (43) found that the degree of stabilization was based on the saccharide sugar mass ratio, which is to be expected if protection is due to hydrogen bonding of the saccharide to the protein in the dried solid. More recently, by studying protein structure in the dried solid with FTIR spectroscopy, Prestrelski et al. (12) found that as the molecular weight of a carbohydrate additive was increased the capacity to inhibit unfolding of interleukin-2 during lyophilization decreased and the level of protein aggregation after rehydration increased. Also, it was clear that protection of the protein did not correlate directly with the formation of a glass (all samples were found to be amorphous) or with the glass transition temperature of the sample (the T_g increased as carbohydrate molecular weight increased). Rather, there was a negative correlation between stabilization and molecular weight, which is to be expected if protection during drying is due to the water replacement mechanism.

Some of the most compelling evidence for the water replacement hypothesis comes from studies on the effects of freeze-drying on a model poly-peptide, poly-L-lysine (8). This peptide assumes different conformations in solution, which have been well characterized with FTIR spectroscopy, depending on the pH and temperature. At neutral pH, poly-L-lysine exists as an unordered peptide. At pH 11.2, the peptide adopts an α-helical conformation. Poly-L-lysine assumes an intermolecular β-sheet conformation (11) in the dried state, regardless of its initial conformation in aqueous solution. The preference for β-sheet in the dried state appears to be a compensation for the loss of hydrogen bonding interactions with water. The β-sheet allows for the highest degree of hydrogen bonding in the dried sample. If poly-L-lysine is freeze-dried in the presence of sucrose, the original solution structure is retained in the dried state because sucrose hydrogen bonds in place of water, obviating the need to form β-sheet.

INFRARED SPECTROSCOPIC STUDIES OF
LYOPHILIZATION-INDUCED STRUCTURAL CHANGES

Until recently, the only way to assess the capacity of an additive to stabilize a protein during lyophilization was to measure activity and/or structural parameters after rehydration. To confound matters further, it was proposed in the protein chemistry literature that dehydration did not alter a protein's conformation (47). Such a claim was clearly counter to the known contributions of water to the formation of the native, folded protein (48,49). Also, it was difficult to reconcile the finding that proteins could be irreversibly inactivated and aggregated after rehydration with the contention that protein structure was not perturbed by dehydration.

Reconciliation of this apparent dilemma was provided by FTIR spectroscopy, which can be used to study protein secondary structure in any state (i.e., aqueous, frozen, dried, or even as an insoluble aggregate). FTIR spectroscopy has long been used for quantitation of protein secondary structure and for studies of stress-induced alterations in protein conformation (50–52). Structural information is obtained by analysis of the conformationally sensitive amide I

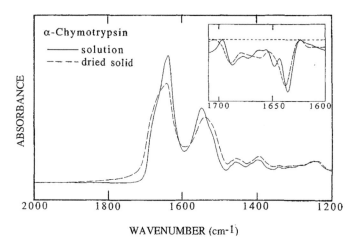

FIGURE 1 Comparison of infrared spectra of α-chymotrypsin in aqueous solution and dried solid state. The inset shows the second derivatives in the amide I region for the spectra in the main panel. *Source:* From Ref. 11.

band, which is located between 1600 and 1700 cm^{-1}. This band is due to the in-plane C=O stretching vibration, weakly coupled with C–N stretching and in-plane N–H bending (50,51,53). Each type of secondary structure (i.e., α-helix, β-turn, and disordered) gives rise to a different C=O stretching frequency (50–54) and, hence, has a characteristic band position, which is designated by wavenumber, cm^{-1}. Band positions are used to determine the secondary structural types present in a protein. The relative band areas (determined by curve fitting) can then be used to quantitate the relative amount of each structural component. Therefore, an analysis of the infrared bands in the amide I region can provide quantitative as well as qualitative information about protein secondary structure (50–54).

To obtain this detailed structural information, it is necessary to enhance the resolution of the protein amide I band, which usually appears as a single broad absorbance contour (Fig. 1). The widths of the overlapping component bands are often greater than the separation between the absorbance maxima of neighboring bands. Because the band overlapping is beyond instrumental resolution, several mathematical band-narrowing methods (i.e., resolution enhancement methods) have been developed to overcome this problem (11,50–52,54). For studies of lyophilization-induced structural transitions, calculation of the second-derivative spectrum is recommended (11). This method is completely objective and alterations in component bandwidths, which are due to protein unfolding, are preserved in the second-derivative spectrum.

For most unprotected proteins (i.e., lyophilized in the presence of only buffer), the second-derivative spectra for the dried solid are greatly altered relative to the respective spectra for the native proteins in aqueous solutions (8–14). For example, Figure 1 compares the original and second-derivative spectra for α-chymotrypsin in solution and in the dried solid. Second-derivative spectra for aqueous and dried lactalbumin and LDH, which are also greatly altered by lyophilization, and granulocyte colony-stimulating factor (GCSF), which is minimally perturbed, are shown in Figure 2. For dozens of proteins studied to date,

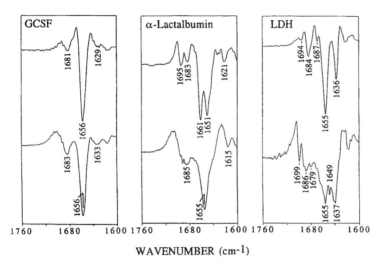

WAVENUMBER (cm⁻¹)

FIGURE 2 Second-derivative amide I spectra of granulocyte colony-stimulating factor (GCSF), α-lactalbumin, and lactate dehydrogenase (LDH) in aqueous solution (*upper spectra*) and dried solid (*lower spectra*) states. *Source*: From Ref. 11, employing data from Refs. 8 and 9.

lyophilization induces varying degrees of shifts in band positions, loss of bands, and broadening of bands.

The lyophilization-induced spectral alterations in the conformationally sensitive amide I region are due to protein unfolding and *not* simply to the loss of water from the protein. The intrinsic effects of water removal on the vibrational properties of the peptide bond, and hence protein infrared spectra, were found to be insignificant by Prestrelski et al. (8). If the direct vibrational effects of water removal were responsible for drying-induced spectral changes, then the infrared spectra of all proteins should be altered to the same degree in the dried solid, which is not the case.

Two different behaviors of proteins unfolded in the dried solid are displayed during rehydration: (*i*) The protein regains the native conformation upon rehydration (reversible unfolding), as observed for α-lactalbumin, lysozyme, chymotrypsinogen, ribonuclease, β-lactoglobulins A and B, α-chymotrypsin, and subtilisin (8,10,11,13,14,55,56). (*ii*) A significant fraction of the protein molecules aggregate upon rehydration (irreversible unfolding), as noted for LDH, PFK, interferon-γ, basic fibroblast growth factor, and interleukin-2 (8–12). It has been documented with several proteins in the latter class that prevention of aggregation and recovery of activity after rehydration correlate directly with retention of the native structure in the dried solid (8–12). Thus, the mechanism by which stabilizing additives (e.g., sugars) minimize loss of activity and aggregation during lyophilization and rehydration is to prevent unfolding during freezing and drying (8–12).

For example, the spectrum for interferon-γ dried in the presence of 1 M sucrose is similar to that for the native aqueous protein, whereas that for the protein dried alone is greatly altered (Fig. 3). For analysis of these data, a

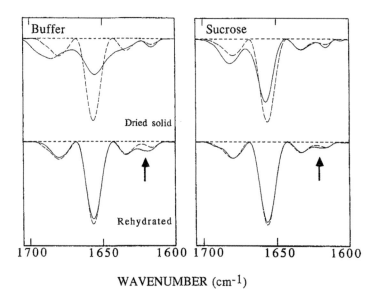

WAVENUMBER (cm⁻¹)

FIGURE 3 Comparisons of second-derivative spectra of interferon-γ in the dried solid and rehydrated states, with or without 1 M sucrose, with the spectrum of the native aqueous state. The spectrum of the native aqueous state is shown with the dashed line. The arrows indicate the band arising from nonnative intermolecular β-sheet. *Source*: From Ref. 11.

baseline was fitted to the second-derivative spectra and have been normalized for total area (11,57). This data presentation is useful because it allows visualization of the relative shifts of area from one component band to another, and, hence, the redistribution from native to nonnative secondary structural elements. For example, for the sample dried without sugar, there is a loss of α-helix as indicated by the decreased absorbance in the 1656 cm⁻¹ band, which is compensated by increased absorbance in bands for β-sheet and turns (∼1640–1645 and 1665–1695 cm⁻¹). These changes are attenuated when the protein is lyophilized in the presence of sucrose, documenting an increased retention of native structure in the molecular population.

After rehydration, the spectra of both samples are very native-like, indicating that the majority of nonnative molecules have refolded (Fig. 3). However, in the spectrum of the sample lyophilized without sucrose, the appearance of a new band near 1625 cm⁻¹, which is assignable to intermolecular β-sheet structure, and the decreased intensities in vibrational bands ascribed to α-helix (1656 cm⁻¹) and turn (1688–1665 cm⁻¹) structures, indicate the formation of protein aggregates upon rehydration (see Ref. 11 for a detailed review of the study of protein aggregation with infrared spectroscopy). In this sample, 18% of the protein formed insoluble aggregates. In contrast, in the sample lyophilized with sucrose, only 9% insoluble aggregate was noted after rehydration. This reduction in aggregation is reflected in a much weaker 1625 cm⁻¹ band in the spectrum of the rehydrated sample. In this case, 1 M sucrose does not provide complete protection during freeze-drying, presumably because it is inadequate at preserving the protein structure during the freezing step (see later).

Also, unfolding of proteins that refold if immediately rehydrated can be inhibited by stabilizing additives (8,10,12–14). It appears crucial that even these proteins should be stabilized against lyophilization-induced unfolding to maintain stability during long-term storage in the dried solid (12,13,15). Thus, an important criterion for a successful freeze-dried formulation of any protein is retention of the native protein structure in the dried solid, which can be readily documented with FTIR spectroscopy.

Although a qualitative visual comparison of second-derivative spectra can be useful to assess the influence of additives on protein structure during lyophilization, a quantitative comparison is often also desirable. For research on lyophilization-induced structural transitions, two approaches can be employed. Occasionally, there is a need to know the secondary structural content. The relative band areas can then be determined with curve fitting (11,50–52,54). For example, the percentage of intermolecular β-sheet can be used to calculate the percentage of aggregated protein in dried samples (11,14).

However, for the general assessment of protein stabilization needed to evaluate formulations, it is usually more meaningful to make an overall global comparison between two spectra. For this analysis, Prestrelski et al. (8,9) originally developed a mathematical procedure to calculate the spectral correlation coefficient (similarity) between two second-derivative spectra. More recent analysis indicated that this method can provide misleading information (57). If the spectra have offset baselines, then the correlation coefficient is much lower than that expected based on a visual assessment of spectral similarity (Fig. 4, top). In contrast, if the spectra are dominated by a large band of high symmetry, the value is too great (Fig. 4, middle). These shortcomings are avoided by simply normalizing the reference (e.g., aqueous native protein) second-derivative spectrum and that for the experimental sample (e.g., unfolded protein in the dried solid), and then determining the fractional area that the spectra share. The method of determining this area of overlap is described in detail in a paper by Kendrick et al. (57), and an example is presented in Figure 4 (bottom). It is important to note here that with some samples a visual impression is still important because band shifts, which are significant in terms of structure, may result in only a relatively small decrease in the area-of-overlap parameter. In such cases, the resolution between fully native and unfolded samples becomes so small that an incremental improvement in structure noted with a stabilizer is not resolved with the area-of-overlap analysis. Then, one must carefully make a qualitative assessment of the spectra to discern what stabilizer concentration and types afford the greatest retention of native-like features in the spectrum of the dried protein, for example, relative intensity and positions of component bands that are most different between spectra for native and unfolded states.

In addition to the quantitating effect of lyophilization on protein structure, we have also used area-of-overlap analysis to quantitate protein unfolding by guanidine HCl (Fig. 5). As can be seen in Figure 5, the changes in a protein's second-derivative spectrum induced by chemical denaturation are very similar to those noted after lyophilization. There are alterations in relative band absorbances, widths, and positions. Furthermore, the unfolding curve generated with infrared spectroscopic data is essentially identical to that based on circular dichroism spectroscopy (Fig. 6). These results further support the contention that FTIR spectroscopy coupled with area-of-overlap analysis can be used to quantitatively assess protein unfolding.

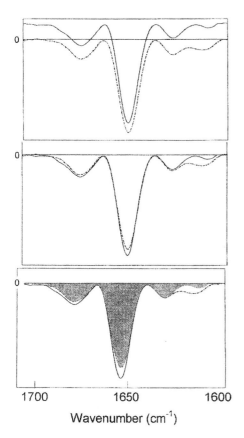

FIGURE 4 Second-derivative amide I spectra of interferon-γ: (*top*) uncorrected spectra, correlation coefficient = 0.80; (*middle*) baseline offset corrected, correlation coefficient = 0.995; (*bottom*) area-normalized spectra, area of overlap = 0.92. Solid lines indicated the native aqueous state, dashed lines indicate rehydrated aqueous state, and the gray fill (bottom panel) indicates the area of overlap. *Source*: From Ref. 57.

It is obvious from the infrared studies to date that the effect of a given additive may vary depending on the protein (8), the presence of other additives, and other specific solution conditions, for example, pH (12). Therefore, the structure of each dried protein in each formulation should be studied with FTIR spectroscopy. Unfortunately, this will not be possible with certain formulations. If albumin is used, then, as is the case with any physical measurement, it will not be possible to separate the albumin contribution to the data from that of the protein drug. If other compounds (e.g., PVP, arginine, histidine, and glycine) that absorb strongly in the amide I region are used in large excess relative to the protein, then they may interfere with the protein spectrum in the dried solid. However, if relatively low concentrations of such additives are used, it may be possible to substract quantitatively their specific absorbances from the protein spectra in both aqueous solutions and dried solids.

There should be few barriers to implementing infrared spectroscopic analysis, at least in industrial laboratories. The instrumentation is available commercially at relatively modest costs. Also, high-quality spectra can be acquired in less than five minutes and with minimal sample preparation. The main disadvantage of the technique is that a minimum protein concentration of 3 to 5 mg/mL is needed to obtain quality spectra of proteins in H_2O solutions. The absolute mass of protein needed is not great because usually less than 50 μL

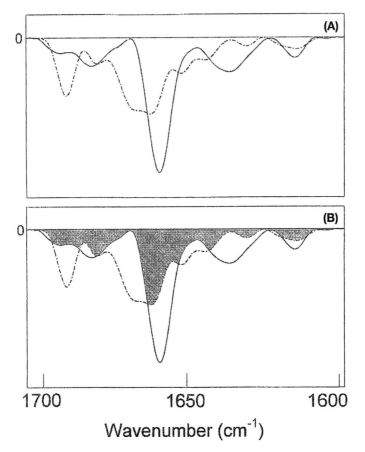

FIGURE 5 Second-derivative amide I spectra of wild-type iso-1-cytochrome c in aqueous solution in the presence and absence of guanidine HCl. (**A**) Baseline offset corrected and area normalized, correlation coefficient = 0.72. (**B**) Baseline offset corrected and area normalized, area of overlap = 0.63. Solid lines indicates zero guanidine HCl, dashed lines indicate 2.5 M guanidine HCl, and the gray fill (panel B) indicates the area of overlap. *Source*: From Ref. 57.

of solution is required to load the sample cell. If solubility is limited, then the protein can be studied at much lower concentrations (around 1 mg/mL) in D_2O. However, the researcher must then be aware of the potential difficulties of data interpretation due to the direct effects of H-D exchange on the vibrational frequencies of amide I component bands (11,56). In some cases, deuteration of the protein makes assignment of bands to different secondary structural types uncertain. This can be a problem if quantitation of secondary structural content is needed. However, if all that is required is a global comparison between a spectrum for an aqueous control sample and that for a freeze-dried protein, then a protein can be studied reliably in D_2O. The only caveat is that sufficient time for H-D exchange must be allowed prior to lyophilization, so that additional exchange does not arise during freezing and drying.

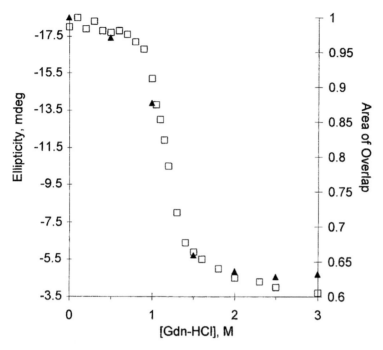

FIGURE 6 Guanidine HCl-induced unfolding of wild-type iso-1-cytochrome c. The unfolding transition is presented as a plot of ellipticity at 22 nm in millidegrees (*open squares*) and infrared spectral area of overlap (*full triangles*) as a function of guanidine HCl concentration. *Source*: From Ref. 57.

MECHANISM FOR STABILIZATION OF MULTIMERIC ENZYMES BY POLYMERS

Hellman et al. (24) found that PVP protected the tetrameric enzyme, L-asparaginase, during freeze-drying and rehydration. In addition, it has been shown that PVP and BSA stabilize tetrameric LDH during freeze-drying (25). Steric hindrance should minimize the ability of PVP or BSA to hydrogen bond effectively to the charged and polar groups on the dried protein's surface. Also, as has already been described, polymer-induced stabilization cannot be ascribed just to the formation of a glassy phase with the proteins during the dehydration step.

To understand further how these polymers protect multimeric enzymes during lyophilization, stress-induced alterations in quaternary structure, especially during freezing, must be taken into consideration. First, simply reducing temperature can foster dissociation of many multimeric proteins. The chilling lability is due to disruption of hydrophobic interactions at the monomer-monomer contact sites and/or an increase in enzyme protonation, and the pKa values of titratable histidines are increased during cooling (58). Second, Chilson et al. (18) demonstrated that LDH dissociates during freeze-thawing and that stabilizers inhibit dissociation. The recovery of enzyme activity correlated directly with the degree to which dissociation was inhibited. We have extended this work and studied the effects of PVP and BSA on LDH dissociation during freeze-thawing and freeze-drying (25). We found that the polymers protected

LDH during both treatments by inhibiting dissociation in the frozen state. Dissociation was not induced by freezing itself but rather during an extended residency time in the frozen state (e.g., due to thawing with slow warming or during the early stages of primary drying) at a relatively high subzero temperature (e.g., $-20°C$). The main factor causing dissociation appeared to be a decrease in the pH of the sodium phosphate buffer system from 7.5 at room temperature to 4.5 in the frozen solution at $-20°C$. Dissociation did not occur when buffers that did not acidify were employed (e.g., Tris and potassium phosphate). PVP and BSA protected the enzyme during lyophilization, at least in part, by inhibiting the reduction in pH in the frozen state. These experiments provide direct evidence that stabilization during freezing is essential for inhibiting protein damage during lyophilization. Further examples are presented below.

EVIDENCE FOR FREEZING-INDUCED UNFOLDING DURING LYOPHILIZATION

If a protein is not adequately protected during freezing, the protein will be unfolded in the final dried solid, no matter how effective the stabilization during the dehydration step (9,36). There is considerable evidence documenting freezing sensitivity of proteins. First, many proteins are irreversibly denatured by freeze-thawing (59). Since this damage is due primarily to freezing, similar damage should also arise during the freezing step of lyophilization. Second, the capacity of an additive to protect during freeze-drying is often directly related to its initial bulk concentration and not to the final mass ratio of additive to protein (60). For example, the data in Figure 7 show that the recovery of PFK activity after lyophilization and rehydration increases as the prefreeze concentration of trehalose increases, even though the same mass ratio of trehalose to protein was used for all the samples. As will be explained later, freezing protection by sugars is governed by initial concentration of the additive, whereas drying protection is related primarily to the mass ratio between the additive and protein.

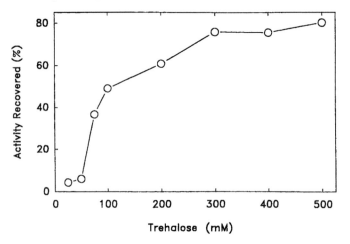

FIGURE 7 The effect of varying concentration of trehalose, while maintaining a constant sugar/protein mass ratio, on recovery of phosphofructokinase (PFK) activity after freeze-drying and rehydration. The protein concentration was adjusted concomitantly with the sugar concentration to maintain a constant sugar/protein mass ratio of 945. *Source*: Data taken from Ref. 59.

Third, Carpenter et al. (36) recently documented more rigorously that freezing-induced denaturation can play an important role in the overall damage to a protein during lyophilization. The impetus for this research was the observation that the disaccharide trehalose was effective at protecting labile enzymes, whereas the constituent monosaccharide, glucose, was not. For example, when PFK is freeze-dried in the presence of 0.2 to 0.4 M trehalose, over 60% of the initial activity is recovered after rehydration (21). In contrast, when similar amounts of glucose are used, the recovery is less than 5%. When considering only the effect of the sugar on the protein during dehydration, these results present a dilemma. This is because if hydrogen bonding of sugar to dried protein in place of water were all that was needed for stabilization, then mono- and disaccharides should provide similar protection. Glucose does not protect during freeze-drying because it provides minimal stabilization during freezing (based on freeze-thawing results), whereas similar concentrations of trehalose are effective at protecting the protein during both freezing and dehydration (21,61).

To examine the separate roles of protein damage and stabilization by freezing and dehydration, Carpenter et al. (36) developed a two-component system for stress-specific stabilization during lyophilization. In this stabilization scheme, polyethylene glycol (PEG) is used as a cryoprotectant and various carbohydrates can be used to protect during dehydration. PEG alone completely stabilizes either LDH or PFK during freeze-thawing. However, it provides little or no protection during dehydration because it crystallizes during lyophilization. When small amounts (e.g., 10–100 mM initial concentration) of trehalose or glucose are added, which alone at the concentrations tested are ineffective at protecting these enzymes during freeze-thawing or freeze-drying, excellent stabilization is noted during freeze-drying. Under conditions where cryoprotection is provided by PEG, glucose is almost as effective as trehalose in stabilizing dried enzymes (i.e., LDH and PFK).

In a complementary structural study of stress-specific stabilization using FTIR spectroscopy, Prestrelski et al. (9) found that the recovery of activity after rehydration correlated directly with the ability of the additives to preserve the native structure of the enzymes in the dried state. Full activity recovery and maintenance of essentially aqueous structure in dried samples were only noted when a combination of PEG and sugar was employed. On the basis of these results, Prestrelski et al. (9) have proposed a model of the conformational events during lyophilization and rehydration, which is shown in Figure 8. Briefly, this model proposes that to recover structure and function after rehydration, the native structure of labile proteins must be retained, both upon freezing and during subsequent dehydration. The appropriate cryoprotectant is required for the initial structural preservation, and a specific stabilizer against drying is needed for the terminal stress during lyophilization. In some instances (e.g., with disaccharides), a single additive can serve both protective functions.

More recently, there has been direct observation of protein structural perturbation in the frozen state using phosphorescence lifetime measurements (62). Reductions in this parameter indicated that freezing perturbed the tertiary structure (at a protein concentration of 3–5 μM) of azurin, ribonuclease, alcohol dehydrogenase, alkaline phosphatase, glyceraldehyde 3-phosphate dehydrogenase, and LDH. The cryoprotectants sucrose and glycerol were tested and found to inhibit the freezing-induced structural perturbations, with almost complete protection noted at a 1 M concentration.

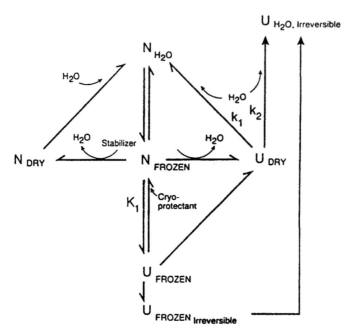

FIGURE 8 Schematic representation of model for conformational changes during freezing, drying, and rehydration. N, native; D, unfolded; K_1, conformation equilibrium upon freezing, which shifts toward the native state in the presence of a cryoprotectant; k_1, rate constant for refolding; k_2, rate constant for formation of irreversibly denatured (aggregated) forms. *Source*: From Ref. 9.

Finally, preliminary FTIR spectroscopic investigations have provided direct evidence for freezing-induced perturbation of protein secondary structure (B. Kendrick, A. Dong, L. Krielgard, and J. Carpenter, unpublished observations). For example, it was found that the structure of LDH was slightly perturbed in the frozen state and that the protein refolded upon thawing. In contrast, recombinant factor XIII dimer was irreversibly unfolded by freezing. In all cases, the degree of structural perturbation noted in the frozen state was intermediate to that noted in dried solid, indicating that freezing-induced unfolding contributes partly to the total protein damage noted during lyophilization. This FTIR method will be valuable for assessing the relative contributions of excipients to stabilization during the freezing and drying steps and, hence, for testing the model presented in Figure 8. The only caveat is that at the concentrations necessary (i.e., about 5mg/mL) to obtain high-quality protein infrared spectra in the frozen state, many proteins that are known to be denatured at lower concentrations (e.g., catalase) are not unfolded during freezing.

PRACTICAL APPROACHES TO MINIMIZING FREEZING-INDUCED DAMAGE

The most damaging stresses to which a protein is exposed during freezing are low temperature and the formation of ice. Cold denaturation has been documented for many proteins and by itself may be sufficient to account for at least some of the damage noted during freezing (63–65). Also, the protein, which partitions into the non-ice phase, is exposed to extremely high solute concentrations as the sample is frozen. If solutes that are destabilizing to the protein are present, then this concentrating effect can contribute to protein denaturation. Finally, as noted earlier, there can be dramatic pH changes during freezing. For example, the dibasic form of sodium phosphate crystallizes in frozen solution, which results in a system that contains essentially solely the monobasic salt and has a very low pH (66,67). Other components in a formulation may inhibit crystallization of dibasic sodium phosphate (25,39). However, such inhibition is not predictable and must be investigated for each formulation with methods such as calorimetry (39) or direct pH measurements on the frozen systems (25). To minimize problems associated with pH changes, whenever possible, sodium phosphate buffer should be avoided. Also, although somewhat obvious, it is important to realize that a sodium phosphate system will be present if one starts with potassium phosphate buffer salts and NaCl, as is the case with phosphate-buffered saline.

Fortunately, to prevent freezing-induced damage to proteins, it is usually not necessary to discern which stresses are responsible for the damage or to target selectively each of these specific stresses. Rather, the most efficient approach is to design a formulation that provides the greatest overall resistance of the protein to denaturing forces. And as noted earlier, the first step in any stabilization process is to choose the specific conditions (pH or ligand) that maximize the stability of the given protein. These specific conditions, or adding a cryoprotectant solute such as sucrose, will protect the protein during freezing, whether the ultimate cause of denaturation is low temperature, high solute concentration, or some combination of stresses. This is because ultimately the stabilization of the protein derives from increasing the free energy barrier between the native and denatured states, which increases resistance to damage by any stress.

A wide variety of compounds have been found to provide nonspecific cryoprotection to proteins. These include sugars, amino acids, polyols, salting-out salts, methylamines, alcohols, other proteins, and synthetic polymers (59,61,68–70). During the initial screening of compounds as cryoprotectants, it is important that a relatively wide range of concentrations be tested for each compound. The range to be tested will be dictated by other formulation concerns (e.g., total excipient mass and tonicity of final rehydrated product) and the effectiveness of the cryoprotectant. Compared to sugars, polymers such as PEG and PVP and other proteins (e.g., albumins) are much more potent cryoprotectants (70). Especially for proteins that must be formulated at low concentration, polymers can be useful as protectants and to minimize loss of active protein on the walls of the vial. Also, if high excipient mass is a concern, polymers are good candidates for cryoprotectants because they are effective at relatively low concentrations (i.e., less than 1% wt/vol). For some proteins, sufficient freezing protection can be obtained by using a disaccharide (e.g., sucrose), which has the added benefit of also protecting the protein during subsequent drying. However, often much higher concentrations (e.g., more than 30% wt/vol) of such low molecular weight solutes are needed to confer adequate protection during freezing.

Finally, at least one aspect of protein stability during freezing is qualitatively unique from that noted in aqueous solution. That is, with numerous proteins it has been found that increasing the initial protein concentration increases protein stability during freeze-thawing. This is usually not the case in unfrozen aqueous solution, except in the special instance in which increasing the concentration of a multimeric protein favors the more stable, fully assembled state. The direct correlation between protein resistance to freezing and initial concentration appears to be related to other observations. First, it has been found with several proteins that increasing the rate of cooling leads to increased damage during freeze-thawing. This effect is attributed to the greater ice surface area associated with the smaller ice crystals that are generated with rapid cooling (71). Strambini and Gabierilli (62) directly documented with phosphorescence lifetimes studies that perturbation of protein tertiary structure in the frozen state was almost twofold greater for protein frozen by cooling at 100°C/min than that for samples cooled at 1°C/min. They and others have interpreted such results to mean that the ice-water interface serves as a stress to protein during freezing. If this is the case, then increasing the initial protein concentration indeed should reduce the percentage of the total sample damaged, assuming that the surface involved will damage only a finite amount of protein. Accordingly, any level of protein exceeding this amount would be spared from damage by the ice-water interface. In addition, any factor that minimized the association of the protein with the interface should increase protein stability during freeze-thawing.

Consistent with this suggestion is the finding that many different nonionic surfactants, at very low concentration (less than 1% wt/vol), have been found to provide complete protection to proteins during freeze-thawing (71,72). This protection has been attributed to the inhibition of damage at the ice-water interface (71). However, it must be stressed that there has yet to be any published direct evidence that surfactants protect proteins during the *freezing* step. Even if surfactants are found to stabilize the native protein structure during freezing, the protection may not necessarily be manifested through competition of protein and surfactant for the ice-water interface. On the contrary, a strong case can be made that under most industrial lyophilization conditions surfactants cannot be protecting in this manner because the ice-water interface is formed much faster than experimentally measured surfactant relaxation times at interfaces. Under typical industrial lyophilization conditions, there is significant supercooling of the liquid before freezing. Once ice nucleates, freezing is nearly instantaneous (less than 0.1 seconds). Surfactant relaxation times at interfaces, in contrast, are on the order of minutes (73). One alternative explanation for the freeze/thawing protection afforded by surfactants is that they may simply be serving to favor refolding over aggregation during thawing (see the next section). In addition, by binding to native protein, surfactants could inhibit interactions between protein molecules themselves and/or between protein molecules and the ice interface, which could be associated with freezing-induced denaturation (see the next section). Much more research is needed to document the role of this interface as a stress vector and to determine at what steps and by what mechanisms surfactants protect proteins during freezing and thawing. Further considerations of the mechanisms by which surfactant protect proteins during freeze-drying and rehydration are presented in the following section.

MECHANISMS FOR PROTEIN PROTECTION BY SURFACTANTS DURING LYOPHILIZATION AND REHYDRATION

Surfactants have often been included in both liquid and lyophilized commercial formulations of pharmaceutical proteins. Most likely, this is because surfactants can protect proteins against surface-induced denaturation, which can arise at air-water interface during vial filling and other processing steps. In general, for lyophilized formulations, it appears that, in addition to the benefits incurred prior to freezing, the presence of a surfactant in the formulation helps minimize the risk of the appearance of undesirable aggregates in the final rehydrated product. Despite the widespread use of surfactants in protein pharmaceuticals, the mechanism(s) by which surfactants protect proteins during lyophilization and rehydration, and even the steps at which this protection is operative, have not been determined. On the basis of the considerations of freezing damage given above, it seems that at least part of the benefit derived from surfactants might be due to inhibition or freezing-induced denaturation (74). But recall that it is not known if the protective effects of surfactants during freeze-thawing are actually manifested during the freezing step. Direct examination of the effects of surfactants on the structure of labile proteins in the frozen state is needed to address this issue.

Surfactants could also serve to increase the resistance of the protein to damage during dehydration, but to date there is no direct evidence documenting improved structure in the solid state of proteins dried in the presence of surfactants. Alternatively, surfactants might only provide protection during rehydration, perhaps by acting as wetting agents and/or by fostering protein refolding. Clearly, there are numerous processing steps at which and mechanisms by which surfactants can be beneficial in freeze-dried formulations. For example, it is known that surfactants increase the resistance of human growth hormone to agitation-induced denaturation in aqueous solution by binding to the native protein and hindering the contact between protein molecules needed for aggregation (75). An increase in protein concentration during both freezing and drying could foster such deleterious intermolecular interactions, which might be inhibited by surfactants. Also, interaction of protein molecules with the ice-water interface might be inhibited if a surfactant were bound to the protein molecule. In addition, a surfactant could serve a "chaperone" function and foster refolding over aggregation during rehydration (76,77). Alternatively, the surfactant, by binding to the native protein more favorably than the denatured state (see below), could simply increase the free energy of denaturation. Much more work is needed to sort out these possibilities and to determine the nature of the interaction of the surfactant with the protein that increases resistance of the protein to damage during freeze-drying and rehydration. And since very specific interactions between protein and surfactants might be important for some proteins but not others, it is clear that the mechanisms by which surfactants protect may vary depending on the specific properties of the protein. Understanding these matters is crucial if the benefits of using these compounds are to be fully exploited for formulation development.

THERMODYNAMIC MECHANISM FOR CRYOPROTECTION OF PROTEINS

Numerous compounds can provide general cryoprotection to proteins, when used at concentrations of several hundred millimolar. These include sugars, polyols, amino acids, methylamines, and salting-out salts (e.g., ammonium sulfate) (59,61,68–70). On the basis of the results of freeze-thawing experiments with LDH and PFK and a review of the literature on protein freezing, Carpenter and Crowe (59) have proposed that this cryoprotection can be explained by the same universal mechanism that Timasheff and Arakawa have defined for solute-induced protein stabilization in nonfrozen, aqueous solution (reviewed in Refs. 4, 70, 78, and 79).

Prior to examining the specifics of the Timasheff mechanism, it is instructive to consider the general effects of ligand binding on protein stabilization. Here we will provide a simplified and qualitative description of the most salient aspects of this relationship, which is referred to as the Wyman linkage function (i.e., in this case, the link between ligand binding and stability of protein states binding the ligand). Rigorous explanations can be found in Wyman (80) and Wyman and Gill (81). Here a two-state model will be considered, in which there is an equilibrium between native and denatured states of the protein (N $\leftrightarrow$ D). At room temperature and in nonperturbing solvent environments, the native state is favored because it has a lower free energy than the denatured state. The magnitude of the difference in free energy between the two states (i.e., the free energy of denaturation) dictates the relative stability of the native state. Any alteration in a system that decreases this difference will reduce stability, for example, reduce the melting temperature of the protein. Conversely, increasing this free energy difference will increase stability of the native state.

Binding of a ligand to either state will reduce the free energy (chemical potential) of that state because thermodynamically binding can only occur if the free energy of the protein-ligand complex is lower than that for the protein alone. The effect of ligand binding on protein stability depends on the difference in binding between the two states. The state that binds the most ligand will have the greatest reduction in free energy. Consequently, if more ligand binds to the native state than to the denatured state, then the free energy denaturation will be increased and the native state will be stabilized. The opposite will be seen if more ligand binds to the denatured state. If binding to the denatured state is sufficiently greater than that to the native state, then the denatured state will have the lowest free energy and will predominate.

Consider next how this general ligand binding argument relates specifically to the Timasheff mechanism for solute-induced protein stabilization and destabilization. Detailed, rigorous reviews of the Timasheff mechanism can be found elsewhere (78,79). For the purpose of the current review, a brief summary, which purposely provides only a simplified explanation, will suffice. First, a descriptive overview will be given, followed by an examination in more detail of the most relevant thermodynamic equations.

In protein cryoprotection, and stabilization and denaturation in nonfrozen aqueous solution, by nonspecific compounds, relatively high concentrations (more than 0.2 M) of ligand (solute) are needed to affect protein stability. This is because the interactions of the solute with the protein are relatively weak. These weak interactions are determined by equilibrium dialysis experiments in which ligand binding is determined by the difference in the ligand concentration in the dialysis bag with the protein and that outside the bag. Binding measured by this method is actually a measure of the relative affinities of the protein for water and ligand. Therefore, the ligand interaction is referred to as "preferential."

Ligand-induced destabilization by denaturants will be considered first because logically it is the easiest to understand in the context of the general ligand binding effects noted above. Timasheff and his colleagues have found that denaturants (e.g., urea and guanidine HCl) are bound preferentially to proteins and that the degree of binding is greatest for the denatured state. The free energy (chemical potential) of the denatured state is decreased more than that for the native state because more surface area for binding is exposed to solvent as the protein unfolds. Therefore, the free energy barrier between the two states is reduced. Consequently, the native state's resistance to stress is reduced (e.g., the melting point of the protein in lowered). If this effect is great enough, the protein will be denatured at room temperature.

Conversely, Timasheff, Arakawa, and their colleagues have observed experimentally that there is a deficiency of stabilizing solutes (e.g., sugars and polyols) in the presence of the protein, relative to that seen outside the dialysis bag. That is, the solutes are preferentially excluded from contact with the surface of the protein. Preferential exclusion, in a thermodynamic sense, means that the solute (ligand) has negative binding to the protein. Thus, there is an increase in the free energy (chemical potential) of the protein. In the presence of preferentially excluded solutes, the native state is stabilized. This is because denaturation leads to a greater surface area of contact between the protein and the solvent and greater preferential solute exclusion. Thus, even though there is an increase in the free energy of the negative state, this effect is offset by the greater increase in the free energy of the denatured state.

Timasheffs preferential interaction mechanism also explains the influence of solutes on the degree of assembly of multimeric proteins. Preferentially excluded solutes tend to induce polymerization and stabilize oligomers since the formation of contact sites between constituent monomers serves to reduce the surface area of the protein exposed to the solvent. Polymerization reduces the thermodynamically unfavorable effect of preferential solute exclusion. Conversely, preferential binding of solute induces depolymerization because there is greater solute binding to monomers than to polymers.

Now, the key thermodynamic aspects of this mechanism (reviewed in Refs. 4, 78, and 79) will be examined in more detail. Setting component 1 = principal solvent (here water), component 2 = protein, and component 3 = solute (e.g., sucrose or PEG), the preferential interaction of component 3 with a protein is expressed, within close approximation, by the parameter $(\delta m_3/\delta m_2)_{\mu1,\mu3}$ at constant temperature and pressure, where μ_i and m_i are the chemical potential and molal concentration of component i, respectively. A positive value of this interaction parameter indicates an excess of component 3 in the vicinity of the protein over the bulk concentration (i.e., preferential binding of the solute). A negative value for the parameter indicates a deficiency of component 3 in the protein domain. Component 3 (the solute) is preferentially excluded and component 1 (water) is in excess in the protein domain.

The preferential interaction parameter is a direct expression of changes in the free energy of the system induced by component 3 and has the relation:

$$\left(\frac{\delta\mu_2}{\delta m_3}\right)_{m_2} = -\left(\frac{\delta m_3}{\delta m_2}\right)_{\mu_1,\mu_3}\left(\frac{\delta\mu_3}{\delta m_3}\right)_{m_2} \tag{1}$$

The term on the left-hand side of the equation defines the change in protein chemical potential as a function of solute concentration. The first term on the right-hand side of the equation is the preferential interaction parameter, which was defined earlier. The second term is the solute self-interaction parameter, which will be described in detail later. Equation (1) indicates that those compounds that are excluded [i.e., $(\delta m_3/\delta m_2)_{\mu 1, \ \mu 3} < 0$)] from the surface of the protein will have positive values of $(\delta \mu_2/\delta m_3)_{m_2}$; they will increase the chemical potential (free energy) of the protein, rendering the system more thermodynamically unfavorable. In the presence of excluded solutes, the exclusion will be greater for the denatured form of the protein than for the native form because the former has a larger surface area, as indicated by $(\delta m_3/\delta m_2)^{D} < (\delta m_3/\delta m_2)^{N} < 0$. Consequently, the increase in chemical potential is greater for the denatured form than for the native form in the presence of a preferentially excluded solute, as indicated by $(\delta \mu_2/\delta m_3)_{m_2}^{D} > (\delta \mu_2/\delta m_3)_{m_2}^{N} > 0$. There is an increase in the free energy difference between the native and denatured forms, thus stabilizing the native state.

The opposite is seen for potent protein denaturants such as urea and guanidine HCl. These solutes bind preferentially to both the native and the denatured form of the protein (reviewed in Refs. 4, 78, and 79), and hence decrease the chemical potential of the protein. Since the number of available binding sites is increased upon unfolding of the protein, an increase in preferential solute binding occurs as indicated by $(\delta m_3/\delta m_2)^{D} > (\delta m_3/\delta m_2)^{N} > 0$. There is a concomitant decrease in protein chemical potential, which is greater for the denatured state: $(\delta \mu_2/\delta m_3)_{m_2}^{D} < (\delta \mu_2/\delta m_3)_{m_2}^{N} < 0$. This serves to lower the free energy difference between the two states, and when the native state becomes the higher energy state, protein denaturation should result.

In more general terms, so long as $(\delta m_3/\delta m_2)^{D} < (\delta m_3/\delta m_2)^{N}$, the native state will be stabilized. Thus, stabilization could also arise if the solute bound preferentially to the native state was excluded from the denatured state, or if the solute was preferentially bound to both states but binding was less for the denatured state. Conversely, in any situation in which $(\delta m_3/\delta m_2)^{D} > (\delta m_3/\delta m_2)^{N}$, the native state will be destabilized.

It is not possible to measure preferential interactions between solutes and proteins in frozen samples. Therefore, it is not known if cryoprotectants are actually preferentially excluded from frozen proteins. However, a recent study by Heller et al. (82) has provided direct evidence that the influence of a solute on protein chemical potential accounts for the solute's effect on protein stability during freezing. First, it was found with infrared spectroscopy that hemoglobin's secondary structure was perturbed in the frozen state. To test the effect of increasing the protein's chemical potential on inhibiting freezing-induced structural perturbation, hemoglobin was prepared in a dextran/PEG mixture. This mixture forms two separate liquid phases, each enriched in one polymer, at room temperature. Hemoglobin partitions into both phases, and the protein's chemical potential will be increased by the presence of polymer in both phases. More importantly, since the system is in equilibrium, the protein's chemical potential is the same in both phases. When samples were removed from each phase and frozen, it was found that the infrared spectra of hemoglobin in the frozen state were identical and much more native-like than that seen in the absence of polymer. These results indicate that increasing protein chemical potential directly equates with increased protein stability during freezing. However, after freeze-drying, the structure of hemoglobin in the dried state was more native-like in the sample taken from the PEG-rich phase than that from the dextran-rich phase. These results indicate that simply increasing protein chemical potential does not necessarily equate with stabilization of a protein during dehydration.

An important and often overlooked aspect of Timasheff's mechanism is the role of solid phase chemical potentials. Adding protective excipients to a protein solution increases the chemical potential of both the protein's native and denatured state. There is a limit to how high the chemical potential can be raised, however. This is because the *solid phase* chemical potentials of the protein are largely unaffected by the addition of excipients to the liquid solution. Thus, once the chemical potential of the native state is increased to a value greater than that of the solid state, precipitation of the protein will be favored. Protein precipitated will be at high concentration; this may pose problems due to aggregation upon drying and/or storage. Thus, there is a clear trade-off between conformational stability of the protein, which is increased with increasing cryoprotectant concentration, and stability against aggregation, which is decreased with increasing cryoprotectant concentration.

Finally, it is important to consider mechanistically how to explain the much greater potency of PEGs as cryoprotectants relative to other compounds such as sucrose. The data for one case, which are shown in Figure 9 and Table 1, illustrate this point. Figure 9 compares cryoprotection of LDH by PEG 8000, PEG 400, and sucrose (MW 342). LDH is completely protected during freeze-thawing

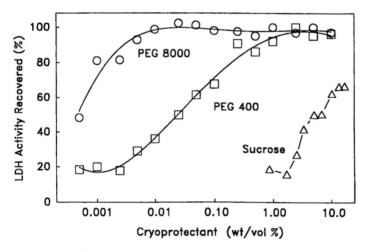

FIGURE 9 Effects of polyethylene glycols (PEG) and sucrose on lactate dehydrogenase stability during freeze-thawing. *Source*: Data taken from Refs. 59 and 70.

TABLE 1 Parameters for Solute Interactions with Chymotrypsinogen

Solute	Concentration	$(\delta m_3/\delta m_2)_{\mu 1, \mu 3}$	$(\delta \mu_3/\delta m_3)_{m2}$[a]	$(\delta \mu_2/\delta m_3)_m$[b]
Sucrose[c]	1.27 m	−10.35	0.56	5.7
PEG 400[d] (0.27 m)	10% wt/vol	−6.87	2.42	16.6
PEG 6000[d] (0.0017m)	1% wt/vol	−0.62	480.00	297.6

[a]kcal/mol of solute/mol of solute in 1000 kg H_2O.
[b]kcal/mol of protein/mol of solute.
[c]Data taken from Ref. 83.
[d]Data taken and calculated from Ref. 84.

by PEG 8000 at concentrations of $\geq 0.01\%$ (wt/vol). In contrast, full protection in the presence of PEG 400 is not realized until the concentration is at least 2.5% (wt/vol). On a weight percentage basis, PEG 8000 is 250-fold more potent as a cryoprotectant. On a molar basis, the higher molecular weight PEG is 5000-fold more potent. Sucrose is much less effective than even PEG 400. Even at sucrose concentrations as high as 10% (wt/vol) the protein is not fully protected.

In the past, we have ascribed these differences in protein stabilization to Timasheff's thermodynamic mechanism. The only protein for which the needed thermodynamic parameters have been measured in the presence of all three cryoprotectants is chymotrypsinogen (83,84) (Table 1). Although these data are not directly applicable to LDH, the general trends shown should be relevant to any protein. The increase in chymotrypsinogen chemical potential, $(\delta\mu_2/\delta m_3)_{m_2}$, in the presence of either of two different molecular weights of PEG (e.g., $M_r = 400$ or 6000) is greater than that noted in the presence of the sucrose, even though the PEGs are excluded to a lesser degree, on a per mole of solute basis. Comparing the two PEG molecules indicates that the larger the PEG the less it is excluded on a mole basis, but the more it increases protein chemical potential.

The basis for these observations can be explained by examining equation (20). The other major component in determining the effect of solute on protein chemical potential is the self-interaction parameter for the solute, $(\delta\mu_3/\delta m_3)_{m_2}$. The value for this parameter is several-fold greater for PEG 400 and almost three orders of magnitude greater for PEG 6000 than that for sucrose. The self-interaction parameter is given by

$$\left(\frac{\delta\mu_3}{\delta m_3}\right)_{m_2} = \left[\left(\frac{RT}{m_3}\right) + RT\left(\frac{\delta\ln\gamma_3}{\delta m_3}\right)_{m_2}\right] \tag{2}$$

where γ_3 is the activity coefficient of the solute and R is the universal gas constant (reviewed in Refs. 4, 78, and 79). The molal concentrations needed for preferential exclusion of PEG are very small and the activity coefficient of PEG is quite large, relative to values for sucrose. Therefore, the self-interaction parameter for PEG is very large compared to that for sucrose. In addition, as the size of PEG increases there is a great increase in such nonideality (Table 1).

This argument does indeed support the contention that on a per-mole basis PEG is much more effective than sucrose at increasing protein chemical potential. And for cases where relatively high concentrations of PEG (e.g., >1% wt/vol) are needed to confer cryoprotection, the Timasheff mechanism may be applicable. However, it seems unlikely that a PEG concentration of 0.01% (wt/vol) would have a significant effect on the thermodynamics of the system. This is because the actual parameter of interest is the transfer free energy of the native versus denatured protein from water into cryoprotectant solution. The difference between the values for the two states determines the magnitude of the effect on the free energy of denaturation. The transfer free energy is obtained by integrating equation (1) from zero to the molal concentration of cryoprotectant of interest. With PEG 8000 at 0.01% wt/vol concentration, the molality is so low that the calculated transfer free energy would be extremely small. Thus, the effect on the free energy of denaturation would be trivial.

So, how can we account for the potent protection afforded by PEGs? The concentration range for cryoprotection is actually very similar to that seen with nonionic surfactants such as Tweens. PEGs have been shown to be surface active, and the magnitude of the decrease in water surface tension has been shown to correlate directly with PEG molecular weight (85). Therefore, the most rational explanation for the results presented in Figure 9 is that PEGs are serving as such potent cryoprotectants because they are operating by the same mechanism(s) as more typical surfactants such as Tweens.

MECHANISMS FOR FAILURE OF DEXTRAN TO PROTECT LYOPHILIZED PROTEINS

In an idealized lyophilization cycle, a solution of proteins, buffers, and excipients is cooled to the solution's freezing point. At this point, it is thermodynamically favorable to form a new solid phase composed of pure ice. Once ice begins to form (not necessarily at the thermodynamic freezing point; substantial supercooling may occur), the remaining components of the solution in the nonfrozen phase become increasingly more concentrated, as shown in Figure 10. The combination of increased concentration and lower temperatures causes the viscosity of the non-ice phase to increase until, at a glass transition point termed T_g', the solution becomes so viscous that further freezing of water is kinetically blocked. Further temperature decreases below T_g' have no additional concentrating effects.

It is important to realize that the preceding is a highly simplified description of the possible phase behaviors that can occur. In actual practice, phase behavior during freezing is rarely so simple. Instead, as the non-ice phase

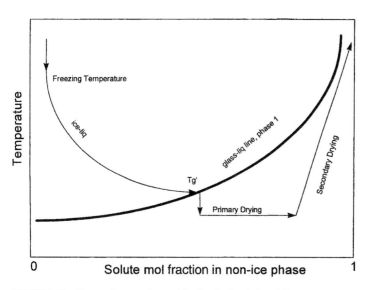

FIGURE 10 Phase diagram for an idealized, simple lyophilization cycle. A liquid sample first is cooled to the freezing temperature. As pure ice is formed, the solute remaining in the non-ice phase is concentrated until the ice-liquid line intersects the glass transition line at T_g'. No further concentration due to cooling occurs. Primary drying occurs under vacuum at a temperature below the glass transition temperature. After primary drying, the temperature is increased to effect secondary drying. Final storage temperature after secondary drying is below the glass transition line.

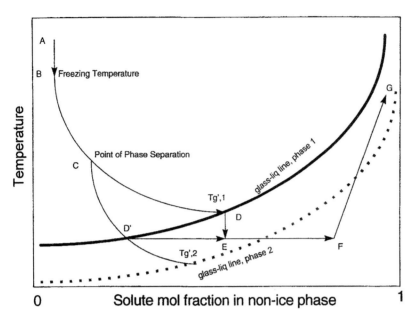

FIGURE 11 Phase diagram for a lyophilization cycle for a phase separation system. Starting with an aqueous solution at point A, the solution is cooled to the freezing point B. Further cooling results in concentration of solutes in the non-ice phase. At point C, the solutes are sufficiently concentrated that a phase split occurs. Additional cooling causes phase I to follow the path CDE, while phase 2 follows path CD'. Primary drying follows the path D'EF for phase 2 and EF for phase I. Secondary drying for both phases begins at F, with the final dried condition at G. Note that for phase 2, the majority of both primary and secondary drying is conducted above its glass transition line, and final storage temperature is also above the glass transition line for phase 2.

is concentrated, the chemical potentials of each solution component increase, until eventually a solubility point or a liquid-liquid miscibility point is reached. At this concentration it will be thermodynamically possible for an additional solid or liquid phase to be formed. Figure 11 shows one such diagram for a hypothetical system that undergoes a liquid-liquid phase split in the freeze-concentrated non-ice phase. That such a phase split can occur should not be surprising, many of the excipients that have been investigated as protein stabilizers (PEG, dextran, PVP, and Ficoll) are well known to form two-phase aqueous systems when present at high concentration (86).

To achieve an optimally stabilized protein and maintain cake integrity, it is important to conduct primary and secondary drying steps at temperatures below the glass transition temperature. For systems that exhibit more complicated phase behavior, such as that shown in Figure 11, the preceding statement should be amended to state that drying steps should proceed at temperatures below *the lowest applicable glass transition temperature*. Systems that undergo liquid-liquid phase splits during freeze concentration will exhibit two or more characteristic glass transition temperatures. These glass transition temperatures may not be easily detected by conventional differential scanning calorimetry, especially if the phase volume of the second liquid phase is small. Electron paramagnetic resonance spectroscopic measurements have indicated the presence of multiple liquid phases in freeze-concentrated dextran/salt samples at $-15°C$, under conditions where only one glass transition temperature could be detected by differential scanning calorimetry (C. Heinen and T. W. Randolph, unpublished observations).

It has been suggested that polymers such as dextran fail to protect proteins because they phase-separate from the protein during the freeze-drying process (87). The dried solid could contain two or more amorphous phases. As Pikal (87) has stated, it is crucial that the stabilizing additive not only remains amorphous but also forms a single phase *with the protein.*

Phase separation during freezing and/or drying may provide a possible reconciliation for the two competing hypotheses to explain protein stabilization by excipients during lyophilization. If a protein phase separates from an excipient on a microscopic scale, the excipient will not afford mechanical protection even if it forms a glass. Likewise, failure of an excipient to hydrogen bond to a protein could be considered a manifestation of local phase separation. Indeed, the experiments of Carpenter et al. (36) showed that when PEG phase separates from PFK or LDH (by forming a solid crystalline phase), it fails to protect the proteins during drying. Thus, one could argue equally well from a "glass mechanics" point of view that because PEG phase separates, it fails to envelop the protein in a competent glass. Or one could argue from the "water replacement" perspective that because PEG phase separates from the protein it cannot supply needed hydrogen bonds. In the end, both theories can be reduced to a single hypothesis: excipients must form a glass that intimately incorporates the protein if the native structure is to be maintained (87).

Polymers such as dextrans could provide many desirable properties (e.g., high T_g' and T_g) to the freeze-dried formulation. Therefore, it is essential that future research address the theoretical and practical aspects of protein/polymer phase separation and develop the mechanistic insight to prevent this phenomenon during lyophilization. Also, as part of this effort, it is important to discern why other polymers (e.g., PVP and BSA) that protect labile proteins apparently do not phase separate from the protein during lyophilization.

ACKNOWLEDGMENTS

We gratefully acknowledge support by grants from the National Science Foundation (NSFBES9505301 and NSFBES9520288); the Whitaker Foundation; Boehringer Mannheim Therapeutics; Genentech, Inc.; Genetics Institute, Inc.; Genencor International; Amgen, Inc.; and Zymogenetics, Inc. We also thank the National Science Foundation and American Foundation of Pharmaceutical Education, the American Pharmaceutical Manufacturing Association, and the Colorado Institute for Research in Biotechnology for providing predoctoral fellowships to our graduate students.

REFERENCES

1. Geisow MJ. Aching for approval, hoping for harmony—biotechnology product regulation. Trends Biotechnol 1992; 10(4):107–108.
2. Wang YJ, Pearlman R. Stability and Characterization of Protein and Peptide Drugs: Case Histories. Vol. 5. New York, Pharmaceutical Biotechnology: Plenum Press, 1993.
3. Talmadge JE. The pharmaceutics and delivery of therapeutic polypeptides and proteins. Adv Drug Deliv Rev 1993; 10:247–299.

4. Arakawa T, Prestrelski SJ, Kinney W, et al. Factors affecting short-term and long-term stabilities of proteins. Adv Drug Deliv Rev 1993; 10:1–28.
5. Thornton CA, Ballow M. Safety of intravenous immunoglobulin. Arch Neurol 1993; 50:135–136.
6. Pikal MJ. Freeze-drying of proteins. Part I: Process design. BioPharm 1990; 3(8):18–27.
7. Pika MJ. Freeze-drying of proteins. Part II: Formulation selection. BioPharm 1990; 3(9): 26–30.
8. Prestrelski SJ, Tedeschi N, Arakawa T, et al. Dehydration-induced conformational changes in proteins and their inhibition by stabilizers. Biophys J 1993; 65:661–671.
9. Prestrelski SJ, Arakawa T, Carpenter JF. Separation of freezing- and drying-induced denaturation of lyophilized proteins by stress-specific stabilization: II. Structural studies using infrared spectroscopy. Arch Biochem Biophys 1993; 303:465–473.
10. Prestrelski SJ, Arakawa T, Carpenter JF. The structure of proteins in lyophilized formulations using Fourier transform infrared spectroscopy. In: Cleland JL, Langer R, eds. Formulation and Delivery of Proteins and Peptides. Washington, D.C.: American Chemical Society Symposium Series No. 567, 1994:148–169.
11. Dong A, Prestrelski SJ, Allison SD, et al. Infrared spectroscopic studies of lyophilization- and temperature-induced protein aggregation. J Pharm Sci 1995; 84:415–424.
12. Prestrelski SJ, Pikal KA, Arakawa T. Optimization of lyophilization conditions for recombinant human interleukin-2 by dried-state conformational analysis using Fourier transform infrared spectroscopy. Pharm Res 1995; 12:1250–1259.
13. Chang BS, Beauvais RM, Dong A, et al. Physical factors affecting the storage stability of freeze-dried interleukin-I receptor antagonist: glass transition and protein conformation. Arch Biochem Biophys 1996; 331:249–258.
14. Allison SD, Dong A, Carpenter JF. Counteracting effects of thiocyanate and sucrose on chymotrypsinogen secondary structure and aggregation during freezing, drying and rehydration. Biophy J 1996; 71:2022–2032.
15. Carpenter JF, Chang BS. Lyophilization of protein pharmaceuticals. In: Avis KE, Wu VL, eds. Biotechnology and Biopharmaceutical Manufacturing, Processing and Preservation. Buffalo Grove, IL: Interpharm Press, 1996:199–264.
16. Ahern TJ, Manning MC. Stability of Protein Pharmaceuticals: Part A. Chemical and Physical Pathways of Protein Degradation. New York: Plenum Press, 1992.
17. Manning MC, Patel K, Borchardt RT. Stability of protein pharmaceuticals. Pharm Res 1989; 6:903–908.
18. Chilson OP, Costello LA, Kaplan NO. Effects of freezing on enzymes. Fed Proc 1965; 24(2):S55–S65.
19. Chen BL, Arakawa T, Hsu E, et al. Strategies to suppress aggregation of recombinant keratinocyte growth factor during liquid formulation development. J Pharm Sci 1994; 83:1657–1661.
20. Brostreaux J, Ericksson-Quensel IB. Etude sur la dessication des proteines. Arch Phys Biol 1935; 23(4):209–226.
21. Carpenter JF, Crowe LM, Crowe JH. Stabilization of phosphofructokinase with sugars during freeze-drying: characterization of enhanced protection in the presence of divalent cations. Biochim Biophys Acta 1987; 923:109–115.
22. Izutsu K, Yoshioka S, Takeda Y. The effects of additives on the stability of freeze-dried β-galactosidase stored at elevated temperatures. Int J Pharm 1991; 71:137–146.
23. Hageman M. Water sorption and solid state stability of proteins. In: Ahern T, Manning MC, eds. Stability of Protein Pharmaceuticals. Part A. Chemical and Physical Pathways of Protein Degradation. New York: Plenum Press, 1992:273–309.
24. Hellman K, Miller DS, Cammack RA. The effect of freeze-drying on the quaternary structure of L-asparaginase from Erwinia carotovora. Biochim Biophys Acta 1983; 749:133–142.
25. Anchordoquy TJ, Carpenter JF. Polymers protect lactate dehydrogenase during freeze-drying by inhibiting dissociation in the frozen state. Arch Biochem Biophys 1996; 332:231–238.
26. Izutsu K, Yoshioka S, Terao T. Stabilizing of, β-galactosidase by amphiphilic additives during freeze-drying. Int J Pharm 1993; 90:187–194.

27. Izutsu K, Yoshioka S, Terao T. Stabilizing effect of amphiphilic excipients on the freeze-thawing and freeze-drying of lactate dehydrogenase. Biotechnol Bioeng 1994; 43:1102–1107.

28. Imai T, Irie T, Otagiri M, et al. Comparative study of antiinflammatory drug fibroprofen with B-cyclodextrin and methylated, β-cyclodextrins. J Inclusion Phenom 1984; 2:597–604.

29. Yoshida A, Arima H, Uekama K, et al. Pharmaceutical evaluation of hydroxyalkyl ethers of, β-cyclodextrins. Int J Pharm 1988; 46:217–222.

30. Hora MS, Rana RK, Smith FW. Lyophilized formulations of recombinant tumor necrosis factor. Pharm Res 1992; 9:33–36.

31. Brewster ME, Hora MS, Simpkins JW, et al. Use of 2-hydroxypropyl-, β-cyclodextrin as a solubilizing and stabilizing excipient for protein drugs. Pharm Res 1991; 8:792–795.

32. Hora MS, Rana RK, Wilcox CL, et al. Development of lyophilized formulation of interleukin-2. Dev Biol Stand 1991; 74:295–306.

33. Carpenter JF, Crowe JH. Infrared spectroscopic studies on the interaction of carbohydrates with dried protein. Biochemistry 1989; 28:3916–3922.

34. Izutsu K, Yoshioka S, Terao T. Decreased protein-stabilizing effects of cryoprotectants due to crystallization. Pharm Res 1993; 10:1232–1237.

35. Izutsu K, Yoshioka S, Terao T. Effect of mannitol crystallinity on the stabilization of enzymes during freeze-drying. Chem Pharm Bull (Tokyo) 1994; 42:5–8.

36. Carpenter JF, Prestrelski S, Arakawa T. Separation of freezing- and drying-induced denaturation of lyophilized proteins by stress-specific stabilization: I. Enzyme activity and calorimetric studies. Arch Biochem Biophys 1993; 303:456–464.

37. Gatlin I, Deluca P. A study of the phase separation in frozen antibiotic solutions by differential scanning calorimetry. J Parent Drug Assoc 1980; 34:398–408.

38. Aldous BJ, Auffret AD, Franks F. The crystallization of hydrates from amorphous carbohydrates. Cryo Letters 16:181–186, 1995.

39. Chang BS, Randall CS. Use of subambient thermal analysis to optimize protein lyophilization. Cryobiology 1992; 29:632–656.

40. Pikal MJ, Dellerman KM, Roy ML, et al. The effects of formulation variables on the stability of freeze-dried human growth hormone. Pharm Res 1991; 8:427–436.

41. Franks F, Hatley RHM, Mathias SF. Materials science and the production of shelf stable biologicals. BioPharm 1991; 4(9):38–55.

42. Angell CA. Formation of glasses from liquids and biopolymers. Science 1995; 267:1924–1935.

43. Tanaka T, Takeda T, Miyajama R. Cryoprotective effect of sachharides on denaturation of catalase during freeze-drying. Chem Pharm Bull (Tokyo) 1991; 39:1091–1094.

44. Levine H, Slade L. Principles of "cryostabilization" technology from structure/property relationships of carbohydrate/water systems—a review. Cryo Letters 1988; 9:21–63.

45. Levine H, Slade L. Thermomechanical properties of small-carbohydrate-water glasses and "rubbers": kinetically metastable systems at sub-zero temperatures. J Chem Soc Faraday Trans 1 1988; 84:2619–2633.

46. Levine H, Slade L. Glass transitions in foods. In: Shartzberg HS, Hartel RW, eds. Physical Chemistry of Foods. New York: Marcel Dekker, 1992:83–221.

47. Rupley JA, Careri G. Protein hydration and function. Adv Protein Chem 1991; 41:37–172.

48. Kuntz ID, Kauzman W. Hydration of proteins and polypeptides. Adv Protein Chem 1974; 28:239–345.

49. Edsall JT, McKenzie HA. Water and proteins II. The location and dynamics of water in protein systems and its relation to their stability and properties. Adv Biophys 16:53–183.

50. Byler DM, Susi H. Examination of the secondary structure of proteins by deconvoluted FTIR spectra. Biopolymers 1986; 25:469–487.

51. Susi H, Byler DM. Resolution enhanced Fourier transform infrared spectroscopy of enzymes. Methods Enzymol 1986; 130:290–311.

52. Surewicz WK, Mantsch HH. New insights into protein conformation from infrared spectroscopy. Biochim Biophys Acta 1988; 953:115–130.

53. Krimm S, Bandekar J. Vibrational spectroscopy and conformation of peptides, poly-peptides and proteins. Adv Protein Chem 1986; 38:181–364.
54. Dong A, Caughey WS. Infrared methods for study of hemoglobin reactions and structures. Methods Enzymol 1994; 232:139–175.
55. Dong A, Meyer JD, Kendrick BS, et al. Effect of secondary structure on the activity of enzymes suspended in organic solvents. Arch Biochem Biophys 1996; 334:406–414.
56. Dong A, Matsuura J, Allison SD, et al. Infrared and circular dichroism spectroscopic characterization of structural differences between j3-lactoglobulin A and B. Biochemistry 1996; 35:1450–1457.
57. Kendrick BS, Dong A, Allison SD, et al. Quantitation of the area of overlap between second-derivative amide I infrared spectra to determine structural similarity of a protein in different states. J Pharm Sci 1996; 85:155–158.
58. Bock PE, Frieden C. Another look at the cold lability of enzymes. Trends Biochem Sci 1978; 3:100–103.
59. Carpenter JF, Crowe JH. The mechanism of cryoprotection of proteins by solutes. Cryobiology 1988; 25:244–255.
60. Carpenter JF, Crowe JH. Modes of stabilization of a protein by organic solutes during desiccation. Cryobiology 1988; 25:459–470.
61. Carpenter JF, Hand SC, Crowe LM, et al. Cryoprotection of phosphofructokinase with organic solutes: characterization of enhanced protection in the presence of divalent cations. Arch Biochem Biophys 1986; 250:505–512.
62. Strambini GB, Gabellieri E. Proteins in frozen solutions: evidence of ice-induced partial unfolding. Biophys J 1996; 70:971–976.
63. Brandts JF, Fu J, Nordin JF. The low temperature denaturation of chymo-trypsinogen in aqueous solution and in frozen aqueous solution. In: Wolstenholme GEW, O'Conner M, eds. The Frozen Cell. London: Churchill, 1970:189–212.
64. Becktel WJ, Schellman JA. Protein stability curves. Biopolymers 1987; 26:1859–1877.
65. Privalov PL. Cold denaturation of proteins. Crit Rev Biochem Mol Biol 1990; 25:281–305.
66. van den Berg L. The effect of addition of sodium and potassium chloride to the reciprocal system: $KH_2-PO_4-Na_2HPO_4-H_2O$ on pH and composition during freezing. Arch Biochem Biophys 1959; 84:305–315.
67. van den Berg L, Rose D. The effect of freezing on the pH and composition of sodium and potassium solutions: the reciprocal system $KHrPO_4-Na_2HPO_4-H_2O$. Arch Biochem Biophy 1959; 81:319–329.
68. Shikama K, Yamazaki I. Denaturation of catalase by freezing and thawing. Nature 1961; 190:83–84.
69. Loomis SH, Carpenter JF, Anchordoguy TJ, et al. Cryoprotective capacity of end products of anaerobic metabolism. J Exp Zool 1989; 252:9–15.
70. Carpenter JF, Prestrelski SJ, Anchordoguy TJ, et al. Interactions of stabilizers with proteins during freezing and drying. In: Cleland JL, Lauger R, eds. Formulation and Delivery of Proteins and Peptides. Am Chem Soc Symp 1994; 567:134–147.
71. Chang BS, Kendrick BS, Carpenter JF. Surface-induced denaturation of proteins during freezing and its inhibition by surfactants. J Pharm Res 1996; 85:1325–1330.
72. Nema S, Avis KE. Freeze-thaw studies of a model protein, lactose dehydro-genase, in the presence of cryoprotectant. J Parent Sci Tech 1993; 47:76–83.
73. Johnson DO, Stebe KJ. Oscillating bubble tensiometry: a method for precisely measuring the kinetics of surfactant adsorptive-desorptive exchange. J Colloid Int Sci 1994; 168:526–538.
74. Izutsu K, Yoshioka S, Kojima S. Increased stabilizing effect of amphiphilic excipients on freeze-drying of lactate of dehydrogenase (LDH) by dispersion into sugar matrices. Pharm Res 1995; 12:838–843.
75. Bam NB, Randolf TW, Cleland JL. Stability of protein formulations: investigation of surfactant effects by a novel EPR spectroscopic technique. Pharm Res 1995; 12:2–11.
76. Cleland JL, Randolph TW. Mechanism of polyethylene glycol interactions with molten globule folding intermediate of bovine carbonic B. J Biol Chem 1992; 267:3147–3153.

77. Horowitz PM. Kinetic control of protein folding by detergent micelles, liposomes, and chaperonins. In: Cleland JL, ed. Protein Folding. In Vivo and In Vitro. Washington, D.C.: American Chemical Series No. 526, 1993:156–163.

78. Timasheff SN. Stabilization of protein structure by solvent additives. In: Ahern T, Manning MC, eds. Stability of Protein Pharmaceuticals. Part B. Vivo Pathways of Degradation and Strategies for Protein Stabilization. New York: Plenum Press, 1992:265–285.

79. Timasheff SN. Preferential interactions of water and cosolvents with proteins. In: Gregory RB, ed. Protein-Solvent Interactions. New York: Marcel Dekker, 1995:445–482.

80. Wyman J. Linked functions and reciprocal effects in hemoglobin: a second look. Adv Protein Chem 1964; 19:223–286.

81. Wyman J, Gill SJ. Binding and Linkage. Functional Chemistry of Biological Molecules. Mill Valley, CA: University Science Books, 1990.

82. Heller M, Carpenter JF, Randolph TW. Effect of phase separating systems on lyophilized hemoglobin. J Pharm Sci 1996; 85:1358–1362.

83. Lee JC, Timasheff SN. The stabilization of protiens by sucrose. J Biol Chem 1981; 259:7193–7201.

84. Bhat R, Timasheff SN. Steric exclusion is the principal source of the preferential hydration of proteins in the presence of polyethylene glycols. Protein Sci 1992; 1:1133–1143.

85. Winterhalter M, Burner H, Marzinka S, et al. Interaction of poly(ethylene-glycols) with air—water interfaces and lipid monolayers: investigations of surface pressure and surface potentials. Biophys J 1995; 69:1372–1381.

86. Albertsson PA. Partition of Cell Particles and Macromolecules. 3rd ed. New York: Wiley-Interscience, 1986.

87. Pikal MJ. Freeze-drying of proteins. In: Cleland JL, Langer R, eds. Formulation and Delivery of Proteins and Peptides. Washington, D.C.: American Chemical Society Symposium Series No. 567, 1994:120–133.

Mechanisms of Protein Stabilization During Freeze-Drying Storage: The Relative Importance of Thermodynamic Stabilization and Glassy State Relaxation Dynamics

Michael J. Pikal
School of Pharmacy, University of Connecticut, Storrs, Connecticut, U.S.A.

INTRODUCTION

Partly because of chemical complexity and partly due to the marginal stability of higher order structure (i.e., conformation), therapeutic proteins often present significant stability problems. While proteins are generally quite stable in aqueous solution for short periods of time, a pharmaceutical product must have adequate stability over storage periods of many months, typically several years. Many proteins do not possess this long-term stability in the aqueous state. Ironically, while the nature of water is an important contributing factor to the conformational stability of a protein, water is a destabilizing factor in the long-term preservation of the chemical and structural integrity of a protein. With proteins, as with most labile molecules, removal of water to form a solid generally improves storage stability. Thus, proteins are typically freeze-dried in an attempt to achieve adequate storage stability. However, some proteins suffer *irreversible* damage during freeze-drying, and even if the protein survives freeze-drying without damage, the freeze-dried solid does not always have the desired storage stability. Stability problems are normally minimized by a combination of proper process control and formulation optimization.

In the context of proteins, "stability" has two distinct meanings. The term pharmaceutical stability refers to the ability of a protein to be processed, distributed, and used without *irreversible* change in primary structure, conformation, or state of aggregation. We refer to pharmaceutical *instability* as "degradation." The phrase "protein stability" is also commonly used to describe the position of the equilibrium between native and unfolded conformations. If a protein formulation requires a high level of chemical denaturant, or a high temperature, to shift the equilibrium between native and unfolded in favor of the unfolded state, the protein is said to be "stable." This meaning of stability is denoted as "thermodynamic stability." The timescale for manifestation of thermodynamic *instability* in aqueous solutions is normally quite short, that is, time constants for unfolding are typically seconds to hours (1,2). While pharmaceutical *instability* during processing may also occur on a short timescale, degradation during storage involves a timescale on the order of years. Thermodynamic *instability* involves physical changes, somewhat analogous to a thermodynamic change of state. Pharmaceutical *instability* may be purely a result of a physical change (i.e., noncovalent aggregation), but may also involve changes in primary structure or "chemical degradation."

Pharmaceutical stability and thermodynamic stability are not necessarily directly related. For example, a protein may exhibit thermodynamic *instability* during freeze-drying and unfold, but if no *irreversible* reactions occur during storage or during reconstitution, the reconstituted protein may refold completely within seconds and therefore display perfect pharmaceutical stability. A protein that remains in the native state and is thermodynamically stable may still degrade via chemical reactions such as deamidation and methionine oxidation over storage times of years, particularly if the reactive moiety is located on the protein surface. Conversely, thermodynamic *instability* may well be a prelude to degradation. Certainly an unfolded protein could expose normally buried and "protected" methionine and asparagine residues to the solution environment and render these residues more reactive. Also, degradation via *irreversible* aggregation is believed to often proceed through unfolded or partially unfolded conformations as intermediates (3).

While the literature does contain some general guidelines for process and formulation optimization (4,5), mechanisms of stabilization are incompletely understood. Consequently, formulation and process development efforts remain somewhat empirical in practice. Mechanisms suggested to rationalize the stabilization of proteins by added formulation components generally fall into two general classifications, "thermodynamic mechanisms" and "pure kinetic mechanisms." A stabilizer that functions by increasing the free energy of denaturation operates via a thermodynamic stabilization mechanism. Stability is conferred by shifting the equilibrium between "stable" native conformation and "unstable" unfolded conformations toward the native state. The "solute exclusion hypothesis," generally believed to be a major factor in stabilization during freezing, and the "water substitute hypothesis," often used to rationalize the stabilizing effect of saccharides during drying, are examples of thermodynamic stabilization mechanisms (6). Conversely, a stabilization strategy that functions by slowing the rate of the degradation process, without significantly affecting the equilibrium constants, operates via a pure kinetic mechanism. That is, if the rate constant for unfolding is sufficiently reduced to prevent unfolding on the experimental timescale, the protein will not unfold regardless of what the free energy of unfolding might become, and stabilization is purely kinetic. The "vitrification hypothesis" (7), which in its simplest form states that a system below its glass transition temperature is stabilized due to immobilization of the reactive entity in a solid-like glassy system, is the primary example of a pure kinetic stabilization mechanism. Thus, since mobility is required for reaction, and if one postulates that mobility is insignificant in the glassy state, the glassy state is a stable state.

Most protein stability studies have focused their interpretation either on a thermodynamic mechanism or on a pure kinetic mechanism, and consequently, there is some controversy and confusion over which mechanism is "correct." Since the direction of a formulation development effort may depend on which "theory" is being followed, clarification of the roles of thermodynamic stabilization and kinetic stabilization in given stability problems would provide some practical benefit. This review is an effort to provide such clarification. To this end, the major stresses, or destabilizing effects, that operate during the freeze-drying process are discussed, selected empirical observations regarding pharmaceutical stability in protein systems are presented, and the structure and dynamics in amorphous protein formulations are discussed.

STRESSES DURING FREEZE-DRYING: THERMODYNAMIC DESTABILIZATION FACTORS

As a sample is cooled during the freezing process, the solution normally supercools to a temperature about 10°C to 20°C below the equilibrium freezing point before ice first nucleates and crystallizes. Once crystallization does begin, the product temperature rises rapidly to near the equilibrium freezing point, decreases slowly until most of the water has crystallized, and then decreases sharply to finally approach the shelf temperature. The process of ice formation is accompanied by an increase in solute concentration between the growing ice crystals and an increase in solution viscosity. Solutes that tend to crystallize easily from aqueous solution, such as sodium chloride and buffer salts, may crystallize, but the protein itself and most carbohydrate excipients do not crystallize. Rather, they remain amorphous, and at the end of the freezing process exist in a glassy state containing a relatively large amount of unfrozen water (i.e., $\approx$ 20% wt/wt).

As an example of freeze concentration and the corresponding viscosity increase, the time profiles of product temperature, solute concentration, and viscosity are illustrated in Figure 1 for the freezing of 3% aqueous sucrose contained in glass vials. These data were calculated from freezing point depression data and viscosity data in the literature (8,9). Here, the solution supercools to about −15°C before ice begins to form, after which the temperature abruptly rises. During the initial portion of the ice growth stage between 10 and about 25 minutes, temperature remains nearly constant as the heat removed

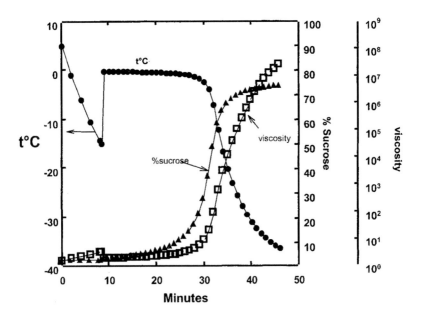

FIGURE 1 Time profile of temperature, concentration, and viscosity during freezing of 3% sucrose. Data are calculated assuming ice crystallization begins at −15°C and that the solution composition follows the equilibrium freezing point depression curve. Viscosities are estimated from a fit of viscosity data over a wide range of composition and temperature to a VTF-type equation. *Source*: Data taken from Refs. 7 and 8.

from the product is nearly balanced by the heat liberated by ice formation. Near the 25-minute mark, as the amount of unfrozen water becomes depleted, the sucrose concentration begins to increase sharply, and the temperature abruptly decreases. At the 30-minute mark, when most of the freezable water has been converted to ice, the sucrose concentration and the viscosity of the freeze concentrate increase very sharply. At about the 45-minute mark, viscosity is about seven orders of magnitude higher than for the starting solution. At this point, the freeze concentrate is nearly a glass.

Since most of the water is removed from the protein phase during freezing, most of the drying is actually accomplished during the freezing process. A number of stresses or perturbations of the free energy of unfolding may develop. First, since it is the unique nature of water that is often credited with stabilization of the native conformation via hydrophobic interactions, one might expect the thermodynamic stability of the native conformation would be decreased as most of the water is transformed to ice. Also, the phenomenon of freeze concentration will increase the protein concentration as well as increase the concentration of any potential reactant, thereby increasing the rate of bimolecular degradation reactions. Thus, as the system freezes, degradation rates may actually increase in spite of the decrease in temperature. Figure 2 gives an illustration of the impact of freeze concentration on degradation rate as well as demonstrates the impact of increasing viscosity on reactivity in the freeze concentrate. The data were calculated for the case of a trace amount of drug in 3% sucrose, using the viscosity and sucrose concentration data from Figure 1 and assuming an Arrhenius activation energy of 20 kcal/mol for the case where rate is independent of viscosity. If one assumes the degradation rate is completely uncoupled from macroscopic viscosity, the rate of degradation is nearly two orders of magnitude higher at the maximum than in the solution at the beginning of the freezing process. Eventually, as the temperature decreases

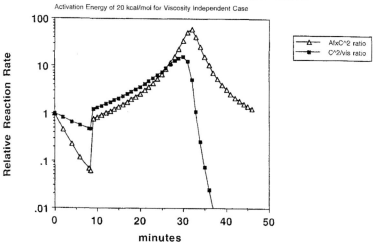

FIGURE 2 Calculated reaction rates for a second-order reaction in freezing 3% sucrose. Key: triangles = rate independent of viscosity with Arrhenius temperature dependence and a 20 kcal/mol activation energy; squares = reaction rate inversely proportional to viscosity.

sharply around 30 minutes, the Arrhenius factor dominates and the rate of reaction decreases. However, since a degradation pathway requires motion or mobility of some kind, most reactions would likely slow as the viscosity increases (7), at least for very high viscosities when the system approaches the glassy state.

The extreme example of coupling is with a diffusion-controlled reaction with the diffusion constant being inversely proportional to viscosity, as in the Stokes–Einstein equation. The curve defined by the square symbols in Figure 2 represents this extreme case. However, even with complete coupling of the reaction rate to viscosity, the degradation rate reaches a maximum of about one order of magnitude higher than in the starting solution. However, once the increase in viscosity dominates, the reaction rate decreases sharply toward extremely low values. While a protein formulation of interest may deviate from the quantitative behavior shown in Figure 2, two generalizations seem valid. Firstly, due to increasing solute concentration, bimolecular processes will be accelerated during the freezing stage in spite of decreasing temperature. Direct experimental observation of this phenomenon has been reported (10). Secondly, at least for reactions that are viscosity dependent, the rates will decrease dramatically near the end of freezing. Indeed, the principle of stabilization by vitrification (7) is based on the concept that viscosity increases near the glass transition temperature depress or eliminate degradation reactions. Experimental studies of three different reactions in frozen maltodextrin systems lend support to this view, though not all reactions are viscosity dependent (11). Rate constants for enzymatic hydrolysis of a substrate, aggregation of a protein, and oxidation of ascorbic acid were obtained in frozen maltodextrin solutions near the T_g' of $-10°C$. The data show about an order of magnitude reduction in rate constant between $-5°C$ and $-10°C$ for enzymatic hydrolysis and for ascorbic acid oxidation, but the protein aggregation reaction studied shows only a small temperature dependence throughout the range studied. Surprisingly, the rate-determining step for aggregation appears not to be viscosity dependent. However, the possibility of significant dilution with ice melt above T_g' confounds the interpretation of these data.

While sucrose and most excipients fail to crystallize during freezing, mannitol, glycine, sodium chloride, and phosphate buffers will crystallize, if present as the major component, once freeze concentration provides sufficient supersaturation. It should also be noted that very high concentrations may be reached before crystallization occurs (i.e., about 6 molal for NaCl), so the ionic strength environment of the protein during freezing may be quite different than in the starting formulation and could present a "stress" for protein stability (4,5). Crystallization of buffer components, resulting in massive pH shifts, may present an even greater stress for proteins. Under equilibrium conditions attained by seeding, the sodium phosphate buffer system shows a dramatic decrease in pH of about 4 pH units due to crystallization of the basic buffer component, $Na_2HPO_4.2H_2O$ (12). Conversely, the potassium phosphate system shows only a modest increase in pH of about 0.8 pH units. Under non-equilibrium conditions (i.e., no seeding) and with lower buffer concentrations, the degree of crystallization is less and the resulting pH shifts are moderated (13). Table 1 shows data accumulated (14,15) during freezing of phosphate buffer solutions in large volumes at cooling rates intended to mimic freezing in vials. For the concentrated buffer solutions (100 mM), the frozen pH values are

TABLE 1 Shifts in pH During Nonequilibrium Freezing with Phosphate Buffer Systems

Concentration (mm)	Initial pH	Frozen pH	Δ pH
Sodium phosphate buffer			
100	7.5	4.1	−3.4
8	7.5	5.1	−2.4
Potassium phosphate buffer			
100	7.0	8.7	+1.7
100	5.5	8.6	+3.1
10	5.5	6.6	+1.1

Source: From Refs. 13 and 14.

close to the equilibrium values. However, lowering the buffer concentration by an order of magnitude considerably reduces the pH shift observed during freezing. It should also be noted that, under some conditions, potassium phosphate buffers also give large pH shifts during freezing. As shown in Table 1, if the initial pH is 5.5, the 100 mM potassium phosphate buffer increases in pH by 3.1 units during freezing. Clearly, if a protein's structural integrity is sensitive to pH shifts, buffer crystallization must be avoided. In our experience, the best solution is to formulate such that the weight ratio of buffer to other solutes is very low (16,17).

It is well-known (18) that protein adsorption to surfaces, such as the air-water interface, may result in a perturbation of the conformation. That is, adsorption is a possible stress. While the surface area of the air-water interface is minimal during a well-designed freezing process, liberation of dissolved air during thawing may generate numerous bubbles, thereby providing a significant surface area for protein adsorption and conformational change. During the freezing process itself, the major interfacial area is the aqueous-ice interface. As the degree of supercooling increases (normally, as the rate of cooling increases), the number of ice crystals increases, thereby increasing the aqueous-ice interfacial area. It is clear that if a protein were to adsorb on the ice crystals and suffer a loss of conformational stability, the formation of ice itself would be a significant stress during freezing. Several observations suggest this speculation has merit. Freezing studies with human growth hormone (hGH) (19) show more insoluble aggregates develop during rapid freezing in a −80°C bath than during freezing procedures that cool more slowly. Classically, one expects less aggregation during a fast-cooling process since the residence time in the potentially reactive freeze concentrate is much less than in a slow-cooling process (i.e., the time required to reach the glassy state is much less). However, since the fast-cooling process will generate a greater aqueous-ice interface, which would maximize the fraction of protein adsorbed, the authors concluded that hGH was denatured by adsorption on ice. We have observed that more rapid freezing to lower temperatures results in more air bubbles on thawing, so an alternate interpretation of the hGH data might be made in terms of denaturation at air bubbles.

A recent study of unfolding during freezing provides strong evidence that proteins can indeed unfold as a result of adsorption to the ice surface (20). A summary of the major findings of this study is given in Figure 3 where the solid circles represent the average phosphorescence lifetime of Trp-48 of azurin in a 1 mM potassium phosphate buffer. All these systems contain ice at subzero temperatures and were formed by seeding with ice at −2°C followed by rapid equilibration to the temperature of interest. The heavy line gives the

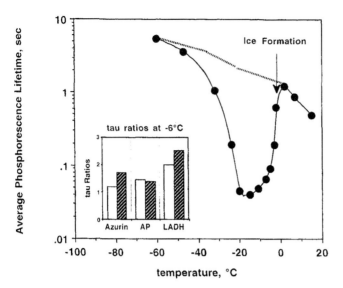

FIGURE 3 The effect of freezing on the average phosphorescence lifetime of Trp-48 of azurin. The circles represent data for protein in 1 mM phosphate buffer. Frozen samples were prepared by seeding a solution at $-2°C$ with ice and quickly bringing the sample to the temperature of interest. The dashed line without symbols gives the behavior of protein in a 1:1 phosphate buffer: ethylene glycol system that does not freeze. The inset shows ratios of lifetimes, low ice surface area divided by high surface area, for the freezing rate variation experiment (*light-shaded bars*) and the annealing experiment (*dark-shaded bars*). The proteins studied were azurin, liver alcohol dehydrogenase (LADH), and alkaline phosphatase (AP). *Source*: Data from Ref. 20.

corresponding lifetime data for a 1:1 mix of buffered protein and ethylene glycol, a system intended to illustrate behavior for a system that does not freeze. It should be noted that such data show a sharp decrease in lifetime when the protein undergoes unfolding. As the temperature is lowered, the lifetime increases reflecting the lower temperature and increased viscosity. However, as soon as ice is introduced, the lifetime decreases abruptly by roughly two orders of magnitude, signaling denaturation of the protein. Of course, as soon as ice forms, a significant amount of freeze concentration occurs and one could argue that several mechanisms other than adsorption of protein to ice are causing the denaturation. However, the authors (20) provide additional data, given in the inset in Figure 3, that are difficult to interpret unless one admits that denaturation at the ice surface is at least a critical factor. Samples of both high and low surface area of ice were prepared by first seeding with ice at $-2°C$, followed by cooling rapidly ($\approx 200°C/min$) or slowly ($\approx 1°C/min$) to a final temperature of $-6°C$. The fast-cooled samples were assumed to have higher surface area than the slowly cooled samples. Additional low surface area samples were prepared by allowing the fast-cooled samples to "anneal" at $-6°C$ for 10 hours to increase the size of the ice crystals and decrease the surface area. The inset in Figure 3 gives ratios of the lifetime of a low surface area sample to the lifetime for the corresponding high surface area sample. A value greater than unity means that the extent of denaturation is less for the low surface area sample. The light-shaded bars represent the freezing rate variation experiment, and the dark bars give the ratios for the annealing experiment. For all six examples, the ratio is greater than unity, indicating that denaturation was indeed less for the low surface area samples. Since all samples were studied at the same temperature, $-6°C$, and the experiments were so well controlled, one is left with the conclusion that, at least with these proteins, denaturation at the ice surface is a significant factor in protein denaturation during freezing. From a mechanistic viewpoint, it is not clear why a protein should denature at the ice surface. One might speculate that the mechanism involves very strong electric fields that can be generated during crystallization of ice via preferential incorporation of one ionic species into the ice lattice (21). However, except for noting that the fields increase with increasing crystal growth rate and that the effect on the protein would obviously be more severe the larger the ratio of ice surface area to protein concentration, there is no evidence known to this author that would link the electric field effect to protein denaturation. In any event, it is clear that ice itself is a stress to protein stability during freezing.

One might speculate that a major role of a surfactant in stabilization during freezing might involve adsorption at the aqueous-ice interface to prevent adsorption of protein with subsequent denaturation. A recent work (22) summarizes the literature precedents for this view and provides a convincing set of data to support the hypothesis. Freezing protocols that should produce a higher specific interfacial area produce greater levels of particulates (i.e., insoluble aggregates), and for a series of proteins, particulate levels after freeze-thaw are well correlated with particulate levels after shaking protein solutions with small teflon beads. Thus, it does appear that formation of insoluble aggregates during freezing does arise from surface denaturation, and the surface involved is the surface of ice. In all cases, addition of low levels ($\approx 0.01\%$) of surfactants greatly retards particulate formation.

The conformational stability of a protein is normally a rather delicate balance between various interactions or "forces," and these interactions may well be modified by changes in the solution environment and/or temperature. Increases in temperature will ultimately decrease the free energy of unfolding to the point where the thermodynamically stable form of the protein is the unfolded or denatured form. At this point, provided the unfolding process is not kinetically hindered, the protein spontaneously unfolds, normally providing a moiety that is much more prone to *irreversible* (i.e., degradation) processes. Solution environments, particularly pH and the presence of chemical denaturants, significantly impact the onset of denaturation. Of course, since freeze-drying is a low-temperature process, high-temperature denaturation is not directly relevant to destabilization during freeze-drying. However, just as proteins undergo thermal denaturation at elevated temperatures, proteins also undergo spontaneous unfolding at very low temperatures, denoted "cold denaturation" (23,24). Estimated cold denaturation temperatures are often well below freeze-drying temperatures and therefore are of questionable relevance to freeze-drying. However, estimates of cold denaturation temperatures are based on thermodynamic parameters measured in dilute aqueous solutions. The impact of perturbations caused by freeze concentration is largely unknown, and therefore the role of cold denaturation in protein inactivation during a practical freeze-drying process is uncertain.

The preceding discussion has focused on stresses that develop during the freezing process. However, since it seems unlikely that these stresses would be relieved during drying, the same stresses must also exist during the drying process. In addition, during drying, the moisture content in the protein phase is reduced from on the order of 20% water to very low levels, often less than 1%. This additional drying may be considered an additional stress. Indeed, the water substitute hypothesis (6) is based on the proposition that a significant thermodynamic destabilization occurs when the hydrogen bonding between protein and water is lost during the last stages of drying. The use of a "water substitute" as a lyoprotectant allows a hydrogen bonding interaction between protein and the water substitute, which thermodynamically stabilizes the native conformation and preserves activity.

During the early portions of freezing, the system is mostly aqueous and of low or moderate viscosity. During the last stages of freezing, and during both primary and secondary drying, the system is a glass or at least not much above the glass transition temperature. These differences are potentially important in the protein's response to a given thermodynamic stress. The timescales of the various stages of freeze-drying are also different. Relative to drying, freezing is relatively fast. Freezing is typically over within a few hours while drying often requires days. However, it must be noted that primary drying, or ice sublimation, constitutes the longest portion of the drying process. Although some water desorption does occur during primary drying, low water content in the solute phase is not achieved until all ice has been removed, and the process enters the secondary drying stage (25). During the early portion of secondary drying, the water content decreases quickly from roughly 10% to 20% to less than 5% within a few hours. Thus, assuming that it is during this drying period that the "drying stress" occurs, the timescale is roughly the same as the freezing timescale.

Since both the last part of freezing and the entire drying stages are normally carried out in a glassy or near glassy state, molecular mobility should be greatly restricted relative to the fluid state that prevails during early freezing, and the dynamic response to a thermodynamic stress will be significantly slower than for a fluid system, assuming the dynamic response depends on viscosity. If the dynamic response is sufficiently slow, thermodynamic *instability* will have no consequence as insufficient time is available for unfolding and subsequent degradation. Recent studies in our laboratories of the kinetics of protein unfolding in highly viscous aqueous systems demonstrate that the protein unfolding rate is strongly coupled to viscosity, at least for the two proteins studied, phosphoglycerate kinase and β-lactoglobulin, in high sucrose content systems. These data (Figs. 1 and 2) suggest that unfolding is on the timescale of months, even at temperatures 10°C to 20°C above the glass transition temperature. The fact that protein inactivation during drying does occur suggests that the mobility needed for inactivation is nearly completely decoupled from viscosity in these systems. Of course, if the mechanism for inactivation does not involve protein unfolding, we do not necessarily expect inactivation kinetics to track with unfolding kinetics.

STABILIZATION OF PROTEINS FOR FREEZE-DRYING: SELECTED EMPIRICAL OBSERVATIONS
In-Process Stability: Freeze-Thaw and Freeze-Dry Stability
While many proteins survive the freeze-drying process with little or no degradation, other proteins exhibit significant degradation and loss of activity during processing. Multimeric proteins seem particularly prone to degradation during freeze-drying (26–29). Degradation during the freeze-drying process may arise

during freezing and/or during drying. As a measure of the degradation during freezing, "freeze-thaw" stability studies are carried out, and to estimate (roughly) the degradation during drying, stability during freeze-drying is compared to stability during freeze-thaw. The basic assumption is that degradation during thawing is comparable to degradation during reconstitution, and therefore the difference in activity between a freeze-dried-reconstituted sample and a freeze-thawed sample is a measure of the loss in activity during drying. This assumption is likely a reasonable approximation for a fast thawing process, at least if air bubbles released during thawing are not a major factor. One observation of particular significance is that some excipients stabilize during both freezing *and* drying (denoted "lyoprotectants"), while others stabilize only during freezing (denoted "cryoprotectants") (26,30,31). Data for phosphofructokinase (PFK), active as a tetramer, illustrate this observation quite well (Fig. 4). With no "stabilizer" added to the formulation, PFK is completely deactivated during freeze-thaw. With all stabilizers at the relatively high level of 0.5 M, some stabilize reasonably well during freeze-thaw but offer no protection during freeze-drying (proline and trimethylamine N-oxide). The disaccharides (trehalose, sucrose, and maltose) stabilize both during freezing and drying, and therefore, at a level of 0.5 M, are effective lyoprotectants. Low levels of disaccharides are not good cryoprotectants (31) and therefore cannot be good lyoprotectants. Of course, if a given formulation offers no protection during freezing, any potential protection during drying will be invisible with the usual experimental design. Polyethylene glycol (PEG) is found to be an exceptionally efficient cryoprotectant for PFK but offers no protection during drying (31). Trehalose and glucose at low levels ($\approx$ 0.1 M) offer essentially no protection during freeze-drying because, at low levels, they are not cryoprotectants. However, when PEG is used in combination with a low level of trehalose (or glucose), nearly complete stabilization during freeze-drying is

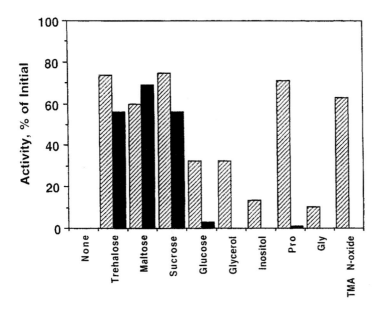

FIGURE 4 Comparison of freeze-thaw stability with freeze-dry stability: phosphofructokinase with additives at 0.5 M. Key: shaded bar = freeze-thaw, solid bar = freeze-dry. *Source*: Data from Ref. 26.

obtained. Studies with lactate dehydrogenase (LDH) give essentially the same results (31). The conclusion is that PEG stabilizes during freezing while the sugar stabilizes during drying, even at relatively low levels. Thus, the combination of stabilizers is a good lyoprotectant. Sugars are effective cryoprotectants only when used at relatively high concentrations, but are generally effective drying stabilizers when used at much lower concentrations. As a general rule, the relevant concentration unit for stabilization during freezing is the molar concentration *in solution*, whereas for stabilization during drying, the relevant concentration unit is the weight ratio of stabilizer to protein (or protein plus buffer) (30–32). Since different formulation strategies are needed for stabilization during freezing than are required for stabilization during drying, it is concluded that the stresses during freezing are different from those during drying, meaning that the mechanisms of destabilization (and stabilization) are different (26,30,31).

Catalase is an example of a multimeric protein that is relatively stable during freeze-thaw, but without a suitable stabilizer suffers significant loss of activity during drying. Without stabilizers, loss of activity during freeze-thaw is only about 20% (33) but loss during freeze-drying is 65% (27). Addition of glucose or sucrose reduces the loss on both freeze-thaw and freeze-drying to about 10%, suggesting that the roughly 45% loss on drying the pure enzyme has been reduced to near zero by these excipients. Mannitol and a variety of saccharides also stabilize during freeze-drying. The degree of stabilization is not correlated with the glass transition temperature of the pure excipient but does appear correlated with the molecular weight of the saccharide. As the molecular weight increases, protein activity remains constant through maltotriose but then decreases, with the high molecular weight dextran (150 kDa) being the least effective of the excipients studied.

L-Asparaginase provides another example of a multimeric protein that suffers severe degradation during drying due to deaggregation of the active tetramer (28). Without stabilizers, L-asparaginase loses about 80% of its initial activity during freeze-drying. Glucose, tetramethylglucose (TMG), mannose, sucrose, and poly(vinylpyrrolidone) (PVP) are all extremely effective lyoprotectants, preserving essentially 100% of the initial activity. Mannitol preserves only about 50% of the initial activity, perhaps due to partial crystallization of the mannitol. Here, as with catalase, a monosaccharide is as effective as a disaccharide. Stability does not correlate with residual water after freeze-drying (28), and since the glass transition temperatures of the effective stabilizers range from 39°C for glucose to about 170°C for PVP, it is obvious that stabilization is not correlated with T_g. The effectiveness of TMG and PVP demonstrates that "sugar-type" hydrogen bonding to the protein is *not* essential for stability.

Storage Stability

Storage stability has generally been the more serious stability issue faced by therapeutic proteins. Storage stability can be extremely formulation specific (16,32,34,35), and even with a knowledge of the major degradation pathways in solution, selection of the optimum formulation for a solid is far from obvious. We illustrate the sensitivity of stability to formulation details with studies of an important protein product, hGH.

hGH is a monomeric 22-kDa protein marketed as a freeze-dried solid with recommended refrigerated storage. While hGH is easily freeze-dried with little or no degradation (16), degradation does occur during storage of the freeze-dried solid at 25°C and 40°C. Chemical degradation proceeds by both methionine oxidation and asparagine deamidation, and aggregation (mostly dimer) develops after storage and reconstitution (16,36,37). Formulation with mannitol or glycine improves storage stability slightly, but a combination of mannitol and glycine (weight ratio of hGH:glycine:mannitol of 1:1:5) provides better stability than an equivalent amount of either mannitol or glycine alone (16), a result attributed to the observation that glycine remains mostly amorphous in the combination formulation. An excipient system of mannitol alone or glycine alone results in nearly 100% crystalline excipient and therefore would not be expected to improve stability greatly, that is, very little excipient could be in the protein phase. Formulation with lactose provides a completely amorphous system that does reduce aggregation dramatically. However, lactose is a reducing sugar, and as might be expected, a massive amount of a degradation product, likely representing an adduct of lactose and hGH, is formed after only one month of storage at 25°C (16). Thus, while a 100% amorphous excipient system offers the best potential for stabilization, a "reactive" excipient system, such as a reducing sugar, is clearly unacceptable.

Chemical and aggregation stability of hGH in several other 100% amorphous systems are compared with corresponding stability in pure protein and the glycine:mannitol formulation in Figure 5 (38). Hydroxyethyl starch (HES),

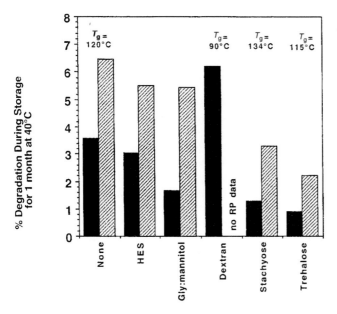

FIGURE 5 The effect of excipients on the storage stability of freeze-dried hGH. Samples were stored for one month at 40°C. Key: solid bars = aggregation (primarily dimer), shaded bar = chemical degradation via methionine oxidation and asparagine deamidation. The glass transition temperatures of the "initial" freeze-dried formulations are given above the bars when a glass transition temperature could be measured by DSC. The glycine:mannitol formulation is a weight ratio of hGH:glycine:mannitol of 1:1:5, the dextran formulation is 1:6 hGH:dextran 40, "none" means no stabilizer, and the others are 1:1 hGH:stabilizer. All formulations contain sodium phosphate buffer (pH 7.4) at 15% of the hGH content. Initial moisture contents are all ≈ 1%. *Abbreviations*: hGH, human growth hormone; DSC, differential scanning calorimetry.

stachyose, and trehalose are formulated in a 1:1 weight ratio of excipient:hGH, whereas the dextran formulation is 6:1 dextran:hGH. While the concept that an excipient system must remain at least partially amorphous to improve protein stability is not in question, it is clear that remaining amorphous is not a sufficient condition for stability. Apparent aggregation in the dextran formulation is greater than in the pure protein. HES shows a slight improvement in stability over the pure protein, but is not nearly as effective as the glycine:mannitol formulation, and increasing the level of HES to 3:1 HES:hGH does not improve stability (38). Conversely, both stachyose and trehalose provide better stability than the glycine:mannitol system, with trehalose superior to stachyose. All systems are glasses at the storage temperature of 40°C, and for those formulations where glass transition temperatures are available, it is clear that storage is well below the T_g, and there is no simple relationship between T_g and stability. While one might speculate that a glass is more "solid," and therefore more stable, the higher the difference between T_g and the storage temperature, the data are not consistent with this speculation. Comparing the stachyose and trehalose formulations, which are both 1:1 formulations with hGH, the T_g of the stachyose formulation is nearly 20°C higher than the trehalose formulation, but trehalose offers slightly better stability than does stachyose. These observations and other similar results (38) suggest that while it is necessary for the formulation to have a T_g well above the highest anticipated storage temperature for both elegance and stability reasons, T_g is not a relevant stability variable for systems stored well below their glass transition temperatures.

STRUCTURE AND DYNAMICS IN AMORPHOUS PROTEIN FORMULATIONS
Protein Formulations as Amorphous Solids

Freeze-dried protein formulations are amorphous systems, at least in part, and the physical and chemical behavior of such products depends on the characteristics of amorphous systems, perhaps as much as their behavior depends on the unique behavior of proteins. Amorphous materials below their glass transition temperatures are termed glasses, and in many respects are solids in the same sense as are crystalline solids. That is, while the long-range order characteristic of crystalline solids is absent, short-range order does exist, and the dynamics in glasses more closely resembles crystalline solids than liquids above their equilibrium fusion temperatures. Glasses differ from liquids in another important respect. The short-range order or structure in liquids represents an equilibrium between possible configurations that responds essentially immediately to changes in temperature. The short-range order in glasses does not represent an equilibrium distribution of configurations. Rather, as a first approximation, the short-range order or configurations characteristic of the liquid at T_g are "frozen in" by cooling quickly through the glass transition, and the resulting glass is in a metastable state. With aging, transitions to lower energy states or enthalpy relaxations occur (39). In short, sufficient mobility exists even well below the glass transition temperature to allow changes in configuration. These relaxations are typically nonexponential in time due to contributions from a number of substates in the glass (39). Dynamics is important in amorphous materials since nearly any degradation reaction will require some degree of motion, or molecular mobility, for the reaction to proceed at a significant rate. The glass transition temperature, T_g, marks the division between mostly solid dynamics and mostly liquid state dynamics. It is important to emphasize, however, that solid dynamics does not mean zero mobility.

A protein dissolved in another glassy component (i.e., sucrose) could behave as a two-phase system with regard to mobility. That is, the protein molecules themselves could undergo internal motion in a rigid matrix and have a pseudo glass transition or mobility transition, which is not strongly coupled with the glass transition of the sucrose matrix. Using Fourier transform infrared (FTIR) spectroscopy to study internal protein motion in carbonmonoxymyoglobin (MbCO) dissolved in aqueous glycerol, protein internal "glass transitions" were determined and compared to the glass transition temperatures of the systems as a whole determined by differential scanning calorimetry (DSC) (40). The protein glass transition temperatures were very close to the corresponding glass transition temperatures determined by DSC. For 65% glycerol and 75% glycerol, the DSC glass transition temperatures are −124°C and −98°C, respectively. The corresponding protein glass transition temperatures are −118°C and −95°C. Thus, at least in these systems, the solvent and the protein dynamics are strongly coupled, likely due to the hydrogen bonding interactions between the solvent and the protein surface (40). While it is perhaps somewhat unusual to refer to the mobility transition in the protein as a glass transition, it should also be noted that the glassy MbCO systems studied also show protein intramolecular relaxation process (i.e., transitions between protein substates) that are both nonexponential in time and non-Arrhenius in temperature dependence, a property characteristic of glasses (40).

Evidence for coupling between glassy matrix dynamics and protein dynamics is not restricted to low-temperature aqueous systems. Studies of the kinetics of ligand binding in MbCO (41) dissolved in *dry* glassy trehalose demonstrates that the glassy trehalose matrix suppresses the equilibrium between protein conformational substates on the timescale of the ligand binding reaction at least up to room temperature. While a protein glass transition temperature was not obtained, the data do demonstrate significant coupling between internal protein motions and the dynamics of the glassy matrix, trehalose.

Coupling between matrix dynamics and internal protein dynamics could have significant pharmaceutical stability implications. While limitations on translational diffusion would be expected to moderate bimolecular reactions regardless of the degree of coupling between protein internal dynamics and the matrix dynamics, degradation processes that depend only on motions within the protein molecule would not necessarily be quenched in the glassy state unless the protein internal dynamics was strongly coupled with the dynamics of the glassy system as a whole. Thus, one would expect optimal stability in those glassy systems that provide effective coupling of the protein dynamics with the dynamics of the glassy matrix.

Molecular Motion, Relaxation, and the Glass Transition

The Stokes–Einstein equation,

$$D = \frac{kT}{6\pi\eta a},\tag{1}$$

predicts that the translational diffusion coefficient, D, is inversely proportional to the coefficient of viscosity, η, where k is the Boltzmann's constant and "a" is the effective hydrodynamic radius of the diffusing species. Thus, while the point at which a material becomes a glass is historically defined in terms of the viscosity (i.e., $\eta > 10^{14}$ poise) (42), if the Stokes–Einstein relationship is valid, the corresponding definition of a glass could also be based on the diffusion coefficient. The translational diffusion coefficient is given in terms of the "jump distance," x, and the diffusional correlation time, τ, or "relaxation time" by

$$D = \frac{x^2}{2\tau} \qquad (2)$$

so the diffusion coefficient is inversely proportional to the diffusional relaxation time. A similar relationship holds for rotational motion. Since the electrical mobility of an ion is directly related to the diffusion coefficient (43), electrical conductance is also inversely proportional to the coefficient of viscosity, given the validity of the Stokes–Einstein equation. We use the term "mobility" in a general sense to refer to translational or rotational diffusion constant, or reciprocal of relaxation time. Assuming validity of the Stokes–Einstein equation, mobility and viscosity are inversely related.

As a liquid is cooled near the glass transition, viscosity increases sharply and the temperature dependence of viscosity becomes distinctly non-Arrhenius. That is, the apparent activation energy increases as the temperature decreases. The Adam–Gibbs equation is a theoretical result describing relaxation behavior in highly viscous systems that was developed using a statistical mechanical analysis of configurational changes in highly viscous systems (44). Configurational changes in systems close to the glass transition temperature take place by highly cooperative motions involving rearrangements in a region whose size is determined by the configurational entropy of the system. As the temperature in a highly viscous system decreases, the configurational entropy decreases, and the size of the cooperatively rearranging region increases, thereby increasing the total free energy barrier to the configurational change and slowing the process. The relationship between configurational entropy, S_c, and the relaxation time is given by

$$\tau(T) = A \exp\left(\frac{C}{TS_c}\right) \qquad (3)$$

where C is a constant which is proportional to the molar (or segmental, for polymers) change in chemical potential for a transition, and A is a pre-exponential factor practically independent of temperature. The configurational entropy must vanish at some temperature, T_0, otherwise one is faced with the "unphysical" conclusion that the configurational entropy will become negative at some low temperature. This result is based on the observation that extrapolations of configurational entropy from temperatures above T_g to lower temperatures predict negative values roughly $50°$ below T_g (i.e., at T_0). Thus, the temperature, T_0, marks the temperature at which the system, at equilibrium, must undergo a second-order phase transition losing the configurational heat capacity. Below T_0 the configurational entropy is zero. With $S_c = 0$ for $T < T_0$, the configurational entropy at temperature T, $S_c(T)$, becomes

$$S_c(T) = \int_{T_0}^{T} \left(\frac{\Delta C_p}{T}\right) dT \qquad (4)$$

where ΔC_p is the configurational part of the heat capacity, frequently taken to be the difference in heat capacity between the equilibrium melt and the glass. In cases where the heat capacity of the glass is significantly larger than the heat capacity of the crystalline phase, ΔC_p should perhaps be taken as the difference between the heat capacity of the melt and the crystal.

Angell (45) takes the configurational heat capacity as inversely proportional to absolute temperature, $\Delta C_p = K/T$, where K is a constant of the material. With this relationship, the relaxation time in equation (3) becomes

$$\tau(T) = A \exp\left(\frac{DT_0}{T - T_0}\right) \tag{5}$$

where D is a constant characteristic of the material $(D = C/K)$. Since "C" is directly proportional to the molar change in chemical potential for a transition, so also is "D." This result is of the same form as another empirical equation known to represent the behavior of highly viscous systems, commonly referred to as the VTF equation (Vogel–Tammann–Fulcher equation). If one insists that the relaxation time for all glasses is (roughly) the same at the glass transition temperature, ≈ 100 seconds, and further insists that τ is the same for all materials at the extreme high-temperature limit (16 orders of magnitude change between T_g and the high-temperature limit), a relationship between D and T_0 results[a] (45):

$$\frac{T_g}{T_0} = 1 + \frac{D}{36.85} \tag{6}$$

Thus, a large value of D means a larger difference between the glass transition temperature and the zero mobility temperature (i.e., larger ratio of T_g/T_0). From equation (5), it is apparent that a larger difference between T_g and T_0 means that as $T \to T_g$, the temperature dependence of relaxation time is smaller. That is, the effective activation energy is smaller, and the deviation from Arrhenius behavior is less for large D. Depending on the nature of the amorphous material, values of D do vary (45). Amorphous materials with large values of D are denoted "strong glasses" while materials with small values of D as "fragile glasses" (45). In short, not all glasses are equivalent in the temperature dependence of relaxation time and therefore are not equivalent in the deviation from Arrhenius behavior.

The traditional "derivation" of the Adam–Gibbs equation as given above assumes the configurations are always in thermal equilibrium, and therefore the results given by equations (4) and (5) do not apply below the glass transition, as the configurations in a glass are *not* in thermal equilibrium. If one assumes that the configurational heat capacity, ΔC_p, is equal to the difference in heat capacity between the melt and the glass at T_g (i.e., no configurational contribution to the heat capacity of a glass), the configurational enthalpy and entropy do not change as the temperature is decreased below the glass transition region. Thus, the configurational entropy of the glass is equal to the configurational entropy of

---

[a]When using the VTF equation in viscosity form, one assumes a difference between the viscosity at T_g and the high-tempearature limit of 17 orders of magnitude (43), which changes the numerical value of 36.85 to 39.14.

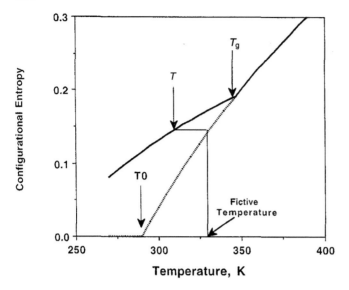

FIGURE 6 A schematic of the temperature dependence of configurational entropy: definition of fictive temperature. Data calculated using input data for sucrose.

the "equilibrium" glass at the glass transition temperature at all temperatures below T_g. In this case, the temperature dependence of the relaxation time becomes Arrhenius below T_g (46). However, the observation that, at least for some materials (47,48), the heat capacity of the glassy phase is significantly higher than the heat capacity of the crystalline phase suggests that one should use the difference in heat capacity between melt and crystalline phase as the configurational heat capacity. With this choice, the configurational heat capacity of the real glass is not identically zero below the glass transition temperature, and the configurational entropy does not remain absolutely constant below T_g, although the configurational entropy curve of the real glass does lie well above the corresponding curve for the equilibrium glass (Fig. 6). The configurational entropy of the real glass at temperature, T, is equal to the configurational entropy of the equilibrium glass at a higher temperature, T_f, termed the "fictive temperature" (49). Thus, we may write

$$S_c(T) = S_c^e(T_f) = \int_{T_0}^{T_f} \frac{\Delta C_p}{T} dT \qquad (7)$$

where "S_c^e" denotes the configurational entropy of the equilibrium glass (i.e., the dotted line in Fig. 6), and the configurational heat capacity, ΔC_p, is the difference between the heat capacity of the melt and the crystalline phase at T_g. Substitution of equation (7) for the configurational entropy of the real glass into the expression for relaxation time, equation (3), and assuming $\Delta C_p T$ is constant, then gives

$$\tau(T, T_f) = \tau_0 \exp\left(\frac{D T_0}{T - (T/T_f)T_0}\right) \qquad (8)$$

From the schematic given by Figure 6, it is clear that the fictive temperature is between the temperature, T, and the glass transition temperature, T_g. A quantitative relationship for fictive temperature may be developed using the expression for the difference in configurational entropy between the real glass and the equilibrium glass, as given by

$$S_c(T) - S_c^e(T) = \int_T^{T_g} \frac{\Delta C_p^{l,g}}{T} dT, \ \Delta C_p^{l,g} = C_p^l - C_p^g \tag{9}$$

where C_p^l is the heat capacity of the liquid or "equilibrium glass," and C_p^g is the heat capacity of the "real" glass. Performing the integration in equation (9), assuming the quantity $T \Delta C_p^{l,g}$ is independent of temperature, and combination with the integrated forms of equation (7) then leads to the desired expression for fictive temperature

$$\frac{1}{T_f} = \frac{(1 - \gamma_c)}{T} + \frac{\gamma_c}{T_g}; \quad T \leq T_g$$

$$\gamma_c \equiv \frac{\Delta C_p^{l,g}}{\Delta C_p} = \frac{C_p^l - C_p^g}{C_p^l - C_p^{xstal}} \tag{10}$$

where C_p^{xstal} is the heat capacity of the crystalline phase. All heat capacities are evaluated at the glass transition temperature. The combination of equations (8) and (10) constitutes a generalization of the Adam–Gibbs theory for the temperature dependence of the structural relaxation time. Above the glass transition temperature, fictive temperature and temperature are identical, and the expression for relaxation time given by equation (8) reduces to the usual expression (i.e., equation 5). Below T_g, equation (10) is used to evaluate the fictive temperature, and the expression for relaxation time differs from the usual expression, the magnitude of the change depending on the value of γ_c. Note that if the configurational heat capacity of the real glass is zero (i.e., the heat capacities of the real glass and the crystal are identical), $\gamma_c = 1$, and the fictive temperature is equal to the glass transition temperature at all temperatures below T_g. In this case, the relaxation time shows Arrhenius temperature dependence below T_g. At the other extreme, if the configurational heat capacities of the real glass and the liquid differ only slightly, $\gamma_c \approx 0$, $T_f \approx T$, and the relaxation time expression is the same both below and above T_g. We note that since enthalpy relaxation is nonexponential, a real glass consists of a number of substates, each having a different configurational entropy and a different fictive temperature. Thus, the results given in equations (8) to (10) refer to an average of substates for temperatures below T_g.

While most treatments of this subject assume that the heat capacities of the crystalline and glassy phases are essentially the same, and therefore Arrhenius temperature dependence is predicted below T_g, experimental heat capacity data for sucrose (47,48,50) indicate that the heat capacity of glassy sucrose is significantly higher than the heat capacity of crystalline sucrose, and $\gamma_c \approx 0.8$. Figure 7 shows calculated relaxation times for two hypothetical amorphous materials with $T_g = 70°C$, but with different fragilities (i.e., $D = 7$ and $\gamma_c = 0.8$ for a fragile glass like sucrose and $D = 23$ and $\gamma_c = 0.94$ for a representative strong

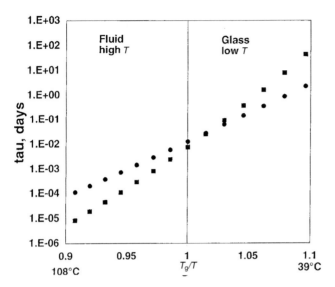

FIGURE 7 Comparison of structural relaxation times for fragile and strong glasses with glass transition temperatures of 70°C. The data were calculated using input data characteristic of sucrose except for the strength parameter, *D*. Key: circles = strong glass (*D* = 23), squares = fragile glass (*D* = 7).

glass). Above the glass transition temperature in the fluid state, the strong glass has the longer relaxation time, but in the glassy state the fragile glass has a longer relaxation time. Thus, assuming that pharmaceutical stability is correlated with structural relaxation, a protein formulation above T_g would be more stable in the strong glass, but below T_g the fragile glass would provide better stability. For the strong glass, temperature dependence is nearly Arrhenius both above and below T_g. For the fragile glass, significant deviations from linearity in the plot indicate non-Arrhenius temperature dependence both above and below T_g, although the non-Arrhenius behavior is more pronounced above T_g. To the extent that the thermal history of the glass will impact the value of γ_c, thermal history will impact the value of relaxation time. Thus, one might speculate that thermal history may well impact pharmaceutical stability in a glassy formulation. Evidence for such an effect is limited, but several studies do suggest greater chemical stability for samples annealed below T_g (51,52).

Structural relaxation times determined from enthalpy relaxation studies with sucrose and trehalose (39,53) are given in Figure 8. The structural relaxation times observed are qualitatively similar to those estimated for the fragile glass example in Figure 7. While there is considerable scatter in the data, it seems clear that the temperature dependence for sucrose is nonlinear and therefore non-Arrhenius. Insufficient data are available for trehalose to judge linearity. The trehalose structural relaxation times are lower than those for sucrose at the same T_g/T, indicating different fragility and/or a different value of γ_c.

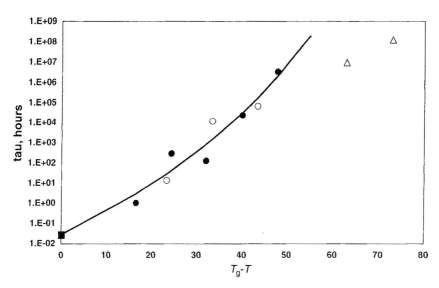

FIGURE 8 Structural relaxation times for quench-cooled glassy disaccharides as determined from enthalpy relaxation data. Structural relaxation times were obtained by a fit of the data to the stretched exponential function (39,52,53). Key: full circles = data for sucrose obtained by differential scanning calorimetry on annealed samples (data from Ref. 39), open circles = data for sucrose obtained by isothermal microcalorimetry (data from Refs. 52 and 53), open triangles = data for trehalose obtained by isothermal microcalorimetry (data from Ref. 53).

Instability and the Glass Transition: Coupling of Reaction Mobility with Structural Relaxation

The vitrification hypothesis assumes that the mobility relevant to pharmaceutical *instability* is strongly correlated with viscosity and structural relaxation time such that high viscosity and large structural relaxation time mean very slow, hopefully insignificant, degradation rate. The mobility required for a given reaction is thought of in terms of a series of diffusional jumps to produce the change in configuration required for the physical or chemical change. These diffusional jumps may involve the entire molecule as in a bimolecular aggregation process, the segmental diffusional motion of amino acid residues as in an unfolding reaction, or the motion on a smaller scale such as formation of the cyclic intermediate in a deamidation pathway. It is further assumed that these diffusional jumps require corresponding motion in the surrounding amorphous medium, otherwise reaction mobility would not couple with (i.e., be dependent on) medium viscosity. Thus, the rate constant for the reaction is proportional to a diffusion constant for the reaction motion, and through equation (2), the rate constant is inversely proportional to the reaction relaxation time. Perfect coupling with viscosity (i.e., validity of the Stokes–Einstein equation) would mean the reaction relaxation time and the structural relaxation time are directly proportional. However, the Stokes–Einstein equation is not entirely appropriate for the high-viscosity systems of interest to protein stability, and not all types of mobility are strongly coupled to viscosity (54).

It is well-known that while values of T_0 determined from ionic conductivity studies are usually identical to those determined from viscosity data on the same systems (9), the values of the strength parameters, D, do differ. Conductivity typically produces slightly smaller D values, meaning the structural relaxation times determined from viscosity "decouple" from the conductivity relaxation times with the differences becoming more marked as temperature decreases (9,54). With some inorganic glasses, the decoupling of conductivity from viscosity becomes extreme with the conductivity relaxation time being nearly 12 orders of magnitude smaller than the (viscosity) structural relaxation time (54). Small-molecule diffusion in glassy polymers also appears to be nearly completely decoupled from structural relaxation (55,56).

A study of mobilities for three different kinds of motion in a polymer having a glass transition at 35°C demonstrates that, even in the same polymer, not all types of motion couple with the viscosity in equal fashion (57). The mobilities are derived from D-NMR correlation times. Complete coupling with the glass transition would imply roughly nine orders of magnitude reduction in mobility between $1000/T = 2.7$ and the glass transition temperature. The data show (Fig. 9) that while chain fluctuation (i.e., chain diffusion) mobility appears to be strongly coupled to the glass transition, chain rotation is only moderately coupled, and ring flip motion of a side chain is nearly completely decoupled. Ring flip mobility shows Arrhenius behavior above and below the glass transition with essentially the same activation energy. The only impact of the glass transition on ring flip is a sharp but small decrease in mobility as the temperature decreases through T_g. It appears that mobility involving motion on a larger scale (i.e., more displacement and/or more generation of free volume needed for a "jump") correlates best with viscosity and the glass transition phenomenon, a conclusion also consistent with the observation made above regarding diffusion of small molecules in polymer glasses.

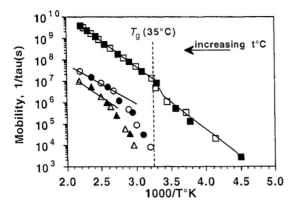

FIGURE 9 Coupling of mobility with the glass transition in a polymer system. Mobility is the reciprocal of the correlation time for the type of motion indicated. Correlation times were evaluated from deuteron NMR relaxation data. Key: triangles = reorientation or "fluctuation" of the chain axis; circles = rotation about the chain axis; squares =180° ring flips of side chain aromatic rings. Open and full symbols refer to different degrees of deuteration. *Source*: Data taken from Ref. 57.

The physical basis of variable coupling between mobility and structural relaxation time likely has its origin in the differences between free volume requirements or, in the context of the Adam–Gibbs theory, differences in the molar chemical potential change between structural relaxation and the relevant diffusional jump (54). Since the strength parameter for structural relaxation, D, is proportional to the molar chemical potential change, one might speculate that the reaction relaxation time could be written in the same form as the structural relaxation time but with a different strength parameter, D_{rx}, where $D_{rx} = gD$, and where the "coupling constant," g, would be expected to be less than unity unless the degradation process required translational diffusion of the entire protein molecule. Thus, with the degradation rate constant, k, being inversely proportional to the reaction relaxation time, one might expect a relationship of the form

$$k = A_k \exp\left(-\frac{gDT_0}{[T - (T/T_f)T_0]}\right) \qquad (11)$$

where the A_k is a preexponential constant depending on the details of the degradation mechanism. One would expect the value of A_k to decrease as the number of "diffusional jumps" needed to complete a reaction increases. Given a distribution of substates in the glass, the values of g and T_f would represent averages for the populated substates. Whether equation (11) is valid is a matter of speculation as no degradation studies have been determined in a series of amorphous systems where D, T_0, and T_f are also available. However, as discussed above, mobility data do provide experimental justification for at least the qualitative features of equation (11). That is, as long as degradation in a highly viscous amorphous material or glass is limited by diffusional motion of some kind, the degradation rate is likely to depend on the factors emphasized in equation (11): (*i*) the zero mobility temperature, T_0; (*ii*) the strength parameter, D, or the glass transition temperature, T_g (i.e., only two of the three parameters, T_g, T_0, and D are independent if equation (6) is valid); (*iii*) for a glass, thermal history through fictive temperature, T_f; and (*iv*) the degree of coupling as expressed in equation (11) via the coupling coefficient, g. Clearly, while $T \ll T_g$ is important for good stability, other material characteristics of the formulation are also important and may dominate.

Instability and the Glass Transition: Experimental Observations

A number of systems show marked deterioration of stability as temperature exceeds the glass transition temperature, but for storage well below the glass transition temperature, stability does not appear to depend directly on the difference between T_g and the storage temperature (Fig. 5). The correlation between reactivity and T_g is most easily discussed in terms of the Williams–Landel–Ferry equation (WLF equation). The WLF equation (58) is an expression for relaxation time, roughly equivalent to the VTF equation, which may be "derived" using free volume concepts. With fixed constants, the WLF equation states that the ratio of relaxation time at any temperature to the relaxation time at T_g depends only on the difference between temperature and the glass transition temperature. Assuming that reaction rate is inversely proportional to structural relaxation time in the amorphous phase, degradation rates normalized to the corresponding rates at T_g should depend only on $T - T_g$. Thus, in an experiment

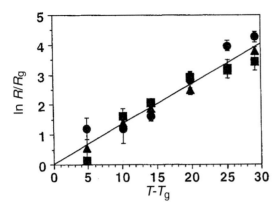

FIGURE 10 The stability of a freeze-dried monoclonal antibody:vinca alkaloid conjugate formulation. Desacetylvinblastine hydrazide is linked to the KS1/4 monoclonal antibody via aldehyde residues of the oxidized carbohydrate groups on the antibody. The formulation is conjugate:glycine:mannitol in a 1:1:1 weight ratio. Storage temperatures are 25°C and 40°C for samples with moisture contents of 1.4%, 3.0%, and 4.7%. Key: circles = dimer formation, triangles = free vinca generation (hydrolysis), squares = vinca degradation, solid line = best fit to the WLF equation. *Source*: From Ref. 59.

where both temperature and glass transition temperature are varied, stability is a function of only one variable, $T - T_g$. Degradation in a monoclonal antibody: vinca alkaloid conjugate system is consistent with the "$T - T_g$" dependence of relative reaction rate (17). Rates of dimerization, hydrolysis of the antibody-vinca linkage, and degradation of the vinca moiety (mostly oxidation) were obtained at two temperatures and three water contents (i.e., three T_g). The rate normalized to the rate at the glass transition temperature is a function only of $T - T_g$, (where here $T > T_g$) as suggested by the WLF equation (Fig. 10). All three degradation reactions form a single curve (a straight line), indicating the WLF constants are a characteristic of the amorphous system and independent of the nature of the degradation pathway, as expected.

The effect of residual water on the chemical degradation of hGH in a trehalose formulation provides another example of a correlation between stability and the glass transition (59). The trehalose formulation shows a well-defined glass transition temperature with DSC, with the expected decrease in T_g as the water content is increased. The pseudo first-order rate constant for chemical degradation is essentially independent of water content while the system remains glassy, but at least for the 50°C data, degradation increases sharply as increasing water content depresses the system glass transition temperature significantly below the storage temperature (Fig. 11). Rates of aggregation show essentially the same behavior. However, since stability is not sensitive to water content in the glassy state, stability is not correlated with $T - T_g$ below the glass transition temperature. The study of formulation effects on hGH stability (Fig. 5) suggests the same conclusion. That is, for glassy systems, stability is not directly related to the precise difference between storage temperature and T_g.

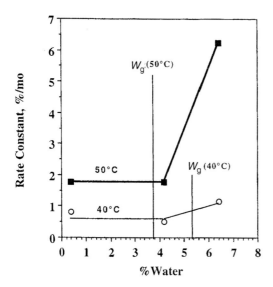

FIGURE 11 Chemical degradation in freeze-dried hGH formulated with trehalose as a function of water content at 40°C and 50°C. The pseudo first-order rate constant for degradation (% per month) is given for the combination of asparagine deamidation and methionine oxidation. The formulation is hGH:trehalose in a 1:6 weight ratio with sodium phosphate buffer (pH 7.4) at 15% of the hGH content. The highest moisture content samples were collapsed after storage at both 40°C (moderate collapse) and 50°C (severe collapse). The water content that reduces the glass transition temperature of the formulation to the storage temperature is denoted by W_g. Key: open circles = 40°C storage, full squares = 50°C storage. *Abbreviation*: hGH, human growth hormone. *Source*: Data from Ref. 60.

Structure of Proteins in the Amorphous Solid State

While it is clear that many proteins may be freeze-dried and reconstituted with little or no loss in activity, it is usually not obvious whether the freeze-dried protein is basically native in conformation or whether the solid-state confor- mation is distinctly "nonnative," with the native and active conformation quickly forming during rehydration. The observation that lysozyme regains enzymatic activity *in the "solid" state* above about 20% water (60) demonstrates that a protein need not be in a predominantly aqueous system to maintain activity and presumably possess native structure. However, this observation does not necessarily imply that the structure is nonnative at lower water contents where the enzymatic activity disappears. The loss of activity at lower water contents (60) could simply be a consequence of greatly slowed kinetic processes (i.e., greatly restricted molecular mobility as the system passes into the glassy state).

Even very dry proteins in the glassy state commonly show denaturation endotherms via DSC (36,61–64), where the heat of denaturation and heat capacity change on denaturation (64) are roughly comparable to solution values. These data provide direct evidence of significant tertiary structure in the dry glass, but do not necessarily mean the structure is fully native. Vibrational spectroscopy perhaps provides the best methodology for study of the con- formational changes induced by freeze-drying. An early Raman study suggested perturbations in the tertiary structure of freeze-dried ribonuclease (65), and FTIR spectroscopy has been employed recently to document changes in the IR spectra of proteins on freeze-drying, presumably reflecting changes in secondary structure (30,66–70), In studies with freeze-dried solids, quantitative comparison between spectra may be made using the spectral correlation coefficient, r, defined by (66,68)

$$r = \frac{\sum_n x_i y_i}{\sqrt{\left(\sum_n x_i^2 \sum_n y_i^2\right)}} \qquad (12)$$

where x_i and y_i are the spectral absorbance values of the reference (i.e., normally the "native" aqueous solution) and the sample of interest (i.e., the freeze-dried solid). Analysis is restricted to the amide I region (i.e., 1720–1610/cm) using smoothed second-derivative spectra (66,68). Changes in the FTIR spectra on freeze-drying vary from slight to dramatic, with corresponding "r" values from greater than 0.9 to less than 0.5. In general, bandwidths are greater for freeze-dried solids, suggesting configurational disorder. New bands may appear in the solid and solution bands may disappear, suggesting major conformational changes. Small shifts in band positions also occur on freeze-drying. Studies with poly-L-lysine (66) suggest that the loss of water during freeze-drying favors formation of β-sheet secondary structure. It is concluded (66) that dehydration alone, without conformational changes, could produce small shifts in protein bands to higher wavenumber, but most of the loss of correspondence between solution and solid spectra arises from conformational changes. The conformational changes involve, at least in part, formation of β-sheet structure. The changes in spectra, and conformation, are not necessarily *irreversible*. In general, the spectral correlation coefficient of a *reconstituted* freeze-dried sample is higher than for the solid. Formulation with good lyoprotectants (i.e., sucrose or trehalose) moderates the change in FTIR spectra on freeze-drying, with a corresponding increase in the spectral correlation coefficient relative to that obtained by freeze-drying from buffer alone. Formulation with disaccharides appears to provide a hydrogen bonding opportunity for the protein that does not involve a shift in conformation to β-sheet, thereby stabilizing the native conformation (66).

While FTIR spectroscopy may have some limitations as a quantitative predictor of the pharmaceutical stability of a protein since only changes in secondary structure are measured, it does seem clear that protein structure is often altered during the freeze-drying process, and the degree of structural perturbation is specific to the protein and quite specific to the formulation.

While one certainly cannot maintain that preserving a protein in the native state is a *sufficient* condition for good storage stability (i.e., otherwise, why freeze-dry?), it does seem intuitive that a native freeze-dried protein will generally be *more* stable than the same freeze-dried protein in an unfolded conformation. Whether the native state is retained via a thermodynamic mechanism or through a purely kinetic mechanism is immaterial. While the volume of relevant data is not large, it does appear that this intuitive concept does have validity for freeze-dried proteins. For example, loss of activity of LDH during freeze-drying from various sugar and polyol formulations does correlate reasonably well with the FTIR spectral correlation coefficients (66). Recently, it has also been demonstrated that storage stability of freeze-dried proteins is

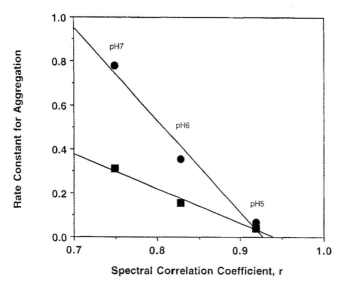

FIGURE 12 The correlation of storage stability with protein structure as determined by Fourier transform infrared (FTIR) spectroscopy: apparent first-order rate constants for aggregation of freeze-dried rIL-2. Key: circles = 45°C storage, squares = 29°C storage. *Source*: Data from Ref. 72.

well correlated with the FTIR spectral correlation coefficient (71,72). Apparent first-order rate constants for aggregation of freeze-dried rIL-2 decrease with decreasing pH in a manner fully consistent with the increase in spectral correlation coefficient with decreasing pH, producing an excellent correlation between storage stability and spectral correlation coefficient (Fig. 12). Therefore, there is some experimental base for the proposition that one requirement for stability during freeze-drying and storage is retention of native conformation in the glassy protein. It should be noted, however, that the correlation between residual activity and spectral correlation coefficient need not necessarily be a direct proportion or even a linear function.

MECHANISMS OF STABILIZATION
Stabilization During Freezing: the "Excluded Solute" Concept
Although many stress factors may operate during freezing, the protein does exist in an aqueous environment during most of the freezing process. Only near the end of freezing does freeze concentration proceed to the point where the protein phase becomes mostly solutes, viscosity becomes high, and mobility becomes slow on the timescale of the experiment (Fig. 1). Assuming the critical stress factors develop before the system becomes a solute-rich high-viscosity system, it is understandable that solutes that stabilize during freezing, or cryoprotectants, are generally those solutes that stabilize the native conformation at more normal concentrations and temperatures (5). Given the chemical diversity

of such stabilizers (i.e., amino acids, polyols, sugars, and PEGs), it is obvious that specific chemical interactions are not the common stabilization mechanism. Such solutes tend to be "excluded" from the surface of the protein and therefore induce "preferential hydration" of the protein. The thermodynamics of this phenomenon is analogous to solute "surface excess" at the air-water interface. A negative surface excess means the solute is partly excluded from the interface, thereby increasing the surface tension of the solution. Indeed, there is a good correlation between those solutes that increase the surface tension of water and those that are excluded from the surface of a protein (6). The thermodynamic consequence of "solute exclusion" and "preferential hydration" is to increase the chemical potential of the protein. The first assumption in relating this thermo-dynamic result to cryoprotection may be stated as: if the increase in chemical potential of the native protein caused by solute exclusion is denoted, $\Delta\mu_N$, the corresponding increase for the unfolded protein is $k\Delta\mu_N$, where $k > 1$. Thus, the free energy of unfolding would be increased by the solute, which is equivalent to stabilization of the native conformation (6). The second assumption is that pharmaceutical stability, or degradation, is directly related to thermodynamic stability. Finally, it is assumed that the solution state concepts are valid throughout the freezing process, or at least valid over the period where unfolding in the absence of a cryoprotectant is both thermodynamically favored and kinetically allowed.

The extent of solute exclusion, $(\partial m_s/\partial m_p)_{T,\mu_w,\mu_p}$, can be measured, and the corresponding effect on chemical potential of the protein, $(\partial\mu_p/\partial m_s)_{T,P,m_p}$, can then be calculated (6,73–76):

$$\left(\frac{\partial\mu_p}{\partial m_s}\right)_{T,P,m_p} = -\left(\frac{\partial m_s}{\partial m_p}\right)_{T,\mu_w,\mu_p}\left(\frac{\partial\mu_s}{\partial m_s}\right)_{T,P,m_p} \tag{13}$$

where the solute exclusion parameter is the partial derivative of the molal concentration of stabilizer in the domain of the protein, m_s, with respect to the molal concentration of protein, m_p, at constant temperature, T, and constant chemical potentials of water, μ_w, and protein, μ_p. The relationship also involves the concentration dependence of the chemical potential of stabilizer, μ_s, which may be written

$$\left(\frac{\partial\mu_s}{\partial m_s}\right)_{T,P,m_p} = \frac{RT}{m_s} + RT\left(\frac{\partial\ln\gamma_s}{\partial m_s}\right)_{T,P,m_2} \tag{14}$$

The first term on the right-hand side is the "ideal solution" contribution while the "nonideal" part involves the concentration dependence of the activity coefficient of the stabilizer in water, γ_s. Since equation (14) involves the recip-rocal of stabilizer concentration, it appears that the response of the chemical potential of the protein to a stabilizer, $(\partial\mu_p/\partial m_s)_{T,P,m_p}$, becomes infinite as the stabilizer concentration approaches zero. However, the molal solute exclusion parameter, $(\partial m_s/\partial m_p)_{T,\mu_w,\mu_p}$, is actually directly proportional to concentration, so the chemical potential derivative, $(\partial\mu_p/\partial m_s)_{T,P,m_p}$ is roughly independent of concentration for most of the systems studied (73–76). The factors impacting the chemical potential derivative are more transparent if we first convert from molal concentration units to mass-based concentrations (i.e., weight ratios, where the weight ratio of component "i" to water is symbolized, g_i); g_i is related to molality and molecular weight, M_i, by

$$g_i = m_i \frac{M_i}{1000} \tag{15}$$

The solute exclusion parameter then becomes, $(\partial g_s/\partial g_p)_{T,\mu_w,\mu_p}$, concentration form. This mass-based solute exclusion parameter may be related to the mass-based preferential hydration parameter, $(\partial g_w/\partial g_p)_{T,\mu_w,\mu_p}$ by the identity (6,74–77),

$$\left(\frac{\partial g_w}{\partial g_p}\right)_{T,\mu_w,\mu_p} = -\left(\frac{1}{g_s}\right) \cdot \left(\frac{\partial g_s}{\partial g_p}\right)_{T,\mu_w,\mu_p} \tag{16}$$

The preferential hydration parameter is nearly independent of concentration, and for most small molecular weight "excluded solutes" is in the range 0.2 to 0.6 (6,73–76). Thus, the chemical potential derivative, in terms of mass-based concentrations, becomes

$$\left(\frac{\partial \mu_p}{\partial g_s}\right)_{T,P,g_p} = RT \frac{M_p}{M_s} \left(\frac{\partial g_w}{\partial g_p}\right)_{T,\mu_w,\mu_s} \left[1 + g_s\left(\frac{\partial \ln \gamma_s}{\partial g_s}\right)\right] \tag{17}$$

The nonideality term, $1 + g_s(\partial \ln \gamma_s/\partial g_s)$, is normally relatively close to unity even in concentrated solutions. The exceptions are some polymers such as higher molecular weight PEGs. For example, for 1 M sucrose, 2 M glycine, and 1 M proline, the nonideality terms are 1.19, 0.67, and 1.1, respectively (77,78), but for 2% PEG 6000, the nonideality term is 2.6 (76). The quantity most directly relevant to stabilization is the free energy of transfer, which is given by the integral of the chemical potential derivative from zero concentration of stabilizer to the concentration of interest, g_s:

$$\Delta \mu_p(\text{transfer}) = \int_0^{g_s} \left(\frac{\partial \mu_p}{\partial g_s}\right) dg_s = g_s\left\langle \frac{\partial \mu_p}{\partial g_s}\right\rangle \tag{18}$$

where $(\partial \mu_p/\partial g_s)$ is the average value of the derivative over the concentration range from zero to g_s. Since the "experimental" values of $\partial \mu_p/\partial g_s$ are relatively insensitive to concentration, the average value may be treated as a constant for a given protein:stabilizer combination, at least as a good first approximation within the range of concentrations studied, typically $g_s < 0.7$. It does not seem reasonable to assume, however, that $\partial \mu_p/\partial g_s$ will continue to remain constant as the system moves toward the high concentrations found in the freeze-concentrated glass, that is, as $g_s \rightarrow 4$. However, since the rate constants for unfolding and folding may slow sufficiently to prevent equilibrium from being maintained on the timescale of a relatively fast freeze (Fig. 7), thermodynamic stabilization may be irrelevant near the glass transition temperature. The free energy of transfer given by equation (18) is the increase in free energy of the protein caused by addition of the stabilizer at concentration g_s, earlier denoted "$\Delta \mu_N$" for the protein in the native conformation. Again, while equation (18) also applies to a protein in the denatured state, data are available only for proteins in their native conformation. Assuming that the corresponding free energy of transfer of the denatured protein is given by $k\Delta \mu_N$, where $k > 1$, the stabilization free energy is given by $(k-1)\Delta \mu_N$, which is greater than zero when a stabilizer increases the chemical potential of the native protein. Thus, in a series of stabilizers, the stability enhancement should correlate with the free energies of transfers evaluated for the protein in the native conformation. Of course, this conclusion is based on the assumption that $k - 1$ is not a major variable in the context of the study.

A selection of data are given in Table 2, where mean values of preferential hydration coefficients, $\partial g_w / \partial g_p$, and chemical potential derivatives, $\partial \mu_p / \partial g_s$, are tabulated for some stabilizers of pharmaceutical significance. The nonideality terms for glycerol and mannitol were assumed to be identical to that for glucose. Since the nonideality is less than the estimated error in $\partial g_p / \partial g_s$, the uncertainty in the nonideality term is of no practical significance. Except for the high molecular weight PEGs, which have very high preferential hydration coefficients, the preferential hydration values are all of the same magnitude, although they are specific to the stabilizer:protein combination. Of the small molecule stabilizers tabulated, glycine is the most effective "preferential hydration inducer," and because of its small molecular weight the chemical potential derivative is by far the highest of the stabilizers in Table 2. Thus, on a weight basis, glycine gives the highest free energy of transfer. The high molecular weight PEGs are exceptional preferential hydration inducers, but because of the inverse relationship of the chemical potential derivative to molecular weight, the chemical potential derivatives and the free energies of transfer are not exceptionally high. Assuming, for the moment, that stability enhancement should correlate with the free energy of transfer, the implication of the data given in Table 2 is that, when compared on a fixed mass-based concentration, high molecular weight PEGs will *not* be outstanding stabilizers. This conclusion seems contrary to implications in the literature (6,73–76). Values of the chemical potential derivative are typically given in the literature in terms of molal concentration, that is, $\delta \mu_p / \delta m_s$ is the reported parameter. While values of $\delta \mu_p / \delta m_s$

TABLE 2 The Effect of Solutes on the Preferential Hydration and Chemical Potential of Proteins

Protein	Stabilizer	$(\partial g_w / \partial g_p)_{T,\mu_w,\mu_s}$	$(\partial \mu_p / \partial g_s)_{T,P,m_p}$ (kcal/mol)
RNase	Sucrose, ≤ 1 M	0.46	12
BSA	Lactose, ≤ 0.4 M	0.30	37
BSA	Glucose, ≤ 2 M	0.23	59
BSA	Mannitol, $\leq 15\%$	0.20	48
BSA	Inositol, $\leq 10\%$	0.40	96
BSA	Glycerol, $\leq 40\%$	0.17	80
BSA	Glycine, ≤ 2 M	0.43	200
BSA	PEG 400, $\leq 10\%$	0.61	70
BSA	PEG 1000, $\leq 10\%$	1.18	74
BSA	PEG 3000, $\leq 6\%$	2.36	56
BSA	PEG 6000, $\leq 4\%$	3.76	67
Lysozyme	Glycine, ≤ 2M	0.58	56
Lysozyme	Proline, 1 M	0.32	26
Lysozyme	PEG 400, $\leq 10\%$	0.26	6.0
Lysozyme	PEG 1000, $\leq 10\%$	0.83	11.3
Lysozyme	PEG 3000, $\leq 4\%$	2.55	12.5
Lysozyme	PEG 6000, $\leq 2\%$	8.6	21

Uncertainties in the data are typically ±10–20%.

for high molecular weight PEGs are extremely high, a comparison based on molal concentration is not practically relevant for pharmaceutical purposes, which is the primary reason why mass-based concentration is used in this presentation. However, stabilization during freeze-thaw is reported to be superior for high molecular weight PEGs, even when comparing stabilizers at constant weight percent (31). Moreover, the trends in stability during freeze-thaw of PFK given in Figure 4 are not consistent with the data given in Table 2. At constant molar concentration of stabilizer (0.5 M), the order of free energies of transfer are (for lysozyme) glycine > proline. For BSA, the order is inositol > lactose > glucose > glycerol. For retention of activity on freeze-thaw (of PFK), the order is proline >> glycine, and disaccharide > glucose = glycerol > inositol. While the correlation with the preferential hydration coefficient is somewhat better, the theory is really based on a correlation with free energies of transfer.

The lack of a quantitative correlation between the thermodynamics of solute exclusion and stability during freeze-thaw described above may have several contributing factors. First, the discussion above is limited to only a few proteins and includes only those excipients that are excluded solutes. That is, a correlation would certainly appear better if one were to include known denaturing solutes such as urea, although the pharmaceutical significance of such a comparison is questionable. Even the sparse data in Table 2 demonstrate that the relative effectiveness of the PEGs varies somewhat with the protein, so a wider protein database could improve the correlation. Also, the data in Table 2 refer to free energies of transfer at room temperature. Free energies are obviously temperature dependent, and whether the rank order of free energies at room temperature will persist at the subzero temperatures encountered during freezing is far from obvious. One must also question the validity of assuming the relationship between the increase in chemical potential of the native form and the denatured form are coupled by a k value that is really a constant for a variety of stabilizers. It must also be recognized that other factors may be controlling *instability* and stability. One possibility that has been largely ignored is the role of denaturation at the aqueous-ice interface and the role of the stabilizer in minimizing protein adsorption on the ice surface. While minimization of protein adsorption is a thermodynamic mechanism, it is a mechanism that does not directly involve thermodynamic stabilization of the aqueous protein. An additional factor, likely important if the stress occurs late in freezing, is the increasing viscosity of the freeze concentrate. Near the end of freezing, the protein is dispersed in an excipient matrix where the viscosity becomes very high (Fig. 1). Both dilution of the protein in the matrix and the low molecular mobility generally associated with high viscosity would tend to slow the rate of denaturation or at least retard aggregation of partially denatured protein, assuming that the mobility critical for degradation is coupled, at least to some extent, to viscosity. Finally, there is at least one example of the stabilizer operating, at least in part, through its ability to prevent crystallization of a buffer component, thereby preventing a large pH shift that destabilizes the protein (79,80). Thus, while solute exclusion is likely a major factor in stabilization during freezing, other stabilization mechanisms are possible and even probable in some cases. It is not the purpose of this discussion to refute the concept of stabilization during freezing via the excluded solute concept, at least as a significant factor. Indeed, in this author's opinion, the concept does have merit and may dominate in many cases. Rather, the intent is to demonstrate that the predictive power of the thermodynamic results, such as in Table 2, for selecting the "optimum" stabilizer system for a "new protein" is quite limited and this limitation should not come as a surprise.

Stabilization During Drying

As discussed earlier, all stresses that develop during freezing are still present during drying, but the normal implicit assumption is that if degradation during freezing did not occur, the freezing stresses are not sufficient to thermodynamically destabilize the protein. Thus, the critical stress during drying is normally assumed to be the removal of the water that is part of the freeze concentrate. This assumption is likely correct as long as degradation occurs during secondary drying where the timescale is similar to that for freezing. Even assuming the critical stress factor in drying is the reduction of water to low levels, thereby removing the assumed stabilizing influence of hydrogen bonding interactions between protein and water, it is not obvious that the role of a stabilizer is to replace water in the hydrogen bonding interactions and thermodynamically stabilize the native conformation. If the stabilizer were to effectively couple the internal motions of the protein to structural relaxation in the glass, thereby reducing the rate of unfolding of a thermodynamically unstable system to insignificant levels, the net result would be a freeze-dried protein in the native conformation. That is, the protein would not unfold, regardless of what the free energy of unfolding might become, and stabilization would be purely kinetic.

Many experimental studies of the relative effectiveness of various solutes in preventing drying damage can be interpreted according to either the water substitute hypothesis, which is a thermodynamic stabilization mechanism, or a purely kinetic stabilization argument based on the effective coupling of the protein motions to the glass. There are studies, however, which do not appear to be consistent with both classes of stabilization mechanisms. For example, the L-asparaginase study discussed earlier demonstrates that the water substitute hypothesis cannot always provide a satisfactory explanation of the data. Neither TMG nor PVP can hydrogen bond as water substitutes but yet are effective in preventing inactivation during drying. The catalase example discussed earlier illustrates that a high glass transition temperature for the excipient is not the critical factor in stabilization during drying. Even glucose ($T_g = 39°C$) and mannitol (which frequently crystallizes) stabilize as effectively as do materials that readily form glasses with glass transition temperatures on the order of 100°C. While the catalase data do appear to be better interpreted in terms of the water substitute hypothesis than by protein immobilization in a glass, it must be acknowledged that mobility of the protein in a glass is not necessarily well measured by the difference between the glass transition temperature of the pure excipient and the sample temperature. Firstly, since the catalase formulations were 1:1 weight ratio mixtures of excipient and catalase, the glass transition temperatures of the formulations will be intermediate between the glass transition temperature of the pure excipient and the glass transition temperature of the protein. Since protein T_g values are normally quite high (63,64), the differences in formulation glass transition temperatures will be much less than the differences in excipient glass transition temperatures. Secondly, glass fragility, thermal history, and perhaps most important, the degree of coupling of the protein motion to motion in the glass are important factors in determining protein mobility at a given value of $T_g - T$.

Stabilization by a thermodynamic mechanism requires that the rates of unfolding and refolding be fast on the timescale of the experiment, which for drying is on the order of hours. Thus, equilibrium is established quickly, and it is the position of the equilibrium (i.e., the ratio of "unfolded" to "native" species) that determines degradation rate and loss of activity. Assuming that drying is carried out close to the glass transition temperature (4), the structural relaxation time is on the order of minutes to tens of minutes (Fig. 7). Therefore, structural relaxation is moderately fast on the drying timescale. However, one can argue that the unfolding time is likely to be significantly greater than the structural relaxation time. Since unfolding involves rather large-scale motion, analogous to polymer chain diffusion, it is likely that coupling to structural relaxation will be strong. Recall that polymer chain diffusion appears to couple well with viscosity and the glass transition (Fig. 9). Further, since a large number of diffusional jumps would be required to complete the unfolding reaction, the total time required for unfolding would be much greater than the time for a single diffusional jump. Thus, one might expect that the unfolding time would consist of a large number of diffusional jump times, each of which is similar in duration to the structural relaxation time. We also note that crystallization of sucrose from an amorphous system near T_g, likely also a diffusion-controlled process, seems to require a reaction time significantly longer than the structural relaxation time. Onset of nucleation and crystallization requires about one day when carried out 10°C above the glass transition temperature of the system (81). The estimated structural relaxation time (Fig. 7) 10°C above the glass transition is $\approx 5 \times 10^{-6}$ days, so at least for sucrose crystallization, the ratio of the reaction time to the relaxation time is on the order of 10^5. Thus, we tentatively conclude that protein-unfolding times in stabilized glassy formulations near T_g are likely of the same order of magnitude as the drying time or longer. Given these considerations, the rates of unfolding and refolding do *not* appear to be fast on the drying timescale, and thermodynamic stabilization mechanisms do not appear plausible. However, uncertainty in the degree of coupling between the diffusional jump process and structural relaxation introduces considerable uncertainty in the above conclusion. Further, it is significant to note that the temperature denoted, $T_{g'}$, may be about 20°C higher than the true glass transition temperature (82). If this interpretation is correct, primary drying and early secondary drying are typically carried out well above the glass transition temperature. For a process carried out at a temperature 20°C above a glass transition temperature, the structural relaxation time is in the range of 10^{-8} to 10^{-5} days, depending on fragility, and one might argue that the rates of unfolding and refolding would be fast compared with the timescale of drying. Thermodynamic stabilization concepts would then become quite viable. However, as mentioned earlier, our data on rates of unfolding in viscous systems suggest unfolding rates can be on the timescale of months even 20°C above a glass transition temperature. To the extent this observation is general, one would not expect unfolding to occur during the usual drying process. However, since *instability* does occur during drying, either the *instability* does not depend on unfolding or the unfolding dynamics is not well coupled to the system mobility in these unstable systems.

Storage Stability

The stresses during storage are exactly those stresses that operate during drying. The major differences between drying stabilization and storage stabilization are the timescales. The timescale available for a stress to produce significant degradation is hours during drying but is months or years for storage. Further, it should also be noted that while the product temperature may closely approach the glass transition temperature during normal processing (4), storage temperatures are typically much below the system glass transition temperature, so relaxation times will also be much longer than during processing. Using the data in Figure 7 as a guide, the structural relaxation time at 40°C may vary from several days to many months depending on glass fragility. The variation could be far more if possible differences in coupling coefficient were also taken into consideration. Thus, kinetic control of degradation via coupling with the structural relaxation time is certainly consistent with large differences in stability between different formulations. Experimental data (Fig. 8) indicate that the structural relaxation times for sucrose at 25°C and 40°C are 570 years and 3 months, respectively, and the structural relaxation time for trehalose at 40°C is over 10,000 years. It is clear that, at least for a protein in a sucrose or trehalose matrix at 25°C, the structural relaxation times are much longer than storage times. Indirect evidence was used earlier to argue that unfolding times are long compared to structural relaxation times. Thus, we conclude that protein-unfolding times in stabilized glassy formulations well below T_g are generally vastly greater than storage times, and any mechanism based on equilibrium between folded and unfolded forms (i.e., thermodynamic mechanism) cannot be correct. Of course, even in stabilized systems, aggregation does occur after storage and reconstitution, which suggests that either some partial unfolding does occur well below T_g, or alternately, aggregation of partially unfolded molecules occurs via small-scale motion that is poorly coupled with structural relaxation.

If we assume that a structurally perturbed protein may have different reactivity than a protein in the native conformation, the observation that proteins do suffer structural perturbations on freeze-drying, which are often highly formulation dependent, suggests that a realistic model of protein degradation in the solid state must recognize the existence of these substates of the protein. Thus, degradation in any given sample is a function of the distribution and reactivities of the substates. Specifically, one might write the rate constant for degradation, k, in terms of substate rate constants, k_i, and the fraction of such substates, w_i, in the general form

$$k = \sum_i w_i k_i = \sum_i w_i A_{k_i} \exp\left(-\frac{g_i D T_0}{[T - (T/T_f)T_0]}\right) \quad (19)$$

where equation (11) has been used to express the substate rate constants, k_i. Presumably, a substate that is partially unfolded would have fewer diffusional jumps to complete the reaction process and its A_{ki} would be larger than the corresponding preexponential factor for the native conformation. It is also quite possible that partial unfolding would mean a lower coupling coefficient, g, for the perturbed state. Both effects would give a larger substate rate constant for the partially unfolded protein. Equation (19) is not necessarily intended to be quantitative or correct in detail. However, the qualitative message is believed valid and consistent with the available data. That is, degradation in a glassy protein formulation depends on both the dynamic properties of the glass and the specific properties of the substates that are created during the freeze-drying process. While thermodynamic stabilization is not required for storage stability, thermodynamic stabilization during processing could well play a significant role in providing stable substate structures.

REFERENCES

1. Tsong TY. Biopolymers 1978; 17:1669–1678.
2. Kiefhaber T, Quaas R, Hahn U, et al. Biochemistry 1990; 29:3053–3061.
3. Brems DN. Biochemistry 1988; 27:4541–4546.
4. Carpenter JF, Pikal MJ, Chang BS, et al. Pharm Res 1997; 14(8):969–975.
5. Pikal MJ. Freeze Drying. In: Swaarbrick J, Boylan JC, eds. Encyclopedia of Pharmaceutical Technology. Vol. 6. New York: Marcel Dekker, 2001.
6. Arakawa T, Kita Y, Carpenter JF. Pharm Res 1991; 8:285–291.
7. Franks F. Cryo Letters 1990; 11:93–110.8.
8. CRC Handbook. Boca Raton, FL: Chemical Rubber Publishing Co, 1957.
9. Angell CA, Bressel RD, Green JL, et al. J Food Eng 1994; 22:115–142.
10. Kiovsky TE, Pincock RE. J Am Chem Soc 1966; 88:7704–7710.
11. Lim M, Reid D. Studies of reaction kinetics in relation to the $T_{g'}$ of polymers in frozen model systems. In: Levine H, Slade L, eds. Water Relationships in Food. New York: Plenum Press, 1991:103–122.
12. Larsen SS. Arch Pharm Chem Sci Ed 1973; 1:41–53.
13. Murase N, Franks F. Biophys Chem 1989; 34:293–300.
14. Gomez G, Rodriguez-Hornedo N, Pikal MJ. Pharm Res 1994; 11:8–265, PPD 7364.
15. Szkudlarek BA, Rodriguez-Hornedo N, Pikal MJ. Mid-Western AAPS Meeting, Chicago, poster #24, 1994.
16. Pikal MJ, Dellerman KM, Roy ML, et al. Pharm Res 1991; 8:427–436.
17. Roy ML, Pikal MJ, Rickard EC, et al. Dev Biol Stand 1991; 74:323–340.
18. Andrade JD, Hlady V. Adv Polym Sci 1986; 79:3–63.
19. Eckhardt BM, Oeswein JQ, Bewley TA. Pharm Res 1991; 8:1360–1364.
20. Strambini GB, Gabellieri E. Biophys J 1996; 70:971–976.
21. Hubel A, Korber C, Cravalho E, et al. J Cryst Growth 1988; 87:69–78.
22. Chang BS, Kendrick BS, Carpenter JF. J Pharm Sci 1996; 85:1325–1330.
23. Franks F. Biophysics Biochemistry at Low Temperatures. Cambridge: Cambridge University Press, 1985.
24. Privalov P. Biochem Mol Biol 1990; 25:281–305.
25. Pikal MJ, Shah S, Roy ML, et al. Int J Pharm 1990; 60:203–217.
26. Carpenter J, Crowe J, Arakawa T. J Dairy Sci 1990; 73:3627–3636.
27. Tanaka R, Takeda T, Miyajima K. Chem Pharm Bull (Tokyo) 1991; 39:1091–1094.
28. Hellman K, Miller D, Cammack K. Biochim Biophys Acta 1983; 749:133–142.
29. Ressing ME, Jiskoot W, Talsma H, et al. Pharm Res 1992; 9:266–270.
30. Arakawa T, Prestrelski S, Kinney W, et al. Adv Drug Delivery Rev 1993; 10:1–28.
31. Carpenter JF, Prestrelski S, Arakawa T. Arch Biochem Biophys 1993; 303:456–464.
32. Pikal MJ. Biopharm 1990; 3(9):26–30.
33. Shikama K, Yamazake I. Nature 1961, 190:83–84.
34. Izutsu K, Yoshioka S. Drug Stability 1995; 1:11–21.
35. Townsend TW, DeLuca PP. J Parenter Sci Technol 1988; 42:190–199.
36. Pikal MJ, Dellerman K, Roy ML. Dev Biol Stand 1991; 74:323–340.
37. Pearlman R, Nguyen TH. Therapeutic peptides and proteins: formulations, delivery, targeting. In: Marshak D, Liu D, eds. Current Communications in Molecular Biology, Cold Spring Harbor Laboratory. New York: Cold Spring Harbor, 1989:23–30.
38. Pikal MJ. Eli Lilly & Co, unpublished observations.
39. Hancock BC, Shamblin SL, Zografi G. Pharm Res 1995; 12:799–806.
40. Iben ET, Braunstein D, Doster W, et al. Phys Rev Lett 1989; 62:1916–1919.
41. Hagen SJ, Hofrichter J, Eaton WA. Science 1995; 269:959–962.
42. Roy R. J Non-Cryst Solids 1970; 3:33–40.

43. Pikal MJ. J Phys Chem 1971; 75:3124.
44. Adam G, Gibbs JH. J Chem Phys 1956; 43:139–146.
45. Angell CA. J Non-Cryst Solids 1988; 102:205–221.
46. Hodge IM. J Non-Cryst Solids 1996; 202:164–172.
47. Finegold L, Franks F, Hatley RHM. Faraday Trans 1989; 89:2945–2951.
48. Shamblin SL, Tang X, Chang L, et al. J Phys Chem 1999; 103:4113–4121.
49. Hodge IM. J Non-Cryst Solids 1994; 169:211–266.
50. Putnam RL, Boerio-Goates J. J Chem Thermodyn 1993; 25:607–613.
51. Mardaleishvili IR, Anisimov VM, Izv Akad Nauk SSSR. Ser Khim 1987; 6:1431–1432.
52. Pikal MJ, Rigsbee DR. Thermometric Seminars on Calorimetry in Materials Sciences. Stockholm, Sweden: Thermometric Inc., May 23, 1996.
53. Liu J, Rigsbee D, Stotz C, et al. J Pharm Sci 2002; 91:1853–1862.
54. Angell CA. Chem Rev 1990:523–542.
55. Oksanen CA, Zografi G. Pharm Res 1993; 10:791–799.
56. Meares P. J Am Chem Soc 1954; 76:3415–3422.
57. Kohlhammer K, Kothe G, Reck B, et al. Phys Chem 1989; 93:1323–1325.
58. Williams ML, Landel RF, Ferry JD. J Am Chem Soc 1955; 77:3701.
59. Pikal MJ, Roy ML, Rigsbee DR. Eli Lilly & Co, unpublished data.
60. Rupley JA, Caren G. Adv Protein Chem 1991; 41:37–172.
61. Rigsbee DR, Pikal MJ. National AAPS Meeting, Orlando, Florida, Abstract #7418, 1993.
62. Bell LN, Hagetnan MJ, Muraoka LM. J Pharm Sci 1995; 84:707–712.
63. Green JL, Fan J, Angell CA. J Phys Chem 1994; 98:13780–13790.
64. Sochava IV, Smirnova OI. Food Hydrocolloids 1991; 6:513–524.
65. Yu NT, Jo BH, Liu CS. J Am Chem Soc 1972; 94:7572–7575.
66. Prestrelski SJ, Tedeschi N, Arakawa T, et al. Biophys J 1993; 65:661–671.
67. Carpenter JF, Crowe JH. Biochemistry 1989; 28:3916–3922.
68. Prestrelski SJ, Arakawa T, Carpenter JF. Arch Biochem Biophys 1993; 303:465–473.
69. Remmele RL, Stushnoff C. Biopolymers 1994; 34:365–370.
70. Dong A, Huang P, Caughey WS. Biochemistry 1990; 29:3303–3308.
71. Prestrelski SJ, Pikal KA, Arakawa T. Pharm Res 1995; 12:1250–1259.
72. Chang BS, Fischer NL. Colorado Protein Stability Symposium, Breckenridge, Colorado, July 17–20, 1994.
73. Lee JC, Timasheff SN. J Biol Chem 1981; 256:7193–7201.
74. Arakawa T, Timasheff SN. Biochemistry 1982; 21:6536–6544.
75. Arakawa T, Timasheff SN. Arch Biochem Biophys 1983; 224:169–177.
76. Bhat R, Timasheff SN. Protein Sci 1992; 1:1133–1143.
77. Robinson RA, Stokes RH. Electrolyte solutions. New York: Academic Press, 1959.
78. Handbook of Biochemistry and Molecular Biology. In: Fasman GD, ed. Physical and Chemical Data. Vol. 1, 3rd ed. Cleveland, OH: CRC Press, 1976:120.
79. Szkudlarek BA, Garcia GA, Rodrigucz-Horncdo N. Pharm Res 1996; 13:8–85.
80. Anchordoquy TJ, Carpenter JF. Arch Biochem Biophys 1996; 332:231–238.
81. Saleki-Gerhardt A, Zografi G. Pharm Res 1994; 11:1166–1173.
82. Shalaev EY, Franks F. J Chem Soc Faraday Trans 1995; 91:1511–1517.

9 Formulation Characterization

D. Q. Wang
Bayer Schering Pharma, Beijing, China

INTRODUCTION

The first and the most important step in developing a lyophilization cycle is to characterize the formulation. This cannot be overemphasized because understanding the formulation will allow us to develop a lyophylization cycle with a scientific rationale, instead of using trial and error (1). Characterization of a lyophilization formulation is nothing new. In 1960, Rey reported the use of differential thermal analysis (DTA) and a freezing microscope to investigate freezing process of biological products (2). In 1964, MacKenzie reported that the first model of a freeze-drying microscope that was constructed by R. J. Williams in 1962 enabled him to develop a second model as an improvement (3). In 1972, MacKenzie discussed how the freeze-drying process could be affected by different formulations by measuring melting temperature (T_m), eutectic temperature (T_e), glass temperature (T_g), and glass transition temperature (T_g') of a water/sucrose system and collapse temperatures (T_c) of NaCl/sucrose system (4). In the 1990s, probably due to numerous publications about the glass transition theory and the wide use of freeze-drying in the biopharmaceutical industry, the importance of characterizing a lyophilized formulation became well recognized. Pikal reported that because eutectic and collapse temperatures vary over an enormous range, determining the maximum allowable product temperature is extremely important and is the first step in formulation and process development (5–7). Franks pointed out that an understanding of a freeze-drying formulation can remove most of the empiricism from the freeze-drying cycle development (1,8).

In addition, the instrumentation used for characterizing formulation has also been improved significantly. Nail et al. developed a further improved model of a freeze-drying microscope that offered better temperature control, faster temperature responses, and sophisticated video recording and image capturing, among many others (9). In the thermal analysis field, TA Instrument (New Castle, Delaware, U.S.) developed modulated differential scanning calorimetry (MDSC), which allows us to extend the thermal analysis to freeze-dried dry cakes to better understand and predict the stability of a lyophilized product. In this chapter, we will focus on techniques that have been used in recent years at Bayer Biotechnology in characterizing lyophilized formulation. Then we will discuss how to interpret data from these characterization studies to develop a lyophilization cycle.

A lyophilized formulation can consist of many different excipients. Sugars such as sucrose are usually used to stabilize protein conformation against denaturation caused by water removal. Polymers and other proteins such as human albumin serum (HAS) also function similarly. These excipients usually remain amorphous during and after lyophilization. A formulation consisting of mainly amorphous excipients is often called "amorphous formulation." On the other hand, bulking agents such as mannitol and glycine are commonly used for providing cake appearance. Such excipients are usually expected to be in crystalline structure after lyophilization. Totally crystalline cakes are not commonly used for protein lyophilization, since the crystallization of all excipients will tend to remove any stabilizing effect of the excipients on the protein (10–12). For pharmaceutical protein products, a formulation termed "crystalline matrix" is widely used. In such a formulation, crystalline components are added at a relatively high level so that a crystalline matrix is formed for the amorphous components to collapse upon. In such a way, the crystalline component provides excellent cake appearance, good reconstitution characteristics, and ease in lyophilization. On the other hand, the amorphous component stabilizes proteins during processing and storage. Lyophilizing such a formulation allows partial collapse (also termed micro-collapse) without affecting cake appearance. As a result, the product can be lyophilized at a relatively higher product temperature if protein activity is not compromised. In this chapter, all of our discussion focuses on the crystalline matrix-type formulation.

A lyophilization process usually includes three steps. First, the aqueous solution that has been filled into containers (such as vials and trays) is frozen to lower than −40°C. Next, in the primary drying, the freeze-dryer chamber is evacuated and the shelf temperature is elevated to sublimate bulk water (also termed free water) out of the system. Finally, the shelf temperature is further increased to remove bound water by desorption. This is called secondary drying. The principles of the lyophilization process have been well described by Pikal (5) and Franks (1).

At Bayer Biotechnology, we usually develop a freeze-drying cycle in five steps. First, the formulation is characterized. Second, based on the understanding of the formulation, the process is optimized. Third, the range of the critical process parameters is found. Fourth, the process is scaled up and transferred to production. Fifth, the process at the production scale is validated and qualified. Here, we focus only on the formulation characterization.

In summary, in this chapter we will discuss the freeze-drying formulation characterization, which is the first and the most important step in developing a freeze-drying cycle. All of the characterization discussion will focus on a crystalline matrix type of formulation. In the discussion, we will follow the order given below:

Utilize differential scanning calorimetry (DSC) to characterize the formulation for freezing and primary drying.

Confirm DSC results with a freeze-drying microscope and determine the maximum allowable product temperature during the primary drying.

Generate a water absorption/desorption curve for characterizing the secondary drying process.

Conduct a moisture optimization study to determine target moisture content for developing secondary drying.

Use MDSC to measure T_g to predict stability of the products.

UTILIZE DSC TO CHARACTERIZE FORMULATION

Basically, DSC measures heat flow as a function of temperature applied to a sample going through freezing, melting, crystallization, and glass transition. The thermal properties that can be measured by using DSC include eutectic crystallization temperature (T_x), eutectic melting temperature (T_e), glass transition temperature (T_g'), and ice melting temperature (T_{im}). Among these critical temperatures, the glass transition temperature T_g' is one of the most important thermophysical properties of the formulation. For a formulation that forms an amorphous cake after being freeze-dried, the T_g' is also the collapse temperature, which is the most critical factor in ensuring the success of the primary drying. For example, MacKenzie measured the collapse temperature of the formulations with different ratios of sucrose to sodium chloride. He demonstrated that at a certain range of ratios, the lyophilization became impractical as the sodium chloride depressed the collapse temperature to a point below −40°C (13–15). Many other reports are available in determining T_g' by DSC (13,14).

In this chapter, as we have mentioned previously, we focus on the crystalline matrix formulation. The example formulation consists of 2.2% glycine, 1.0% of sucrose, 0.02 M of histidine, 0.03 M of sodium chloride, and 0.0025 M of calcium chloride. The concentration of therapeutic protein is in the range of 50 to 200 μg/mL, which is negligible.

Figure 1 shows a DSC thermogram of warming for the formulation. The sample was frozen to −60°C and then warmed up to 20°C. During the warming, T_g can be observed between −30°C and −35°C. Under these conditions, the highest allowable product temperature during primary drying is about −35°C. In this case, we do not believe that the glycine crystallizes out and, as a result, we must freeze-dry the product below this maximum allowable temperature. The maximum allowable temperature here is also the collapse temperature. In other words, the cake will collapse if the product temperature is higher than this

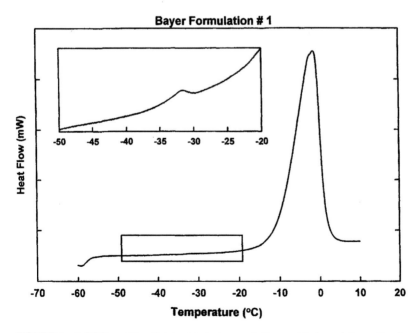

FIGURE 1 A DSC warming thermogram of a crystalline matrix formulation without annealing. *Abbreviation*: DSC, differential scanning calorimetry.

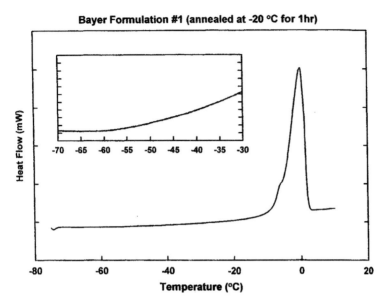

FIGURE 2 A DSC warming thermogram of a crystalline matrix formulation after annealing. *Abbreviation*: DSC, differential scanning calorimetry.

maximum allowable temperature during sublimation. Generally speaking, the lower the allowable product temperature, the less efficient the freeze-drying cycle. It may even become impractical for large-scale production if the highest allowable temperature approaches −40°C.

For the crystalline matrix-type formulation, however, adding an annealing step in freezing may make a significant difference in the freeze-drying process. Figure 2 shows a DSC thermogram of the same formulation described above, but it has been annealed during freezing. As before, the sample was first frozen to −60°C. Then the sample temperature was slowly increased to −20°C and held for one hour at this temperature. We call this step annealing or heat treatment. After holding at −20°C for one hour, the sample was again slowly frozen down to −60°C to a fully frozen state. The sample was then warmed to 20°C. The thermogram shown in Figure 2 is from the last warming. As a result of the annealing, as shown in the thermogram, the thermal event observed between −30°C and −35°C disappeared. Instead, there is a new thermal event, like a shoulder, which occurred at a temperature above −10°C. We believe the new thermal event is the eutectic melting of the glycine. It appears like a shoulder on the thermogram because the peak of glycine melting was quickly integrated into the peak of the ice melting. This also indicates that annealing has made glycine crystallize out as evidenced by the eutectic melting of glycine during the warming. The complete crystallization of the crystalline component makes the thermal transition shift to −10°C. In other words, it is evident that annealing will allow primary drying to occur at a much higher product temperature because, for the crystalline matrix formulation, the eutectic melting temperature, T_e, is the collapse temperature, T_c. This is a clear advantage of the crystalline matrix-type formulation. However, care must be taken to prevent protein activity and stability from being compromised by annealing. We will discuss this in detail in a later section.

We stated in the previous paragraph that glycine was completely crystallized out as a result of annealing. The evidence was the eutectic melting of glycine shown in Figure 2. Another way to verify the crystallization of glycine is to conduct an X-ray powder diffraction study on the freeze-dried cake. Figure 3 is a diffractogram of the lyophilized crystalline matrix formulation. Figure 4 shows a diffractogram of a lyophilized glycine-only formulation. Comparing

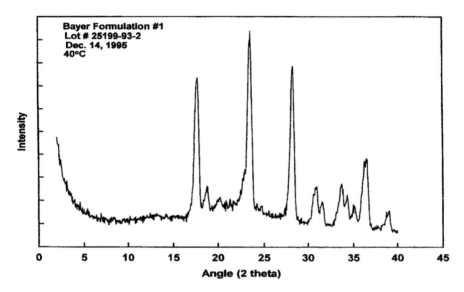

FIGURE 3 X-ray diffractogram of the crystalline matrix formulation.

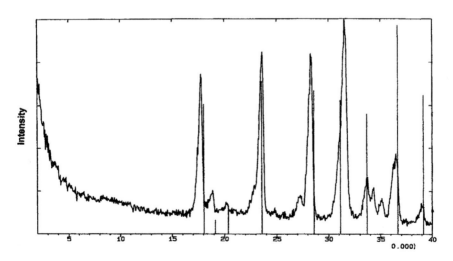

FIGURE 4 X-ray diffractogram of glycine.

these two figures, we conclude that glycine was indeed crystallized when the annealing step was incorporated into the freezing cycle as observed by the sharp peaks in the diffractogram at the specific angles corresponding to standard crystalline glycine.

In addition to DSC, other instruments, such as DTA and electrical resistance analysis (ERA), are also commonly used in determining the thermophysical properties of a lyophilized formulation, such as the glass transition temperature, the collapse temperature, etc. One of the examples using DTA and ER has been recently reported by Ma et al. on a similar crystalline matrix formulation (14). Other literature on the application of DTA (16–18) and ER (3,16,17, 19–21) in formulation characterization is also very informative. However, DSC is the most common and reliable means, and also the easiest.

CONFIRM DSC RESULTS WITH A FREEZE-DRYING MICROSCOPE

A freeze-drying microscope provides real-time images of freezing, melting, crystallization, collapse, and melt-back during the freezing and lyophilization processes. A freeze-drying microscope that we have in our laboratory is shown in Figure 5, consisting of two major parts, an optical microscope with a Physitemp FDC-1 freeze-drying microscope stage. To investigate the freeze-drying behavior of the crystalline matrix formulation described previously, a small aliquot (10 μL) of formulation was frozen to −48°C between quartz coverslips in the freeze-drying stage, either with or without annealing. The behavior of the material during sublimation drying was then observed at temperatures between −48°C and −11°C using an Olympus BX50P polarizing microscope. Micrographs were recorded using a Javelin Smartcam CCD video camera mounted on the microscope and a Data Translation DT3153 frame grabber card. The magnification used for observation of the freeze-drying front was 150×. More details about this type of freeze-drying microscope can be found in Nail (9). Studies of freeze-drying microscopy have also been reported by others (3,4,18, 22–24).

FIGURE 5 A freeze-drying microscope.

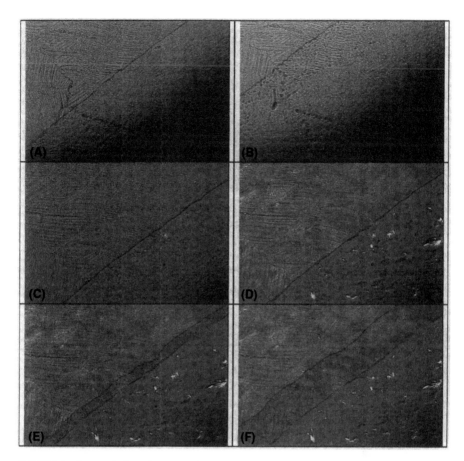

FIGURE 6 Freeze-drying microscope images.

The freeze-drying microscope has become an excellent tool to confirm DSC results. The advantage of a freeze-drying microscope is that it provides real-time images, which can be seen by our own eyes, during freezing and freeze-drying. Figure 6 shows six freeze-drying microscopy images of the crystalline matrix formulation. Figure 6A shows the formulation was initially frozen at −48°C with no annealing. Figure 6B shows that the sample was then freeze-dried with a stepwise increase of drying temperature (i.e., the sample sublimation temperature) and we can see that significant amounts of structure are lost if samples are freeze-dried at temperatures as low as −42°C. Figure 6C shows that since the amorphous structure does not appear to be fully retained during freeze-drying even at −42°C, we stopped freeze-drying and froze the sample down to −48°C again. Then we slowly warmed the sample to −20°C and held it there for about one hour for annealing. Figure 6D shows that after annealing some purple shadows appear between the ice crystals. We believe these purple shadows are crystallized glycine. Figure 6E shows that after annealing, freeze-drying restarted. The sample temperature was first set at −32°C and the cake structure is perfectly preserved. Figure 6F shows the retention of crystalline cake structure when lyophilization was performed at a sample temperature at −15°C. Clearly, the observation agrees with the DSC results described above, that is, with an annealing step, glycine crystallizes out and freeze-drying can be done at a much higher temperature.

So far we have described how to use DSC and a freeze-drying microscope to determine (i) the necessity of annealing in freezing and (ii) the highest allowable product temperature in primary drying. With these two different methods, we can confidently design the freezing cycle and primary drying cycles. For the crystalline matrix-type formulation discussed in this chapter, it is clear that annealing is necessary to crystallize glycine. The annealing temperature should be at −20°C. After the glycine crystallizes out, the highest allowable product temperature is elevated and, therefore, primary drying can be done with a much higher efficiency. Determining the highest allowable product temperature is the most critical factor in developing a primary drying cycle. During the primary drying the product temperature must be kept lower than the highest allowable temperature to avoid collapse and melt-back of the cake. The melt-back here includes both the melt-back of a solid, such as glycine described above, and the melt-back of ice, which occurs at a much higher temperature than that of glycine. In addition, during primary drying, both heat transfer and mass transfer are important in affecting product temperature. In other words, we can achieve a product temperature lower than the highest allowable temperature by using many different combinations of shelf temperature and chamber pressure. Figure 7 demonstrates the possible combinations for the crystalline matrix-type formulation. This provides very important information not only for designing a primary drying process, but also for bracketing process parameters during validation. For example, if −16°C is the highest allowable temperature for the formulation, Figure 7 illustrates that all combinations of shelf temperature and chamber pressure resulting in a product temperature lower than −16°C would be acceptable process parameters. In practice, we usually take at least 2°C to 5°C as a safety margin, and we always optimize the process by looking for the combinations that result in the highest sublimation rate and, consequently, the shortest drying time.

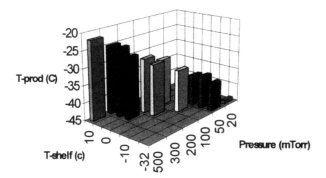

FIGURE 7 Relationship of product temperature as a function of shelf temperature and chamber pressure.

STUDY WATER SORPTION

Water sorption, including desorption and adsorption, provides the degree of "hygroscopicity" of the formulation by plotting the increase in water content of a dry cake as a function of relative humidity to which the sample has been exposed at a given temperature. The curve indicates water vapor sorption characteristics; in other words, the degree of ease in secondary drying. Although the desorption process is not a straightforward reversal of the adsorption, the trends are predictive. In general, the easier the water adsorption, the easier the water desorption.

To conduct a water sorption study, the product was stored in desiccators with solid salt or saturated salt solutions for 48 hours of equilibration at an ambient temperature (Fig. 8). The salts included phosphorus pentoxide, lithium chloride, potassium acetate, magnesium chloride, potassium carbonate, and sodium chloride, which generated relative humidities of approximately 0%, 11%, 23%, 33%, 43%, and 75%, respectively. The vials were sealed immediately after equilibration. The moisture in the lyophilized product was determined by the Karl Fischer method.

Figure 9 shows a typical water adsorption, curve for a typical crystalline matrix-type formulation. Such characteristics of a freeze-dried formulation provide information on the affinity of water for a dried product. It illustrates a progressive increase in water content as the relative humidity is increased. When samples were exposed to relative humidities from 6% to 75%, the resultant moisture content ranged from approximately 1.2% to 17%. This indicated that the lyophilized product was moderately hygroscopic in the relative humidity range of 6% to 43%, and more hygroscopic at relative humidities above 43%. The crystalline material is often "drying friendly" because only the surface of crystals is available for water vapor sorption. In addition, the degree of hygroscopicity of the formulation also provides guidance for product handling, for example in the case of measuring moisture content for product release. Figure 9 also shows that in an environment where the relative humidity was

FIGURE 8 A demonstration of conducting a moisture sorption study with desiccators.

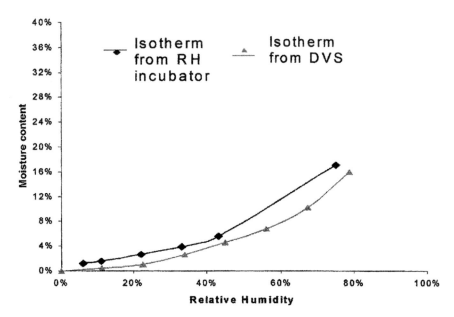

FIGURE 9 Water adsorption curves generated from desiccator method and DVC methods.

greater than 75%, which is common under a noncontrolled laboratory environment during a wet day, the moisture content could increase to as high as 17%. This suggests how important it is to control the environmental relative humidity during the moisture testing. Another tool for determining water sorption is Dynamic Vapor Sorption (DVS, Surface Measurement System, London, U.K.). It measures weight changes due to moisture gain at different relative humidity conditions. This instrument also provides information on water sorption kinetics. Figure 9 also includes the results from the DVS method. The DVS data were comparable to those obtained by incubating the material in desiccators and indicate that moisture could be removed rapidly during secondary drying.

OPTIMIZE MOISTURE CONTENT

Lyophilization is a process to remove water so that the water activity, or the mobility of the water, in the dried product can be reduced to an optimal level. This is achieved by using a secondary drying cycle. As a result, fully understanding the effects of the moisture content on product stability becomes the most important information necessary to develop a secondary drying. Generally speaking, for any formulation there is an optimal range of moisture content that would result in the best stability of the lyophilized product. Hsu et al. reported that there is an optimum residual moisture range for a lyophilized recombinant protein (25). Overdrying may result in opalescence in the product upon reconstitution, while underdrying leads to a greater protein activity loss upon storage under temperature stress conditions. Greiff (26) and Liu et al. (27) revealed that aggregation or insoluble proteins can be induced by a moisture content that was higher than the optimal range.

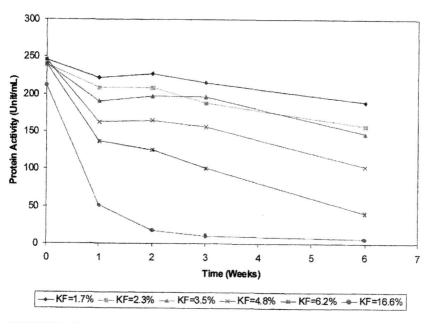

FIGURE 10 Protein activity as a function of storage time and moisture content when samples are stored at 40°C.

Samples for a moisture optimization study are usually generated in the same desiccators that incubate samples for the water adsorption study. That is, samples were exposed to different relative humidities for 48 hours to allow them to adsorb various amounts of moisture. An accelerated stability program was then conducted at 40°C storage temperature. For a crystalline matrix-type formulation, Figure 10 shows how protein activities are affected by moisture levels. It clearly demonstrates that as moisture increased, protein activity decreased accordingly. It should also be noted that the greatest loss occurred in the first week, particularly for the high moisture content samples. The percentage monomer of another protein product, shown in Figure 11, was measured by the size-exclusion high-performance liquid chromatography (SEC-HPLC) method. Data stored for only six weeks had already shown a clear trend: as moisture increased, percentage monomer dropped rapidly, meaning protein aggregation increased drastically. The data demonstrate a trend of "the dryer, the better," which suggests having a long secondary drying cycle that can make a "bone-dry" product. In addition, partial production loads can be comfortably bracketed in the validation because "overdrying" is not a concern for such a formulation.

It should be pointed out that Greiff proposed that the thermostability of freeze-dried protein products could be a bell-shape function of the residual moisture content (26). In other words, both overdrying and underdrying are detrimental. Hsu et al. also concluded that the generally accepted concept the drier, the better is not necessarily appropriate for their product (25). It is noted that their data were generated from excipient-free proteins. In this study, however, our results did not demonstrate a bell-shape function for the

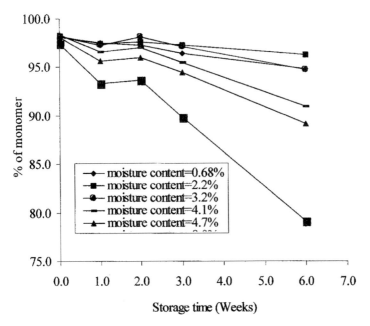

FIGURE 11 Percentage monomer measured by HPLC for product samples stored at 40°C. *Abbreviation*: HPLC, high-performance liquid chromatography.

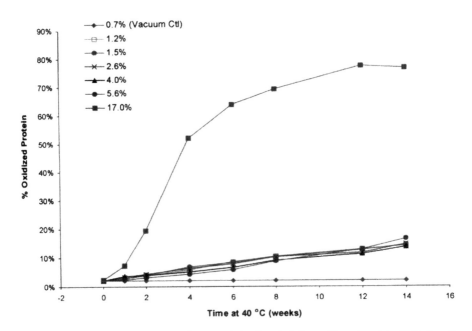

FIGURE 12 Result from another moisture optimization study for a different protein. Oxidation of the protein was studied.

crystalline matrix formulation, but a trend of the drier, the better. In another study on the same protein but in a different formulation, we found that overdrying was indeed detrimental and resulted in an increase of insoluble aggregates and a decrease of protein activity. Therefore, the range of optimal moisture content and the shape of the curve are formulation dependent.

Figure 12 shows results from a moisture optimization for a different protein product but with a similar crystalline matrix-type formulation. However, as oxidation is one of the instability mechanisms, this figure shows the percentage of oxidized protein versus storage time and moisture content. The assay of oxidation used here is reversed phase HPLC. The percentage of oxidation was calculated as the area under the peak for oxidized protein divided by the sum of the areas under the oxidized and nonoxidized peaks. The percentage of oxidized protein increased dramatically at a moisture content of 17%. However, the oxidation rate was comparable for moisture contents of 1.2% to 5.6%. Note that the percentage oxidation is lowest for the vacuum control because of the headspace vacuum. Unsealing the other vials and exposing the contents to atmospheric pressure when samples were incubated in the desiccators increased the percentage of oxidation.

MEASURE T_g OF DRY CAKES

Another way to determine the optimal moisture content for stability is to develop a relationship between moisture content and T_g of lyophilized dry cakes. T_g can be measured using either DSC or MDSC. T_g is a measure of the molecular motion of water, which is a major factor in denaturing proteins (28,29). To increase the long-term storage stability of a pharmaceutical protein, the motion of water can be limited (frozen storage) or water can be removed (lyophilization). However, even for a lyophilized biopharmaceutical product, residual moisture in the dry cake could be mobile depending on the level of the residual moisture content and storage temperature. High residual moisture content and high storage temperature may result in more mobility of water (29), and as a consequence may damage the product protein significantly as well as harm the cosmetic properties of the lyophilized cake. The glass transition temperature of a freeze-dried cake (T_g) marks an increase in the mobility of the remaining water (28). Freeze-dried products stored below their T_g will exhibit much greater stability and structural integrity than those stored above T_g. Because of the cost of storing a product at low temperatures, it is economically more desirable for the T_g of a lyophilized product to be relatively high. Each product formulation will have a different relationship between moisture content and glass transition (29). MDSC can be used to characterize the glass transition of a dry cake (30,31). By analyzing cakes of different moisture contents, the relationship between T_g and moisture content can be quantified. In the following paragraphs, the relationship between dry cake moisture content and T_g is reported for the crystalline matrix-type formulation.

Again, samples for T_g measurement are prepared similarly to those for adsorption study using desiccators. Each desiccator maintained a different relative humidity by using different saturate salts as described previously in the water adsorption study. As a result, different levels of moisture content were achieved for the samples in different desiccators. The driest samples were obtained by keeping them sealed after freeze-drying until sample preparation.

The moisture content of each sample was determined by near-infrared (NIR) content analysis. The NIR used here is provided by a FOSS Model 5000 Rapid Content Analyzer.

All sample preparation for T_g measurement using MDSC took place within a glove box. The relative humidity of the glove box was matched to that of the desiccator each vial had been stored in. This was done to avoid any change in the moisture content of the sample. Approximately 5 to 10 mg of each freeze-dried cake were sealed into a hermetic aluminum DSC sample pan. All DSC work was performed on a TA Instruments 2920 DSC in MDSC mode. Thermogram analysis was carried out with TA Instruments Universal Analysis software. Helium was used as the purging gas. A typical thermogram and testing conditions are given in Figure 13. The scanning rate was 2°C/min, from −40°C to 120°C, modulated at ±0.6°C every 100 seconds. Glass transition values were determined from the reversing heat flow and then corrected to the total heat flow. The relationship of T_g and moisture content for the crystalline matrix formulation is given in Figure 14. It shows if the relationship between T_g and the moisture content of a dry cake is known, an optimum moisture content can be chosen based on the desired storage conditions. Product stability at a higher storage temperature can be achieved by lowering the moisture content of the cake as much as possible. For example, to achieve a product stability at room temperature storage, we have to keep moisture content lower than 2%. In the pharmaceutical industry, it is common practice to use 40°C storage as an accelerated stability test to predict product stability using Arrhenius kinetics. Figure 14 also shows that for this formulation such practice may not be appropriate because it is hardly possible to have T_g above 40°C with a reasonable moisture content. Instead, WLF (Williams-Landel-Ferry) kinetics may better predict the stability as suggested by Franks (1).

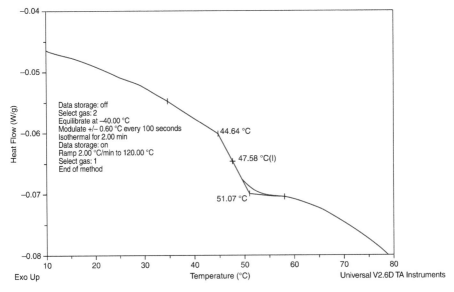

FIGURE 13 A typical thermogram of T_g measurement using MDSC. *Abbreviations*: T_g, glass temperature; MDSC, modulated differential scanning calorimetry.

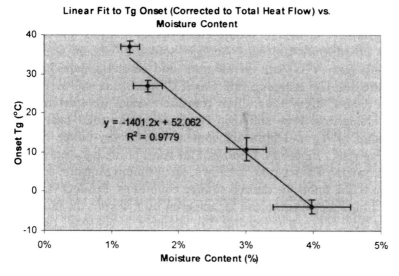

FIGURE 14 T_g of dry cakes as a function of moisture content. *Abbreviation*: T_g, glass temperature.

INTERPRET DATA FOR CYCLE DEVELOPMENT

With a thorough understanding of the formulation, we can reduce the amount of trial and error in cycle development. Results from DSC and the freeze-drying microscope, as we described in previous sections, indicate an annealing step is necessary to crystallize glycine in the formulation. Consequently, this allows us not only to conduct the primary drying aggressively at a relatively high product temperature but also to obtain a good cake appearance. In addition, it also provides a scientific rationale to determine freezing temperature, annealing temperature, primary drying shelf temperature, and primary drying pressure. Data from the adsorption study and moisture optimization studies, on the other hand, demonstrate a characteristic of the drier, the better for this formulation. As a result, it guides us to confidently design secondary drying processes such as determining optimal drying temperature, pressure, and time to provide uniform moisture content in the final containers. The information from the formulation characterization studies also assists in the validation of the cycle, particularly in bracketing the process parameters.

A typical lyophilization cycle for a crystalline matrix-type formulation, such as the one discussed here, is illustrated in Figure 15. The cycle includes the following:

1. An annealing step in freezing to slowly warm the product to $-20°C$ and then hold for one hour.
2. An aggressive primary drying with a shelf temperature of $5°C$ and a chamber pressure of 300 mTorr.
3. An overdrying secondary drying with a ramped shelf temperature of $25°C$ and a chamber pressure of 100 mTorr for a total of 24 hours.

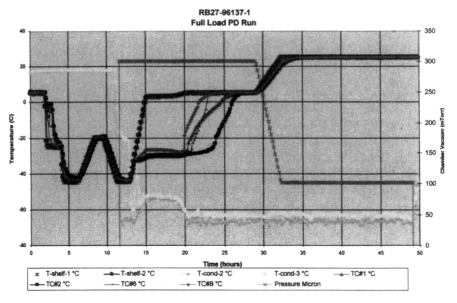

FIGURE 15   A typical freeze-drying cycle.

CONFIRM THE HYPOTHESIS

As described previously, lyophilizing a crystalline matrix type of formulation allows "micro-collapse" of the amorphous components. This allows primary drying to be performed at a relatively high product temperature without loss of the cake structure. The hypothesis we made is that the micro-collapse would not compromise product quality, particularly in stability. Here we discuss this issue and particularly a study that we performed to confirm the hypothesis.

In the case of pharmaceuticals, collapse of the cake structure during freeze-drying results in a pharmaceutically unacceptable product. There are several obvious potentially detrimental effects of collapse on the stability and desirability of a freeze-dried product. The collapse of a system during freeze-drying will reduce the surface area of the cake and so will prevent efficient secondary drying. Also, the collapse may remove the pores or channels left by the sublimated ice, and so may increase the resistance to water vapor moving out of the product, thereby inhibiting primary drying to a greater extent. Since the secondary drying is less efficient, the final product may contain higher levels of moisture than is optimal and the higher moisture content of the cake can then lead to increased levels of instability of the protein being lyophilized. The collapse of the cake may also promote protein aggregation (29). A further complication brought about by collapse is that the reconstitution step may take longer to complete since the surface area of the cake is greatly reduced. Although some investigations (10,29–36) have been published on the effects of collapse on freeze-dried materials, little has been published on the effects of collapse on the long-term storage stability of freeze-dried proteins. Therefore, it is necessary to conduct an investigation of the long-term storage stability. The previously described protein in the crystalline matrix formulation was freeze-dried using three different freeze-drying protocols. These protocols were chosen to produce a collapsed cake, a micro-collapsed cake, and a noncollapsed cake. This investigation aims to test the impact of the micro-collapsed cake on the protein stability.

The three freeze-drying cycles were selected to produce cakes with different physical properties. The first, "no collapse" (NC), was a very gentle cycle, in which the samples were annealed to allow crystallization of the glycine eutectic to occur, and the primary drying temperature was very low. The second cycle, "micro-collapse" (MC), also included an annealing step, but the primary drying temperature was much higher. The third cycle, "collapse" (C), did not include an annealing step, and the temperature during primary drying was also kept low to inhibit the crystallization of the glycine eutectic. Following freeze-drying, the products were placed in stability chambers at 5°C, 25°C, and 40°C to assess their long-term stability during storage at each of these temperatures. Sample vials were removed from the stability chambers periodically and assayed for protein activity, moisture content, and aggregate by SEC-HPLC.

Figure 16 is a photograph of samples freeze-dried by each of the three methods, and shows the level of physical, macroscopic collapse that typically resulted from using the collapse freeze-drying method.

Scanning electron microscopy (SEM) showed that the C cakes are significantly different on a microscopic scale from the other cakes, as shown in Figure 17. The magnification of these pictures is 760×. The C cake shows much less structure than is observed for the NC or MC cakes, which at this magnification are indistinguishable. No evidence was detected by SEM for any level of collapse in the MC samples.

Figure 18 shows the average protein activity measurements for each freeze-drying protocol up to the 18-month time point of the stability study. This plot shows data from each of the storage temperatures, 5°C, 25°C, and 40°C, as well as for liquid samples that were taken immediately prior to freeze-drying, and zero time point (T_0) samples taken immediately after freeze-drying. The

FIGURE 16 Photograph of noncollapsed (NC), micro-collapsed (MC), and collapsed (C) α-amylase samples after freeze-drying.

Non-Collapse   Micro-Collapse

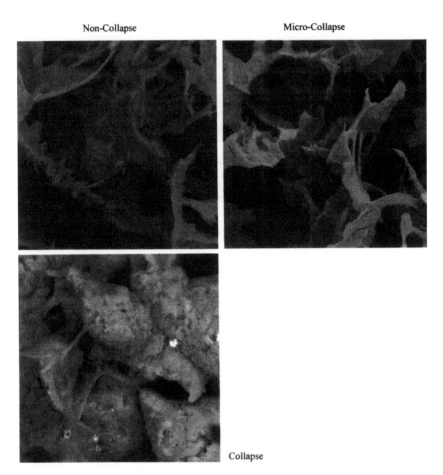

Collapse

FIGURE 17 Scanning electron micrograph of each of the freeze-dried cakes, noncollapsed (NC), micro-collapsed (MC), and collapsed (C). Magnification is 760×.

protein activity is calculated as the percentage of the T_0 activity. As expected, the activity decreases most rapidly at the higher storage temperatures. However, this loss of activity at 40°C appears to be slightly greater for the NC and MC samples than for the C samples. At the lower storage temperatures, the loss of activity appears to be slightly greater in the C samples than in the MC or NC samples after the 18 month storage time point.

The results of this study suggest that collapse is not necessarily detrimental to either the activity or stability of freeze-dried protein. Particularly, data support the hypothesis that the more aggressive, shorter primary drying cycle permitted by the MC cycle does not compromise the stability of the protein product. The similarity between the results of the stabilities of the samples lyophilized by the NC and MC cycles indicates that using the more aggressive cycle is not harmful to the protein. This is important because it permits a more efficient freeze-drying cycle to be used, without decreasing the stability of the protein. However, it is important to point out that such a conclusion may only

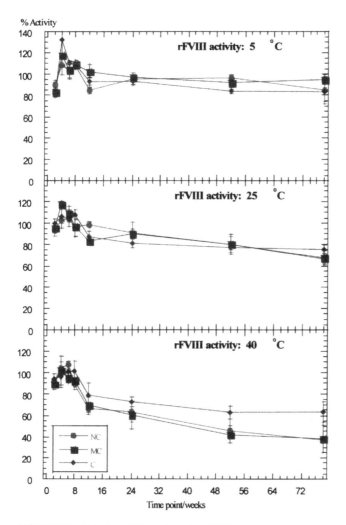

FIGURE 18 Protein activity versus storage time.

apply to the protein and formulation studied here. For other proteins and formulations, the behavior may be totally different.

CONCLUSIONS

This chapter described how to characterize a crystalline matrix-type formulation to successfully develop a lyophilization cycle. Techniques used for characterizing the formulation include DSC, lyo-microscopy, water adsorption study, and moisture optimization. We found that DSC and lyo-microscopy are excellent tools to determine the thermophysical properties of a formulation such as freezing, melting, crystallization, and glass transition temperatures. This information allows us to precisely design freezing and primary drying parameters, such as annealing temperature, shelf temperature, and pressure. We also described how to generate a water adsorption curve and how to conduct a moisture optimization study. We demonstrated that this information is critical for optimizing the secondary drying process, such as drying temperature, pressure, and time. In addition, we also reported a confirmation study that demonstrated after annealing that we could lyophilize protein product at a high temperature without compromising protein activity and stability. The techniques and methods used can be used in other protein formulation characterization studies. In summary, we have demonstrated that a better understanding of formulation characteristics allows us to develop an optimal lyophilization cycle with a scientific rationale.

ACKNOWLEDGMENTS

The author wants to thank Dr Jeff Hey, Dr Xinghang Ma, and Bruce Gardner from Bayer Corporation for their hard work in generating some of the data reported in this chapter. The author also wants to thank Dr Steve Nail from Purdue University for his work on some of the DSC and X-ray diffraction data.

REFERENCES

1. Franks F. Freeze-drying: from empiricism to predictability. The significance of glass transitions. In: May J, Brown, eds. Developments in Biological Standardization. Vol. 74. Basel: Karger, 1991:9–38.
2. Rey L. Thermal analysis of eutectics in freezing solutions. Ann N Y Acad Sci 1960; 85 (pt 2):510–534.
3. MacKenzie AP. Apparatus for microscopic observations during freeze-drying. Biodynamica 1964; 9(186):213–222.
4. MacKenzie AP. Freezing, freeze-drying and freeze-substitution. Proceedings of the Workshop on Biological Specimen Preparation Techniques for Scanning Electron Microscopy, IIT Research Institute, April 1972.
5. Pikal MJ. Freeze-drying of proteins. Part I: Process design. BioPharm 1990; 3(8):18–27.
6. Pikal MJ. Freeze-drying of proteins. Part II: Formulation selection. BioPharm 1990; 10:26–30.
7. Pikal M, Shah S. The collapse temperature in freeze-drying: dependence on measurement methodology and rate of water removal from the glassy phase. Int J Pharm 1990; 62:165–186.
8. Franks F. Improved freeze-drying: an analysis of the basic scientific principles. Proc Biochem 1989; 2:3–7.
9. Nail SL, Her LM, Christopher PB, et al. An improved microscope stage for direct observation of freezing and freeze drying. Pharm Res 1994; 11(8):1098–1100.
10. Izutsu K, Yoshioka S, Kojima S. Physical stability and protein stability of freeze-dried cakes during storage at elevated temperatures. Pharm Res 1994; 11(7):995–999.
11. Arkawa T, Prestrelski SJ, Kenney WC, et al. Factors affecting short- term and long-term stabilities of proteins. Adv Drug Delivery Rev 1993; 10:1–128.
12. Carpenter JF, Prestrelski SJ, Arkawa T. Separation of freezing- and drying- induced denaturation of lyophilized proteins using stress-specific stabilization. Arch Biochem Biophys 1993; 303(2):456–644.
13. Her LM, Nail SL. Measurement of glass transition temperatures of freeze-concentrated solutes by differential scanning calorimetry. Pharm Res 1994; 11(1): 54–59.
14. Ma X, Wang DQ, Bouffard R, et al. Characterization of murine monoclonal antibody to tumor necrosis factor (TNF-MAb) formulation for freeze-drying cycle development. Pharm Res 2001; 18(2):196–202.

15. Hatley RHM. The effective use of differential scanning calorimetry in the optimization of freeze-drying processes and formulations. Dev Biol Stand 1991; 74:105–122.
16. Rey LR. Thermal analysis of eutectic in freezing solution. International Institute of Refrigeration and Laboratoire de Physiologic. Paris, France: Ecole Normale, 1975:510–534.
17. Rey L. Study of the freezing and drying of tissues at very low temperatures. In: Parkes A, Smith A, eds. Recent Research in Freezing and Drying. Oxford: Blackwell, 1960:40–620.
18. MacKenzie AP. Factors affecting the transformation of ice into water vapor in the freeze-drying process. Ann N Y Acad Sci 1965; 125:522–547.
19. Jennings T. Thermal-analysis instrumentation for lyophilization. Med Devices Diagn Instrum 1980; 11:49–57.
20. DeLuca P, Lachman L. Lyophilization of pharmaceuticals, I: Effect of certain physical-chemical properties. J Pharm Sci 1965; 54:617–624.
21. MacKenzie AP. Changes in electrical resistance during freezing and their application to the control of the freeze-drying process. In: Refrigeration Science and Technology, IIF-IIR Commission C Tokyo, 1985:155–163.
22. MacKenzie AP. A current understanding of the freeze-drying of representative aqueous solutions. In: Refrigeration Science and Technology, IIF-IIR Commission C Tokyo, 1985:21–34.
23. MacKenzie AP. Collapse during freeze-drying—qualitative and quantitative aspects. In: Goldblith S, Key L, Rothmayr W, eds. Freeze-Drying and Advanced Food Technology. New York: Academic Press, 1975:277–307.
24. Rey L. Dispositif pour l'examen microscopique aux basses temperatures. Experientia 1957; 13:201–202.
25. Hsu CC, Ward CA, Pearlman R, et al. Determining the optimum residual moisture in lyophilized protein pharmaceuticals. Dev Biol Stand 1991; 74:257–271.
26. Greiff D. Protein structure and freeze-drying effects of residual moisture and gases. Cryobiology 1971; 8:145–152.
27. Liu WR, Langer R, Klibanov AM. Moisture-induced aggregation of lyophilized protein in the solid state. Biotechnol Bioeng 1991; 37:177–184.
28. Yoshioka S. Aso Y, Kojirna S. The effect of excipients on the molecular mobility of lyophilized formulations, as measured by glass transition temperature and NMR relaxation-based critical mobility temperature. Pharm Res 1999; 16(1):135–140.
29. Hancock B, Zografi G. The relationship between the glass transition temperature and the water content of amorphous pharmaceutical solids. Pharm Res 1994; 11(4):471–477.
30. Miller D, de Pablo J, Corti H. Thermophysical properties of trehalose and its concentrated aqueous solutions. Pharm Res 1997; 14(5):135–140.
31. Royal P, Craig D, Doherty C. Characterisation of the glass transition of an amorphous drug using modulated DSC. Pharm Res 1998; 15(7):578–595.
32. Lueckel B, Helk B, Bodmer D, et al. Effects of formulation and process variables on the aggregation of freeze-dried Interlcukin-6 (IL-6) after lyophilization and on storage. Pharm Dev Technol 1998; 3(3):337–346.
33. Adams GDJ, Ramsay JR. Optimizing the lyophilization cycle and the consequences of collapse on the pharmaceutical acceptability of Erwinia L-Asparaginase. J Pharm Sci 1996; 8606(12):1301–1305.
34. Adams GDJ, Irons LI. Some implications of structural collapse during freeze-drying using Erwinia Caratovora L-Asparaginase as a model. J Chem Biotechnol 1993; 58:71–76.
35. Lueckel B, Bodmer D, Helk B, et al. Formulations of sugars with amino acids or mannitol—influence of concentration ratio on the properties of the freeze-concentrate and the lyophilizate. Pharm Dev Technol 1998; 3(3):325–336.
36. Jiang S, Nail SI. Effect of process conditions on recovery of protein activity after freezing and freeze-drying. Eur J Pharm Biopharm 1998; 45(3):249–257.

10

Practical Aspects of Freeze-Drying of Pharmaceutical and Biological Products Using Nonaqueous Cosolvent Systems

Dirk L. Teagarden and Wei Wang
Pfizer Biotherapeutics Pharmaceutical Sciences, Pfizer, Inc., Chesterfield, Missouri, U.S.A.

David S. Baker
Pharmaceutical Sciences, QLT Plug Delivery Inc., Menlo Park, California, U.S.A.

INTRODUCTION

Freeze-drying of pharmaceutical and biological solutions to produce an elegant stable powder has been a standard practice employed to manufacture many marketed pharmaceutical and biological products for over 50 years. The vast majority of these products are lyophilized from simple aqueous solutions. Water is typically the only solvent of significant quantity that is present, which must be removed from the solution via the freeze-drying process. However, frozen water (ice) is not the only frozen liquid that can sublime under reduced pressures. Numerous organic and mineral solvents have been shown to possess this property (1). It is also noteworthy that during the freeze-drying of pharmaceutical or biological products it is not unusual for small quantities of organic solvents to be present in either the active pharmaceutical ingredient or one of the excipients. These low levels of organic solvent are commonly found because they may be carried through as part of the manufacture of these individual components since the ingredient may be precipitated, crystallized, or spray dried from organic solvents. Therefore, many freeze-dried products may be dried from solutions, which contain low levels of a variety of organic solvents. Additionally, there may be instances where freeze-drying from substantial quantities of organic solvents or mixtures of water and organic solvent may offer the formulation scientist advantages over simply drying from an aqueous solution. An example of at least one pharmaceutical product on the market that has utilized an organic cosolvent system during freeze-drying is CAVERJECT® Sterile Powder (2,3). This particular product has been successfully manufactured by freeze-drying from a 20% vol/vol *tert*-butanol/water cosolvent system. Another example is the anticancer drug Imexon. This drug, which is undergoing clinical testing, utilized dimethyl sulfoxide (DMSO) as a formulation vehicle prior to lyophilization (4).

There are many reasons why it may be beneficial to both product quality and process optimization to select a lyophilization process that employs a strictly organic or organic/water cosolvent system. A list of some of these potential advantages is summarized in Table 1. However, the development scientist must be aware that use of these organic/water cosolvent systems causes a variety of issues that must be properly addressed. A list of some of these potential disadvantages is summarized in Table 2. When developing a new freeze-dried product it is critical to evaluate and gain a proper understanding of the fundamental interrelationships

TABLE 1   Potential Advantages of Use of Cosolvents in Freeze-Drying

- Increases rate of sublimation and decreases drying time
- Increases chemical stability of the pre-dried bulk solution
- Increases chemical stability of the dried product
- Facilitates manufacture of bulk solution by increasing drug wettability and solubility in solution
- Improves reconstitution characteristics (e.g., decreases reconstitution time)
- Improves dissolution behavior of lipophilic drugs
- Offers potential alternative to manufacture of products requiring powder filling
- Enhances sterility assurance for pre-dried bulk solution

TABLE 2   Potential Disadvantages of Use of Cosolvents in Freeze-Drying

- Toxicity concerns
- Operator safety concerns due to high degree of flammability or explosion potential
- Lack of compendia grades or monographs
- May require special manufacturing facilities/equipment or storage areas
- Possess difficult handling properties
- Requires high-purity solvent with known impurities at low levels
- Must reach acceptable residual solvent in final product
- High cost to use
- Potential for splash/spattering of product in vial neck
- Lack of regulatory familiarity
- Potential adverse environmental impact

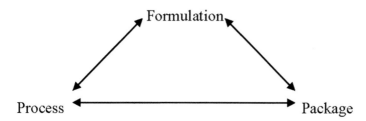

FIGURE 1   Interrelationships between formulation, process, and package.

between the formulation, the process, and the package (Fig. 1), since all must work in unison for a successful product to be developed. The knowledge gained from the interrelationships enables optimization of the formulation, which can be successfully manufactured and packaged in a production-scale setting. These same principles apply to the use of organic solvents in freeze-drying. The advantages and disadvantages for a particular organic cosolvent system must be carefully weighed before they are selected for use in the manufacture of a pharmaceutical or biological product, especially one that is an injectable dosage form.

A list of some of the solvents that have been evaluated in one form or another in freeze-drying studies is provided in Table 3. Included in this table is a summary of some of the critical physical/chemical properties for each of these solvents. Several pharmaceutical and biological products or drugs in various stages of formulation and/or clinical development have been manufactured via a process that required freeze-drying from organic cosolvent systems. These

TABLE 3 Properties of Organic Solvents Evaluated in Freeze-Drying

Solvent	Solubility in water (%[a])	Vapor pressure (mm Hg at 20°C)	Freezing point (°C)	Boiling point (°C)	Flash point (°F/°C)	Flammability		
						Autoignition temperature (°F/°C)	Lower flammability limit (in air vol. %)	Upper Flammability limit (in air vol. %)
tert-Butanol	100	26.8	24.0	82	52/11	892/478	2.4	8.0
Ethanol	100	41.0	−114	78.5	62/16	793/423	3.3	19
n-Propanol	100	14.5	−127	97.1	59/15	760/404	2.1	13.5
n-Butanol	7.7	5.6	−90	117.5	95/35	689/365	1.4	11.2
Isopropanol	100	31.0	−89.5	81	53.6/12	750/398	2.5	12
Ethyl acetate	8.7	64.7	−84	77.1	24/−4	800/426	2.2	11.5
Dimethyl carbonate	9.5	72	2	90	65/18	–	4.2	12.9
Acetonitrile	100	69.8	−48	80.1	45/8	975/524	4.4	16.0
Dichloromethane	1.3	343.9	−97	40	None	1033/556	14	22
Methyl ethyl ketone	27.0	76.2	−87	79.6	26/−3	885/474	1.7	10.1
Methyl isobutyl ketone	2.0	5.1	−80	117	56/13	860/460	1.2	8
Acetone	100	160.5	−94	56.2	1/−17	1000/538	2.6	12.8
1-Pentanol	2.7	1.8	−78	138	120/49	572/300	1.2	10
Methyl acetate	25	148.7	−98	57	15/−9	935/502	3.1	16
Methanol	100	87.9	−98	65	52/11	835/446	6.0	36
Carbon tetrachloride	0.08	78.9	−23	76	None	None	None	None
Dimethyl sulfoxide	100	0.5	18.4	189	188/87	572/300	3.5	42
Hexafluoroacetone	100	5.0[b]	−129	−26	None	None	None	None
Chlorobutanol	0.8	–	97	167	>212/>100	–	–	–
Dimethyl sulfone	100	–	107	248	290/143	–	–	–
Acetic acid	100	11.6	16.2	118.5	103/39	960/516	6.6	19.3
Cyclohexane	0.008	66.4	6.5	81	−1/−18	500/260	1	9
Tetrahydrofuran	100	143	−108	66	−14	610/321	2	11.8

[a]100% = Miscible.
[b]25°C.

TABLE 4 Examples of Drug Preparations Freeze-Dried from Cosolvents

Drug	Cosolvent system	Reference
Alprostadil (CAVERJECT® Sterile Powder)	20% vol/vol *tert*-Butanol/water	2
SarCNU	Neat *tert*-Butanol	49
Aplidine	40% vol/vol *tert*-Butanol/water	6
Rhizoxin	40% *tert*-Butanol/water	47
Amoxicillin sodium	20% vol/vol *tert*-Butanol/water	8
Bendamustine	30% *tert*-Butanol/water	9
Tobramycin sulfate	*tert*-Butanol/water	10,11,67
Nomodipone proliposome	*tert*-Butanol/water	12
Liposome + ketoprofen	*tert*-Butanol/water	13
Gentamicin sulfate	*tert*-Butanol/water	14
Amyloid β-peptide	30% *tert*-Butanol/water	15
N-cyclodexyl-N-methyl-4-(2-oxo-1,2,3,5-tetrahydroimidazo-[2,1-b]quinazolin-7-yl)oxybutyramide with ascorbic acid	50% vol/vol *tert*-Butanol/water	16
HPβCD + ketoprofen or nitrendipine	*tert*-Butanol/water	17
Cyclohexane-1,2-diamine Pt (II) complex	*tert*-Butanol	18
Annamycin	*tert*-Butanol/DMSO/water	19
Edaravone	*tert*-Butanol/water	20
Diazepam	*tert*-Butanol/water	21
Zinc peptide	*tert*-Butanol/water	22
Fructose-1,6-diphosphate	*tert*-Butanol/water	23
Fenofibrate	*tert*-Butanol/water	24
Cephalothin sodium	*tert*-Butanol/water	25
Cephalothin sodium	5% wt/wt Isopropyl alcohol/water	26
Cephalothin sodium	4% Ethanol, 4% methanol or 4% acetone/water	27
Prednisolone acetate/polyglycolic acid	CCl$_4$/hexafluoroacetone sesquihydrate	28
Gabexate mesylate	Ethanol/water	29
Piraubidin hydrochloride	Ethanol/water	30
Progesterone, coronene, fluasterone, phenytoin	Chlorobutanol hemihydrate/Dimethyl sulfone	31
Poly(lactide-co-glycolide)	Acetic acid	32
Phospholipid + clarithromycin	Acetic acid	33
Dioleoylphosphatidylcholine and dioleoylphophatidylglycerol	Cyclohexane	34
Vecuroniumbromide	Acetonitrile	35
Bovine pancreatic trypsin inhibitor	DMSO/1% water	36
Imexon	Neat DMSO	4,37
Danazol/HPβCD	Tetrahydrofuran/water	38

types of solvent systems were chosen for one or more of the advantages described earlier. Table 4 contains a list of examples of a few drug preparations that have been evaluated.

Additional uses for the technique of freeze-drying from organic cosolvent systems, other than that in the manufacture of pharmaceuticals and/or biologicals, includes the preparation of biological specimens, the preparative isolation of excipients such as lecithin, the preparation of aerogels, and the increase in the catalytic activity of certain imprinted enzyme systems. The biological specimens can be prepared by lyophilization from organic cosolvent systems to improve specimen preservation for scanning electron microscopy (SEM) examinations (39–42). *tert*-Butanol appears to be the major organic solvent selected for this use. The surface structure of the specimen remains intact when employing rapid freezing followed by freeze-drying from an appropriate organic solvent such as *tert*-butanol (43). The phospholipid, lecithin, has been shown to be readily prepared in a solvent-free form via lyophilization from cyclohexane (44). The freeze-drying of resorcinol-formaldehyde gels with *tert*-butanol produced more mesopores than when freeze-drying from water (45). Certain imprinted enzymes such as thermolysin, subtilisin, and lipase TL were shown to have significantly enhanced enzymatic activity after freeze-drying the enzyme plus imprinter from 30% *tert*-butanol/water (46).

FACILITATING MANUFACTURE OF BULK SOLUTION

The first step in the manufacture of most freeze-dried products is the formation of a solution of the ingredients to be dried. Typically these solutions are sterile filtered, aseptically filled into containers, and freeze-dried. Some hydrophobic ingredients (e.g., the bulk drug or excipients) may be difficult to wet or may require large amounts of water to adequately solubilize. The use of organic cosolvents can greatly facilitate the wetting of the hydrophobic substance, decrease the time to achieve a solution or uniform dispersion, and decrease the amount of solvent that needs to be removed during the drying process. All of these attributes can potentially have a positive effect on the consistency and ease of product manufacture. Several examples of this increased drug solubility in the presence of organic cosolvents targeted for lyophilization include alprostadil formulated in a *tert*-butanol/water solution (2); cardiotonic phosphodiesterase inhibitors complexed with vitamins formulated in a *tert*-butanol/water solution; however, other alcohols such as ethanol, *n*-propanol, or isopropyl alcohol are also claimed to provide further increases in solubility (16); aplidine (6) formulated in *tert*-butanol/water solution; and rhizoxin formulated in *tert*-butanol/water (47). The actual *tert*-butanol concentration (i.e., 40% vol/vol in water) selected for aplidine produced a greater than 40-fold increase in solubility compared to that in pure water. Rhizoxin formulated in 40% *tert*-butanol produced an increase in solubility, which was greater than 500-fold compared to water (47). The antitumor agent EO-9 exhibited a significant increase in solubility (~8- to 16-fold) when using 30% to 50% vol/vol *tert*-butanol/water compared to water (7). However, one must be aware of the possible negative impact of organic cosolvents on the solubility of hydrophilic excipients. The precipitation potential must be evaluated to determine whether the organic level chosen will allow the solution components to remain in solution throughout the processing steps, especially as temperature changes occur or nucleation sites become available.

STABILIZATION OF BULK SOLUTION

A major challenge in developing a sterile injectable product can be its instability in solution. Most freeze-dried products are developed as this dosage form to circumvent poor stability whether it is chemical or physical instability. The manufacture of a freeze-dried product necessitates that the product is usually first manufactured as a solution, filtered to sterilize, aseptically filled, and finally lyophilized to remove the solvents. All of these unit operations require that the product be held in the solution state for a defined period of time. However, as the product is held in the solution phase it can experience various levels of degradation that are dependent on the kinetics of the degradation mechanism. The presence of the various levels of organic solvent can have a profound effect on the chemical stability. Those drug candidates that are very labile in aqueous solutions may require improved stability to achieve an acceptable level of degradation during manufacture.

Early efforts to freeze-dry an antineoplastic agent (1,3-bis(2-chloroethyl)-1-nitroso-urea) from an ethanol/water solution were initiated because of the rapid degradation in aqueous solution and improved solution stability in ethanol/water solutions (48). Unfortunately freeze-drying this product in the ethanol/water cosolvent system proved to be unsuccessful due to potency losses and unacceptable clarity. Flamberg et al. (48) suggested that an alternative process to freeze-drying solvent systems containing ethanol would be to use low-temperature vacuum drying.

The antitumor drug (SarCNU) was so unstable in aqueous solution that it needed to be freeze-dried from neat *tert*-butanol to prevent degradation during processing (49). The first-order degradation rate noted for SarCNU was 57 times faster in water compared with that in neat *tert*-butanol. Other solvents (e.g., acetic acid, DMSO, or *tert*-butanol/water mixtures) could also be used to freeze-dry SarCNU; however, the solution stability in these solvents was significantly less stable than in neat *tert*-butanol (49). DMSO has been used to stabilize solutions of imexon sufficiently to minimize degradation prior to lyophilization from this solvent system (4). The use of *tert*-butanol (40% wt/vol) was also evaluated for its impact on imexon solution stability; however, neat DMSO was shown to be superior (4).

Another anticancer agent (Rhizoxin), which was freeze-dried from *tert*-butanol (40%), benefited from an enhanced stability in the cosolvent system compared with water (47). A similar observation of improved solution stability in a *tert*-butanol/water cosolvent prior to producing freeze-dried product was noted for the bladder cancer agent (EO-9) (7). Both of these examples enabled manufacture of the product with minimal degradation during processing.

Alprostadil (the active ingredient in CAVERJECT Sterile Powder) has been successfully freeze-dried from a *tert*-butanol/water solution. The first-order degradation rate constant of alprostadil in 20% vol/vol *tert*-butanol/water ($k = 0.0011$/day at 25°C) was significantly reduced compared with water buffered at the same pH value ($k = 0.0041$/day at 25°C). This data is consistent with the claims of extraordinary stability of prostaglandins in *tert*-butanol (50). This decreased degradation rate enables the manufacturing unit operations to be performed at ambient conditions without requiring cooling of the solution during manufacture. Additionally, it adds flexibility in scheduling these various operations because the solution degradation has been minimized.

The formulation of trecetilide fumarate, a sterile injectable in clinical development for treatment of arrhythmias, also involved freeze-drying from a *tert*-butanol/water mixture (51). Kinetic analysis showed solution degradation occurred by a process of defluorination through SN_1 substitution and E_1 elimination, both proceeding through the same carbonium ion intermediate. Since factors such as ionic strength, buffer type, solution pH, and drug and buffer concentrations did not significantly affect degradation rate, destabilization of the fluoride leaving group was one of the few methods left to control this reaction. Use of tertiary butyl alcohol as a cosolvent slowed solution state degradation by a factor of approximately 4 to 5. This significantly increased the probability of

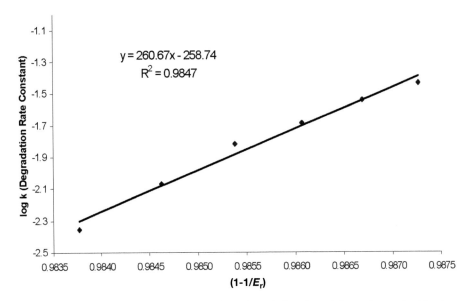

FIGURE 2 Effect of relative permitivity (E_r) on the solution degradation rate constant for tre-cetilide fumarate, a compound undergoing SN_1 substitution and E_1 elimination through the same carbonium ion intermediate.

being able to scale up the manufacturing process while maintaining tight control of the level of degradation. The rate constant (k) for drug degradation was decreased substantially as the *tert*-butanol content was increased. The work required to separate two charges to infinite distance is related to the function $(1 - 1/E_r)$ where E_r is the relative permitivity of the medium. The linear relationship observed between log k and this function (Fig. 2) indicates that the decreased relative permitivity of the solvent system (i.e., the increased work required to remove the fluoride group) was the major effect for the improved solution state stability of trecetilide fumarate. The decreased ability of *tert*-butanol (relative to water) to solvate and stabilize the two ions appeared to be less of a factor. The use of *tert*-butanol allowed formulation and filling on a production scale over a 24-hour period for this compound. The resulting freeze-dried product was predicted to have an acceptable shelf life of at least two years at ambient temperature. This was a dramatic improvement compared with a frozen aqueous solution that had to be stored at −80°C and required use within three hours of thawing and admixture preparation. This type of effect would be expected to be observed for many other drug products that are degraded in the presence of water.

IMPACT ON THE FREEZE-DRYING PROCESS
Effect on Freezing
The first stage of freeze-drying involves freezing the solution to remove solvent (typically water) from the drug and excipients through the formation of ice. The resulting semi-frozen system is cooled further to transform all components into a frozen state. A selected time-temperature profile is achieved by placing the solution, which is commonly held in glass vials or syringes, onto cooled shelves.

Suspended impurities in the solution or imperfections in the walls of the container initiate heterogeneous nucleation during freezing. This event almost always involves supercooling where upon crystallization occurs below the equilibrium freezing point of the solution. Consequently, when freezing does occur, crystal growth tends to be rapid and results in a complex mixture of crystalline, amorphous, and metastable materials. The impact of the presence of organic solvents on the various phases of freeze-drying has been discussed in detail (52–56). Not surprisingly, the type and concentration of the organic solvent present affects the freezing characteristics of the solution prior to initiation of drying. The resulting frozen or semi-frozen solution significantly impacts the crystal habit of the ice, the drying rates, the collapse temperatures, the appearance of the dried cake, the surface area of the dried cake, and reconstitution properties, etc. The choice of solvent can also affect the degree of crystallinity of the drug. It has been demonstrated that incorporation of isopropyl alcohol readily results in highly crystalline cefazolin sodium (26). Scaling up this process required incorporation of a heat treatment step to insure complete crystallization of the drug. Use of cosolvents can sometimes have deleterious effects during freezing. The use of volatile organic solvents has been reported to result in drug precipitation in the latter parts of freezing due to solvent evaporation. This can lead to an increase in drug concentration above its saturation level (55). Care should be taken to select excipient concentrations such as buffer salts so that they do not exceed their saturation solubility. This is particularly important for phosphate buffers since they have very low solubility products with certain cations such as aluminum, calcium, or iron (57–60). As a result, salt precipitation can produce a haze upon reconstitution. Additionally, the preferential precipitation of one of the forms of phosphate can also cause a significant pH shift for the frozen solution. Therefore, it is critical to select buffer components that can maintain pH in both the solution and frozen state. These precipitation problems can be exacerbated in cosolvent systems due to the decreased solubility and higher association constants for such systems.

The size and shape of the ice crystals has been found to vary with different organic solvents. Trace impurities of isobutyl alcohol have been shown to significantly alter the crystal habit of ice crystals (61). The presence of small levels of organic solvents (e.g., DMSO, ethanol, dichloromethane, or acetone) in an aqueous solution has been shown to influence the nucleation of ice crystals and their subsequent growth (62). These organic impurities can also potentially reduce the collapse temperature for the frozen solution (63,64). The presence of ethanol was shown to cause the inhibition of crystallization of solutions of mannitol during freeze-drying unless an annealing step was utilized (65). The presence of high melting point solvents, such as *tert*-butanol, results in solvent crystallizing between the ice matrix as the temperature is decreased. The presence of the *tert*-butanol altered the crystal habit of the ice as it formed. The size of the ice crystals (i.e., large vs. fine) changed depending on the quantity of *tert*-butanol present in the system. The other materials present with the *tert*-butanol/water system can influence the freezing behavior of the alcohol. It has been reported that crystallization of *tert*-butanol hydrate was inhibited by the presence of excipients such as sodium bicarbonate or mannitol (7). Sucrose can also influence the crystallization of *tert*-butanol during cooling such that as the ratio of *tert*-butanol to sucrose was less than 0.2 the *tert*-butanol does not crystallize (66). Investigators demonstrated that *tert*-butanol crystallizes more readily in the presence of tobramycin sulfate than in the presence of tobramycin base (67). Thermal analysis studies (via differential scanning calorimetry and freeze-dry microscopy) have been used to evaluate the various stable and metastable states that form for *tert*-butanol/water systems during freezing (68). The authors were able to apply various annealing techniques to eliminate the metastable states and were able to construct the true phase diagram. Although this phase diagram agreed well with other *tert*-butanol/water phase diagrams reported in the literature (69,70), it was claimed that the slight differences could be explained by the presence of metastable events that thermal treatments eliminated. These data suggested that *tert*-butanol levels in the range of 3% to 19% caused the ice to form needle-shaped crystals. As these large needle-shaped crystals sublimed, they created a more porous, less resistant matrix, which facilitates drying. Other solvents such as acetic acid, formic acid, or dimethyl carbonate also appear to freeze under production freeze-dryer conditions and can be adequately lyophilized. However, most of the organic solvents investigated (56) such as methanol, ethanol, *n*-propanol, *n*-butanol, acetonitrile, methyl ethyl ketone, dichloromethane, and methyl isobutyl ketone do not freeze in typical commercial freeze-dryers but remain as liquid residues within the frozen matrix. The following appears to occur when using conventional freeze-dry equipment: (*i*) solutions containing 8% ethyl acetate, 10% dimethyl carbonate, or 10% *n*-butanol appeared to dry rapidly; (*ii*) solutions containing 10% ethanol, 10% *n*-propanol, or 10% methanol appeared to dry slowly; and (*iii*) solutions containing up to 20% ethanol experienced collapsed cakes and were near impossible to dry [56]. However, other investigators stated that mannitol, in ethanol concentrations up to 30%, may be freeze-dried and produce an elegant cake appearance (65). The ability to achieve this acceptable cake would be a function of various parameters such as ethanol ratio, cycle design, mannitol concentration, and characteristics of other excipients present. Some of these more hydrophilic solvents such as ethanol and methanol retained significant amounts of associated water that only partially froze as the temperature decreased. Samples dried with organic solvents that do not completely freeze may produce a product that is heterogeneous with respect to residual solvent. Use of appreciable levels of solvents, which do not freeze, usually result in unacceptable cake appearance. However, in those products that produce a resistant surface skin during the drying process, a small level of unfrozen organic can cause discontinuities in the skin sufficiently to potentially facilitate the removal of the frozen water vapor (71).

In those systems that completely froze (e.g., *tert*-butanol), the ice and frozen solvent grew upward until reaching the solid surface and formed a eutectic skin. However, other investigators noted that solutions of sulfobutylether 7-β-cyclodextrin in 5% *tert*-butanol froze by having the ice crystals nucleate, grow from the vial bottom, float to the top, and result in fast top-down freezing (72). Hydrophilic solvents that retained large volumes of water formed thick liquid skins containing ice whereas less hydrophilic solvents containing less water formed thinner skins with less ice.

It should also be noted that the time between filling the cosolvent solution and the freezing of this solution should be carefully controlled. The volatility of the organic portion of the solution can be such that a significant portion of the organic solvent can be lost due to evaporation. The loss of solvent may affect content uniformity among different vials. One should be aware of the potential for a reflux type phenomenon when using highly volatile solvents such as *tert*-butanol. This situation can happen when the evaporating *tert*-butanol condenses near the top of the vial and forms a stream of solvent returning to the solution. The dissolved substances in the solution can diffuse in this stream. After freeze-drying has been completed, the vial can contain spots of powder near the neck of the vial. The presence of dried powder near the neck of a vial is not desired because of both a poor appearance and the possibility of negatively impacting the seal with the rubber closure. This problem can be decreased by shortening the time period between the filling and the freezing of the solution.

An interesting study was performed (31) to investigate the feasibility of using organic solvents that are solid at room temperature to lyophilize drug products without the use of conventional freeze-drying equipment. These organic solvents were selected on the basis of their ability to solubilize hydrophobic drugs, increase the solution stability of water-sensitive drug molecules, be readily removed via vacuum drying, and produce elegant dried cakes that were easily reconstituted with acceptable solvents. It was found that the chlorobutanol hemihydrate–dimethyl sulfone eutectic was an optimum solvent system based on its low toxicity, excellent solubilizing capability, and ease of removal via sublimation under vacuum. Lyophilization was accomplished without refrigeration and only required modest heating under vacuum. The resulting dried cakes contained less than 1% residual solvent. Use of this method should be undertaken with caution since the practicalities of heating potentially flammable solvents and keeping the solutions in the liquid phase during filtration, filling, etc. may require additional development time and cost to make this work at a production scale.

Acceleration of Sublimation Rate
The freeze-drying process is a unit operation that typically involves a long and expensive process. Improvements in the rate of mass transfer of solvent through the partially dried cake layer will increase the rate of sublimation and hence decrease the time for the primary drying phase of the freeze-dry cycle. Mathematically it has been shown that

$$\text{Sublimation rate} = \frac{\text{pressure difference}}{\text{resistance}}$$

The resistance term is an additive term that reflects the sum of the dried product resistance, the vial/stopper resistance, and the chamber resistance. However, the predominant resistance of the three terms is typically the product resistance, which usually accounts for about 90% of the total resistance (73). The mass transfer in the dry layer occurs via two general mechanisms: bulk flow (the movement of material in the direction of a pressure gradient, which may be molecular or viscous) or diffusive flow (the movement of material by molecular motion from higher concentration to lower concentration or partial pressure) (74). Typically the resistance to mass transfer increases with cake depth. What happens during the freezing phase can have a profound impact on the resistance of the dried cake to mass transfer during primary drying. Those cosolvents that can alter the ice crystal habit and size of the ice crystals, such as promoting the formation of large needle crystals, can dramatically increase the rate of bulk flow. This is because the permeability of the dried layer would increase proportionally (to the third power for molecular flow and to the fourth power for viscous flow) to the average diameter of the pore created by the sublimation process (74). Many of the cosolvents selected for freeze-drying increase sublimation rate because they have higher vapor pressures than water and hence an expected larger driving force for sublimation because the latter depends on this vapor pressure difference (75).

The potential acceleration of the freeze-drying rates of aqueous solutions of lactose and sucrose with 5% and 10% aqueous solutions of *tert*-butanol has been studied (76). It was found that both lactose and sucrose solutions could be successfully freeze-dried in the presence of *tert*-butanol at considerably higher shelf temperatures than corresponding aqueous solutions of either lactose or sucrose. The drying rates were significantly increased when using the *tert*-butanol as a cosolvent. The drying times were decreased by approximately 50% when drying sucrose in the presence of *tert*-butanol. The collapse temperature for the frozen solutions appeared to increase when *tert*-butanol was present. The *tert*-butanol readily froze and remained frozen during the primary drying phase. The *tert*-butanol sublimed during primary drying and created a porous structure that facilitated the mass transfer of water vapor due to decreased cake resistance. The resulting dried cakes exhibited significantly increased surface areas. Freeze-drying of a similar amorphous carbohydrate such as a lactose base formulation from a *tert*-butanol/water cosolvent (e.g., CAVERJECT Sterile Powder) also produces a very porous cake structure, as illustrated by the SEM shown in Figure 3. Alternatively, it was demonstrated that using other organic cosolvents at a 5% level (e.g., methanol, ethanol, isopropanol, acetone, *n*-butanol, or dioxane) that do not freeze under operating conditions for conventional commercial freeze-dryers produced unacceptable freeze-dried cakes of either lactose or sucrose due to boiling of solvent and cake collapse.

FIGURE 3 SEM picture of CAVERJECT Sterile Powder that has been lyophilized from a 20% vol/vol *tert*-butanol/water solution. *Abbreviation*: SEM, scanning electron microscopy.

A more in-depth study was later completed evaluating the use of *tert*-butanol as a mass transfer accelerator during the freeze-drying of a 5% wt/vol sucrose solution (77,78). Again it was demonstrated that the primary drying phase (i.e., sublimation) proceeded more rapidly when 5% wt/wt *tert*-butanol was present, thereby resulting in an approximately 10-fold reduction in drying time. This subsequent decrease in sublimation time was confirmed by others studying the effect of *tert*-butanol used in freeze-drying and concluded that the drying times were comparable to what was observed with agitated vacuum contact dryers (79). This increased drying rate was caused by the formation of needle-shaped ice crystals that dramatically lowered the product resistance of the dried cake. The resulting surface area of the dried cake increased by approximately 13-fold when the 5% *tert*-butanol was used. The presence of the *tert*-butanol did not impact the collapse temperature; however, the rapid sublimation prevented the product from reaching the collapse temperature. The rationale for this was postulated to be because the water content in the partially dried layer decreased faster in the presence of the *tert*-butanol, which resulted in an increased viscosity and thereby prevented collapse. The rate of sublimation of both the water and the *tert*-butanol was impacted by the ratio of the two solvents. Water appeared to sublime faster at ratios of less than 20% wt/wt *tert*-butanol/water. *tert*-Butanol appeared to sublime faster at ratios of greater than 20% wt/wt *tert*-butanol/water. Both solvents sublimed at equal rates at 20% wt/wt *tert*-butanol/water. The latter data suggested a strong association at this concentration. This data is consistent with the sublimation of water and *tert*-butanol from the frozen matrix of CAV-ERJECT Sterile Powder during freeze-drying, as illustrated in Figure 4. Others noted an increased drying rate for sulfobutylether-7-β-cyclodextrin lyophilized from 5% *tert*-butanol (72). They also noted a top-down freezing mechanism for this system. It was postulated that, in addition to the large ice crystals that form in solutions of *tert*-butanol/water, the ice structure formed by the top-down freezing mechanism might help increase the sublimation rate (72). The addition of the 5% *tert*-butanol for this system reduced the T_g' by 10°C; however, it only decreased the collapse temperature by 2°C.

Mixtures of *tert*-butanol/water were also used to increase the rate of freeze-drying for solutions for a model drug, gentamicin sulfate (14). Simple freeze-drying of the gentamicin with the cosolvent system was not sufficient to produce an acceptable cake. The investigators had to apply a statistical experimental design to achieve the final formulation. All components of the formulation studied, that is, the active ingredient, *tert*-butanol/water ratio, and the amount of the maltose bulking agent were optimized to achieve a reduction in drying time by approximately 40% and yet produce a freeze-dried cake of acceptable porosity. The addition of the maltose was a key component for this particular formulation and process. It should be readily apparent that optimization of the formulation and process parameters are necessary to maximize the impact of the cosolvent system on drying rates.

IMPACT ON STABILITY OF FREEZE-DRIED PRODUCT
Positive Impact on Stability
It is noteworthy that the chemical stability of the freeze-dried product can be positively impacted by the type of solvent system from which it is lyophilized. The product, CAVERJECT Sterile Powder, is significantly more stable when it is

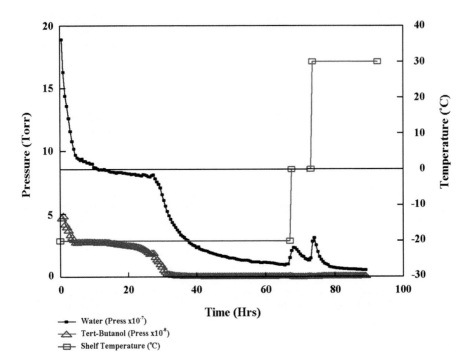

FIGURE 4 Sublimation data (via Turboquad mass spectrometer) for CAVERJECT Sterile Powder (freeze-dried from 20% vol/vol *tert*-butanol/water).

freeze-dried from a 20% vol/vol *tert*-butanol/water cosolvent system compared to a simple aqueous solution. The reason for the improved stability appears to be related to the marked increase in dried cake surface area (~5-fold), which is produced when using this solvent system. Figure 5 illustrates that the

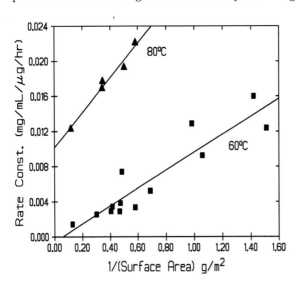

FIGURE 5 Impact of surface area of CAVERJECT Sterile Powder on degradation rate constant.

degradation rate constant for this product is related to the reciprocal of the dried cake surface area. Since the degradation kinetics of the active ingredient, alprostadil, appear to fit an apparent second-order mechanism with respect to the alprostadil concentration (i.e., the rate of formation of the major degradation product, PGA_1, increases by the square of the alprostadil concentration), the improved chemical stability is consistent with a larger cake surface area, which might increase the distance between the alprostadil molecules.

Another example of improved chemical stability for the lyophilized product includes the use of isopropyl alcohol in the freeze-drying of cefazolin sodium (26). The presence of the isopropyl alcohol helped induce crystallization of the amorphous cefazolin sodium during the freezing phase. The presence of the isopropyl alcohol decreased the glass transition temperature of the system and lowered the temperature of crystallization (80). Use of the isopropyl alcohol with a thermal treatment phase enabled a freeze-dried crystalline form of the drug to be produced, which possessed superior stability (26). Use of this cosolvent system also enabled the product to be more effectively processed with shorter lyophilization times and fewer instances of cake collapse. However, when mannitol was added to the system, the presence of the mannitol prevented the crystallization of the cefazolin sodium even with a thermal treatment (80). Other short chain alcohols, for example, methanol, ethanol, and *n*-propanol or acetone are also claimed to provide similar improvements to freeze-drying of cefazolin sodium (27). Crystalline lyophiles of cefazolin sodium could be obtained using a *tert*-butanol/water cosolvent system (25). A thermal treatment with high initial concentration of cefazolin sodium was necessary to induce crystallization.

Freeze-drying of sucrose solutions with isopropanol has been shown to produce a cohesive cake, which is more physically stable at higher temperatures (81). Heating of the dried amorphous sucrose under the right conditions can induce the sucrose to become crystalline as the residual alcohol evaporates.

A very unique technique was employed in stabilizing tobramycin sulfate by use of freeze-drying from a 20% *tert*-butanol/water system (10). Initially the product was freeze-dried using this cosolvent system. The resulting cakes contained amorphous tobramycin sulfate. However, prior to unloading the dryer, humidified nitrogen was pumped into the freeze-dry chamber. The increasing moisture level caused the glass transition temperature of the tobramycin sulfate to sufficiently decrease and allow crystallization to occur. This was followed by rapid release of the residual *tert*-butanol. The resulting product was an in situ crystallized form of tobramycin sulfate with very low level of residual *tert*-butanol (0.008%).

Spray freeze-drying of Δ^9-tetrahydrocannabinol in inulin from a *tert*-butanol/water cosolvent system was shown to help stabilize the labile drug (82). It was noted that spray freeze-drying of above produced more stable product than simple freeze-drying in a vial. It was postulated that the spray freeze-drying enables faster freezing and hence less time for phase separation to occur. It was recommended to use spray freeze-drying for the production of solid dispersions.

Freeze-drying of plasmid DNA from 20% vol/vol *tert*-butanol/water with calcium chloride produced a condensed shear resistant form of DNA with improved stability (83).

Negative Impact on Stability

Conversely, there are cases where the lyophilization from cosolvents can produce a less stable system. An example of this occurrence is illustrated by the lyophilization of the protein, bovine pancreatic trypsin inhibitor from 1% water/DMSO (36). The data appear to support the premise that the protein dissolving DMSO denatures the protein sufficiently to reduce its enzymatic activity after reconstitution. Additionally, it is important to evaluate the impact of residual organic solvent remaining at the end of primary drying since the combination of the solvent and higher product temperatures during secondary drying may lead to undesirable chemical reactions (71). The use of cosolvents can also have a negative impact on the stability of cyclodextrin-bound stabilized drug preparations since the cosolvent may compete for the binding of the drug molecule to the cyclodextrin and lessen the percentage of bound drug at the end of freezing prior to lyophilization.

IMPACT ON RECONSTITUTION PROPERTIES FOR STERILE PRODUCTS

The ability of the freeze-dried cake to be readily reconstituted upon addition of an appropriate pharmaceutical solvent is dependent on several factors. The structure of the dried product, the degree of cake collapse or melt-back that has occurred during drying, the surface area of the cake, the presence of hydrophobic coatings, and the homogeneity of the dry matrix are all factors that can influence the reconstitution properties of the dried product. Depending on the organic cosolvent selected and processing conditions used to freeze-dry, the product may or may not readily reconstitute. Therefore, one will need to evaluate this property on a case-by-case basis. However, there are examples of freeze-drying sucrose and lactose solutions from *tert*-butanol/water solutions with the proper drying cycle where amorphous cakes with large surface areas were produced (2,76). These cakes tended to reconstitute extremely rapidly upon addition of reconstitution vehicle. Proper freeze-drying of tobramycin sulfate from *tert*-butanol/water systems produced a friable easily reconstitutable cake; however, freeze-drying the same drug with less than 10% ethanol or isopropanol produced a hard, difficult to reconstitute cake (10).

It is also important to recognize the importance of extractables from the filling line and package system (i.e., the glass vial and rubber closure or the filter and tubing) used to filter, fill, and hold the solution being lyophilized. The use of siliconized vials or stoppers can cause problems when organic cosolvents are used because the organic solvent can solubilize or extract silicone oil from the package component. This can be especially problematic when freeze-drying in syringes since they are typically siliconized to facilitate their use. This same problem can also occur if silicone tubing is used to transfer solution during filtration, filling, etc. The extracted silicone oil can impede the wetting of the affected portions of the cake, result in the cake being difficult to reconstitute, cause the reconstituted solution to become hazy, or exceed the particulate matter and/or clarity specification limits. A high organic content also requires a judicious choice of sterilizing filters. Fortunately there is now a reasonable choice of high-quality solvent-resistant filters available. However, it is recommended that the filter manufacturer be consulted on their selection and that appropriate compatibility studies be performed.

IMPACT ON DISSOLUTION RATE FOR NONSTERILE PRODUCTS

Studies have also been completed that demonstrated that freeze-drying from cosolvents such as *tert*-butanol can improve the dissolution behavior of lipophilic drugs such as fenofibrate (24). This work illustrated how to produce nanocrystalline particles via controlled crystallization during freeze-drying. Crystallization could occur during either the freezing or the drying phase. When crystallization occurred during freezing, faster freezing or use of solution with lower water/*tert*-butanol ratios produced smaller crystals. However, when crystallization occurred during drying, faster freezing or use of higher water/*tert*-butanol ratios resulted in the formation of smaller crystals. Other workers demonstrated that cryogenic spray freezing of danazol/hydroxypropyl β-cyclodextrin complex in tetrahydrofuran/water followed by lyophilization produced high surface area powders with improved dissolution characteristics (38). Other investigators freeze-dried hydrophobic drugs such as budesonide or salmeterol xinafoate with either β- or hydroxyl propyl β-cyclodextrins from *tert*-butanol/water cosolvent systems (17). The resulting complex exhibited an enhanced dissolution rate. Model drugs, ketoprofen and nitrendipine, and their hydroxyl propyl β-cyclodextrin complexes were prepared by freeze-drying from *tert*-butanol. The resulting dried complex exhibited markedly enhanced dissolution rate in simulated gastric and intestinal fluids (84,85). Drug cyclodextrin complexes prepared by this method can be further manufactured into injection powders, tablets, or capsules that exhibit improved in vivo absorption (86). The lipophilic drug diazepam, when freeze-dried with sugars from *tert*-butanol/water systems and later tabletted, produced anomalous dissolution behavior, which was governed by how it crystallized (21). The latter study noted that the fastest dissolution occurred for low drug loads or when using slow dissolving sugars. Others found that freeze-drying from a *tert*-butanol/water mixture is an excellent method to produce solid dispersions of lipophilic drugs in sugar glasses (87). The dispersions prepared by this method were usually amorphous and had improved dissolution rates. The residual *tert*-butanol trapped after freeze-drying could be reduced by the exposure to moist air. Spray freeze-drying using *tert*-butanol/water as a solvent was used to produce solid dispersions of Δ^9-tetrahydrocannabinol in inulin (82). The dispersions produced fine particles with improved dissolution characteristics that were suitable for inhalation therapy.

ENHANCEMENT OF STERILITY ASSURANCE

Since many freeze-dried products tend to be injectable formulations, they are required to be manufactured under sterile conditions. The growth-promotive properties of the bulk solution prior to freeze-drying can impact the facilities and processes required to manufacture the sterile freeze-dried product (e.g., conventional aseptic manufacturing vs. advanced aseptic manufacturing such as barrier-type suites). Many of the potential organic cosolvents proposed for lyophilization possess some form of microbicidal properties (88). Ethanol and isopropanol at concentrations of greater than 20% exhibit excellent antimicrobial activity against both gram-positive and gram-negative bacteria, fungi, yeasts, and molds (89). However, neat ethanol is less bactericidal than mixtures with water because the combination of alcohol plus water promotes the loss of cell proteins through a solvent damaged cell membrane. Neat *tert*-butanol exhibits excellent bactericidal properties against *Pseudomonas aeruginosa, Escherichia coli, Staphylococcus aureus,* and *Staphylococcus epidermidis.* There are also reports of several organic solvent-resistant organisms (90,91). The presence of an organic solvent, therefore, does not guarantee that the bulk solution will have adequate microbicidal properties. Each formulation must be evaluated as to its growth-promotive nature to determine the appropriate aseptic manufacturing requirements. If the solution is sterilized by filtration then the prefiltration bioburden must be evaluated and an appropriate holding time for the unfiltered solution qualified.

CONTROL OF RESIDUAL SOLVENT LEVELS
The retention of volatile components such as organic solvents during freeze-drying has been well described in the literature (5,92–100). Many of the organic compounds in the food aroma have a vapor pressure higher than water. However, a significant number of these compounds remain with the freeze-dried cake after lyophilization. Two theories that are used to explain this phenomenon have been described as "selective diffusion" (100) and "microregion entrapment" (98). The retentive behavior for model amorphous carbohydrate systems, which were freeze-dried from various volatile organic solvents, was studied by Flink and Karel (94). They found that the volatile retention for organic solvents (acetone, methyl acetate, ethanol, *n*-propanol, isopropanol, *n*-butanol, *tert*-butanol, and 1-pentanol), which were freeze-dried in maltose, sucrose, or lactose ranged from 1.4% to 3.3%. The retention levels were similar for volatiles of different vapor pressures. The microregion entrapment theory postulates that the retention is not due to adsorption to the dried material. The solvent retention appears to occur in localized regions where the volatile was initially frozen. These regions occur on a microscale. Increases in the secondary drying conditions of temperature or reduced pressure do little to decrease the volatile retention level. The volatile retention shows no competition with water vapor for sorption sites, which might be indicative of different modes of interaction with the amorphous carbohydrate. The previous statement appears to be true up to a point as long as the samples are kept dry. Upon humidification of the dried carbohydrate containing the entrapped volatile organic, there appears to be a critical humidity condition that results in a corresponding moisture level where the volatile organic is rapidly released. An example of this effect is illustrated by the humidification of maltose, which had been freeze-dried from either isopropanol/water or *tert*-butanol/water. Both exhibited a trend whereby as the moisture content reached approximately 8% to 9%, the residual alcohol content dropped significantly (96). Structural changes in the cake due to moisture absorption, especially during cake collapse, can result in rapid loss of the entrapped volatile. Others noted that residual *tert*-butanol in sugars (e.g., sucrose, trehalose, and inulin) could be reduced by exposure to moisture vapor (87). The *tert*-butanol removal from the sucrose and trehalose was accompanied by crystallization of the sugars; however, the glassy inulin remained amorphous. The volatile retention appears to be related to hydrogen bonding to the amorphous carbohydrates. As the hydrogen bonds are broken, such as at certain moisture levels, the volatile loss accelerates. The selective diffusion model predicts that as the water content for carbohydrate systems decreases, the ratio of

the diffusion coefficient of the volatile organic to the diffusion coefficient of water decreases. This ratio becomes so small for low water/organic concentrations in dry amorphous carbohydrates that the system is only permeable to water (100,101).

The impact of formulation and process variables on residual solvent levels for formulations that were freeze-dried from *tert*-butanol/water cosolvent systems has been critically evaluated (102). The physical state of the solute (i.e., amorphous vs. crystalline), the initial *tert*-butanol concentration, freezing rate, cake thickness, and temperature during secondary drying were examined. It was noted that when a crystalline matrix was used (e.g., glycine), the residual *tert*-butanol was very low (0.01–0.03%), regardless of freezing rate or initial *tert*-butanol concentration. Freeze-drying of SarCNU from neat *tert*-butanol produced crystalline SarCNU, which was very low in residual *tert*-butanol (0.001%) (49). Interestingly, the residual *tert*-butanol data when a D-mannitol matrix was used was reported to be approximately 0.8% (6). Although the authors claimed to have produced crystalline mannitol via annealing at −20°C, it is possible that the thermal treatment was only annealing some unfrozen *tert*-butanol hydrate and that part of the mannitol remained uncrystallized. This may explain the higher than expected residual *tert*-butanol levels for a totally crystalline matrix. Considerably higher residual *tert*-butanol levels were noted when freeze-drying an amorphous sugar such as sucrose from *tert*-butanol/water systems. However, processing conditions had a profound impact on the residual solvent level present at the end of drying. Low levels of *tert*-butanol/water (1–2% wt/wt) resulted in high levels of *tert*-butanol residuals in the dried sucrose amorphous cake ($\approx$10–18%). Higher levels of *tert*-butanol (3–5% wt/wt of an aqueous solution) resulted in lower residual *tert*-butanol levels in the dried amorphous cake ($\approx$2%). Freeze-drying tobramycin sulfate from various levels of *tert*-butanol (5–9%) produced residual solvent levels ranging from 0.6% to 1.0% (67). Similar results (1–2% residual levels) were obtained when freeze-drying lactose solutions from 20% vol/vol *tert*-butanol/water mixtures (2). This latter matrix would also be expected to produce an amorphous cake. It was postulated that, when using *tert*-butanol levels above the threshold concentration required for eutectic crystallization of the solvent, lower residual *tert*-butanol levels in the freeze-dried cake are obtained. The reverse is true when the starting *tert*-butanol concentration is below this threshold. Freezing rate appeared to impact the residual *tert*-butanol level for amorphous systems in that fast freezing (e.g., with liquid nitrogen) produced cakes with higher residual *tert*-butanol. Examples of the latter observation were supported by the higher residual alcohol levels when flash freezing *tert*-butanol solutions of tobramycin sulfate or sucrose (10,102) or liquid nitrogen freezing of aqueous solutions of *n*-butanol (99). These data appeared to contradict the data reported for isopropanol retention in freeze-dried maltose or dextran (98) and other *n*-alcohols in maltodextrin (100). It also contradicts data for residual *tert*-butanol retention in freeze-dried lactose. It would, therefore, appear that the impact of freezing rate on residual solvent content requires evaluation on a case-by-case basis.

Evaluation of the factors that influenced the residual *tert*-butanol in cyclodextrin complexes freeze-dried from *tert*-butanol/water revealed that the initial *tert*-butanol concentration should be higher than the crystal formation concentration, the appropriate cyclodextrin should be selected, and one should employ an annealing technique (103). Increases in both time and temperature for

TABLE 5 Effect of Variable Residual *tert*-Butanol in CAVERJECT Sterile Powder on T_g and the Degradation Product (PGA$_1$) Formation After Nine Months at 25°C

% *tert*-Butanol	Estimated[a] T_g (°C)	PGA$_1$ (μg/mL) after 9 months at 25°C
1.4	97	1.0
1.4	97	0.9
4.0	79	3.1
4.0	79	3.2
5.1	71	4.0
5.2	71	3.8

High residual *tert*-butanol caused decreased T_g and increased degradation.
[a]Estimated using Gordon–Taylor equation (104).

the secondary drying phase did not significantly reduce the residual *tert*-butanol level in the freeze-dried sucrose cakes. Similar results for lactose dried from *tert*-butanol/water have been observed. Other solvent systems may be behaving differently than *tert*-butanol since freeze-drying of imexon from neat DMSO produced crystalline imexon; however, the residual DMSO content was relatively high at 4.6% (4).

It should be noted that the level of residual solvent may have an impact on the stability of the freeze-dried product. The residual solvent can act as a plasticizer at low levels, decrease the glass transition temperature (T_g) of the dry matrix, increase its molecular mobility, and thereby affect stability of the lyophilized powder. CAVERJECT Sterile Powder, which was freeze-dried from *tert*-butanol/water, could produce variable residual *tert*-butanol levels for non-annealed samples. The different levels of residual *tert*-butanol in the dried matrix would be expected to affect the T_g, as predicted by the Gordon–Taylor equation (104). Table 5 and Figure 6 show data that illustrate the impact of different residual *tert*-butanol levels on the T_g and stability of CAVERJECT [as evidenced by the formation of the degradation product (PGA$_1$)]. The increase in residual *tert*-butanol appeared to plasticize the dry cake, decrease the T_g, and increase the degradation rate.

Thermal Treatment (Impact of Annealing)

It is noteworthy that the level of residual solvent remaining at the end of the freeze-dry cycle can be significantly impacted by use of thermal treatments (e.g., annealing) of the frozen solution prior to initiation of the drying phase. This effect is illustrated by studying the impact that process conditions had on the residual *tert*-butanol levels remaining in CAVERJECT Sterile Powder. This product contains a predominantly lactose base and is lyophilized from a 20% vol/vol *tert*-butanol/water solution. Normal freezing by loading on precooled (i.e., −40°C) shelves followed by lyophilization produced a bimodal distribution of residual *tert*-butanol levels (Fig. 7). The majority (94–97%) of a typical lot contained residual *tert*-butanol levels in the range of 1% to 2%. However, the remaining 3% to 6% of the lot contained *tert*-butanol levels ranging from 3.4% to 5.5%. A tighter control of the residual alcohol level was achieved for this product through the addition of a thermal treatment step (i.e., annealing) during the freezing phase to control metastable forms of solvents that might form. The differential scanning calorimetry thermogram of the CAVERJECT frozen

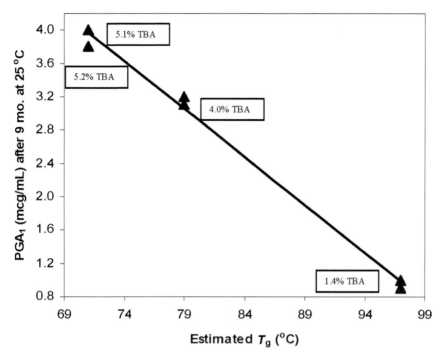

FIGURE 6 The impact of the estimated glass transition temperature on degradation product (PGA$_1$) formed for CAVERJECT Sterile Powder after nine months at 25°C. *Abbreviation*: TBA, tertiary butyl alcohol.

solution (unannealed and annealed) is illustrated in Figure 8. The annealing likely enabled any remaining unfrozen *tert*-butanol hydrate to crystallize and produce a more uniform frozen product. The presence of the amorphous solute such as the lactose probably was the cause of the presence of the metastable unfrozen *tert*-butanol since it would be expected to inhibit the crystallization of the *tert*-butanol during cooling (80,105). The effect of this thermal treatment on the residual *tert*-butanol was rather dramatic (Fig. 7). The resulting *tert*-butanol levels were on average slightly lower (0.8–1.1%), much tighter, and no values exceeded 1.3%. A similar annealing technique was employed during the freezing phase for aplidine in a 40% vol/vol *tert*-butanol/water system and for tobramycin sulfate in a 5% *tert*-butanol/water system prior to lyophilization (6,10,67). Annealing of cefazolin sodium in solutions isopropanol/water, ethanol/water, and *tert*-butanol/water (followed by lyophilization) all produced lower residual solvent levels than non-annealed samples (25). However, only the *tert*-butanol was able to truly crystallize during the thermal treatment. It also produced the lowest residual solvent level (<0.2%). It was postulated that, when annealing compounds that crystallize as hydrates, one must be aware that the dehydration that occurs during the lyophilization process may produce a crystalline hydrate to amorphous anhydrate transition and negate the utility of the cosolvent (25).

   It is clear from the above discussion that the formulation scientist must evaluate the impact of the process conditions on the developing freeze-dried formulation to adequately control the residual volatile organic present at the end

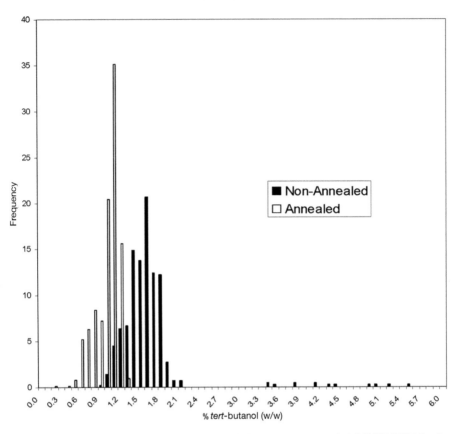

FIGURE 7 Residual *tert*-butanol levels in annealed versus non-annealed CAVERJECT Sterile Powder.

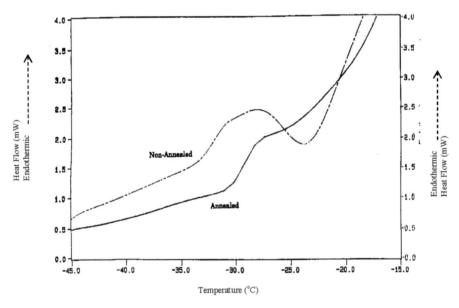

FIGURE 8 Differential scanning calorimetry thermograms for annealed versus non-annealed bulk solution used in CAVERJECT Sterile Powder.

of freeze-drying. It should be noted that the use of annealing can impact the overall drying cycle time. As shown by Chouvenc et al. (106), annealing can accelerate sublimation rates by increasing ice crystal sizes and hence decreases the dried mass transfer resistance. Therefore, annealing can typically decrease the primary drying time. However, annealing can also decrease the desorption kinetics and lead to longer secondary drying times. The resulting consequence is that optimization of the annealing may be necessary to achieve an appropriate freeze-dry cycle.

Purity of Solvent (Impact of Impurities, etc.)

A critical component in the selection of an organic cosolvent is the evaluation of the purity of the solvent, which is intended for use. Since many of the organic solvent impurities may not freeze or sublime, they can be retained in the freeze-dried cake. Typically they are either not removed or are only partially removed during the freeze-drying process. Therefore, it is possible that the impurities can become concentrated in the freeze-dried cake. Knowing the type and level of impurities in the selected organic solvent is critical to determining whether the grade or supplier is acceptable. A list of the significant organic impurities found in a lot of *tert*-butanol is summarized in Table 6. The organic impurities retained in a lactose cake prepared from freeze-drying from a 20% vol/vol *tert*-butanol/water cosolvent system are summarized in Table 7. It is readily apparent that many of the organic impurities in the parent solvent are only partially removed via the lyophilization process. It is therefore critical that proper quality control specifications are established for the selected organic solvent. Table 8 contains a list of potential specification assays, which one might use to determine raw material quality. Once specifications are established, each lot of organic solvent used for freeze-drying should be tested to confirm acceptable purity.

TABLE 6 Organic Impurities Found in *tert*-Butanol

Impurity	Concentration (% wt/wt)
Methanol	0.02
Ethanol	0.01
Isopropanol	0.26
2-Butanol	0.02
2-Methyl-1-propanol	0.04
tert-Butyl peroxide	<0.01
tert-Butyl ether	0.01

TABLE 7 Residual Organic Impurities in Lactose Freeze-Dried from 20% vol/vol *tert*-Butanol/Water

Impurity	Concentration (% wt/wt)
tert-Butanol	1.1–2.0
Ethanol	0.01
Isopropanol	0.12–0.24
2-Butanol	0.01
2-Methyl-1-propanol	0.03–0.04

TABLE 8 Potential Specification Assays for
Organic Solvents Used in Freeze-Drying

- Description
- Identification
- GC impurities
- Nonvolatile residue
- Potency
- Water
- Specific gravity
- Acidity
- Alkalinity
- Refractive index
- Microcount
- Pathogens

MISCELLANEOUS PHARMACEUTICAL USES
Liposomes
Several authors have utilized freeze-drying of liposome forming lipids from *tert*-butanol or mixtures of *tert*-butanol/water to prepare preliposomes and liposomes. Various drugs could be incorporated into the liposome either during the preparation or after reconstitution (12,34,107–113). Certain preliposomes could be prepared by dissolving liposome forming lipids and water-soluble carriers such as sucrose in the *tert*-butanol/water mixtures, sterile filtering, and freeze-drying to remove the solvents (13). Reconstitution of these dried materials enabled the spontaneous formation of homogeneous liposomes. Drugs such as ketoprofen could be passively loaded into the liposome by dissolving it in the *tert*-butanol lipid mixture prior to lyophilization (13). Entrapment efficiencies of approximately 44% were achieved. Other drugs such as ciprofloxacin lactate and propanolol hydrochloride were actively loaded into the liposome after reconstitution in response to pH gradients (13). Entrapment efficiencies of 77% and 94% were achieved, respectively. The homogeneity and size of the liposome was controlled by adjustment of the lipid/sucrose ratio. Proliposomes of nimodipine could be prepared by freeze-drying lipids and nimodipine from *tert*-butanol/water (12). The lipid concentration and surface charge were the major factors influencing the entrapment efficiency. If the drug to be encapsulated has good solubility in *tert*-butanol/water and good affinity to phospholipids, one can use passive loading. Those drugs that can be loaded in response to some type of gradient (e.g., pH gradient) can use the active loading approach. This latter approach can be utilized to prepare drug-encapsulated liposomes just prior to clinical use. Other investigators evaluated the impact of freeze-drying preformed liposomes from *tert*-butanol/water (114). They found that freeze-drying of liposomes from cosolvent systems was acceptable for hydrogenated soybean phosphatidylcholine liposomes but not for soybean phosphatidylcholine liposomes. A novel method has been reported that involved preparation of an oil solution containing both poorly water-soluble and poorly oil-soluble drugs (33). A phospholipid and a poorly oil-soluble drug (clarithromycin) were dissolved in acetic acid and lyophilized. Addition of this dried material to oil produced a clear oil solution.

Microparticles

Spherical microparticles of a zinc peptide were shown to form during freeze-drying from a *tert*-butanol/water cosolvent system (22). Parameters such as the peptide to zinc ratio, the percentage of *tert*-butanol used, the cooling technique employed, and the use of annealing or not had an impact on whether spherical or irregular-shaped microparticles formed. A unique mechanism for the spherical microparticle formation was proposed that involved the formation of a solid emulsion, which after removal of the solvent generated spherical micro-particles.

ULTRALOW TEMPERATURE FREEZE-DRYING

Rey (115) pioneered the use of ultralow temperature freeze-drying by evaluating solvents such as benzene and solid ammonia to lyophilize unstable compounds such as phospholipids. Ammonia is described as an exceptional solvent, on the basis of its remarkable properties, which are somewhat analogous to water. It is easily frozen by liquid nitrogen and can be lyophilized by sublimation between $-130°C$ and $-110°C$. Liquid ammonia is also a reaction medium that allows the study of new chemical entities and includes the storage of unstable and reactive elements containing free radicals. Liquid ammonia allows the lyophilization of tissues since it completely dissolves the glycerol used to coat and protect the tissue during freezing. Truly freeze-drying low freezing solvents (i.e., those with freezing points ranging from $-50°C$ to $-120°C$) would require very specialized freeze-dry equipment. A description of the specialized equipment required to achieve these ultralow temperature freeze-drying conditions has been provided by various authors (63,64). Since these solvents, like water, will supercool prior to freezing, a shelf that can reach very low temperatures such as that achieved with liquid nitrogen would be required to enable the solvent to properly freeze. Additionally, external radiation would need to be controlled to achieve uniform drying (63). Rey (115) also describes the use of liquid carbon dioxide to extract organic compounds. Subsequent lyophilization is achieved at low temperature ($-78.8°C$) and atmospheric pressure. Solidification of CO_2 is achieved either by using liquid nitrogen or by releasing some CO_2 as a gas with subsequent cooling and freezing of the remaining CO_2. Carbon tetrachloride is also described as a good solvent for lipids, with the use of glycol distearate to modify its solubilizing characteristics. Other solvents used include dioxane and chloroform. Ultralow temperature freeze-drying was found to mimic that of aqueous systems such that with a suitable choice of the drying temperature and pressure, sublimation of the crystallized solvent resulted in preservation of the structure of the frozen interstitial phase (116). Freeze-drying more complex systems was studied by depositing and freezing thin films of two immiscible solvent systems. Two sets of solvents were studied water/dioxane and benzene/chloroform/carbon tetrachloride/cyclohexane. Freeze-drying was described as stepwise if the vapor of one of the solvents is eliminated preferentially and complex if the vapors of both solvents are eliminated together at comparative rates.

TOXICITY ISSUES

As discussed in the previous section, the volatile organic solvent component will be retained to a certain extent by the freeze-dried cake. The amount retained will be governed by the solvent used, the formulation which is lyophilized, and the

processing conditions selected. Because most residual solvents offer no therapeutic benefit, they should be removed to low levels to meet product specifications, good manufacturing practices, and other quality-based requirements. A
major concern that must be addressed is whether the residual solvent present is
at an acceptable level on the basis of safety considerations. The acute toxicity
(i.e., the LD_{50} for several species) for many of the potential solvents investigated
as a freeze-dry solvent has been summarized previously (117). These data, plus
other toxicity data that reside in the literature, provide some level of assurance
of the relative safety of the solvent. However, an additional resource to use to
gauge the acceptability of a particular solvent and its acceptable residual level
has been provided in the International Conference on Harmonization guidance
document on impurities: residual solvents (118). The guidance ranks various
solvents into classes on the basis of a safety risk assessment. Class 1 solvents
should be avoided since they can be known carcinogens, possess a high toxicity
potential, or pose an environmental hazard. An example of a class 1 solvent
investigated in freeze-drying is carbon tetrachloride (28). Class 2 solvents are
those solvents that cause reversible toxicity. They can also be nongenotoxic
animal carcinogens or be possible causative agents of other irreversible toxicity.
Their content should be limited. Examples of class 2 solvents found in pharmaceutical products are illustrated in Table 9 along with their permitted daily
exposure and concentration limits. Several of the class 2 solvents listed in Table 9

TABLE 9 Class 2 Solvents in Pharmaceutical Products

Solvent	Permitted daily exposure (mg/day)	Concentration limit (ppm)
Acetonitrile[a]	4.1	410
Chlorobenzene	3.6	360
Chloroform	0.6	60
Cyclohexane[a]	38.8	3880
1,2-Dichloroethene	18.7	1870
Dichloromethane[a]	6.0	600
1,2-Dimethoxyethane	1.0	100
N,N-dimethylacetamide	10.9	1090
N,N-dimethylformamide	8.8	880
1,4-Dioxane	3.8	380
2-Ethoxyethanol	1.6	160
Ethylene glycol	6.2	620
Formamide	2.2	220
Hexane	2.9	290
Methanol[a]	30.0	3000
2-Methoxyethanol	0.5	50
Methylbutyl ketone	0.5	50
Methylcyclohexane	11.8	1180
N-methylpyrrolidone	48.4	4840
Nitromethane	0.5	50
Pyridine	2.0	200
Sulfolane	1.6	160
Tetralin	1.0	100
Toluene	8.9	890
1,1,2-Trichloroethene	0.8	80
Xylene	21.7	2170

[a]Organic solvents evaluated in freeze-drying.

TABLE 10 Class 3 Solvents That Should be Limited by GMP or Other Quality-Based Requirements

Acetic acid[a]	Heptane
Acetone[a]	Isobutyl acetate
Anisole	Isopropyl acetate
1-Butanol[a]	Methyl acetate[a]
2-Butanol[a]	3-methyl-1-butanol
Butyl acetate	Methylethyl ketone[a]
tert-Butylmethyl ether	Methylisobutyl ketone[a]
Cumene	2-Methyl-1-propanol
Dimethyl sulfoxide[a]	Pentane
Ethanol[a]	1-Pentanol[a]
Ethyl acetate[a]	1-Propanol[a]
Ethyl ether	2-Propanol[a]
Ethyl formate	Propyl acetate
Formic acid	Tetrahydrofuran

[a]Organic solvents evaluated in freeze-drying.

have been used in freeze-drying investigations. The residual content of all of these class 2 solvents should be limited in pharmaceutical products because of their inherent toxicity. The allowable residual level should be less than the permitted daily exposure limit. Class 3 solvents exhibit low toxicity potential to man. Examples of class 3 solvents found in pharmaceutical products are illustrated in Table 10. Several of these class 3 solvents have been used in freeze-drying investigations. Permitted daily exposure limits of 50 mg/day or less would be acceptable for these solvents without additional justification. The most used organic solvent in freeze-drying, that is, *tert*-butanol (2-methyl-2-propanol) is not listed in the guidance document. However, it is likely to fall in the category of a class 3 solvent based on the similarity of acute LD_{50} toxicity data for other class 3 solvents. It should be noted that the ICH guidance does not apply to either potential new drug substances, excipients, and drug products used during the clinical research stages of development or to products marketed prior to July 1997. The guidance also notes that higher levels of residual solvent may be acceptable in cases such as short-term exposure (30 days or less). However, this would require justification on a case-by-case basis.

HANDLING, SAFETY, AND STORAGE ISSUES

After an appropriate organic solvent has been selected for use in the freeze-drying process, it is important to consider how the solvent can be safely handled and properly stored. The package in which the solvent is received must be inert and not provide any additional extractables that may contaminate the freeze-dried product. Usually stainless steel or glass containers are preferred to minimize contaminants. Most of the organic solvents are highly flammable so proper care must be exercised when handling and storing to prevent fires or explosions. Manufacture of the bulk solution may need to take place in an explosion proof manufacturing module. Use of electrical mixers should be avoided. Equipment should be appropriately grounded to make intrinsically safe where possible. Transport of the manufactured solution should occur in sealed tanks (preferably stainless steel vessels). Use of a nitrogen blanket above the solution headspace is recommended as an additional safety feature. The tanks must have a pressure relief valve, and there must be a shutoff valve on the tank to turn off liquid flow, should there be an emergency. If pressure filtration is employed, nitrogen as opposed to air should be used. Hard piping of the transfers is recommended as opposed to use of flexible tubing. Appropriate care must be taken during the filling and loading of the freeze-dryer. Distance of electrical equipment used during the filling operation from the manufacturing tank must comply with appropriate National Electrical and Fire safety codes for flammable materials. Local ventilation should be used near vessel connection points to prevent vapor buildup should a leak occur. The vacuum sensors for the freeze-dryer that are hot wire resistance variety (thermal conductivity-type gauge, Pirani, etc.) should be replaced with capacitance manometer–type diaphragm gauges to minimize the hazards associated with these volatile organics (71). It may even be necessary to work in explosion proof environments or at least explosion proof the electric motors on the vacuum pumps. Depending on the type of organic solvent used it may be necessary to add a liquid nitrogen cold trap prior to the vacuum pumps if the condensers are not cold enough to remove the migrating vapors. The freeze-drying of ethanol containing products has been such that it is sufficient to install a tube that is in thermal contact with the condenser coils and that can be drained at the end of drying (62). One must be aware that contamination of the vacuum pump oil with organic solvents may decrease pump efficiency and lead to the inability to adequately control the vacuum in the drying chamber. The proper disposal of the solvent and its environmental impact must be appropriately assessed.

The most commonly used organic solvent in freeze-drying, *tert*-butanol, presents an additional unusual handling requirement. This solvent freezes close to room temperature with the pure solvent creating a negative volume change on freezing. The pure material may be completely liquid, a solid, or a mixture of both at room temperature. To use this solvent, any solid material must be liquefied by warming, taking care to not break the container as the solution expands. The total container should be mixed after warming to produce a uniform solution since the unfrozen impurities typically tend to concentrate in the liquid portion because they possess lower freezing points. This solvent is flammable so caution must be used to warm it.

However, an alternative method to use this key freeze-drying solvent is to have the supplier provide the *tert*-butanol as a slightly diluted mixture with water (e.g., 95% vol/vol *tert*-butanol/water). The appropriate quality of water (e.g., water for injection) should be selected to make the dilution if the solvent mixture is destined for use in the manufacture of an injectable product. This addition of water lowers the freezing point from room temperature to approximately 2°C. This mixture is much easier to handle because it remains liquid at room temperature. The *tert*-butanol is commonly used as a mixture with water anyway during freeze-drying so the additional water will not have any negative impact on the drying process.

SUPPLIER SELECTION AND QUALIFICATION

As discussed previously, the purity of the organic cosolvent must be closely monitored since the freeze-drying process may not remove unwanted impurities that can be retained in the freeze-dried cake. It is wise to select a supplier who has a drug master file established for the manufacture, packaging, and storage of

GMP-grade organic solvent. If a drug master file is not available, the supplier needs to work to establish one. It is recommended to audit the supplier to confirm that appropriate documentation of the production process is in place to satisfy GMP requirements. This would especially include verification of cleaning validation and records to assure that manufacturing equipment, packages used, and transport tankers, etc. do not add unwanted contaminants. A sufficient number of lots should be evaluated to confirm that the quality will consistently meet targeted specifications and enable the manufacture of reproducible freeze-dried product lots.

COST

All factors impacting the cost of goods need to be assessed before one should choose to use organic solvents in freeze-drying. Those parameters such as potentially increased drying rates, increased solution stability, ease of drug wetting, increased product stability, etc. can have a positive influence in reducing manufacturing costs. However, there are numerous factors that can add significantly to the cost. First, there are several one-time development costs that must be factored into the expense estimates. Typically a new solvent supplier may need to be located, qualified, and materials tested/evaluated. Additional safety studies may need to be completed to satisfy potential regulatory concerns of residual solvents in the product. The manufacturing facilities may require modification to handle the storage and use of flammable solvents in a safe manner. The continual fixed cost of high-purity GMP-grade solvents must be evaluated. It is not unusual for the GMP-grade solvent expense to exceed 10 to 20 times the cost of the same corresponding analytical grade solvent. The cost of disposal of the used solvent and its environmental impact must also be taken into consideration. The positive and negative influences of the use of organic cosolvents on the cost to manufacture product must be weighed together to determine the overall cost acceptability.

REGULATORY CONCERNS

The majority of the pharmaceutical regulatory agencies throughout the world have limited experience with the review or approval of freeze-dry processes using organic cosolvents. Therefore, increased scrutiny of the regulatory file should be expected. However, there has been some recent precedence with the worldwide regulatory approval of CAVERJECT Sterile Powder. It has also been noted that there are other drug molecules entering clinical testing that have been lyophilized from cosolvents. Additionally, since it is quite common for most active pharmaceutical ingredients and some excipients to be precipitated and/or dried from organic solvents, the same type of safety evaluation and/or justification for the residual organic solvent levels should be employed. It is important to demonstrate that the product has been carefully evaluated for all possible residual solvents since it is quite possible that impurities in the starting solvent can concentrate during the drying process. Appropriate qualification of the solvent vendor (ideally with a drug master file), the solvent storage package and conditions would appear a prudent course. This is especially critical when using noncompendia grades of solvents. The ICH guidelines on residual solvents discussed earlier should be used to justify the proposed specification limits.

CONCLUSIONS

Nonaqueous cosolvent systems have been used in the freeze-drying of pharmaceutical and biological products, preparation of biological specimens, or preparative isolation of excipients. The reasons for freeze-drying from these nonaqueous solvent systems for pharmaceutical products include increased drug wetting or solubility, increased sublimation rates and hence decreased drying time, increased pre-dried bulk solution or dried product stability, decreased reconstitution time, increased dissolution rate, presence of a potential alternative to powder filling, and possible enhancement of sterility assurance of the pre-dried bulk solution. However, the practicalities of use of these cosolvent systems must be properly assessed before they should be considered for use. This especially applies when using them in the manufacture of a pharmaceutical or biological product. The issues that must be evaluated include the proper safe handling and storage of flammable and/or explosive solvents, the special facilities or equipment that may be required, the determination and control of residual solvent levels, the toxicity of the remaining solvent, the qualification of an appropriate GMP purity and supplier, the overall cost benefit to use of the solvent, a possible adverse environmental impact, and the potential increased regulatory scrutiny when used in pharmaceutical or biological products.

The cosolvent system that has been most extensively evaluated was the *tert*-butanol/water combination. The *tert*-butanol possesses a high vapor pressure, freezes completely in most commercial freeze-dryers, readily sublimes during primary drying, can increase sublimation rates, and has low toxicity. This cosolvent system is being used in the manufacture of a marketed injectable pharmaceutical product. When using this solvent system, both formulation and process control required optimization to maximize drying rates and to minimize residual solvent levels at the end of drying.

Other cosolvent systems that do not freeze completely in commercial freeze-dryers were more difficult to use, do not sublime but rather boil during the evacuation process, often resulted in unacceptable freeze-dried cakes, and may bypass condensers and contaminate vacuum pumps. Their use appears limited. Many of these types of solvents were near impossible to dry when levels used exceed 10%.

REFERENCES

1. Rey L. Potential prospects in freeze-drying. In: Rey L, May JC, ed. Freeze-drying/ Lyophilization of Pharmaceuticals and Biological Products. 1st ed. New York: Marcel Dekker, 1999:470–471.
2. Teagarden DL, Petre WJ, Gold PM. Stabilized prostaglandin E_1. US patent 5,741,523. April 21, 1998.
3. Teagarden DL, Petre WJ, Gold PM. Method for preparing stabilized prostaglandin E_1. US patent 5,770,230. June 23, 1998.
4. Den Brok M, Nuijen B, Lutz C, et al. Pharmaceutical development of a lyophilized dosage form for the investigational anticancer agent Imexon using dimethyl sulfoxide as solubilizing and stability agent. J Pharm Sci 2005; 94(5):1101–1114.
5. Lambert D, Flink J, Karel M. Volatile transport in frozen aqueous solutions. I: Development of a mechanism. Cryobiology 1973; 10:41–45.
6. Nuijen B, Bouma M, Henrar REC, et al. Pharmaceutical development of a parenteral lyophilized formulation of the novel antitumor agent aplidine. PDA J Pharm Sci Technol 2000; 54(3):193–208.

7. van der Schoot S, Nuijen B, Flesch F, et al. Development of a bladder instillation of indoloquinone anticancer agent EO-9 using tert-butyl alcohol as a lyophilization vehicle. AAPS PharmSciTech 2007; 8(3):E1–E10.

8. Tico Grau JR, Del Pozo Carrascosa A, Salazar Macian R, et al. Determinacion de disolventes residuales en materias primas farmaceuticas: Aplicacion en amoxicilinas sodicas liofilzadas. Cienc Ind Farm 1988; 7(11):325–330.

9. Brittain J, Franklin J. Lyophilized bendamustine pharmaceutical compositions for treatment of autoimmune and neoplastic diseases. WO patent application 2006076620. July 20, 2006.

10. Needham G. Formulation and processing factors affecting residual solvent levels in TBA/water formulations. Freeze-Drying of pharmaceuticals and biologicals. Breckenridge, CO, August 1–4, 2001.

11. Kwok K, Yang K. Free-flowing lyophilized tobramycin formulation. US patent 20080221049. September 11, 2008.

12. Wang Z, Deng Y, Zhang X, et al. Preparation of proliposome nimodipine and quality evaluation. Shenyang Yaoke Daxue Xuebao 2006; 23(11):681–685.

13. Li C, Deng Y. A novel method for the preparation of liposomes: freeze-drying of monophase solutions. J Pharm Sci 2004; 93(6):1403–1412.

14. Baldi G, Gasco M, Pattarino F. Statistical procedures for optimizing the freeze-drying of a model drug in tert-butyl alcohol-water mixtures. Eur J Pharm Biopharm 1994; 40(3):138–141.

15. Breitenkamp K, Skaff H, Sill K. Polymer-encapsulated amyloid-beta peptides. WO patent application 2008106657. September 4, 2008.

16. Benjamin EJ, Visor GC. Cardiotonic phosphodiesterase inhibitors complexed with water soluble vitamins. US patent 4,837,239. June 6, 1989.

17. Wang Z, Deng Y, Sun S, et al. Preparation of hydrophobic drugs cyclodextrin complex by lyophilization monophase solution. Drug Dev Ind Pharm 2006; 32:73–83.

18. Tanno N, Ae N, Ishizumi K. Freeze-dried pharmaceuticals of fat-soluble platinum (II) complexes. Japanese patent 03255025. March 2, 1990.

19. Zou Y, Priebe W, Perez-soler R. Submicron liposome suspensions obtained from preliposome lyophilizates. US patent 5,902,604. May 11, 1999.

20. He J, Li X, Huang S, et al. Freeze-dried preparation of edaravone and preparation process thereof. China patent 101288650. October 22, 2008.

21. vanDrooge D, Hinrichs W, Frijlink H. Anomalous dissolution behaviour of tablets prepared from sugar glass-based solid dispersions. J Control Release 2004; 97:441–452.

22. Qian F, Ni N, Chen J, et al. Formation of zinc-peptide spherical microparticles during lyophilization from tert-butyl alcohol/water co-solvent system. Pharm Res 2008; 25(12):2799–2806.

23. Sullivan BW, Marangos PJ. Partially lyophilized fructose-1,6-diphosphate (FDP) for injection into humans. US patent 5,731,291. March 24, 1998.

24. deWaard H, Hinrichs W, Frijlink H. A novel bottom-up process to produce drug nanocrystals: controlled crystallization during freeze-drying. J Control Release 2008; 120:179–183.

25. Telang C, Suryanarayanan R. Crystallization of cephalothin sodium during lyophilization from tert-butyl alcohol water cosolvent system. Pharm Res 2005; 22(1):153–160.

26. Koyama Y, Kamat M, De Angelis RJ, et al. Effect of solvent addition and thermal treatment on freeze drying of cefazolin sodium. J Parenter Sci Technol 1988; 42(2):47–52.

27. Cise MD, Roy ML. Improvements in or relating to freeze-drying cephalothin sodium. British patent 1,589,317. May 13, 1981.

28. DeLuca PP, Kanke M, Sato T, et al. Porous microspheres for drug delivery and methods for make same. US patent 4,818,542. April 4, 1989.

29. Kamijo S, Imai A, Kodaira H. Preparation of crystalline lyophilized gabexate mesilate with high stability. Japanese patent 63270623. April 30, 1987.

30. Kaneko S, Kishikawa M, Sato T, et al. Manufacture of freeze-dried anthracycline glycoside preparations with improved solubility. Japanese patent 07076515. June 30, 1993.

31. Tesconi MS, Sepassi K, Yalkowsky SH. Freeze-drying above room temperature. J Pharm Sci 1999; 88(5):501–506.

32. Meredith P, Donald AM, Payne RS. Freeze-drying: in situ observations using cryoenvironmental scanning electron microscopy and differential scanning calorimetry. J Pharm Sci 1996; 85:631–637.

33. Qin L, He H, Tang X. Phospholipid solubilization of both poorly water-soluble and poorly oil-soluble drug into oils using freeze-drying technology. Asian J Pharm Sci 2006; 1(2):103–109.

34. Felgner PL, Eppstein DA. Stable liposomes with aqueous-soluble medicaments. European patent 0172007B1. May 22, 1991.

35. Jansen FHJ. Process to prepare pharmaceutical compositions containing vecuronium bromide. US patent 5,681,573. October 28, 1997.

36. Desai UR, Klibanov AM. Assessing the structural integrity of a lyophilized protein in organic solvents. J Am Chem Soc 1995; 117(14):3940–3945.

37. Opitz H, Woog H, Gruber W. Parenteral forms of administration of imexon and method for the production by lyophilization. WO patent application 2006042662. April 27, 2006.

38. Rogers T, Nelsen A, Hu J, et al. A novel particle engineering technology to enhance dissolution of poorly water soluble drugs: spray-freezing into liquid. Eur J Pharm Biopharm 2002; 54:271–280.

39. Inoue T, Osatake H. A new drying method of biological specimens for scanning electron microscopy: the t-butyl alcohol freeze-drying method. Arch Histol Cytol 1988; 51:53–59.

40. Akahori H, Ishii H, Nonaka I, et al. A simple freeze-drying device using tert-butyl alcohol for SEM (scanning electron microscopy) specimens. J Electron Microsc (Tokyo) 1988; 37(6):351–352.

41. Hojo T. Specimen preparation of the human cerebellar cortex for scanning electron microscopy using a t-butyl alcohol freeze-drying device. Scanning Microsc Suppl 1996; 10:345–348.

42. Fujii Y, Ohno N, Li Z, et al. Morphological and histochemical analyses of living mouse livers by new "cryobiopsy" technique. J Electron Microsc (Tokyo) 2006; 55 (2):113–122.

43. Herman R, Müller M. Limits in high-resolution scanning electron microscopy: natural surfaces? Scanning 1997; 19:337–342.

44. Radin NS. The preparative isolation of lecithin. J Lipid Res 1978; 19(7):922–924.

45. Tamon H, Ishizaka H, Yamamoto T, et al. Influence of freeze-drying conditions on the mesoporosity of organic gels as carbon precursors. Carbon 2000; 38:1099–1105.

46. Rich J, Mozhaev V, Dordick J, et al. Molecular imprinting of enzymes with water-insoluble ligands for non-aqueous biocatalysis. J Am Chem Soc 2002; 124(19): 5254–5255.

47. Stella V, Umprayn K, Waugh W. Development of parenteral formulations of experimental cytotoxic agents I. Rhizoxin (NSC-332598). Int J Pharm 1988; 43:191–199.

48. Flamberg DW, Francis DL, Morgan SL, et al. Low temperature vacuum drying of sterile parenterals from ethanol. Bull Parenter Drug Assoc 1970; 24(5):209–216.

49. Ni N, Tesconi M, Tabibi S, et al. Use of pure t-butanol as a solvent for freeze-drying: a case study. Int J Pharm 2001; 226:39–46.

50. Monkhouse DC. Tertiary alcohol stabilized E-series prostaglandins. US patent 3,927,197. December 16, 1975.

51. Baker DS. Unpublished results.

52. Seager H, Taskis CB, Syrop M, et al. Freeze drying parenterals containing organic solvent—Part 1. Manuf Chem Aerosol News 1978; 49(11):43–44,47–48.

53. Seager H. Freeze drying parenterals containing organic solvent—Part 2. Manuf Chem Aerosol News 1978; 49(12):59–60, 64.

54. Seager H. Freeze drying parenterals containing organic solvent—Part 3. Manuf Chem Aerosol News 1979; 50(1):40, 42–45.

55. Seager H. Freeze drying parenterals containing organic solvent—Part 4. Manuf Chem Aerosol News 1979; 50(2):41–42, 44–45, 48.

56. Seager H, Taskis CB, Syrop M, et al. Structure of products prepared by freeze-drying solutions containing organic solvents. J Parenter Sci Technol 1985; 39(4):161–179.

57. Hasegawa K, Hashi K, Okada R. Physiochemical stability of pharmaceutical phosphate buffer solutions I—complexation behaviour of Ca(II) with additives in phosphate buffer solutions. J Parenter Sci Technol 1982; 36:128–133.
58. Hasegawa K, Hashi K, Okada R. Physiochemical stability of pharmaceutical phosphate buffer solutions II—complexation behaviour of Al(III) with additives in phosphate buffer solutions. J Parenter Sci Technol 1982; 36:168–173.
59. Hasegawa K, Hashi K, Okada R. Physiochemical stability of pharmaceutical phosphate buffer solutions IV—prevention of precipitation in parenteral phosphate buffer solutions. J Parenter Sci Technol 1982; 36:210–215.
60. Hasegawa K, Hashi K, Okada R. Physiochemical stability of pharmaceutical phosphate buffer solutions V—precipitation behaviour in phosphate buffer solution. J Parenter Sci Technol 1983; 37:38–45.
61. Hallet J. Nucleation and growth of ice crystals. In: Rey L, ed. Advances in Freeze-Drying Lyophilisation. Paris: Hermann, 1966:33.
62. Willemer H. Experimental freeze-drying. In: Rey L, May JC, eds. Freeze-Drying/Lyophilization of Pharmaceuticals and Biological Products. New York: Marcel Dekker, 1999:79–119.
63. Willemer H. Low temperature freeze drying of aqueous and non-aqueous solutions using liquid nitrogen. In: Goldblith SA, Rey L, Rothmayr WW, eds. Freeze Drying and Advance Food Technology. New York: Academic Press, 1975:461–477.
64. Rey L. Orientations nouvelles de la lyophilisation. In: Rey L, ed. Research and Development in Freeze-Drying. Paris: Hermann, 1964:630–643.
65. Takada A, Nail S, Yonese M. Subambient Behavior of Mannitol in Ethanol-Water Co-solvent System. Pharm Res 2009; 26(3):568–576.
66. Zuo J, Hua Z, Liu B, et al. Assay study on glass transition temperature of solutions in freeze-drying. Shipin Kexue 2006; 27(2):58–60.
67. Wittaya-Areekul S, Needham G, Milton N, et al. Freeze-drying of tert-butanol/water cosolvent systems: a case study in formation of friable freeze-dried powder of tobramycin sulfate. J Pharm Sci 2002; 91(4):1147–1155.
68. Kasraian K, DeLuca PP. Thermal analysis of the tertiary butyl alcohol-water system and its implications on freeze-drying. Pharm Res 1995; 12(4):484–490.
69. Ott JB, Goates JR, Waite BA. (Solid + liquid) phase equilibria and solid-hydrate formation in water + methyl, + ethyl, + isopropyl, + tertiary butyl alcohols. J Chem Thermodyn 1979; 11:739–746.
70. Woznyj M, Lüdemann HD. The pressure dependence of the phase diagram t-butanol/water. Z Naturforsch 1985; 40a:693–698.
71. Murgatroyd K. The freeze drying process. In: Cameron P, ed. Good Pharmaceutical Freeze-Drying Practice. Buffalo Grove, Illinois: Interpharm Press Inc., 1997:55–57.
72. Liu J, Viverette T, Virgin M, et al. A study of the impact of freezing on the lyophilization of a concentrated formulation with a high fill depth. Pharm Dev Tech 2005; 10:261–272.
73. Pikal M. Freeze-drying of proteins. Part I: Process design. Biopharm 1990; 3:18–27.
74. Nail SL, Gatlin LA. Freeze-drying: principles and practice. In: Avis KE, Lieberman HA, Lachman L, eds. Pharmaceutical Dosage Forms: Parenteral Medications. Vol. 2. New York: Marcel Dekker, 1993:163–233.
75. Wittaya-areekul S. Freeze-drying from nonaqueous solution. Mahidol University J Pharm Sci 1999; 26(1–4):33–43.
76. DeLuca PP, Kamat MS, Koida Y. Acceleration of freeze-drying cycles of aqueous solutions of lactose and sucrose with tertiary butyl alcohol. Congr Int Technol Pharm 1989; 1:439–447.
77. Kasraian K. The utilization of tertiary butyl alcohol (TBA) as a mass transfer accelerator in lyophilization. Diss Abstr Int B 1994; 54(7):3570.
78. Kasraian K, DeLuca PP. Effect of tert-butyl alcohol on the resistance of the dry product layer during primary drying. Pharm Res 1995; 12(4):491–495.
79. Daoussi R, Vessot S, Andrieu J, et al. Experimental sublimation kinetics and modeling during freeze-drying of pharmaceutical active principle with organic co-solvent formulations. The Freeze-Drying of Pharmaceuticals and Biologics Conference, Breckenridge, Colorado, August 6–9, 2008.

80. Pyne A, Suryanarayanan R. The effect of additives on the crystallization of cefazolin sodium during freeze-drying. Pharm Res 2003; 20(3):283–291.
81. te Booy MPWM, de Ruiter RA, de Meere ALJ. Evaluation of the physical stability of freeze-dried sucrose-containing formulations by differential scanning calorimetry. Pharm Res 1992; 9(1):109–114.
82. vanDrooge D, Hinrichs W, Dickhoff B, et al. Spray freeze-drying to produce a stable Δ^9-tetrahydrocannabinol containing inulin-based solid dispersion powder suitable for inhalation. Eur J Pharm Sci 2005; 26:231–240.
83. Knight J, Adami R. Stabilization of DNA utilizing divalent cations and alcohol. Int J Pharm 2003; 264:15–24.
84. Wang Z, Zhang X, Deng Y, et al. Complexation of hydrophobic drugs with hydroxypropyl-β-cyclodextrin by lyophilization using a tertiary butyl alcohol system. J Inclusion Phenom Macrocyclic Chem 2007; 57(1–4):349–354.
85. Wang Z, Deng Y, Zhang X. The novel application of tertiary butyl alcohol in the preparation of hydrophobic drug-HP beta CD complex. J Pharm Pharmacol 2006; 58(3):409–414.
86. Wang Z, Deng Y, Zhang X. Method for manufacturing hydroxypropyl β-cyclodextrin inclusion complex of lipophilic drug. China patent 1709514. December 21, 2005.
87. vanDrooge D, Hinrichs W, Frijlink H. Incorporation of lipophilic drugs in sugar glasses by lyophilization using a mixture of water and tertiary butyl alcohol as solvent. J Pharm Sci 2004; 93(3):713–723.
88. Olsen WP. Volatile solvents for drying and microbial kill in the final container—a proposal and a review, with emphasis on t-butanol. Pharm Eng 1997; 112–117.
89. Wallhausser KH. Appendix B. Antimicrobial preservation used by the cosmetic industry. In: Kabara JJ, ed. Cosmetic and Drug Preservation. New York: Marcel Dekker, 1984:605–745.
90. Inoue A, Horikoshi K. A Pseudomonas thrives in high concentrations of toluene. Nature 1989; 338:264–266.
91. Pinazo A, Puigjamer L, Coderch L, et al. Butanol resistant mutants of clostridium acetylbutylicum: isolation and partial characterization. Microbios 1993; 73, 93–104.
92. Flink JM. Loss of Organic Volatiles in Freeze-Dried Carbohydrate Solutions (PhD thesis). Massachusetts: Massachusetts Institute of Technology; 1970.
93. Flink J, Gejl-Hansen F. Retention of organic volatiles in freeze-dried carbohydrate solutions: macroscopic observations. J Agric Food Chem 1972; 20(3):691–694.
94. Flink JM, Karel M. Retention of organic volatiles in freeze-dried solutions of carbohydrates. J Agric Food Chem 1970; 18(2):295–297.
95. Flink JM, Karel M. Effects of process variables on retention of volatiles in freeze-drying. J Food Sci 1970; 35:444–447.
96. Flink JM, Karel M. Mechanisms of retention of organic volatiles in freeze-dried systems. J Food Technol 1972; 7:199–211.
97. Flink JM, Labuza TP. Retention of 2-propanol at low concentrations by freeze-drying carbohydrate solutions. J Food Sci 1972; 37:617–618.
98. Flink JM. The retention of volatile components during freeze-drying: a structurally based mechanism. In: Goldblith SA, Rey L, Rothmayr WW, eds. Freeze Drying and Advance Food Technology. New York: Academic Press, 1975:351–372.
99. Lambert D, Flink J, Karel M. Volatile transport in frozen aqueous solutions II: influence of system parameters. Cryobiology 1973; 10:52–55.
100. Thijssen HAC. Effect of process conditions in freeze drying retention of volatile components. In: Goldblith SA, Rey L, Rothmayr WW, eds. Freeze Drying and Advance Food Technology. New York: Academic Press, 1975:373–400.
101. Rulkens WH, Thijssen HAC. Retention of volatile compounds in freeze-drying slabs of malto-dextrin. J Food Technol 1972; 7(1):79–93.
102. Wittaya-areekul S, Nail SL. Freeze-drying of tert-butyl alcohol/water cosolvent systems: effects of formulation and process variables on residual solvents. J Pharm Sci 1998; 87(4):491–495.
103. Wang Z, Deng Y, Zhang X, et al. Factors influencing the content of residual tert-butyl alcohol in cyclodextrin complex prepared by lyophilization cosolvent system. Yao Xue Xue Bao 2007; 42(3):314–317.

104. Schneider H. Glass transition (theoretical aspects). In: Salamone JC, ed. Polymeric Materials Encyclopedia. Boca Raton: CRC Press, 1996:2779–2781.
105. Oesterle J, Franks F, Auffret T. The influence of tertiary butyl alcohol and volatile salts on the sublimation of ice from frozen sucrose solutions: implications for freeze-drying. Pharm Dev Tech 1998; 3(2):175–183.
106. Chouvenc P, Vessot S, Andrieu J. Experimental study of the impact of annealing of ice structure and mass transfer parameters during freeze-drying of a pharmaceutical formulation. PDA J Pharm Sci Tech 2006; 60(2):95–103.
107. Evans JR, Fildes FJT, Oliver JE. Process for preparing freeze-dried liposome compositions. US patent 4,311,712. January 19, 1982.
108. Amselem S, Gabizon A, Barenholz Y. Optimization and upscaling of doxorubicin containing liposomes for clinical use. J Pharm Sci 1990; 79(12):1045–1052.
109. Yang Z, Yang M, Xiahou G. Preparation of zedoary tumeric oil proliposomes and its quality evaluation. Chin Pharm J 2008; 43(19):1488–1499.
110. Zuo J, Hua Z. Study on preparation of proliposomes by freeze-drying. International Congress of Refrigeration: Refrigeration Creates the Future, 22nd, Beijing, China, August 21–26, 2007, c1737/1–c1737/6.
111. Chen X, Cai K. Pharmaceutical liposomes for injection containing antitumor agent oxaliplatin. China patent 101103972. January 16, 2008.
112. Li C, Deng Y, Cui J. Preparation of liposomes and oily formulations by freeze-drying of monophase solutions. In: Gregoriadis G, ed. Liposome Technology. 3rd ed., Vol. 1. New York: Informa Healthcare, 2007:35–53.
113. Li C, Deng Y. Freeze drying of liposomes using tertiary butyl alcohol-water cosolvent systems. Shenyang Yaoke Daxue Xuebao 2005; 22(4):241–244.
114. Cui J, Li C, Deng Y, et al. Freeze-drying of liposomes using tertiary butyl alcohol/water cosolvent systems. Int J Pharm 2006; 312:131–136.
115. Rey L. Un dévelopment nouveau de la lyophilisation: la cryodessication des systèmes non-aqueux. Experientia 1965; 21:241–246.
116. Rey L, Chauffard F, Dousset M. Les lyophilisations complexes. In: Rey L, ed. Advances in Freeze Drying, Lyophilisation Recherches et Applications Nouvelles. Paris: Hermann, 1966:89–94.
117. Teagarden DL, Baker DS. Practical aspects of lyophilization using non-aqueous cosolvent systems. Eur J Pharm Sci 2002; 15:115–133.
118. Note for Guidance on Impurities: Residual Solvents. International Conference on Harmonization Topic Q3 C (R3) Impurities: Residual Solvents. March 1998.

11

Regulatory Control of Freeze-Dried Products: Importance and Evaluation of Residual Moisture

Joan C. May

Center for Biologics Evaluation and Research, United States Food and Drug Administration, Rockville, Maryland, U.S.A.

INTRODUCTION

The residual moisture content of the freeze-dried biological product is usually near 1% (wt/wt) to 3% (wt/wt). Optimum residual moisture limits are set for each freeze-dried biological product on a case-by-case basis. Residual moisture limits for the freeze-dried product are supported with stability data that demonstrate that at the recommended moisture level the safety, purity, and potency of the product is maintained throughout the product's dating period. The Center for Biologics Evaluation and Research (CBER) of the U.S. Food and Drug Administration (FDA) regulates freeze-dried biological products in its section pertaining to residual moisture, as published in Title 21 of the Code of Federal Regulations for Food and Drugs. The regulation requires that each lot of dried product be tested for residual moisture and meet and not exceed established limits as specified by an approved method on file in the product license application. Specific information about test methods and illustrative residual moisture limits for vaccines and biological product is addressed in the CBER Guideline for the Determination of Residual Moisture in Dried Biological Products. Many recent articles report studies that have focused on residual moisture as it relates to potency, formulation, aggregation, and container closure and stopper considerations. Methods used for the measurement of residual moisture in the freeze-dried final container include the traditional loss on drying, Karl Fischer, and gas chromatographic methods as well as recent advances in coulometric Karl Fischer methods, thermogravimetry (TG), and thermogravimetry/mass spectrometry (TG/MS). Special applications for near-infrared reflectance (NIR) spectroscopy and tritium isotope methods are described. Recent advances in vial headspace moisture methodology (vapor pressure moisture) research have the potential to shed light on moisture and processes occurring within the freeze-dried vial over time, for example, redistribution of moisture between container closure, container headspace, and freeze-dried cake. In addition, there is the redistribution of moisture between surface moisture and bound water in the various chemical constituents such as protein and hydrated salts in the freeze-dried cake.

Residual Moisture

Residual moisture is the low level of water, usually in the range of less than 1% to 3% (wt/wt), remaining in a freeze-dried product after the freeze-drying (vacuum sublimation) process (1–8) is complete. Nail and Johnson (9) have described in-process methods to monitor the endpoint of freeze-drying using residual gas analysis, pressure rise, comparative pressure measurement, and

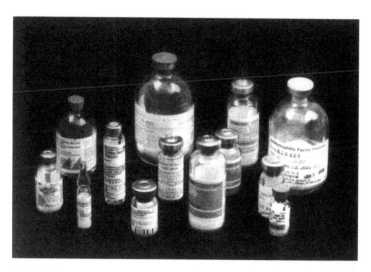

FIGURE 1 The variety found in freeze-dried final containers includes flame-sealed glass ampoules and many sizes of glass vials with container closures.

product temperature measurement. Roy and Pikal (10) used an electronic moisture sensor inside the lyophilization chamber. Residual moisture (11) content is important in the final freeze-dried product because it affects the potency of the product, its long-term stability, and the official shelf life of the product.

Examples of freeze-dried biological products (Fig. 1) include yellow fever vaccine, thrombin, BCG vaccine, intravenous immune globulin, α-interferon, measles virus vaccine live, antihemophilic factor (human), honey bee venom allergenic extract, streptokinase, and meningococcal polysaccharide vaccine groups A and C combined. These freeze-dried products could contain live bacteria, attenuated virus, or therapeutic proteins. The final container could be a single- or multidose glass vial with a rubber container closure or a flame-sealed glass ampoule.

The freeze-dried product contains the freeze-dried biological material, residuals of the manufacture of the product such as buffer, and any excipient such as lactose, sucrose, mannitol, sodium chloride, or sorbitol that has been added to the product to optimize the freeze-drying process and protect product potency. The types of water present in the freeze-dried product cake may be different for each product depending on the residuals of manufacture and the excipients present; the water could be surface water, bound water, or trapped water.

The levels of residual moisture in the biological products should be low and optimized so that, depending on the type of biological product, properties such as the viability, immunological potency, and integrity of the product are retained over time. Final overdrying could kill living cells (12–15). If the level of water remaining in the freeze-dried product is not enough, the excess water may promote structural changes in protein molecules, which in turn may alter the product's immunological properties or foster heat instability. Optimum residual moisture limits are set for each freeze-dried biological product on a case-by-case basis. Residual moisture limits for the freeze-dried product are supported with stability data that demonstrate that at a certain moisture value the safety, purity, and potency of the product are maintained throughout the product's dating period. Potency and residual moisture are monitored when freeze-dried product stability is being determined.

Regulations

Freeze-dried biological products are regulated by the specification for residual moisture that is contained in the Code of Federal Regulations (16) revised in 1990. The specification states that

> 21 CFR 610.13 (a) Purity
> "Each lot of dried product shall be tested for residual moisture by an approved method on file in the product license application."

A guideline (17) is available that summarizes pertinent technical considerations. Prior to 1990, in the previous regulation (18), the gravimetric or loss-on-drying moisture method was described and specified as the method to be used with limits obtained by the gravimetric method listed for several products. The regulation was very specific. It stated

> 21 CFR 610.13 Purity
> "Products shall be tested as provided in paragraphs (a) of this section

> a. Test for residual moisture. Each lot of dried product shall be tested for residual moisture and other volatile substances.
>   1. *Procedure.* The test for dried products shall consist of measuring the maximum loss of weight in a weighed sample equilibrated over anhydrous phosphorus pentoxide at a pressure of not more than 1 mm of mercury, and at a temperature of 20°C to 30°C for as long as it has been established is sufficient to result in a constant weight.
>   2. *Test results*, standards to be met. The residual moisture and other volatile substances shall not exceed 1.0 percent except that, (i) they shall not exceed 1.5 percent for BCG Vaccine, (ii) they shall not exceed 2.0 percent for Measles Virus Vaccine Live, Measles Live and Smallpox Vaccine, Rubella Virus Vaccine Live, and Antihemophilic Factor (Human); (iii) they shall not exceed 3.0 percent for Thrombin and Streptokinase, and (iv) they shall not exceed 4.5 percent for Antibody to Hepatitis B Surface Antigen for the Reverse Passive Hemaglutination Test."

Alternate methods (19) to the gravimetric moisture method could be used if appropriate comparative data were submitted to the FDA, showing the alternate method to be equal or superior to the gravimetric method.

The gravimetric method most accurately measures surface moisture and loosely bound waters of hydration. Changes in technology prompted the change in the FDA regulation. The Karl Fischer, gas chromatographic, and TG moisture methods are rapid tests and are preferred by testing laboratories. Evaluation is required as these methods may measure not only surface moisture but also bound water. This fact has led to different moisture limits by different moisture methods (20).

Recently Reported Studies Focused on Residual Moisture and Its Relation to Potency, Formulation, Aggregation, and Container Closure/Stopper Considerations

Product Potency and Residual Moisture

Evidence continues to accumulate in the literature attesting to loss in product potency or activity associated with elevated moisture content in freeze-dried products.

Of particular interest is work by Pitombo et al. (21). They studied the effect of moisture content on the invertase activity of freeze-dried *Saccharomyces cerevisiae* with respect to monolayer moisture content. Samples with water activity above the monolayer moisture content lost at least 60% of their invertase activity; samples with water activity at about the monolayer moisture content retained at least 85% of their original invertase activity.

Ford and Dawson (22) used alkaline phosphatase as a model and found that alkaline phosphatase activity decreased with increasing moisture content. The study involved treated stoppers, flame-sealed glass ampoules, and untreated rubber stoppers. The lowest alkaline phosphatase activity was found in vials with untreated stoppers and relatively high moisture content. The moisture appeared to have originated from the stopper.

Bell et al. (23) found, via differential scanning calorimetry, that the denaturation temperature of lyophilized bovine somatotropin and lysozyme decreased, and therefore stability decreased with increasing moisture content. The denaturation temperature decreased with increasing moisture irrespective of the excipient. The magnitude of the decrease in denaturation temperature was, however, dependent on the type of excipient.

Herman et al. (24) found that increased water content in the microenvironment of freeze-dried methylprednisone sodium succinate decreased the glass transition temperature (T_g) of the amorphous phase, resulting in an increased rate of decomposition reaction.

Other articles have stressed the importance of residual moisture content. Oliyai et al. (25) evaluated the chemical stability of an Asp-hexapeptide (Val-Tyr-Pro-Asp-Gly-Ala) in lyophilized formulations as a function of multiple formulation variables, including pH, temperature, moisture content, and amorphous or crystalline bulking agent. Analysis-of-variance (ANOVA) calculations of the main effects indicated that while the influence of pH of the starting solution was not statistically significant, residual moisture level, temperature, and, especially, type of bulking agent had a significant impact on the solid-state chemical reactivity of the hexapeptide. Furthermore, residual moisture level and temperature could be important stability variables depending on the type of excipient used.

Oliyai et al. (26) studied an Asn-hexapeptide (Val-Tyr-Pro-Asn-Gly-Ala) in lyophilized formulation as a function of pH of the bulk solution, temperature, and residual moisture content in a factorial study. A statistically significant two-factor interaction indicated that the pH of formulation solution determined the extent to which this peptide stability depends on moisture level and temperature.

Chang and Fischer (27) developed an efficient single-step freeze-drying cycle for protein formulations, namely, human interleukin-1 receptor antagonist (rhIL-1ra) formulations. The process resulted in successful drying of 1 mL of rhIL-1ra formulation to 0.4% moisture content within six hours.

Adams and Irons (28) freeze-dried the enzyme *Erwinia caratovora's* L-asparaginase assessing criteria of dried cake appearance, moisture content, or ease of reconstitution.

Baffi and Garnick (29) considered

> "the most common modes of degradation to evaluate the stability of" proteins ... to be "oxidation, deamidation, aggregation, disulfide rupture and rearrangement, and fragmentation" ... caused by exposing candidate lyophile to heat, light, agitation, or freeze-thaw cycle.... "Oxidation may be detected by reversed-phase high-performance liquid chromatography (RP-HPLC), hydrophobic interaction chromatography (HIC), or peptide mapping. Deamidation ... detected by isoelectric focusing (IEF) or high-performance ion-exchange chromatography (HPIEC). Aggregation and fragmentation may be detected by sodium dodecylsulfate polyacrylamide gel electrophoresis (SDS-Page) or high-performance size-exclusion chromatography (HPSEC).... These methods" are "integrated into stability-monitoring protocols used to evaluate multiple batches of product." The residual moisture in the cake must be defined and controlled. "The desired moisture level can be obtained by the development and validation of a reproducible lyophilization cycle. If the moisture level is too high, the cake" might collapse.... "Degradation by deamidation may continue in the presence of small traces of residual moisture. Overdrying of the protein is yet another concern. If the residual moisture is too low aggregation, inadequate reconstitution, and/or loss of activity may occur."

Lai et al. (30) studied the mechanistic role of water in the deamidation of a model asparagine-containing hexapeptide (Val-Tyr-Pro-Asn-Gly-Ala) in freeze-dried formulations containing poly(vinylpyrrolidine) (PVP) and glycerol. Glycerol was used to vary T_g of the formulation without any significant changing water content or activity. Residual water appeared to assist deamidation in the solid PVP formulations investigated by promoting molecular mobility, solvent/medium interaction, and by acting as a chemical reactant in the breakdown of the cyclic imide.

Formulation and Residual Moisture

Pikal et al. (31) studied effects of formulation variables on the in-process and shelf life stability of freeze-dried human growth hormone at low moisture levels. A combination of mannitol and glycine, where the glycine remains amorphous, provided the greatest protection against decomposition via methionine oxidations and asparagines deamidation and aggregation.

Severe aggregation was observed at high pH.

Skrabanja et al. (32) reviewed the lyophilization of biotechnology products and emphasized that the final quality of a protein product is determined by an interplay between the proper choice of excipients and the freeze-drying process; the glass temperature that defines that state of the freeze-dried cake can be influenced by the moisture content and the choice of the excipient.

Mattern et al. (33) investigated sugar-free L–amino acid systems as stabilizers in protein formulations. Increase in moisture content during storage reduced the T_g enhancing crystallization and decreasing protein stability.

Aggregation and Residual Moisture

Katakam and Banga (34) studied moisture-induced aggregation of solid-state albumin and γ-globulin. Moisture-induced (2–10 μL added to 10 mg) aggregation of solid-state albumin and γ-globulin was investigated by incubation at 37°C for 24 hours. Aggregation was observed with increasing moisture content and was especially prominent for bovine serum albumin. When mixed with carbohydrate excipients in a 1:1 ratio, aggregation was reduced for both bovine serum albumin and γ-globulin by Emdex, dextrose, trehalose, and hydroxypropyl β-cyclodextrin excipients. The mechanism of the aggregation was indicated to be covalent linkages formed due to intermolecular thiol disulfide interchange.

Bell et al. (35) studied lyophilized recombinant bovine somatotropin (rbSt) and found an increasingly significant contribution of exothermic aggregation at higher moisture contents. In the presence of moisture they identified hydrophobic aggregation as being most prominent. In the dry state, covalent modifications including polymerization into compounds of higher molecular weight were more prominent, whereas in the presence of moisture, hydrophobic aggregation was most prominent. This can be explained by the increasingly significant contribution of the exothermic aggregation at higher moisture contents.

Hekman et al. (36) linked elevated moisture content to the formation of aggregates for a conjugated I_gG lyophilized with maltose and citrate buffer. The increase in molecular size was a function of both the moisture content in the vial and the amount of time for which the sample was stressed thermally. The data suggest that the increase in molecular size as a function of thermal stress is due to attachment of maltase, which is a glucose disaccharide present in the lyophile as an excipient. This degradation pathway was only observed in the lyophile.

Prestrelski et al. (37) using Fourier transform infrared (FTIR) spectroscopy and accelerated stability studies examined interleukin-2 (IL-2) with respect to pH and stabilizers that provide optimal storage stability for the lyophilized product. IL-2 prepared at pH 5 is approximately an order of magnitude more stable than at pH 7 with regard to formation of soluble and insoluble aggregates.

Pikal et al. (38) also looked at effects of moisture and oxygen on the formulation and stability of freeze-dried human growth hormone evaluating the formation of irreversible aggregates.

Prestrelski et al. (39) demonstrated that FTIR is a rapid and useful method for studying protein conformation in the dried state and can aid in determining the optimum conditions for stabilization of proteins during freeze-drying.

Dong et al. (40) indicated that a successful lyophilized protein formulation is the preservation of the native conformation in the dried solid. They used FTIR as a tool to study lyophilization-induced unfolding and aggregation of proteins, namely, through the bands at 1620 cm^{-1} to 1685 cm^{-1}, common IR spectral features indicative of lyophilization- and temperature-induced protein aggregation, used to monitor and quantify aggregation even in the dried solid.

Lueckel et al. (41) assessed the residual moisture, T_g, and excipient physical state of different formulations in relation to the in-process and shelf life stability of freeze-dried IL-6. The amorphous state of the excipients and a high T_g can be considered necessary condition for preventing aggregation of freeze-dried IL-6. Sarciaux et al. (42) compared effects of buffer composition and processing conditions on aggregation of bovine I_gG during freeze-drying. The data were supportive of a mechanism of aggregation formation involving denaturation of the I_gG at the ice/freeze concentrate interface that is reversible upon freeze-thawing, but becomes irreversible after freeze-drying and reconstitution. Breen et al. (43) studied the effect of residual moisture in the range of 1% to 8% on the stability of a lyophilized humanized monoclonal antibody formulation. They found that high-moisture cakes had higher aggregation rates than drier samples if stored above their T_g values. Intermediate moisture-containing cakes were more stable to aggregation than low-moisture-containing cakes.

Container Closure/Stoppers and Residual Moisture
Pikal and Shah (44) studied moisture transfer from stopper to product and the resulting stability implications. They found that the product moisture content increases with time and reaches an apparent equilibrium value characteristic of the product, the amount of product, and the stopper treatment method. Templeton et al. (45) and Donovan et al. (46) conducted further studies relating stoppers, stopper processing, and product final moisture content.

Crist (47) studied lyophilized pharmaceuticals sealed under reduced pressure and demonstrated that vial pressure is primarily due to desorption of water vapor from the stopper into the headspace of the vial. The presence of a hydrophilic product decreases the rate of pressure rise, and pressure increases more rapidly in smaller vials. Results are discussed in terms of testing for seal integrity and for stopper effects on moisture in the product.

Pikal and Shah (48) studied intravial distribution of residual moisture in dextran, human serum albumin, and bovine somatotropin. In general, the residual moisture in the top of a core sample of the freeze-dried product was less than the moisture content in the bottom core section of the product. The section closest to the vial wall was consistently found to be lowest in moisture content.

RESIDUAL MOISTURE MEASUREMENT IN THE FREEZE-DRIED FINAL CONTAINER
Current Methods
The methods for the determination of residual moisture currently used at the CBER at the FDA are the gravimetric (loss-on-drying) method, the Karl Fischer method, and the TG and TG/MS method. Current work in progress involves the use of vapor pressure moisture measurements to provide additional information about residual moisture content and its interaction with the components of the freeze-dried final container and its contents.

Gravimetric Method (Loss on Drying)
The gravimetric method measures surface moisture and loosely bound water of hydration (20). The gravimetric method was described in the U.S. Code of Federal Regulations (18) before 1990. In the CBER/FDA laboratory, the gravimetric (loss on drying) test (2,49) is performed in a humidity- and temperature-controlled room. In actuality the room is a walk-in incubator converted to maintain the 20% to 25% relative humidity range and 20°C to 25°C temperature range required by the test. The room contains a five-place analytical balance on a marble balance table, vacuum pump, Pirani vacuum gauge, hygrometer, desiccators, and sulfuric acid scrubbers for the air used to release desiccator

FIGURE 2 Vacuum pump, Pirani vacuum gauge, and hygrometer in low-humidity room for gravimetric moisture test.

vacuum (Figs. 2 and 3). Figure 4 shows the typical dry box configuration of weighing bottle, sample vials, hygrometer, Petri dishes containing anhydrous phosphorus pentoxide to maintain a low humidity, spatula, and pliers. This equipment is necessary to perform the transfer (and pooling if necessary) of the freeze-dried sample from the vials. The dry box and its contents equilibrate for over three hours to ensure that the phosphorus pentoxide has absorbed excess moisture and dried the box as indicated by the portable hygrometer in box.

FIGURE 3 Desiccator in desiccators guard and sulfuric acid scrubbers to remove water from air used to release vacuum in desiccators in gravimetric moisture test.

FIGURE 4 Dry box with sample vials, weighing bottles, hygrometer, and Petri dishes filled with anhydrous phosphorus pentoxide for sample preparation for gravimetric moisture test.

Sample is transferred from the vials to the preweighed weighing bottles. The weighing bottles plus sample are capped, removed from the dry box, weighed on the five-place Mettler balance, and placed in a desiccator containing anhydrous phosphorus pentoxide. The weighing bottle tops are opened and the desiccator is closed. A vacuum of less than 1 mmHg measured by the Pirani gauge is drawn on the desiccator by the vacuum pump. The desiccator is sealed and remains in the controlled room for three days. At that time the desiccator vacuum is released by allowing air to enter after passing through three gas-washing bottles filled with concentrated sulfuric acid, which removes water from the air. This process takes approximately 1.5 hours. The desiccator is then reopened, and the weighing bottles are capped and reweighed. The loss in sample weight divided by the initial sample weight and multiplied by 100 yields the percent moisture (wt/wt) in the original sample.

Karl Fischer Method

Iodine in the presence of pyridine, sulfur dioxide, and methanol reacts quantitatively with water. Karl Fischer (50) developed this quantitative method for water determination (51) in 1935. In the coulometric method iodine is electrically generated at the surface of the electrode emersed in pyridine, sulfur dioxide, and methanol to react with water. Coulometric Karl Fischer measurements (52) are conducted in a Plexiglas glove box that is located in a chemical fume hood. A low relative humidity is maintained in the glove box with anhydrous phosphorus pentoxide. The relative humidity is monitored by a portable hygrometer. The Karl Fischer instrument (Aquatest 8 Coulometric Moisture Analysis System, Photovolt, Indianapolis, Indiana, U.S.) is placed on top of the dry box (Fig. 5) to minimize corrosion of the electrical wiring. Custom elongated wires (Photovolt) connect the titration vessel inside the dry box to the instrument outside through rubber-stoppered ports in the Plexiglas. Samples are vortexed to render the freeze-dried cake into a powder; vials are scraped free of labels and glue. The vial is placed inside the dry box at approximately 15% to 20% relative humidity, and the vial's vacuum, if present, is released by quickly opening and then closing the stopper. Not releasing the vacuum could cause a significant erroneous weight due to buoyancy when the vial or ampoule contains vacuum

FIGURE 5 Karl Fischer dry box configuration with coulometric processing unit outside and titration vessels inside the dry box. Elongated wires connect the instrument to electrodes in the titration vessel in the dry box. Unit on the right uses pyridine solvent; unit on the left uses pyridine-free solvent.

FIGURE 6 Close-up of Karl Fischer titration vessel containing Karl Fischer reagent, electrodes, automatic stirrer, and portable hygrometer.

instead of air. The vial's weight is determined to four decimal places with a Mettler balance and then the vial is placed back into the dry box. Approximately 20 to 30 mg of sample is poured into the pyridine-containing vessel (Fig. 6) for titration after a 0 μg of water background reading is obtained by the instrument. After the sample is stirred in the vessel for 1.5 minutes (or another optimized time) the sample is titrated for moisture content. The instrument readout indicates the micrograms of water in the sample. The vial is reweighed to determine the exact amount of sample delivered for titration. The micrograms of water (converted to milligrams) determined by the coulometric titrator divided by the milligrams of sample multiplied by 100 yields the percent water in the sample. Standards used for this method included sodium tartrate dihydrate or known amounts of water carefully and accurately delivered into the titration solution with a microsyringe.

Thermogravimetry
TG analysis is described in the U.S. Pharmacopeia (53) as a method for the determination of the weight of a substance as a function of temperature. TG measurements for biological products (52) are carried out with the electro-balance (with furnace) (Thermal Analyst ASI, TA Instruments, Wilmington, Delaware, U.S.) in a Plexiglas glove box (Fig. 7) at a low humidity maintained by phosphorus pentoxide and monitored by a portable hygrometer. The quartz tubes surrounding the samples and balance counter weights are painted with gold paint (20) to minimize the effect of static electricity on sample handling and the balance in the dry box. Wires connect the gold layer to the electrobalance ground (Fig. 8). Pulverized samples ranging from 5 to 11 mg are placed on the TG pan for analysis. Thermograms are collected by the Thermal Analyst 2220 (TA Instruments) with IBM Personal System 12 color display (Fig. 9). The IBM LaserPrinter 10 by LexMark (Lexington, Kentucky, U.S.) prints out the ther-mograms. In the TG profile method sample loses weight as the furnace

FIGURE 7 TG electrobalance (with furnace) enclosed in dry box/glove box maintained at low humidity by anhydrous phosphorus pentoxide.

FIGURE 8 Close-up of electrobalance showing sample pan, gold-coated quartz tubes, thermocouple, and furnace before sample is placed on pan.

temperature increases from room temperature to 400°C at a programmed heating rate of 20°C/min. Figure 10 shows a typical thermogram for freeze-dried varicella virus vaccine live. The weight of the residual moisture is taken as the difference of the initial sample weight and the sample weight at constant weight, usually the first horizontal plateau of the thermogram after the initial weight loss. In this case, the residual moisture weight loss is small. There is not a

FIGURE 9   TG data processor displaying thermogram of sodium tartrate dehydrate standard.

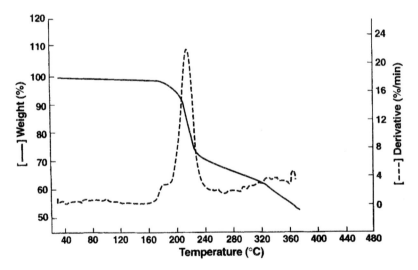

FIGURE 10   TG of varicella virus vaccine live.

well-defined plateau for the residual moisture weight loss. However, the DT curve (dashed line) clearly indicates that the weight loss ends at approximately 140°C. The temperature varies with each product type. The ratio of the lost residual moisture weight to the initial sample weight multiplied by 100 is taken as the percentage residual moisture in the sample. For varicella virus vaccine live, lot A, the TG moisture result was calculated to be 0.92% (Table 1). The Karl Fischer moisture result for this lot was 0.95%. Sodium tartrate dihydrate is used as a standard. The waters of hydration, shown by well-defined plateaus, are

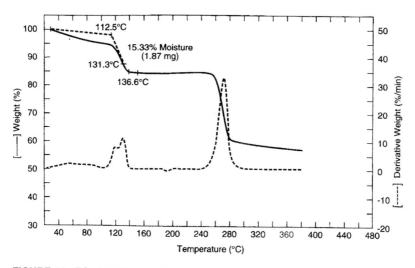

FIGURE 11 TG of standard sodium tartrate dihydrate.

TABLE 1 TG and Karl Fischer Residual Moisture Data for Varicella Virus Vaccine Live and Limulus Amebocyte Lysate

	Residual moisture (%)[a]			
Sample	TG method	TG/MS method	Karl Fischer method	Relative error (%)[b]
Varicella virus vaccine live lot A	0.92 ± 0.03	–	0.95 ± 0.18	3.3
Limulus amebocyte lysate lot A[c]	–	7.84 ± 0.22	7.97 ± 0.33[d]	1.6

[a]Arithmetic mean and standard deviation of two determinations unless otherwise indicated.
[b]Relative error from the Karl Fischer value.
[c]Moisture results are both over the 5.0% limit for this product.
[d]Arithmetic mean and standard deviation of three determinations.
Abbreviation: TG/MS, thermogravimetry/mass spectrometry.

measured thermogravimetrically and are accurately determined as shown in Figure 11.

Thermogravimetry/Mass Spectrometry

When a clearly defined plateau is not in evidence for the residual moisture TG transition, as is shown in the data for a limulus amebocyte lysate (LAL) illustrated in Figure 12, TG/MS is employed to determine the endpoint of the evolution of the residual moisture. TG/MS has been shown to elucidate the TG transitions attributable to residual moisture in freeze-dried biological products (20,52,54,55). The combination of TG and MS has proven to be effective by providing precise TG heating conditions and weight-loss information along with mass spectral identification of volatiles evolved during the weight-loss process. As shown in the Figure 12 composite of TG and MS data for LAL, mass spectra are taken of the TG off-gases continuously while the weight loss and rate of weight loss [differential thermogram (DTG)] scans are recorded. The ion intensities of mass peaks 18 and 44 are monitored to show the changes in the

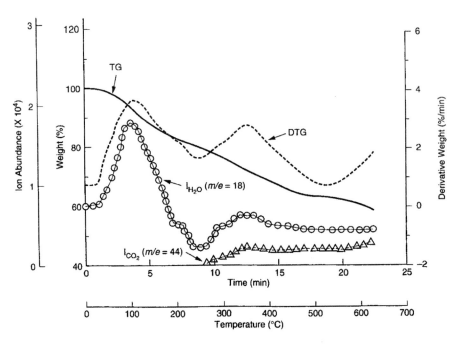

FIGURE 12 Plot of TG/MS data for limulus amebocyte lysate.

amounts of water and carbon dioxide, respectively, in the TG off-gases. When superimposed on the respective TG data, the mass spectral ion intensities verify the transition caused by moisture in the sample by differentiating between the water content of the sample and the water evolved from thermal decomposition of the sample, which coincides with the evolution of carbon dioxide (20).

In the TG/MS data obtained for the LAL, the TG curve does not display a clearly defined weight-loss transition that could be attributed to residual moisture (Fig. 12). The presence of carbon dioxide (shown by the ion abundance of $m/e = 44$) coinciding with the evolution of water after 260°C would indicate that the water evolved (after 260°C) resulted from sample decomposition. Since carbon dioxide is not evolved during the evolution of the first water peak, the TG weight loss attributed to residual moisture is indicated by the ion abundance for water ending at approximately 260°C. The loss in weight of LAL to 260°C is 7.84%. The LAL TG/MS residual moisture result is in close agreement with the residual moisture results obtained by the Karl Fischer method (7.97%) (Table 1). This particular lot of LAL has test results that are outside the acceptable limit of 5% for LAL residual moisture by both the Karl Fischer and TG methods.

Figure 13 displays the TG/MS data for U.S. standard pertussis vaccine lot 8. The TG has a well-defined plateau for the weight loss attributable to residual moisture. The DTG curve clearly indicates the end of the residual moisture evolution at approximately 160°C. The ion intensity for carbon dioxide also begins at approximately 160°C, indicating that the moisture before this temperature is residual moisture, not moisture resulting from product decomposition. Earlier TG/ MS data for residual moisture in freeze-dried biological products were obtained by the continuous monitoring method of Chiu and Beattie using a DuPont 990 thermal

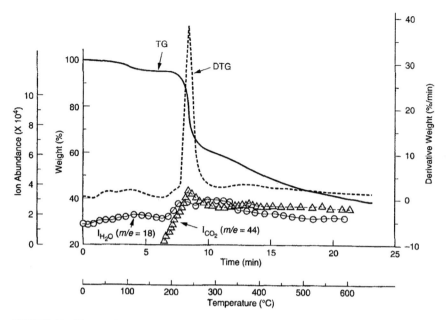

FIGURE 13 Plot of TG/MS data for U.S. standard pertussis vaccine lot 8. *Abbreviations*: TG/MS, thermogravimetry/mass spectrometry; DTG, differential thermogram.

FIGURE 14 TG/MS-jet separator interface.

analysis system interfaced with a glass tee to a DuPont 21-104 mass spectrometer and the methodology of May et al. (54) in which a DuPont 1090 thermal analysis system was interfaced to a Hewlett-Packard 5995B quadrupole mass spectrometer. Figure 14 shows the DuPont 1090 TG analyzer (Wilmington, Delaware, U.S.) and Hewlett-Packard 5995B quadrupole mass spectrometer (Rockville, Maryland, U.S.)

configuration in the laboratory. The TG/MS interface consists of a glass tube of ¼ in. of outer diameter, that is, l0 in. in length. One end connects to the TG. Swagelock connections are used in both cases. This straightforward TG/MS interface utilizes the characteristics of the jet separator to reduce TG effluent pressure (about 1 atm) to the low pressure necessary for MS operation (10^{-6} Torr). While discriminating against lower molecular weight species (helium) and therefore eliminating helium carrier gas, the jet separator increases the relative concentration of thermal decomposition products in the flow.

The TA Instruments thermal analyst 220 was interfaced (51) to the Hewlett-Packard 5972 series mass selective detector (Fig. 15) equipped with a hyperbolic quadrupole mass filter and vapor diffusion high-vacuum pump used in conjunction with a LaserJet 4 Plus printer. The TG analyzer's effluent tube was modified to terminate in a straight ¼-in. OD glass tube. A ¼ to ⅙ in. tube reducing union (Swagelok) was used to connect the TG effluent tube to a 0.53-mm ID fused silica capillary tube (Fig. 16). An OSS-2 variable outlet splitter (Scientific Glass Engineering Pty. Ltd., Ringwood, Australia) (Fig. 17) was plumbed

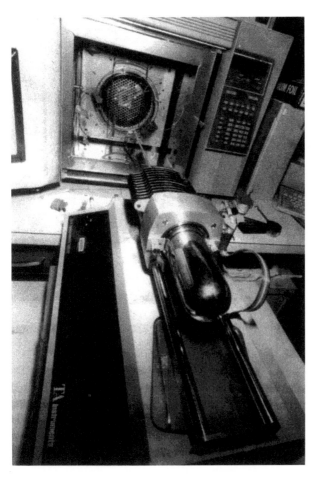

FIGURE 15   TG/MS-capillary interface.

FIGURE 16 Capillary interface: Swagelock adapter connecting TG effluent to capillary tubing.

FIGURE 17 Capillary interface: variable outlet splitter.

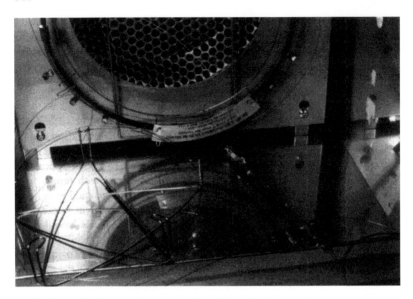

FIGURE 18 Capillary interface: variable outlet splitter in place in spectrometer.

between the TG and the mass spectrometer (Fig. 18). This effluent splitter reduces the flow into the mass selective detector by venting a portion of the flow to the atmosphere. All of the 0-rings and connections in the TG balance were tightened so that excessive amounts of oxygen and nitrogen did not enter the system. This capillary interface allows continuous monitoring of the ion intensities of mass peaks $m/e = 18$ (water) and $m/e = 44$ (carbon dioxide). Comparable results for TG/MS plots were obtained with both interfaces indicating continuity.

TG, TG/MS, and differential scanning calorimetric (DSC) measurements were also performed with a Netzsch STA 409PC Luxx (Selb, Germany) TG analyzer interfaced to a Pfeiffer Vacuum Technology (Asslar, Germany) ThermoStar mass spectrometer. In the accurate measurement of residual moisture by TG methods this instrumentation is used to identify interferences. Using this instrumentation ethanol was detected in a polysaccharide used in vaccine production and H_2S was detected in an antivenin. When residual moisture was calculated in these products using TG/MS, interferences from ethanol and H_2S were avoided in these cases by using in the moisture calculation the weight loss corrected for the ethanol or H_2S weight loss that appeared in the thermogram (56).

Prevention of Sample Contamination by Ambient Humidity Peaks Occurred

The sample must be protected from contamination by ambient humidity in both the Karl Fischer and TG moisture methods as well as in the gravimetric method. This is necessary for every sample analyzed. Normally the sample is manipulated in the Karl Fischer dry box maintained at a low humidity (~15–20% relative humidity) monitored by a portable hygrometer. If the sample appears to take on or give up water rapidly, the relative humidity in the dry box is lowered or raised, respectively, for the analysis. Lowering the relative humidity involves adding sufficient phosphorus pentoxide to the dry box to have the hygrometer read less than 5% relative humidity and working quickly to transfer the sample from the vial to the Karl Fischer solution to minimize its exposure. At this point, the sample in question is analyzed for moisture content by the Karl Fischer method (at lower dry box relative humidity) and by the TG method. The moisture results should be in good agreement if the adjustment to the relative humidity in the dry box has successfully dealt with preventing sample contamination.

Samples with lactose as the excipient are usually run at between 0% and 5% relative humidity in the dry box. Single-dose vials containing approximately 5 mg or less per vial are pooled by consecutive delivery into the vessel solution of the contents of four or five vials.

To increase the accuracy and reliability of results, Karl Fischer results are usually compared to TG results. The moisture results should be nearly equivalent from these two methods since they are both measures of total bound and surface moisture in the sample cake (20). Vial-to-vial moisture variability in the samples is usually inherent in freeze-dried samples since each vial is unique with respect to freeze-drier shelf and position on the shelf. Both the Karl Fischer and TG methods are frequently capable of measuring the moisture content of one vial and therefore vial-to-vial variability for one lot since one test requires either 20 or 5 mg of sample, respectively. For these two methods also the relative standard deviation is near 10%. Low vial-to-vial variability is produced by a well-controlled sample lyophilization process. After the freeze-drying of U.S. National Reference Preparation for α-fetoprotein in mid-pregnancy maternal serum (57), 85 samples were chosen at random. The mean residual moisture content of the 85 samples was 0.55% with a standard deviation of $\pm 0.19\%$. These residual moisture results are illustrative of an excellently freeze-dried product with relatively low vial-to-vial residual moisture variability as determined by the Karl Fischer test method.

Comparative Results

Evaluation is required as the gravimetric, Karl Fischer, and TG methods may measure not only surface moisture but also bound water. This fact has led to different moisture limits by different moisture methods. For example, measles virus vaccine has a limit of 3% by the Karl Fischer method and 2% by the gravimetric method. This was set by correlating the data collected from both assay methods on the same lots (20).

Gravimetric and Karl Fischer results are in excellent agreement for certain manufacturers' anistreplase, mite allergenic extract, fibrinolysin and desoxyribonuclease combined, and digoxin immune FAB (ovine) (17) with results in the 0.48% to 2.13% moisture range.

Karl Fischer and TG or TG/MS results have been shown to be in good agreement for varicella virus vaccine and LAL (Table 1) with relative errors between the different moisture test method results of 3.3% and 1.6%, respectively. Excellent agreement between Karl Fischer and TG or TG/MS moisture test results have also been demonstrated for certain manufacturers' typhoid vaccine, meningococcal polysaccharide vaccine groups A and C combined, honey bee venom allergenic extract, measles virus vaccine live attenuated (58), antibody to hepatitis B surface antigen, Anti-Jk[a] blood grouping serum (55), antihemophilic factor (human), and BCG TheraCys (52) in the range of 0.65% to 3.18% residual moisture.

Data for accuracy and assessment of error or uncertainty for residual moisture methods are presented in detail (56).

Vapor Pressure Moisture Methodology

Vapor pressure moisture methodology has added a new piece of information to the evaluation of residual moisture in the freeze-dried final container. Rey (personal communication) used water vapor pressure methodology to determine the moisture content of the headspace in several freeze-dried biological products using and electro-optical dew point measurement instrument (Fig. 19). This information has agreed with results from gravimetric, Karl Fischer, TG, and TG/MS testing for both high and low residual moisture levels in vials. Verifying high residual moisture levels in freeze-dried biologicals is important since excessive residual moisture levels have led to decreases in product potency and therefore decreases in product stability.

Figure 20 shows the condensation temperatures for three lots of α-interferon with residual moisture values near 1.0% (Table 2), which are within the moisture limit for the product (3.0%). The condensation temperatures are very low (between $-11.8°C$ and $-24.3°C$) and are relatively close to one another. The corresponding water vapor pressure moisture values are low, between 2.05 and 6.67 µg of water per vial, indicative of only small amounts of moisture in the vial headspace. Antihemophilic factor (Table 2) similarly has a low residual moisture (1.05%), a condensation temperature of $-42°C$, and a very low vapor pressure moisture within the vial headspace (0.2 µg water/vial). In contrast, the condensation temperature graphs for the sequence of U.S. pertussis vaccines, lots 8,

FIGURE 19 Instrument for the measurement of vapor pressure moisture within sealed vials (JMD Electronique, Montelier, France).

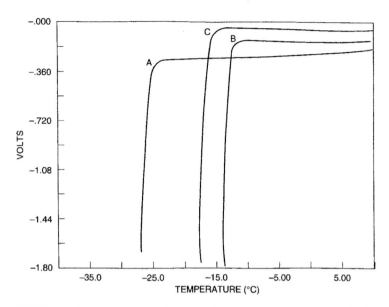

FIGURE 20 Vapor pressure moisture condensation temperature curves for three lots (A, B, and C) of α-interferon.

TABLE 2 Freeze-Dried Cake Residual Moisture Values Compared with Corresponding Vapor Pressure Moisture Values and Whether the Product Met Its Dating Period Stability Requirement

Sample	Residual moisture TG method (%)	Vapor pressure moisture (μg/vial)	Stability
α-Interferon lot A	1.19	2.05	+
α-Interferon lot B	0.98	6.67	+
α-Interferon lot C	1.28	4.76	+
AHF	1.05	0.20	+
Pertussis vaccine lot 8	2.44	9.5	+
Pertussis vaccine lot 9	4.75	26	−
Pertussis vaccine lot 10	2.14	10.2	+

+ indicates product meets dating period requirement; − indicates product does not meet dating period requirement.
Abbreviations: AHF, antihemophilic factor; TG, thermogravimetric.

9, and 10 shown in Figure 21, show condensation temperatures near −6°C and −7°C for lot 8 and lot 10 and a condensation temperature near +4.5°C for lot 9. This corresponds to vapor pressure moisture values of 9.5 and 10.2 μg water/vial for lots 8 and 10, respectively. Lot 9 has a vapor pressure moisture value of 26 μg water/vial, indicating high headspace water vapor and therefore moisture content. This agrees with the higher cake residual moisture value of 4.75% for lot 9 compared to 2.44% and 2.14% of lots 8 and 10, respectively. These are high-moisture values for both cake and headspace for lot 9. Lot 9 failed the product stability requirement. This is illustrative of high-moisture values in a freeze-dried product leading to loss of product potency and therefore stability over

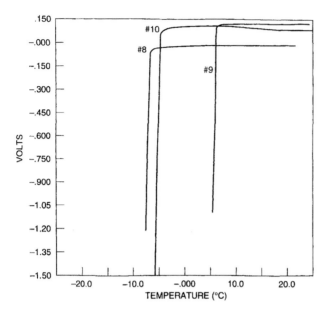

FIGURE 21 Vapor pressure moisture condensation temperature curves for three lots (8, 9, and 10) of U.S. standard pertussis vaccine.

time. This vapor pressure moisture methodology is being applied to the study of vial-to-vial variability within one lot and redistribution of moisture between cake, headspace, and stopper or headspace and cake over time.

Methodology in Use by Manufacturers of Biological Products
The methods approved for analyzing residual moisture in freeze-dried biological products licensed by the FDA are the gravimetric (loss-on-drying) method, many variations of the Karl Fischer method, TG, gas chromatography, and a modification of the moisture evolution analyzer (MEA).

Gravimetric Methods
Flosdorf (1) described the gravimetric method in 1949. There are several variations on the basic gravimetric method in use. The test may be carried out in a relatively large dry box rather than in the humidity-controlled room. The sample may be heated in a drying oven. The Abderhalden method uses a small sample size and heat is provided by a refluxing solvent. The sample is dried in the Abderhalden apparatus over refluxing liquid that boils near room temperature. There are other variations in terms of time, temperature, and vacuum that have been licensed for a particular product since, on a case-by-case basis, data demonstrated that at a chosen temperature a constant weight loss was obtained without decomposing the product.

Karl Fischer Methods
The approved variations (17) in the Karl Fischer method include volumetric titration methods to either a visual (excess iodine or addition of an indicator) or

voltametric endpoint detection method. The visual or voltametric endpoint methods usually require 30 to 40 mg of sample for analysis for freeze-dried biological products containing from 1.0% to 3.0% residual moisture. Coulometric Karl Fischer instruments generate the iodine from potassium iodide for water titration at the electrodes. Only 10 to 20 mg of freeze-dried sample is required for accurate analysis.

Methods of sample addition include the direct addition of pulverized (vortexed) freeze-dried cake to the Karl Fischer vessel solution using a four-place analytical balance to determine the weight of the sample delivered. This is the most accurate sample addition method.

A second method involves the delivery of an aliquot of the suspension of the freeze-dried cake in a methanol solvent. The methanol solvent is usually added to the sample vial by a syringe through the stopper. The test aliquot is usually withdrawn from the stoppered vial with the syringe and delivered into the vessel solution through a stopper on the top of the vessel solution cover. This type of procedure has been operated on the open laboratory bench with the intent that the syringe and rubber closure system keeps out moisture contamination. However, the syringe volumetric delivery of a portion of the solvent-suspended cake has not yet been demonstrated to be as accurate as a four-place analytical balance determination of sample weight delivered to the Karl Fischer titration vessel for analysis. The methanol solvent that is usually used is volatile and its evaporation increases the delivery error. In addition, methanol picks up water from its surroundings. It is difficult to accurately compensate for this moisture contamination in the blank. In a similar variation the sample is titrated inside the vial using burettes with needle tips that pierce the freeze-dried final container rubber closure.

In a third variation of the Karl Fischer method, a sample is heated and the evolved moisture is taken by a carrier gas from the sample to the vessel solution for titration. Careful validation data must be collected that ensure that the heating temperature does not decompose the biological material in the sample. This decomposition would evolve carbon dioxide and water. The water of decomposition would be mislabeled as residual moisture.

To prevent contamination of the sample and reagents by moisture in the surrounding air, the Karl Fischer apparatus is enclosed in a dry box or other apparatus. Anhydrous phosphorus pentoxide is placed in the box as a desiccant. A portable hygrometer monitors humidity in the dry box.

Both pyridine and nonpyridine Karl Fischer solvent systems are approved for licensed biological products. The major advantage of the nonpyridine solvent is that its use eliminates the hazard of pyridine fumes. The solubility properties of the two different Karl Fischer reagents are not always the same.

For a very limited number of biological product sample types, the Karl Fischer method cannot be used. There may be a substance present in the product that interferes chemically with the Karl Fischer reaction such that a substance that binds iodine, or the sample may not dissolve adequately into the Karl Fischer reagent, or the sample moisture may not adequately extract into the Karl Fischer solvent.

Thermogravimetric Methods

TG has been used for measuring residual moisture in freeze-dried viral and bacterial vaccines.

Gas Chromatographic Methods

Robinson (59) described a gas chromatographic method for measuring residual water in freeze-dried smallpox vaccine in 1972. The method was developed to optimize quality control of a tissue culture smallpox vaccine. Water is extracted from the sample with benzene and determined by gas chromatography with thermal conductivity (hot wire) detection and columns packed with Chromosorb 102.

Moisture Evolution Analyzer

The TA Instruments MEA has been adapted for use for determining residual moisture in freeze-dried allergenic extracts. Jewell et al. (60) applied the method to several allergenic extracts including freeze-dried mold, ragweed, and house dust allergenic extracts. All determinations were performed at low humidity in a controlled humidity dry box using phosphorus pentoxide as desiccant. In the MEA, the sample is heated in a controlled oven. The moisture driven off by the heat applied is carried by a dry nitrogen purge gas to an electrolytic cell where it reacts with phosphorus pentoxide. The current required to regenerate the phosphorous pentoxide is converted to micrograms of water and is identified as the amount of water evolved from the sample.

OTHER METHODOLOGY

A tritium isotope technique and an NIR technique have been reported in the literature but have not been formally approved for use of moisture methods for freeze-dried biological products.

Tritium Isotope

In 1973, Kassai and Sikos (61) described the determination of moisture content in freeze-dried products by tritium isotope. This method involves adding tritium oxide to 5 to 10 ampoules of the dissolved product in experimental lots containing 2000 to 8000 ampoules prior to lyophilization.

After lyophilization, the dry material is assayed for radioactivity using liquid scintillation counting and the water content is calculated. A comparison was made between the tritium method and the gravimetric method. The sensitivity of the tritium method was stated to be 1 μg of water; it measures the water content of 1 mg of dry material with an accuracy of ±0.1%. It was noted that small amounts of tritium were found in the nontest ampoules after lyophilization. In addition, tritium oxide was used to measure the exchange of water molecules between stopper and material. Kassai and Sikos stated that "the product distributed into the vial takes up water from the stopper until a state of equilibrium ensues."

NIR Spectroscopy

Last and Prebble (62) developed an NIR method for the determination of moisture in an experimental freeze-dried injection product. NIR spectra were collected through the bases of unopened product vials using a horizontal instrument accessory. The samples in these vials were then used for Karl Fischer analysis to generate a standard curve for the analysis. The NIR data must be correlated with an accepted residual moisture technique to yield a meaningful result. This article states that NIR accuracy and precision in this application are not consistent with allowing the use of the current method in anything but a screening role. A major problem with the NIR reflectance spectroscopic method is that only a millimeter or less of the sample is analyzed as the NIR beam is reflected from the sample and residual moisture content particularly in large cakes may not be uniform throughout the cake.

Lin and Hsu (63) assessed the value of using NIR spectroscopy to analyze residual moisture in lyophilized protein pharmaceuticals sealed in glass vials. They found that doubling or halving the concentration of a disaccharide used as a lyoprotectant caused significant deviation between NIR and Karl Fischer data because the NIR absorbance of the disaccharide overlapped with the moisture signal. Bai et al. (64) studied the ability of NIR spectroscopy as a noninvasive method to detect protein conformation damage during lyophilization due to elevated temperatures and other freeze-drying stresses, thereby complementing information provided by FTIR spectroscopy.

FUTURE DEVELOPMENT

The current challenges to moisture measurement include devising methods to deal with smaller and smaller sample sizes in single-dose final containers, working with samples that are very sensitive to ambient humidity even in the dry box, and understanding the interaction between water in the freeze-dried cake, the vial headspace, the container closure, and the changes that occur over time.

REFERENCES

1. Flosdorf EW. Freeze-Drying. New York: Reinhold, 1949.
2. Seligmann EB Jr., Farber JF. Freeze-drying and residual moisture. Cryobiology 1971; 8:138–144.
3. Pikal MJ. Freeze-drying of proteins. Part1: Process design. BioPharm 1990; 3:18–27.
4. Pikal MJ. Freeze-drying of proteins. Part II: Formulation selection. BioPharm 1990; 4:26–30.
5. Rey LR. Basic aspects and future trends in the freeze-drying of pharmaceuticals. In: International Association of Biological Standards, ed. Developments in Biological Standardization. Basel: Karger, 1992:3–8.
6. Nail SL, Jiang S, Chongprasert S, et al. Fundamentals of freeze-drying. Pharm Biotechnol 2002; 14:281–360.
7. Tang X, Pikal MJ. Design of freeze-drying processes for pharmaceuticals: practical advice. Pharm Res 2004; 21(2):191–200.
8. Pikal MJ, Cardon S, Bhugra C, et al. The nonsteady state modeling of freeze drying: in process product temperature and moisture content mapping and pharmaceutical product quality application. Pharm Dev Technol 2005; 10(1):17–32.
9. Nail SL, Johnson W. Methodology for in-process determination of residual water in freeze-dried products. In: International Association of Biological Standards, ed. Developments in Biological Standardization. Vol. 74. Basel: Karger, 1992:137–151.
10. Roy ML, Pikal MJ. Process control in freeze-drying: determination of the end point of sublimation drying by an electronic moisture sensor. J Parent Sci Technol 1989; 43(2):60–66.
11. Greiff D. Protein structure and freeze-drying: the effects of residual moisture and gases. Cryobiology 1971; 8:145–152.
12. Nei T, Araki T, Souzu H. Studies of the effect of drying conditions on residual moisture content and cell viability in the freeze-drying of microorganisms. Cryobiology 1965; 2:68–73.

13. Nei T, Souzu H, Araki T. Effect of residual moisture content on the survival of freeze-dried bacteria during storage under various conditions. Cryobiology 1966; 2:276–279.
14. Greiff D. Stabilities of suspensions of influenza virus dried by sublimation of ice in vacuo to different contents of residual moisture and sealed under different gases. Appl Microbiol 1970; 20:935–938.
15. Sparkes JD, Fenje P. The effect of residual moisture in lyophilized smallpox vaccine on its stability at different temperatures. Bull World Health Organ 1972; 46:729–734.
16. Code of Federal Regulations. Test for residual moisture, 21 CFR61 0.13(a). Washington, D.C.: U.S. Government Printing Office, 2009.
17. May JC, Wheeler RM, Etz N, et al. Measurement of final container residual moisture in freeze-dried biological products. Dev Biol Stand 1992; 74:153–164.
18. Code of Federal Regulations. Test for Residual Moisture, 21 CFR61 0.13(a). Washington, D.C.: U.S. Government Printing Office, 1988.
19. Code of Federal Regulations. Equivalent methods and processes, 21 CFR610.9. Washington, D.C.: U.S. Government Printing Office, 2009.
20. May JC, Grim E, Wheeler RM, et al. Determination of residual moisture in freeze-dried vial vaccines: Karl Fischer, gravimetric and thermogravimetric methodologies. J Biol Stand 1982; 10:249–259.
21. Pitombo RN, Spring C, Passos RF, et al. Effect of moisture content on the invertase activity of freeze-dried S. cerrevisiae. Cryobiology 1994; 31:383–392.
22. Ford AW, Dawson PJ. Effect of type of container; storage temperature and humidity on the biological activity of freeze-dried alkaline phosphatase. Biologicals 1994; 22:191–197.
23. Bell LN, Hageman MJ, Muraoka LM. Thermally induced denaturation of lyophilized bovine somatotropin and lysozyme as impacted by moisture and excipients. J Pharm Sci 1995; 84:707–712.
24. Herman BD, Sinclair BD, Milton N, et al. The effect of bulking agent on the solid-state stability of freeze-dried methylprednisolone sodium succinate. Pharm Res 1994; 11:1467–1473.
25. Oliyai C, Patel JP, Carr L, et al. Chemical pathways of peptide degradation. VII. Solid state chemical instability of an aspartyl residue in a model hexapeptide. Pharm Res 1994; 11:901–908.
26. Oliyai C, Patel JP, Carr L, et al. Solid state chemical instability of an asparaginyl residue in a model hexapeptide. J Pharm Sci Technol 1994; 48:167–173.
27. Chang BS, Fischer NL. Development of an efficient single-step freeze-drying cycle for protein formulations. Pharm Res 1995; 12:831–837.
28. Adams GD, Irons LI. Some implications of structural collapse during freeze-drying using Erwinia caratovora L-asparaginase as a model. J Chem Technol Biotechnol 1993; 58:71–76.
29. Baffi RA, Garnick RL. Quality control issues in the analysis of lyophilized proteins. Dev Biol Stand 1992; 74:181–184.
30. Lai MC, Hageman MJ, Schawen RL, et al. Chemical stability of peptides in Polymers. 2. Discriminating between solvent and plasticizing effects of water on peptide deamidation in poly(vinylpyrrolidone). J Pharm Sci 1999; 88:1081–1089.
31. Pikal MJ, Dellerman RM, Roy, ML, et al. The effects of formulation variables on the stability of freeze-dried human growth hormone. Pharm Res 1991; 8:427–436.
32. Skrabanja AI, de Meere AL, de Ruiter RA, et al. Lyophilization of biotechnology products. J Pharm Sci Technol 1994; 48:311–317.
33. Mattern M, Winter G, Kohnert U, et al. Formulation of proteins in vacuum dried glasses. II. Process and storage stability in sugar-free amino acid system. Pharm Dev Technol 1999; 4:199–208.
34. Katakam M, Banga KA. Aggregation of proteins and its prevention by carbohydrate excipients: albumins and gamma-globulin. J Pharm Pharmacol 1995; 47:103–107.
35. Bell LN, Hageman MJ, Bauer JM. Impact of moisture on thermally induced denaturation and decomposition of lyophilized bovine somatotropin. Biopolymers 1995; 35:201–209.
36. Hekman C, Park S, Ten WY, et al. Degradation of lyophilized and reconstituted Macroscint. J Pharm Biomed Anal 1995; 13:1249–1261.

37. Prestrelski SJ, Pikal MJ, Arakawa T. Optimization of lyophilization conditions for recombinant human interleukin-2 by dried-state conformational analysis using Fourier-transform infrared spectroscopy. Pharm Res 1995; 12:1250–1259.
38. Pikal MJ, Dellerman K, Roy ML. Formulation and stability of freeze-dried proteins: effects of moisture and oxygen on the stability of freeze-dried formulations of human growth hormone. Dev Biol Stand 1992; 74:22–38.
39. Prestrelski SJ, Arakawa T, Carpenter JF. Separation of freezing- and drying-induced denaturation of lyophilized proteins using stress-specific stabilization. Arch Biochem Biophys 1993; 303:465–473.
40. Dong A, Prestrelski SJ, Allison SO, et al. Infrared spectroscopic studies of lyophilization- and temperature-induced protein aggregation. J Pharm Sci 1995; 84:41–424.
41. Lueckel B, Helk B, Bodmer O, et al. Effects of formulation and process variables on the aggregation of freeze-dried interleuking-6 (IL-6) after lyophilization and on storage. Pharm Dev Technol 1998; 3:337–346.
42. Sarciaux JM, Mansour S, Hageman MJ, et al. Effects of buffer composition and processing conditions on aggregation of bovine IgG during freeze-drying. J Pharm Sci 1999; 88:1354–1361.
43. Breen ED, JG Curley, Overeashier DE, et al. Effect of moisture on the stability of a lyophilized humanized monoclonal antibody formulation. Pharm Res 2001; 18: 1345–1353.
44. Pikal MJ, Shah S. Moisture transfer from stopper to product and resulting stability implications. Dev Biol Stand 1992; 74:165–177.
45. Templeton AC, Placek J, Xu H et al. Determination of the moisture content of bromobutyl rubber stoppers as a function of processing: implications for the stability of lyophilized products. PDA J Pharm Sci Technol 2003; 57(2):75–87.
46. Donovan PD, Corvari V, Nurton MD et al. Effect of stopper processing conditions on moisture content an ramifications for lyophilized products: comparison of "low" and "high" moisture uptake stoppers. PDA J Pharm Sci Technol 2007; 61(1):51–58.
47. Crist B. Time-dependence of pressure in lyophilization vials. J Pharm Sci Technol 1994; 48:189–196.
48. Pikal MJ, Shah S. Intravial distribution of moisture during the secondary drying stage of freeze drying. PDA J Pharm Sci Technol 1997; 51:17–24.
49. May JC, Wheeler RM, Grim E. The gravimetric method for the determination of residual moisture in freeze-dried biological products. Cryobiology 1990; 26:277–284.
50. Scholz E. Karl Fischer Titration: Determination of Water Chemical Laboratory Practice. Berlin: Springer-Verlag, 1984.
51. United States Pharmacopeia/National Formulary (USP32/NF 27) 2009 Water Determination <921>. Rockville, MD: Pharmacopeial Convention, Inc., 2009:388–390.
52. May JC, Del Grosso AV, Wheeler RM, et al. TG/MS capillary interface: applications to determination of residual moisture in BCG vaccine and other freeze-dried biological products. J Therm Anal 1997; 49:929–936.
53. United States Pharmacopeia/National Formulary (USP32/NF 27) 2009. Thermal Analysis <891>. Rockville, MD: U.S. Pharmacopeial Convention, Inc., 2009:380–381.
54. May JC, Del Grosso A, Wheeler R. TG/MS interface: applications to the determination of moisture in polysaccharides and freeze-dried biological products. Thermochim Acta 1987; 115:289–295.
55. May JC, Wheeler RM, Del Grosso A. Compositional analysis of drugs and injectable biological products by thermogravimetry. In: ASTM Special Technical Publication 997. Philadelphia: American Society for Testing and Materials, 1988:48–56.
56. May JC. Highlights on the Control of Freeze-Dried Biologicals with Special Reference to the Role of Residual Moisture. In: Refrigeration Science and Technology Proceedings. New Ventures in Freeze-Drying, Strasbourg, France, November 7–9, 2007.
57. Reimer CB, Smith SJ, Wells TW. The U.S. National Reference Preparation for alphafetoprotein in mid-pregnancy maternal serum. Clin Chem 1982; 28:709–716.
58. May JC, Wheeler RM, Grim E. The determination of residual moisture in several freeze-dried vaccines and a honey bee venom allergenic extract by TG/MS. J Thermal Anal 1986; 31:643–651.

59. Robinson LC. A gas chromatographic method of measuring residual water in freeze-dried smallpox vaccine. Bull World Health Organ 1972; 47:7–11.
60. Jewell JE, Workman R, Zelenznick LD. Moisture analysis of lyophilized allergenic extracts. Dev Biol Stand 1977; 36:182–189.
61. Kassai L, Sikos K. Determination of moisture content of freeze-dried products by tritium isotope. Ann Immunol Hung 1973; 17:259–264.
62. Last IR, Prebble KA. Suitability of near-infrared methods for the determination of moisture in a freeze-dried injection product containing different amounts of the active ingredient. J Pharm Biomed Anal 1993; 2:1071–1076.
63. Lin TP, Hsu CC. Determination of residual moisture in lyophilized protein Pharmaceuticals using a rapid and non-invasive method: near infrared spectroscopy. PDA J Pharm Sci Technol 2002; 56:196–205.
64. Bai S, Nayar R, Carpenter JF et al. Noninvasive determination of protein conformation in the solid state using near infrared (NIR) spectroscopy. J Pharm Sci 2005; 94:2030–2038.

12 Freeze-Drying of Biological Standards

Paul Matejtschuk, Michelle Stanley, and Paul Jefferson
*National Institute for Biological Standards and Control, Health Protection Agency,
Potters Bar, U.K.*

INTRODUCTION
Biological Medicines and Biological Standards

Biological medicines due to their inherent complexity and heterogeneity cannot be adequately characterized solely by physical or chemical means. They include bacterial and viral vaccines, blood and serum products, and other immunological, endocrinological, and cell-based medicines. Biologicals play a significant role in medicine and public health also featuring in transplantation and cell therapy programs. In addition, the growth in biotechnology-derived products has introduced the need for new reference standards for those materials that although not batch-released in a formal manner do undergo postmarketing testing and surveillance, and indeed high-quality reference materials may be invaluable in the comparison of biosimilar/biogeneric products that are now coming to the market.

Many national regulatory authorities make special provisions for the control of biological medicines, reflecting their complex nature and production processes. The functional activity of biologicals in most cases cannot be determined in absolute units; it has to be measured against some reference preparation (standard) of the same material. Usually, the standard is a single large batch of well-characterized biological material dispensed into suitable containers with minimal between-container variation, stored under conditions that ensure good stability prior to use. The large majority of biological standards are therefore freeze-dried for long-term stability and ease of distribution.

The definitive reference material for most biological medicines is the appropriate World Health Organization (WHO) International Biological Standard. These International Standards are the primary benchmark for the relevant biological and thus are the biological equivalent of the kilogram or meter. They are established by WHO after extensive international collaborative studies and meet demanding requirements for consistency and stability. International Biological Standards are generally assigned potency values expressed in terms of International Units of biological activity. WHO International Reference Reagents are biological materials usually selected for their qualitative value and have been studied less extensively than International Biological Standards.

The WHO Expert Committee for Biological Standards (ECBS) assesses and if "appropriate" approves proposals for materials to be recognized as WHO International Standards or Reference Reagents. The criteria reflect the suitability for purpose of the material, stability, reproducibility, and, if freeze-dried, the residual moisture (see sect. "Processing of Biological Standards" for details). There is no requirement for sterility but any microbial contamination should not cause interference in the assay system in which the reference material is to be used, which may include cell culture or in vivo assays.

The National Institute for Biological Standards and Control (NIBSC) is a center of the Health Protection Agency in the United Kingdom, reporting to the U.K. Department of Health. Its mission is to safeguard and enhance public health through the standardization and control of biological medicines. The provision of these standards is a central feature of the work of NIBSC (1–4).

Candidate Biological Standards and Reference Materials Processed at NIBSC

Most of the materials are single batch products, indeed the specific combination of biological material and formulants may not be encountered exactly again until the reference material is replaced. Typically, a batch comprises a 0.5- to 1-g fill weight of material in some 100 to 20,000 containers, usually heat-sealed, type I neutral glass ampoules, although glass vials may be used for some reference materials. In general, freeze-drying is tailored to yield a residual moisture of less than 1% weight of the dry weight, and dispensing is performed with a coefficient of variation (CV) of 1% or less for formulations with a plasma-like viscosity or 0.25% or less for aqueous-type materials. Table 1 shows some of the biological materials developed at NIBSC and established by WHO ECBS as International Biological Standards and International Reference Materials during the years 2006/2007 (Table 1).

Typical biological standards are shown in Figure 1.

NIBSC holds and distributes over 500 reference items in catalogue and makes over 5000 shipments to laboratories around the world per year. Figure 2A shows the analysis of the preparation of freeze-dried biological reference materials in Centre for Biological Reference Materials (CBRM) during 2008/2009 according to product type, and Figure 2B shows the distribution by area of application.

NIBSC distributes biological standards and reference materials with a "handling" charge to account for the costs of storage and distribution by general mail. For materials, which have to be shipped by other means, for example, infectious materials via special carrier, this is at the cost of the requestor. Current charges and the catalogue of materials available are available from NIBSCs web site (http://www.nibsc.ac.uk).

Quality Management System

NIBSC processing facilities operate under a quality management system independently certified to the international standard ISO 9001:2000. This quality system is a visible sign of NIBSCs commitment to quality for the preparation of reference materials, as is the Institute's independent accreditation to ISO 17025 for its regulatory batch release testing of biological medicines and other related testing activities. Other materials comply with the In Vitro Diagnostic (IVD) Directive and are *Conformité Européenne* (CE) marked in compliance with ISO 13485. Compliance is facilitated by the use of an Institute-wide specific Reference Material Quality System and other ISO 9001 quality systems that control the reference material development from laboratory analytical work through to the process scale filling and distribution.

PROCESSING OF BIOLOGICAL STANDARDS
Requirements for Processing—Acceptance Criteria

Each new standards project is managed under the auspices of the Reference Materials Quality Manual. Each project is managed by a scientist with relevant specialization with representation of both statisticians, development scientists, production processing, and dispatch staff ensuring that all aspects of the development and production of the standard is adequately addressed. The particular requirements for processing (e.g., batch size, fill volume, and

TABLE 1 List of Freeze-Dried Candidate International Reference Materials Submitted by NIBSC to WHO and Endorsed in (**A**) 2006 and (**B**) 2007, Listing Filling Accuracy (CV) and Residual Moisture Content of Lyophilized Materials

Code	Reference material	CV of fill (%)	Residual moisture content (% wt/wt)	Comments
(A) Lyophilized Materials Submitted to WHO in 2006				
04/176	1st IS *Plasmodium falciparum* DNA for NAT assay	2.385	ND	Vials filled and dried at external contractor
97/750	2nd IS hepatitis B virus DNA for NAT assays	0.6	ND	Vials filled and dried at external contractor
03/190	1st IS human antibody to human platelet antigen 3a	0.05	0.18 (0.31% wt/wt prior to desiccation)	Further desiccated over phosphorus pentoxide
02/112	Anti-HIV-1 A human reference preparation solvent detergent inactivated. Filled in vials (one of seven panel members)	0.3	0.8 (1.77% wt/wt prior to desiccation)	Further desiccated over phosphorus pentoxide
03/188	Blood grouping anti-A minimum potency reagent	0.08	0.8	
03/164	Blood grouping anti-B minimum potency reagent	0.08	1.2	
03/178	1st IS for serum folate	0.08	0.8	
03/192	1st IS for thyroid-stimulating hormone, recombinant human for bioassay	0.07	3.95	
03/228	2nd IS for protein S in plasma human	0.09	0.06 (FD only sample not taken)	Further desiccated over phosphorus pentoxide
03/242	2nd IS protein C in human plasma	0.06	0.09 (0.41% wt/wt prior to desiccation)	Further desiccated over phosphorus pentoxide
05/162	1st IS for α-1-antitrypsin	0.15	0.66	
03/200	WHO reference reagent for IL-18	0.1	1.94	
01/420	WHO reference reagent for IL-17	0.07	0.39	
82/585	3rd IS for anti-polio antiserum types 1,2,3	0.53	ND	
06/116	2nd IS smallpox vaccine	N/A	1.52	Supplied by commercial vaccine supplier — infectious
94/532	4th IS whole-cell pertussis vaccine	0.26	1.69 (1.45% prior to desiccation)	Further desiccated over phosphorus pentoxide

(Continued)

TABLE 1 List of Freeze-Dried Candidate International Reference Materials Submitted by NIBSC to WHO and Endorsed in (**A**) 2006 and (**B**) 2007, Listing Filling Accuracy (CV) and Residual Moisture Content of Lyophilized Materials (*Continued*)

Code	Reference material	CV of fill (%)	Residual moisture content (% wt/wt)	Comments
(B) Lyophilized Materials Submitted to WHO in 2007				
04/200	1st IS parathyroid hormone 1-34 recombinant human	0.06	1.31	
05/112	2nd IS low molecular weight heparin for molecular weight calibration	0.15	0.29 (FD only not sampled)	Further desiccated over phosphorus pentoxide
06/100	3rd IS hepatitis C virus for NAT assay	0.62	ND	Vials filled and dried at external contractor, no RMD as infectious
05/132	1st IS syphilitic antibodies plasma IgG and IgM	0.17	0.69	
05/122	1st IS syphilitic antibodies plasma IgG and IgM	0.34	0.41	
04/150	2nd IS tetanus toxoid for flocculation test	0.04	0.92	
02/176	2nd International reference reagent diphtheria toxoid for flocculation test	0.06	0.28	
05/134	1st IS antibodies to HPV16	0.09	1.36	
04/252	1st IS protein C concentrate human	0.32	0.09	Further desiccated over phosphorus pentoxide
94/730	1st IS tissue plasminogen activator	0.13	0.24 (1.33% wt/wt prior to desiccation)	Further desiccated over phosphorus pentoxide
06/166	3rd IS antithrombin concentrate human	0.21	0.12	
05/106	1st IS for human platelet antigen 1a (minimum potency)	0.18	0.62	

See WHO web site at http://www.who.int/biologicals/expert_committee/en/.
Abbreviations: ND, not determined; CV, coefficient of variation; IS, international standard; NAT, nucleic acid amplification technique; RMD, residual moisture determination.

container) are agreed between Standards Processing Division and the responsible scientists for individual fills, prior to processing requests being accepted. The fundamental criterion is that data submitted to the ECBS demonstrates fitness for intended use. In addition, every container (ampoule or vial) should be identical, in terms of quantity, potency, composition, and stability.

The general criteria set by WHO have been published (5), although they are applied on a case-by-case basis. These criteria have been incorporated by Standards Processing Division into its **Quality** Management System (ISO 9001) and are a routine requirement for all fills undertaken. These criteria can be reduced for individual fills where they are not deemed necessary. Some examples of how these criteria are met are shown below:

WHO criteria: "the CV of the fill weights should be less than 0.25% for aqueous materials." Control of the variation of fill is important since a defined volume of typically water is normally used to reconstitute the freeze-dried

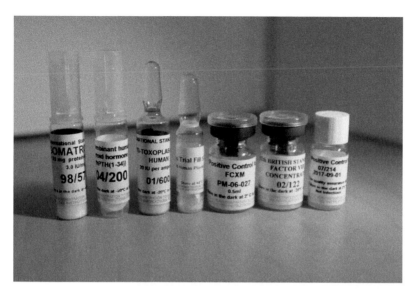

FIGURE 1 Lyophilized biological reference materials produced at NIBSC, showing 5-mL and 3-mL ampoules, crimp topped, and screw-capped vials.

material irrespective of the volume of material originally dispensed into that particular container. The between-container variation in fill weight must be insignificant compared to the uncertainty of the assay in which it will be used.

Within the Quality Management System, the target for variability of fill weight for a 1-mL fill of aqueous (or similar viscosity) material in ampoules is a CV of 0.25% or less, for a 1-mL fill of more viscous materials it is 1% or less. In addition, material filled into vials has a target CV of 1%, due to the lesser precision of the peristaltic pump on the vial-filling machine compared with the piston dispenser used for ampoules.

During any filling run at least 1% to 2% of the containers filled are selected at intervals for "check-weighing," as a measure of the variability of the material dispensed over the period of the fill (see sect. "Monitoring of the Dispensing Process").

Some reference materials, for instance those that are qualitative or "positive/negative" controls, may not require such tight limits and for other materials (e.g., cellular suspensions or adjuvanted vaccine materials) the nature of the material, in particular its viscosity, may make such a tight CV of fill difficult to achieve.

WHO criteria: "standards must be stable, it is WHO policy not to set expiry dates for International Biological Standards and Reference Reagents." Stability is determined by various factors including the intrinsic properties of the material, the moisture content of the freeze-dried material, and the oxygen and moisture content of the headspace gas within the container.

Since, in general, WHO standards are produced without a defined shelf life they are processed using heat-sealed glass ampoules (in preference to

2008/9 production of freeze dried reference material by product type

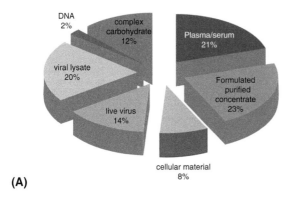

(A)

2008/9 Production of freeze dried reference materials by area of application

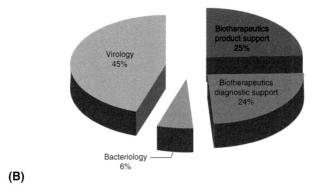

(B)

FIGURE 2   Distribution of freeze-dried products produced by NIBSC in 2008/2009 in terms of (**A**) product type and (**B**) area of application.

rubber-stoppered vials). These containers currently offer the most secure method of preventing ingress of moisture and oxygen. The superiority of ampoules over vials in maintaining a constant environment for reference materials was originally shown by Ford and Dawson (6) and has been more recently demonstrated to still hold true despite improvements in closure technology (7). When a model formulation (lyophilized human albumin in hepes saline buffer) was stressed through repeated cycles of −30°C and +56°C storage over a period of several weeks, the moisture content and oxygen content of the ampoules remained unchanged but the moisture content of vials rose significantly due to moisture released from the closure

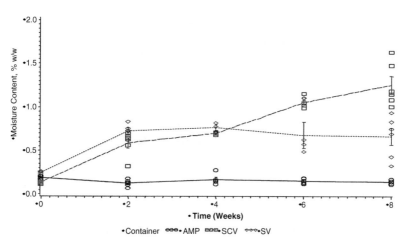

FIGURE 3 Comparison of trends in moisture content of a model lyophilized materials on stress testing of ampoules against screw-capped and crimp-topped vials. *Source*: From Ref. 7.

(Fig. 3). In addition, to maintain stability over long periods of time, freeze-dried materials are normally stored in the dark at −20°C until shipped.

Candidate WHO materials are subjected to accelerated degradation tests (see sect. "Accelerated Degradation Studies" for further details) to predict the stability of the material at the temperature of storage. This prediction supplements real-time measurements.

WHO criteria: "residual oxygen is determined using at least 3 containers to confirm that the atmosphere within the container is inert and the material processed is protected against oxidative change. Oxygen levels below 45 μmol/litre have been shown to ensure adequate long-term stability." Since the presence of oxygen may result in loss of activity, air must be excluded from ampoules and vials at the end of processing. Vials are stoppered under vacuum or dry nitrogen to preserve the internal environment and then capped. For ampoules, which are sealed by flame-induced fusion, expanding gas can result in poor seals or even blow-out failures such that previously a capillary labyrinth (8) was used for many years to facilitate sealing. However, following extensive trials, a specially designed ampoule has been used from 2006 that allows the ampoules to be sealed more simply, without the need for the capillary labyrinth. A working limit of 1.1% oxygen has been used at NIBSC and with the new automated sealing capabilities of the Bausch and Strobel filling machine (capable of processing ampoules at up to 6000/hr) this limit is easily met in routine processing.

WHO criteria: "the moisture content of the material shall be determined in order to verify that drying is adequate but not so excessive that the nature of the material has been changed." If the formulation has been freeze-dried previously, the conditions required and achievable residual moisture level will be known. However, for formulations not previously freeze-dried, a series of laboratory tests and pilot studies are performed (see sect. "Process Design and Troubleshooting") on the formulation. Results of these tests and trials are

Variation in Mean Check solution moisture content over 35 months of coulometric Karl Fischer determinations

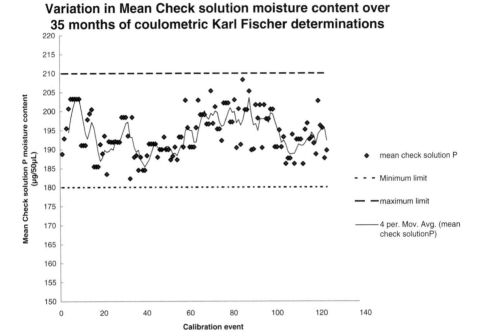

FIGURE 4 Reproducibility over 35-months use of Check solution P run as a callibrant for coulometric Karl Fischer testing in CBRM.

used to determine the most suitable freeze-drying conditions to achieve the desired product including optimum residual moisture level.

As part of this development work, residual moisture in the freeze-dried material is routinely measured by the Karl Fischer method (in a dry box as recommended by May et al. (9), see sect. "Use of Product Temperature Probes"), calibrated with a control liquid sample of known moisture. Table 1 shows residual moistures of some of the materials established by ECBS in the year 2006/2007. Figure 4 shows the reproducibility of the pre-assay check of the moisture control liquid monitored over a period of 35 months. This check must be within the range 180 to 210 μg per 50 μL with a CV of <5% before any product testing proceeds.

WHO criteria: "potency tests are essential to confirm there has been no undue loss of activity/potency during processing." Following processing of the candidate materials, the responsible scientist assays the material for activity/potency. In addition, at the end of the dispensing and before freeze-drying, "frozen baseline" samples of the material are sealed and stored in the vapor phase above liquid nitrogen at approximately −150°C to allow comparative testing between finished freeze-dried material and frozen material, to indicate any loss of potency attributable to freeze-drying.

WHO criteria: "full records should be kept of all procedures and tests used during processing." All materials processed are assigned a unique code used to identify the material throughout its existence. All information, including the original request, agreed criteria, documentation, anomalies in processing, and any tests undertaken, is included in a batch record referred to as the product record. Product records are stored for at least the lifetime of the material.

Additional measurements/criteria carried out at NIBSC: Processing is carried in a controlled environment with respect to temperature and environmental cleanliness. Although there is no formal specification for a maximum permissible microbial content or for sterility per se (materials distributed by NIBSC are not for administration to humans), a high microbial content could interfere with the final assay procedure in which the material is to be used or cause increased degradation of the active material.

The microbial content of the bulk material is determined on arrival for processing, on samples removed during processing, and on the final product.

The dry weight of the freeze-dried material is measured and is compared with the expected dry weight of the material in the formulation.

The final freeze-dried material must completely reconstitute within a reasonable period (typically less than 2 minutes) with occasional gentle agitation.

Process Description
Equipment Used*
The Institute has three pharmaceutical grade filling machines for different uses, a Bausch and Strobel AVF5090 ampoule dispensing/sealing machine, a Bausch and Strobel FVF5060 vial-filling machine, and a Schubert Paxal ARN584 vial-dispensing/capping machine. There is also other small-scale filing equipment used for pilot studies and laboratory work.

Currently, we use three production freeze dryers, manufactured by Serail (Argenteuil, France), a CS100 (usable capacity 20,000 5-mL ampoules, shelf area 6 m^2), a CS15 (capacity 3500 5-mL ampoules or shelf area 1 m^2), and a CS150 (capacity 24,000 5-mL vials, shelf area 12 m^2). This latter dryer is set up with a negative pressure isolator and dedicated filling machine to allow its use for the freeze-drying of infectious materials (see sect. "Recent Developments").

For development work we have two laboratory-scale Virtis Genesis machines, which are fully PC controlled and so can reliably model the production units, when devising and assessing potential drying cycles.

We have various measuring and monitoring equipment, including balances, temperature and vacuum measuring devices, which are used during processing to monitor the various stages of the process. All equipment has a defined maintenance and where necessary calibration program (traceable to national standards) within the Division's formal **Quality** Management System. In addition, all critical equipment has alarms to indicate malfunction and where possible automatic reversion to conditions "safe" for the product.

---

*Note: Details of any named equipment used should not be taken as a recommendation or endorsement for use but is purely for information.

Consumables Used

For candidate WHO standards 5 mL or 3 mL, type I neutral glass, DIN ampoules are used for freeze-drying, fitted with 13-mm igloo-style halobutyl closures. In addition to DIN ampoules, various vials are also used, ranging from 5 to 20 mL with crimped, tear-off, aluminum caps and 14- to 20-mm diameter halobutyl cruciform-style closures. A screw capped 5-mL vial is also used with cruciform 14-mm diameter halobutyl closure and a plastic overcap that can be color coded.

Processing Environment

The environment in which materials are processed is tightly controlled in terms of temperature and microbial contamination. All reusable equipment parts such as vessels, needles, and pump heads are washed and then autoclaved before use, and disposable material such as tubing is autoclaved before use and disposed of after use. The microbial contamination of the environment is monitored on a weekly basis to ensure that low bioburden conditions are maintained. This is achieved through the placing of settle plates and of swab samples throughout the processing areas. There are defined alert and action limits for the number of colonies detected in each area. All products for dispensing are also tested as bulk material, post filling and post freeze-drying. The testing includes methods for bacterial contaminants and for moulds and yeasts.

Pilot Scale Work (see sect. "Process Design and Troubleshooting")

In most cases, small-scale trials are carried out well in advance of the large-scale batch to determine the best processing conditions; this is not always necessary where past experience can be used, but in all cases the results of the pilot studies and any previous batches of the same formulation are reviewed and recorded well in advance of processing the definitive batch to allow time for any potential problems to be identified and addressed. Previous lyophilization experience, obtained with different dryers and cycles, should only be used with caution.

Preparation

All critical consumables are quarantined on receipt until checked and released, if acceptable, for use in the processing. All glass consumables are cleaned in a nondetergent system. Normally dilute acetic acid is used for glassware and alcohol for cleaning of rubber/plastics. Closures are washed in a commercial stopper-washing machine and siliconized. Freeze-drying stoppers for use with vials are subjected to a drying process to remove absorbed moisture. Prior to use, all filling vessels, tubing, and needles are autoclaved after cleaning.

Dispensing

Prior to processing, the bulk material is stored, if required, at the appropriate temperature using controlled, calibrated, and monitored storage facilities. Dispensing is carried out under controlled, monitored conditions, as demanded by the material, typically 2°C to 8°C, with gentle stirring. The dispensing machine is adjusted and verified to give the nominal fill weight required.

After dispensing, ampoules and vials are automatically fitted with the closure by the filling machine to the part-stoppered depth and are packed into stainless steel tins with removable bases. They are then stored at the same temperature as required by the bulk product until the entire batch has been filled.

FIGURE 5 Laser-etched ampoules showing pump head used and order of dispensing.

Monitoring of the Dispensing Process

Ampoules on the AFV 5090 are laser-etched with the unique batch code (Fig. 5) as part of the filling process. This enables them to be identified at all stages of subsequent processing prior to application of the final label, and also allows the pump head used and the sequence within the filling process to be identified.

The Bausch and Strobel vial-filling machine FVF5060 weighs each vial before and after filling and records the net weight. Vials that are out of range cause an alarm and are automatically rejected. The fill weight is adjusted manually to maintain an optimum weight.

The ampoule-filling machine AVF5090 uses three pumps and weighs a set of three ampoules before and after filling. During setup mode it automatically weighs all the ampoules and adjusts each pump to bring the weights in range. Once acceptable weights are achieved, dispensing is allowed to proceed and 3 in every 90 ampoules are checked routinely throughout the rest of the fill. Small variations will prompt the machine to self-adjust to optimize the fill weights. A weight that is out of range stops the machine and reverts it to setup mode.

The older Paxal Schubert filling machine uses a peristaltic pumping mechanism and relies on manual check weighing and adjustment. Typically 1% to 2% of the batch is checked manually.

The CV of the filling process is calculated across a number of ampoules/vials in the filling operation and the data is used to assess the accuracy of fill across the run (Fig. 6)

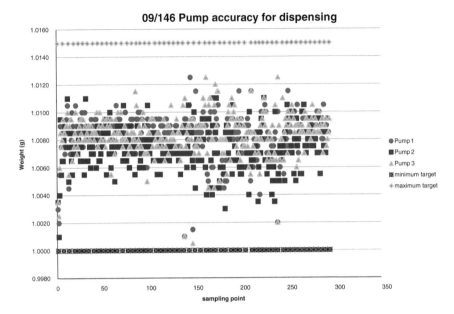

FIGURE 6 Accuracy of fill for typical ampouled reference material with targets set +/− 3CV. Mean fill = 1.0080 g with CV of 0.16%. *Abbreviation*: CV, coefficient of variation.

A minimum of six of the ampoules/vials used are marked and numbered at intervals throughout the fill to determine the dry weight (by weighing before filling and after freeze-drying) of the final material. These samples are used to determine the percentage residual moisture content of the freeze-dried material by the coulometric Karl Fischer method.

Other samples are also taken during/after filling, for assessing the microbial contamination.

Freeze-Drying

The appropriate freeze-drying cycle (see typical cycle from trial lyophilization in Fig. 7), as determined previously, is selected and the freeze-dryer started. When the shelves reach the set temperature (typically either +4°C or −50°C) and filling is complete, the entire batch of steel tins containing the filled ampoules/vials is loaded onto the shelves and the bases of the tins are removed. The position of each tin within the freeze-dryer is recorded on the Product Record and predetermined loading patterns are defined for all sizes of fill.

Resistance thermometers are inserted in one to six of the ampoules/vials (depending on the specified need) and the location of each is recorded on the Product Record. Once the temperature of the thermometers in the ampoules/vials reaches the shelf temperature, the freeze-drying cycle is started. The information gained from these thermometers is used in comparisons between batches and sometimes as an indicator as to the speed of freeze-drying, in the knowledge that the containers with the probes are not typical of the batch.

The freeze-drying parameters are continuously monitored and recorded by the freeze-dryer control system, and are checked and recorded twice daily

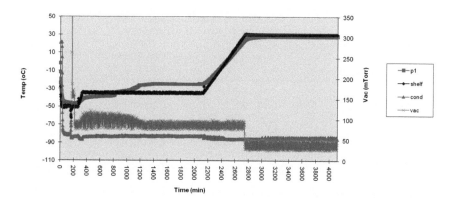

FIGURE 7 Typical lyophilization cycle developed at pilot scale showing product temperature probe inflection and drop in pressure as indicators of end of primary drying. *Source*: From Ref. 10.

by the processing team to detect and deal with any deviations or potential problems. Typical parameters measured are shelf temperature, product temperature, condenser temperature, condenser, and chamber vacuum; these are compared to the set points in the freeze-drying cycle used. All recordings are attached to the Product Record.

At the end of freeze-drying, the chamber is restored to atmospheric pressure with pure, dry nitrogen, and the ampoules/vials are fully stoppered by contracting the shelf stack and then removed. The ampoules/vials are then sealed or overcapped (see sect. "Sealing/Capping").

Further Desiccation

Further desiccation over phosphorus pentoxide following the freeze-drying was practiced at NIBSC for some reference materials until 2005 to ensure very low moisture contents. However, by optimizing freeze-drying cycles accordingly residual moisture levels of <1% wt/wt can typically be achieved without resorting to such additional processing. Indeed in comparative studies of labile biological activities (factor V and factor VIII activity in plasma), Hubbard et al. (11) showed that not only did the nondesiccated material have adequate stability similar to the further desiccated material but that stability was also dependent as much on appropriate formulation as ultralow moisture content. Addition of glycine and in particular glycine plus hepes buffer gave greater stability of the factor VIII activity at 20°C storage without further desiccation of the product than for the desiccated product (Table 2).

Sealing/Capping

Whereas manual sealing may be suitable for small numbers of ampoules [e.g., using a semiautomated machine from Adelphi Tubes Ltd., (Haywards Heath, W Sussex, U.K.)], to deliver the consistency and quality required for International Standards an automated flame sealer is required (Fig. 8).

TABLE 2 Residual Moisture Content and Predicted Loss of Factor VIII Activity from Accelerated Degradation Testing of Formulation Options for Lyophilized Human Plasma

Formulation	Residual moisture (% wt/wt) with CV (%) in parentheses ($n = 6$)	Mean % loss per year at $-20°C$ (and 95% upper confidence limit)	Mean % loss per year at 20°C (and 95% upper confidence limit)
Plasma FD	0.47 (5.8)	0.75 (1.00)	36.4 (38.7)
Plasma FD & SD	0.15 (9.9)	0.12 (0.30)	20.7 (27.8)
Plasma + Hepes FD	0.38 (15.4)	0.80 (0.09)	22.0 (22.4)
Plasma + Hepes FD & SD	0.10 (5)	0.07 (0.14)	16.3 (20.4)
Plasma + glycine FD	0.37 (15.9)	0.03 (0.04)	8.8 (9.7)
Plasma + glycine/Hepes FD & SD	0.13 (16.5)	0.13 (0.26)	10.33 (12.9)
Plasma + glycine/Hepes FD	0.61 (14.5)	0.03 (0.08)	6.3 (8.3)
Plasma + glycine/Hepes FD & SD	0.90 (10.3)	0.06 (0.06)	7.54 (7.6)

Abbreviations: FD, freeze-dried only; *FD & SD*, freeze-dried and secondary desiccated; CV, coefficient of variation.
Source: From Ref. 11.

FIGURE 8 Flame sealing of ampoules on Bausch & Strobel filling line.

Stoppered vials are removed and oversealed with either a screw cap or an aluminum foil tear-off cover. This prevents ingress resulting if any stoppers were dislodged or "popped up" during storage.

Leak Testing of Containers

Prior to the introduction of DIN ampoules, bespoke "test-tube" type ampoules were used with a diameter of 10 mm. The large diameter caused some problems in sealing and therefore each batch of ampoules was tested for leaks by immersion in a dye. With the introduction of industry standard DIN ampoules, this practice continued until a review of one year's production showed no leaks, and the practice was no longer necessary. Now noninvasive sensing methodologies are available that can assure the quality of the internal atmosphere (see sect. "Oxygen Headspace Measurement").

Since 2005 the capillary leak adaptor (8) has been replaced by specially sourced ampoules with a parallel-sided neck for the last centimeter. These are stoppered with standard 13-mm diameter igloo closures and the seal is sufficiently good for the headspace gas to remain unchanged with respect to its oxygen content for at least 24 hours if undisturbed. On sealing, the closure is slightly lifted a matter of seconds before the ampoules are flame sealed. This process has been shown routinely to deliver final oxygen headspace well below 1% residual oxygen.

Labeling

Generally ampoules/vials are labeled to indicate the unique batch identifier, product name, manufacturer, and storage conditions. An example of a typical label is shown in Figure 9A.

On establishment by WHO, the ampoules/vials are overlabeled to indicate their new international status, the new label has opaque backing to occlude the previous label, as in the example below (Fig. 9B). Overlabeling is preferred to removal of the previous label for quality assurance reasons.

Test and Inspection

Visual inspection and monitoring of product quality is carried out by the processing team at all stages of the process; this is to detect any potential problems, for example, splashing during dispensing, broken or damaged ampoules, poor freeze-drying (collapse), and poor sealing or capping. All comments are noted in the Product Record.

In addition, the following are recorded/tested:

Fill weight and CV
Dry weight and CV
Residual moisture and CV
Residual oxygen content

All of the above are recorded in the Product Record, which is reviewed by a team consisting of production, processing, and development personnel, prior to release of the batch. In addition, reconstitution time, biological activity/potency, and stability are determined subsequently by the relevant scientist.

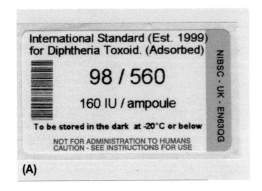

FIGURE 9   (A) Typical NIBSC label for candidate reference material; (B) typical WHO label for endorsed International Reference material.

The whole process is reviewed to ensure it has been carried out in accordance with the agreed criteria and the freeze-drying is reviewed to determine if the chosen conditions were satisfactory.

Storage and Dispatch

Biological standards and reference materials are stored at the appropriate temperature in controlled, calibrated, monitored, and alarmed storage facilities. The freeze-dried materials are normally stored at $-20°C$ to maintain stability over the period of the material's availability, which may be a decade or more. Freeze-dried materials should be sufficiently stable to withstand short-term shipping at ambient temperatures without deterioration to its intended use.

The ampoules/vials are removed from storage on the day of dispatch and packed in accordance with national/international regulations. Shipment is typically by post (mail) except for countries where import via the post is a known problem or the shipment is urgent when couriers are used.

Infectious standards, for example, hepatitis B surface antigen, and frozen materials that require solid carbon dioxide (dry ice), are shipped via specialized carriers and according to International Air Transport Association (IATA) and air security regulations, using UN certified packaging. Shipments sent with solid carbon dioxide include a "tell-tale" monitor to indicate thawing of the contents (even if refrozen). Special training and certification is required for staff involved in shipping these substances.

Biological standards and reference materials are supplied for immediate use, and prior to use should be stored at the temperature indicated on the label. Once freeze-dried material is reconstituted, users must determine the stability of the material according to their own methods of preparation, storage, and use. In general, NIBSC follows the WHO policy in not setting expiry dates for freeze-dried biological standards and reference materials.

PROCESS DESIGN AND TROUBLESHOOTING
Cycle Design
The cycle design process used at NIBSC has resulted from different priorities than those facing most users of freeze-drying in the pharmaceutical or food industries. Unlike these sectors, there is little requirement for repeat batches of the same material. The normal constraints, such as the number of batches that have to be processed per week and the cost of the freeze-drying process per batch, are not considered as the highest priority. Unlike most pharmaceutical freeze-drying, our control programs must take into account that active processing occurs in a "9 a.m. to 5 p.m." weekday-only environment.

The key factors in determining successful freeze-drying are long-term stability, preservation of sufficient activity, and rapid facile reconstitution. The product residual moisture content and atmospheric composition (in terms of oxygen content) are adjuncts to this. In the WHO guidelines, there are no set limits for minimal moisture level. We have an in-house limit of <1% to reduce the likelihood of water-catalyzed hydrolytic and other degradative processes. However, in practice, the level of residual moisture is usually well below this for most of our materials.

In common with most freeze-drying operations there are three basic stages in the process—freezing, primary drying, and secondary drying.

Freezing
All constituent water needs to be totally immobilised, either by crystallization to ice or incorporation into a glass of super-concentrated biological material, before a vacuum is drawn and freeze-drying can commence. The product must be maintained at a sufficiently low temperature to be below the eutectic (T_{eu}) point(s) of the major crystallizing electrolyte present and the glass transition temperature (T_g') of the protein/carbohydrate glass for sufficient time to ensure this. NIBSC is unusual in facing different freeze-drying challenges in terms of disparate formulations with almost every fill. The start material may be any of a wide range of biologicals, by nature protein, carbohydrate, glycoprotein, nucleic acid, or a complex mixture of several different components. Indeed, we are now encountering whole-cell and whole-virus preparations where integrity is required after reconstitution. There may be a wide variety of excipients required for the preservation of the biological activity following reconstitution but which may help or hinder the freeze-drying process.

It is vital to know the composition of the material to be freeze-dried so as to determine the suitable freezing conditions. Some salts such as tris, hepes, or calcium chloride with low T_g' values may lower the glass transition to such an extent that the freezing temperature may be below the operational capabilities of the freeze drier; the lowest practical temperature achievable with our production-scale freeze driers is −50°C.

The freezing process will also be important; snap freezing by placing product on precooled shelves can result in heterogeneous crystal formation and final product appearance (12). This has been particularly apparent to us when freeze-drying plasma where shelf freezing on precooled shelves (−50°C) resulted in up to 60% of ampoules having freeze-dried cakes with a markedly striated appearance and some having a heterogeneous appearance, partially uniform (noncrystalline) and partially striated.

For plasma, to prevent this heterogeneity, we routinely use shelf freezing from 4°C down to −50°C at a modest rate of cooling (0.2–1°C/min) and then perform primary drying from shelf temperatures of −30°C to −40°C. This controlled freezing results in a homogeneous cake without marked crystalline appearance.

However, if there is a tendency for freeze-induced damage in a biological then there may be no alternative to rapid freezing, achieved by loading product on to precooled shelves at −50°C Historically, snap freezing in liquid nitrogen was used but this is neither safe nor practical at the scale of current manufacture, often resulting in a high proportion of ampoules breaking and leading to the crystalline problems described above. Materials that contain adjuvant suspension or cellular components are also loaded on a precooled shelf to minimize the time taken to achieve a frozen state and so the likelihood of sedimentation. There is evidence that rapid freezing of cellular or viral materials may also minimize the damage caused by ice crystal formation as the crystals will be smaller (13).

Primary Drying

The aim of primary drying is to remove the crystalline water ice. As freeze-drying commences, the products must be held below these temperatures to avoid collapse as drying commences and a visually unacceptable cake or, worse still, loss of functional activity. The initial product temperature during primary drying must not exceed the T_g' or T_{eu} points while not being so low that the sublimation rate is too slow. Product temperatures 0°C to 10°C below the T_g' or T_{eu} are chosen and maintained for sufficient time for an inflexion in the temperature profile to be evident and the product temperature to rise to equal or above that of the shelf (Fig. 7). During the early stages of freeze-drying, the product temperature is at a temperature lower than the shelves due to sublimative cooling; the heat loss from the product exceeds the heat flow into the product, primarily from the shelf. The inflexion in the product temperature profile represents the stage at which this cooling loss finishes and the product temperature rises and so gives a coarse indication of the end of primary drying, at least for the ampoules being monitored.

During primary drying, should a processing equipment failure occur, it is important that the shelf temperature and therefore the product temperature be rapidly lowered to ensure that the product retains biological activity. Routinely, low shelf temperatures have been used at NIBSC, with prolonged primary drying periods. Under such conditions, although sublimation rates are slow, the likelihood of product temperature exceeding the product T_g' is low. This, although not optimal in terms of the duration of the freeze-drying process, does allow the processing of a wide variety of formulations with minimal cycle development. Chamber vacuum is maintained typically at 30μbar to 100 μbar during the primary drying process in most of our operations.

It has been our practice to extend the length of the process stages well beyond the minimum time indicated from the probe-containing ampoules/vials during development studies; this helps to prevent problems during scale-up without the need for further validation runs. Although there are a number of end-point determination technologies now commercially available and indeed the smaller production dryer is fitted with a manometric pressure rise tool, we normally do not aim to modify the cycles, developed at laboratory scale, for the scale-up to manufacturing scale.

Secondary Drying

Following primary drying, secondary drying is used to remove most of the amorphous water component of the glass. The shelf temperature used is as high as can be applied to the biologicals to promote rapid drying without loss of functional activity. Given that some of our standards are labile biological factors there might well be loss of activity if too high a secondary drying temperature is used. We typically use 25°C to 30°C with a 30 μbar chamber pressure across a range of biological materials. Typically, the product temperature probes do not quite attain the shelf temperature in our experience even after prolonged secondary drying periods. Residual moisture is a particular concern for biological standards due to the requirement for a prolonged shelf life. Levels of moisture that may be acceptable for some freeze-dried products such as 2% to 5% wt/wt may not be satisfactory for reference materials and, routinely, we dry to 1.0% or less residual moisture.

There is evidence to support the view that products can be overdried. Some water may be needed to maintain the structural stability of the biological, and removal of this may lead to destabilization. Extended drying of insulin resulted in a rise in the level of degradation products detected by reverse phase high-performance liquid chromatography (HPLC) analysis, as shown in Table 3 (14). Other researchers have noted loss of activity in influenza virus with excessive drying (15). The optimum level of dryness should be determined for each material to be dried (16) where feasible, given the number of formulations to be dried.

Use of Product Temperature Probes

It is well accepted that containers housing product temperature probes are uncharacteristic of the containers that do not have probes; nucleation is enhanced by the heterogeneous surface introduced by the probe (17) and the probe leads present a heat source. Probes in containers cannot be used as an absolute indication of the conditions occurring in those containers without a

TABLE 3 Damage Caused by Overdrying of Insulin

Preparation method	Predicted degradation rate at −20°C % per annum	Residual moisture (%)
Freeze-dried only	0.105	0.46
Freeze-dried then further desiccated 1 wk over phosphorus pentoxide	0.475	<0.1

Degradation rates based on Arrhenius calculations from degradation at elevated temperatures and measurement of degradation products by reverse phase HPLC.
Source: Adapted from Ref. 14.

probe during a freeze-drying run and should be regarded as a source of information and not used to control the system. For the processing of infectious materials, probes are not included due to decontamination concerns.

Given this proviso, temperature probes and vacuum probes can and do provide a useful means of monitoring the progress of the freeze-drying process and to some degree indicate the consistency of process changes within and across the shelves of the freeze-drying chamber. Probes are also useful indicators of equipment noncompliance and system failures—for instance, in terms of pressure leakages, inadequate heating, or cooling steps, etc.

Evidence of ice crystallization (and the degree of supercooling) and the crystallization of excipients can be observed (for the probe-containing ampoules at least) from the temperature profile as inflections in the product temperature profile during the progress of the freezing step (Fig. 10).

Most commercial freeze driers allow for a number of product temperature probes that are usually distributed so as to provide details of temperature profiles across the shelves and between shelves. Even then there may be only six probes (in the case of the Serail CS-100) for instance to monitor the freeze-drying of 10,000 ampoules/vials. Of the types of probes available simple K-type or T-type thermocouples are small, convenient, and inexpensive. We use resistance thermometers, although these are larger (3–4 mm in length and 2-mm wide), as they give more accurate readings with a more linear response than thermocouples. Typically we would aim to place at least one probe in product on each shelf of a run covering multiple shelves.

The primary concern with the location of probes is the reproducibility in terms of the location within the container (centrally or to the side) and the degree to which contact is made with the material to be freeze-dried (see sect. "Impact of Process Scale-up").

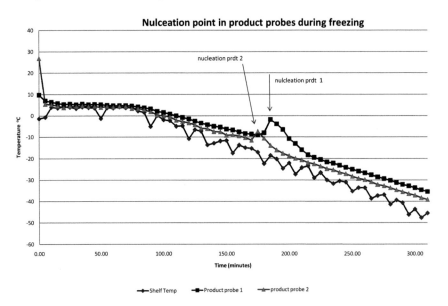

FIGURE 10 Inflection in signal from product temperature probe on nucleation of ice formation during ramped freezing step.

Although the drying process progresses on the basis of the pre-programmed profile of shelf temperature and chamber pressure with time, there can be variation in terms of temperature across a shelf and across the chamber itself. The mapping of the variation in temperature across the freeze drier is an important part of the qualification of a new plant and should be repeated periodically, especially if problems are suspected in coolant flow path.

Formulation and Freeze-Drying Conditions

At its simplest, a product for freeze-drying might contain only water as the solvent and the biological to be dried as the solute. The establishment of suitable freezing conditions would be straightforward, and once the water had been eliminated by sublimation the product would be stable and remain active. However, in reality a variety of nonvolatile excipients will be present either to stabilize the product during the drying process or to ensure activity once the product is reconstituted. Other excipients may supply bulk, as often the quantity of the biological standard is minute and to ensure good cake formation at least 2% wt/wt solids should be maintained. These excipients will themselves influence the freeze-drying conditions and may influence the activity and stability of the product as they are concentrated on sublimation of the water ice. For instance buffers may be added to preserve biological activity, but the pH range created by these buffer components may vary due to differential crystallization during freezing, and indeed some buffer component may be volatile and so pH may shift on sublimation.

Typically we would avoid high levels of salts, such as sodium chloride and buffers (such as phosphate-based systems) that might shift pH on freezing (18). Preferred excipients would be lyoprotectants such as nonreducing sugars, especially trehalose (19), and albumin (20) or glycine as bulking agents (21).

To determine the freeze-drying parameters, a number of critical parameters are required. The first among these is the T_{eu} of formulation components that crystallize on freezing and the T_g' of those that form an amorphous state; these should be determined to set the maximum safe product temperature for primary drying and can be ascertained by one of a number of methods.

Modulated Differential Scanning Calorimetry—mDSC

The determination of T_g' of frozen liquids and the T_g of freeze-dried solids has been widely reported (22). We have recently used modulated DSC (for a review of mDSC see Ref. 23) as a routine primary tool in determining the freeze-drying conditions to apply to standards being processed. The glass transition is observed as an inflexion in the modulated reversed heating profile and the T_g' is derived by application of the manufacturer's software (24). In our experience, the use of large (100 μL) sample pans has helped to enhance what can be very weak signals. In addition clearer glass transitions can be observed at high concentrations of sample. For some samples, at least, the determined T_g' at a high concentration holds true also at the lower concentrations being freeze-dried. For example, low molecular weight heparin analyzed at concentrations of 100, 50, and 10 mg/mL (the latter being the concentration at which freeze-drying was to be performed) all gave a similar T_g' value. However, the inflexion in the DSC signal was far less distinct at the concentration at which freeze-drying was to occur (Fig. 11).

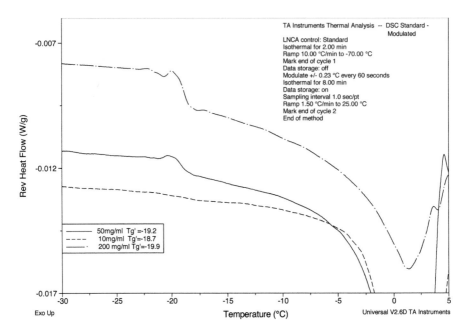

TA Instruments Thermal Analysis -- DSC Standard - Modulated
LNCA control: Standard
Isothermal for 2.00 min
Ramp 10.00 °C/min to -70.00 °C
Mark end of cycle 1
Data storage: off
Modulate +/- 0.23 °C every 60 seconds
Isothermal for 8.00 min
Data storage: on
Sampling interval 1.0 sec/pt
Ramp 1.50 °C/min to 25.00 °C
Mark end of cycle 2
End of method

50mg/ml Tg' =-19.2
10mg/ml Tg'=-18.7
200 mg/ml Tg'=-19.9

FIGURE 11 Modulated DSC profile of heparin run at three different concentrations showing amplitude of the T_g' response with concentration [run on 2920 DSC (TA Instruments, Crawley, U.K.)] at 1.5°C/min, 0.23°C/min modulation in steel high-volume pans.

The limitations of this methodology are that some samples (e.g., plasma and albumin) appear to give no glass transition, even at high concentration (10–20% wt/vol protein), and often the standards to be dried are only available in small quantities and dilute concentration making T_g' determination by mDSC difficult. For example, in our experience no specific T_g' was detectable by DSC for a preparation of tRNA at 1 mg/mL. In the literature, a value for modified RNA was reported using DSC (25) but with analysis on a sample at 100-mg/mL concentration—conditions that would be unlikely to be available for most nucleic acid samples we encounter. In other samples containing sodium chloride the T_{eu} event dominates the profile and weak T_g' may be missed or masked.

In summary, DSC is a rapid technique and can be automated but requires a significant financial outlay for the equipment and does not provide T_g' values for all samples under consideration for freeze-drying. However, it offers the advantage of also allowing the determination of the T_g of the dried state that may be of benefit when studying formulation selection and storage stability (see chap. 8).

Electrical Resistance
The changes in electrical resistance that occur as a sample undergoes incipient melting can be followed by a number of methods (26). At NIBSC the SLTT—solid-liquid transition temperature—was measured for many years [using an in-house technique, performed with equipment developed in conjunction with

Brunel University, London based on the design of Lachman (27)] to indicate the temperature at which the electrical conductivity of a sample varies during controlled heating after snap freezing in liquid nitrogen to equal or below −60°C. Values obtained were derived from extrapolation of the tangent to the conductivity curve at the water mobilization point. The values obtained were not identical to those derived from DSC, and this technique was replaced with a commercially available device in 2003. Resistivity measurement is an easily performed technique and gives useful data with a wide variety of formulations in our experience (Fig. 12A).

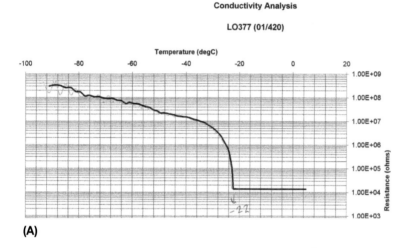

(A)

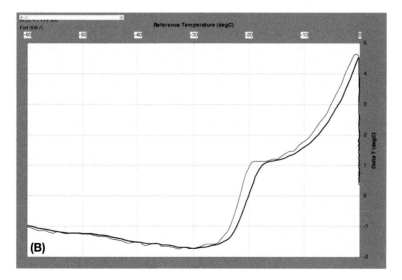

(B)

FIGURE 12 (**A**) Electrical resistance(conductivity) profile for reference material formulation showing critical temperature at −22°C. (**B**) Differential thermal analysis profile of common reference material formulation showing two thermal events.

Differential thermal analysis permits more detailed analysis of thermal events comparing the heat changes in a sample against that of a reference material (often water) and can be used to determine equilibrium freezing temperature, phase changes, and onset temperature for ice melting. An advantage is that analysis can be performed in the same containers in which lyophilization will occur, but the interpretation of the profile, especially in complex formulations, may be difficult (Fig. 12B) as multiple thermal events may occur and other means will be needed to determine which are important to freeze-drying. Other thermal analysis techniques can also be used (see chap. 5).

Freeze-Drying Microscopy

Freeze-drying microscopy (FDM) mimics the freezing and thawing process under the microscope by means of a controlled temperature microscope stage with a sample mounted under vacuum in a thin film (24,28,29). It yields a collapse temperature based on visual inspection of collapse of crystal structure as the sample is gradually warmed from frozen. Although initially a home-built technology, the equipment is less expensive than DSC; cryostages are available from microscopy suppliers (such as Linkam, Epsom, U.K.) and an integrated freeze-drying microscope is also available (e.g., Lyostat, Biopharma Technology Limited, Winchester, U.K.).

Although FDM should generate T collapse values for all samples (Fig. 13), these figures are derived from the collapse in a 2 to 5 μL of sample on the slide whereas freeze-drying will be of containers possibly filled to a height of several centimeters. For this reason, broad margins for error are applied. It has been suggested to perform freezing at temperatures 2°C to 5°C below the T_g' value (24,29).

In conclusion, it is necessary to have at least one and ideally several different techniques available to get the most accurate T_g' information for use during freeze-drying cycle development (30).

Impact of Process Scale-Up

The scalability of freeze-drying cycles can be a problem as cycles devised for a pilot machine may not necessarily work in a process scale dryer with 10-fold or greater surface area, different pumps, and cooling systems. As part of the qualification of our new 12 m² freeze-dryer at NIBSC we adopted a strategy of investigating in the new dryer the performance of our cycles devised at laboratory scale and used successfully in smaller 1- and 6 m² dryers. Tests were performed when using only a single shelf and also when operating under full occupancy. Two run cycles were evaluated, one for human plasma and one for formulated influenza antigen. Both materials were frequently processed and so were typical of the operating demand on the new dryer. Following lyophilization, assessment of the product appearance and of the residual moisture content and product functional activity were made at a number of key locations across the batch. For a one-shelf run, this was the center tray and each corner tray of vials and when assessing a full load (of six shelves) the midpoint on each shelf and the midpoint and corners on top and bottom shelves were assessed. Good reproducibility across a single shelf and across a 5-shelf 20,000 run was shown for influenza antigen. However, for plasma, although moisture and activity were

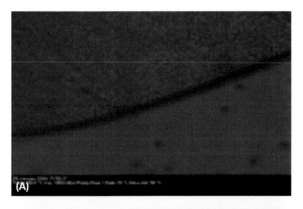

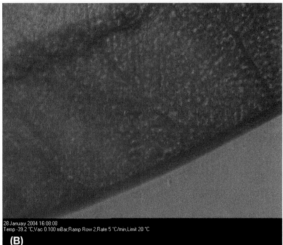

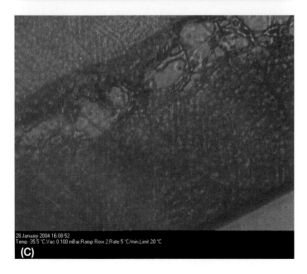

FIGURE 13 Freeze-drying microscopy images on a sample showing (**A**) frozen, (**B**) drying, and (**C**) collapse. Images taken using a 5-μL sample on a Linkam FDCS 196 cryostage.

Shelf 1		
0.26 (5)		0.28 (6)
	0.30 (5)	
0.22 (5)		0.30 (6)

Shelf	Mean % w/w moisture
1	0.30 (5)
2	0.30 (5)
3	0.30 (6)
4	0.30 (6)
5	0.24 (6)
6	0.29 (5)

Shelf 6		
0.26 (6)		0.23 (6)
	0.29 (5)	
0.26 (6)		0.24 (6)

Figure showing the mean moisture content and number of vials tested (in brackets) at each sampling point from shelves 1 and 6, (left) and the mean moisture content from the vials tested from the centre of each tin (right).

FIGURE 14 Performance qualification profile for lyophilization of plasma over 25,000 vials on a 6-m^2 freeze-dryer (06-040-PM)—residual moisture was monitored across the 6 shelves (central point) and across shelves 1 and 6 (5 points). Mean moisture across all points was 0.27% wt/wt with range from 0.15% to 0.36% wt/wt.

maintained, a more crystalline appearance was seen in some vials in one corner of one of the shelves. This problem was investigated both by our own scientists and with external consultants, and it was concluded finally that a chill blast effect was being seen from the condenser, which was located to the side of the chamber on this dryer. This was causing a difference in the freezing of the product in these locations from a gradual controlled freeze to an uncontrolled more rapid and stochastic event, the end result being similar to that already described in section "Freezing" for plasma frozen on a precooled shelf. This problem was successfully addressed by programming a closure of the isolation valve during loading and freezing stages and opening it only when vacuum was being drawn. After this good uniformity of product properties was observed across the dryer (see Fig. 14).

Residual Moisture Determination

The Karl Fischer solutions are methanolic and in some cases the reconstitution of samples—plasma in particular—can be problematic. We overcome solubility problems by the injection of entire vial/ampoule contents (redissolved in 1–4 mL of anolyte) into the electrolytic cell. Also, we perform Karl Fischer determinations within a dry box (9) to avoid interference from atmospheric moisture.

There is a practical limit to the number of samples that can be measured before the electrode solutions are replaced. The limit of detection in our hands is

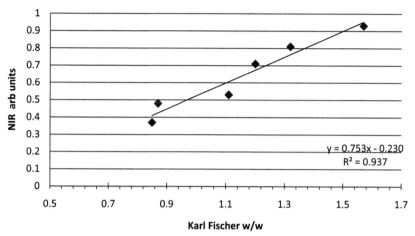

FIGURE 15 Comparison of residual moisture in a lyophilized plasma (0.5-mL filling volume) in 2-mL crimp top vials: arbitary reflectance NIR units plotted against residual moisture content by coulometric Karl Fischer analysis. *Source*: From (Ref. 34).

approximately 10 µg of water, but we have found that where the dry weight of samples is very low (say <10 mg) the apparent moisture rises and we think this is the result of measuring at the accuracy limits of the coulometric method. In theory, it is possible to pool multiple samples and perform analysis on the pool, and this may be required on occasions in future as the frequency of standards with very low dry weight increases.

Moisture determination by nondestructive infrared reflectance methods has been demonstrated by a number of groups (31–33) in some pharmaceutical situations but has proved more difficult to apply in our situation as calibration of the near infra red (NIR) output against a standard curve of known moistures is required for each product and presentation size. This is attractive where there are repeated batches of a few products, but is less straightforward for NIBSC where our standards differ so much from preparation to preparation and the generation of standard curves is time consuming to perform.

At NIBSC, one potentially useful application may be in the assurance of moisture content for reference standards of infectious materials where to perform destructive coulometric Karl Fischer analysis would require containment conditions (Fig. 15).

Oxygen Headspace Measurement
WHO guidelines stipulate very low oxygen content (5) to minimize the risk of loss of activity due to oxidation of the biological materials, particularly proteins. We have used a working value of 1.1% vol/vol oxygen and have used two methods to determine this. A destructive method [using a electrochemical oxygen sensor—such as the Orbisphere Pharmapack (Hach Ultra, Chesterfield,

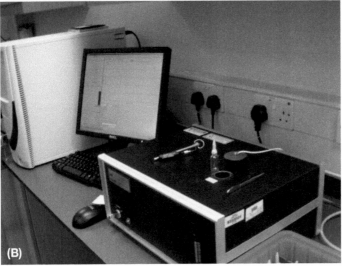

FIGURE 16 Residual oxygen testing methods: (**A**) destructive analysis by Orbisphere Pharm-apack and (**B**) noninvasive analysis using laser frequency modulated spectroscopy.

U.K.)] and a nondestructive method (frequency modulated spectroscopy using laser infrared light, for instance the FMS 760 by Lighthouse Instruments, Charlottesville, U.S.) (Fig. 16).

The destructive method involves the controlled breakage of the inverted ampoule base or piercing of the vial closure using a specially designed needle through which nitrogen saturated water can be injected so displacing the headspace gas past a calibrated electrochemical oxygen electrode. This method can detect oxygen content to below 1% vol/vol oxygen. However, contamination of the injection needle or electrode chamber by reconstituted product can be problematic and regular servicing of the container-piercing assembly is required. Noninvasive oxygen measurement involves the use of the frequency-modulated signal absorption of near infrared laser light at 760 nm by oxygen in the headspace gas. This method is nondestructive, is independent of the formulation and indeed appearance of the freeze-dried cake (providing it has not broken up to such an extent to contaminate the sides of the container), and so is more versatile and easier to operate. A limitation is that a standard curve is required for each diameter of container used (as the signal will depend on the path length of the gas and so the diameter of the container). These can be supplied by the manufacturer traceable to NIST certified oxygen/nitrogen mixtures. The method is rapid and eliminates the risks associated with breaking open a container for the destructive method. A comparative study between the two methods was performed and similar results were obtained.

Product Stability

Of prime importance as a test of the success of the freeze-drying of standards is the retention of good reactivity after reconstitution. In practice, some loss of activity may be acceptable on freeze-drying but the resultant product should maintain its activity once dried. Within NIBSC, this assessment is often by specialist functional assays, some of which may have confidence limits that are wider than others. However, we offer the option of laying down snap frozen sample stored in liquid nitrogen (frozen baselines) as a means of checking whether it is the freezing or drying process that is causing the deterioration. Activity of the standards after freeze-drying is generally well preserved as can be seen from the illustrations of four biologicals in Table 4.

Vials may be readily stoppered within the freeze-drier and so do not pose the same degree of problem in maintaining the internal environment during sealing. However, these rubber closures may prove less suitable in terms of long-term storage, particularly at subzero temperatures, compared with flame-sealed ampoules that are impervious to gas exchange.

TABLE 4 Preservation of Biological Activity in Some Typical Biological Reference Materials After Freeze Drying

	Activity (IU/mL)		
	Activity pre-drying (defined as 100%)	Activity post-drying	Percent
01/592 Heparin[a]	1312 IU/mL	1270 IU/mL	96
01-037-MA factor VIII in plasma	0.5 IU/mL	0.46 IU/mL	92
01/586 Thrombin[b]	24 IU/mL	22.3 IU/mL	93
04/150 Tetanus toxoid[c]	701 Lf/mL	646 Lf/mL	92

[a]Measuring factor Xa activity.
[b]Measured by chromogenic substrate assay.
[c]Flocculation activity units, measured by ELISA.
Abbreviation: LF, limes flocculationis.
Source: Data Courtesy of Drs Elaine Gray, Anthony Hubbard, Colin Longstaff, and Thea Sesardic, National Institute for Biological Standards and Control.

ACCELERATED DEGRADATION STUDIES
Overview
Degradation at the selected temperature of storage often is too slow for real-time studies to be practical before determining whether a material can be used as a reference material, although real-time studies can continue in parallel with the initial use of the materials (35).

In principle, material is stored at a range of elevated temperatures and also at some reference temperature at which degradation is assumed to be insignificant. The activity of these samples is determined at known time intervals relative to the activity of the reference material stored for the same duration. It is important to always include a sample from the reference temperature at every testing and sufficient material should be placed in storage for this purpose.

The reference temperature should be less than the intended temperature of storage. These studies yield no direct information on the stability of the material at the reference temperature, although some indirect information can be obtained from the magnitude of the assay response of the reference material over time.

Typical Testing Protocol
At first, samples from the higher temperature storage sites (e.g., 56°C) are removed periodically and tested against samples from the reference storage temperature until significant degradation (20–30%) is observed. It is important not to use samples from the lower temperature storage sites early in the study. Such samples probably would show no significant degradation and would waste irreplaceable material (Table 5).

Attention is then focused on the next highest temperature storage sites (e.g., 45°C and 37°C). These are removed periodically and tested until significant degradation (15–25%) is observed. The study proceeds for a suitable period until degradation (5–10%) is observed at the lower storage temperatures (e.g., 20°C and 4°C). Subsequently, samples are assayed from all the available storage temperatures against samples from the reference storage temperature.

The frequency of testing will depend on the rate of degradation observed at the higher storage temperatures. Initially, in the absence of any prior knowledge of the degradation rate, a suggested protocol is one month, two months, three months, six months, and one year. On removal from storage

TABLE 5 Typical Testing Protocol for an Accelerated Thermal Degradation Study

Samples removed from storage site at time	Sample removed from given storage temperature					
	Reference temperature	4°C	20°C	37°C	45°C	56°C
1 mo	✓			✓	✓	✓
2 mo	✓			✓	✓	✓
3 mo	✓		✓	✓	✓	(✓)
6 mo	✓		✓	✓	(✓)	
12 mo	✓	✓	✓	✓		
Subsequent times	✓	✓	✓	✓		

Source: From Ref. 35.

temperature, if the material cannot be assayed without delay, it should be stored at the reference temperature.

Models

It is usual to model the degradation of materials using the Arrhenius equation

$$\ln(K\{t\}) = A + B/T,$$

or the Heyring equation

$$\ln(K\{t\}) = A + B/T + \ln(T),$$

where

- $K\{t\}$ is the degradation rate at the absolute temperature T (Kelvin), relative to that at the reference temperature; and
- A and B are constants.

The Heyring equation is said to have a slightly stronger theoretical basis. Both models assume a unimolecular, single mechanism of degradation of first-order kinetics.

By determining the relationship between the degradation rate and temperature using samples stored at a range of higher temperatures of storage, the degradation rate at lower, that is, conventional, storage temperature can be predicted using these models. For the computational methods used refer to the work of Kirkwood et al. (36,37).

Confidence Limits of Prediction

The random error in potency estimates is assumed to be log-normally distributed. The upper 95% confidence limit is derived from the equation

$$K\{t\}'' = K\{t\} + (C.\text{se}Kt),$$

where

- $K\{t\}''$ is the upper 95% confidence limit of the degradation rate $K\{t\}$;
- C is a constant; and
- $\text{se}K\{t\}$ is the standard error of $K\{t\}$.

It is not possible to determine a value of C to be used in all cases. The value of C depends on the total statistical weight of the study (the reciprocal of the variance of $\log_{10}$ estimates of relative potency, from which the mean relative potency is obtained), that is,

- for a total statistical weight of 30,000 or more, and where at least 25% degradation has occurred at temperatures of 37°C or greater, a value for C of 4 (with 3 elevated temperatures) or of 3 (with 4 elevated temperatures) or of 2 (with 5 elevated temperatures) should be used, and
- for a total statistical weight of less than 30,000, a value for C of 5 should be used.

The precision of estimates of degradation rate from accelerated degradation tests is much improved by

- increasing the range of temperatures at which the samples are stored, even if the total statistical weight is kept constant;
- increasing the total statistical weight by having replicate samples tested;
- increasing the duration of the study.

A separate analysis of the measured response from the reference samples (e.g., if a radial diffusion assay, the ring diameter; if an ELISA assay, the absorbance; or if an HPLC assay, the peak height) will indicate the between-assay variation. The analysis of duplicates of the reference samples will indicate the within-assay variation.

Common Problems Seen with Accelerated Degradation Studies
Reduction in Degradation Rate with Time
This may be seen with freeze-dried preparations and is likely to be caused by an irreversible consumption of residual oxygen or moisture present as a contaminant of the atmosphere within the container or within the freeze-dried material itself.

Discontinuity in the Relationship Between Degradation Rate and Temperature
The analytical models assume that the rate of molecular diffusion (i.e., viscosity) in the stored samples does not significantly alter over the range of temperatures studied.

A significant change in viscosity can occur in the accelerated degradation testing of freeze-dried preparations containing a large proportion of substances that do not crystallize on freezing but form a glass, for example, some proteins and carbohydrates, used as bulking materials or cryoprotectants. If, during subsequent storage, the temperature of the freeze-dried material exceeds the T_g' of the glassy matrix, for example, by storage at some elevated degradation storage temperature, there is progressive collapse of the previously stable glass by the water released from the glass. Subject to the magnitude of the water released, the glass progressively collapses into a deformable rubber, a viscous syrup fluid, and finally to a mobile fluid.

Over the temperature range where this progressive transformation occurs, there will be a marked change in the rate of diffusion and the Arrhenius/Heyring equations are not valid above this point of discontinuity. This may be revealed by a discontinuity in the relationship between ln(relative potency) or degradation rate, and the reciprocal of absolute temperature.

For information, at temperatures near the glassy transformation point, the kinetics is said to be better described by William Landel Ferry equation

$$\log(k) = \frac{C1(T - T_g)}{C2 + (T - T_g)},$$

where

- k is the rate constant,
- T the absolute temperature (Kelvin),
- T_g the glass transition absolute temperature (Kelvin), and
- $C1$ and $C2$ are constants.

Additional Degradation Processes at Higher Temperatures

At higher storage temperatures, additional degradation processes can become significant that are irrelevant at lower and conventional temperatures of storage. These additional degradation processes can include Amadori products arising from reactions between protein and carbohydrates and often resulting in a discoloration of the material stored at elevated temperatures. If statistically significant, these additional processes can prevent the experimental data from being fitted to the degradation equation since there is no longer a linear relationship between ln(degradation rate) and the reciprocal of absolute temperature.

When not statistically significant, such additional processes can lead to an overestimation of the rate of degradation at conventional storage temperatures by using the combined degradation at the higher temperatures. This is particularly the case if the degradation study has not continued long enough for significant degradation to occur at the lower storage temperatures, when undue weight is placed on the data from the higher temperatures.

Temperature Effects on the Integrity of the Container

If vials are used as the container of the material, high or low temperatures may affect the physical properties of the stopper and thus the integrity of the seal it makes with the neck of the container. This should be investigated prior to their use (6,7).

Example of Accelerated Degradation Studies on a Biological Standard

The use of accelerated degradation studies is illustrated in Figure 17 by reference to factor VIII concentrate stored at $-20°C$: potency was tested using the European Pharmacopoeia chromogenic assay.

Data was fitted using the Arrhenius equation

$$\ln(K_{\{T\}}) = A + B/T,$$

where

$K_{\{T\}}$ is the degradation rate at the absolute temperature T, relative to the rate at the reference temperature, and

A and B are constants.

Maximum likelihood estimates of the constants A and B are as follows:

A	24.24
Asymptotic error of A	1.61
B	−7244.98
Asymptotic error of B	502.27
Asymptotic covariance of A and B	−810.90

The Chi-square test statistic for the predicted versus observed activity remaining at the various time points is 4.41 for 5 degrees of freedom. At the 5% level, the predicted remaining activities are not significantly different from the observed remaining activities.

The predicted potency loss is shown in Figure 17, indicating that at a storage temperature of $-20°C$, the standard had good stability with low degradation rate.

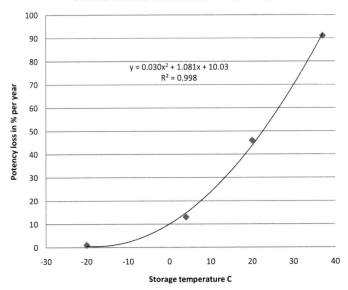

Stability profile of a Factor VIII concentrate reference material

$$y = 0.030x^2 + 1.081x + 10.03$$
$$R^2 = 0.998$$

Potency loss in % per year

Storage temperature C

FIGURE 17   Predicted stability at $-20°C$ for factor VIII concentrate reference material based on accelerated elevated temperature degradation studies.

RECENT DEVELOPMENTS
Infectious Filling Capability
NIBSC has a new reference materials production unit, the Center for Biological Reference Materials, which is fully operational. In particular, this facility offers capabilities for processing both infectious liquid and freeze-dried reference materials, in addition to improved facilities for the preparation of noninfectious standards.

The capability to freeze-dry up to 10 L of containment level 3 materials comprises a freeze-drier interfaced with a negatively pressured pharmaceutical-grade isolator containing vial-filling/capping equipment. The freeze-dryer is steam sterilizable and the isolator decontaminated by formaldehyde gassing. This should significantly increase the availability and ease of supply of these key reference materials that currently are available only as liquid preparations. Product can be filled on the automated vial-filling line and then is transferred to the dryer by an operator, working in a half-suit, which is an integral part of the isolator (Fig. 18). Following lyophilization, vials can be removed while maintaining containment through the use of transfer mobiles and reinserted and screw capped prior to the product being decontaminated ready for removal and storage.

Trends in Reference Materials
As predicted in the second edition, there has been over the past few years an increase in the range and variety of reference materials processed by NIBSC. In

FIGURE 18 System for infectious filling of vials within negative pressure isolator/freeze dryer; note half-suited operator transfers filled vials stacked in trays into the "pizza door" of the freeze-dryer at rear of isolator.

particular, a number of panels of genetic reference materials have been produced to some common genetic conditions either as International Standards (38) or as CE-marked reagents, this illustrates the potential for reference materials to provide improved quality assurance in the genetic diagnostic testing market. Also, the use of intact nonviable cells as a matrix for the presentation of a reference material has been demonstrated for the leukemia-associated BCR-ABL gene (39), and studies using reference materials based on mixtures of cells with different phenotypes are progressing well.

Polymerase chain reaction (PCR) has become a dominant assay format and the number of reference standards required for this area continues to rise. Some of the challenges that have arisen from these projects have caused us to improve our analytical testing methods and revise the formats in which materials are presented, as the quantities of material required and the ways in which they are used change. To continue to lead in the areas of reference materials will require continually improving lyophilization techniques with which to yield robust stable reference material as well as flexibility and vision with regards to the demands of reference materials to serve the new challenges of the 21st century medicine.

ACKNOWLEDGMENTS

We thank all the staff of the Standards Processing Division and Standardization Science development team for their expertise and commitment that have made the work described possible, and especially to Dr Peter Phillips (who has retired since the second edition was published) for his contributions to the earlier edition of this chapter, in particular for the section on accelerated degradation studies that has been retained unaltered. We also thank numerous NIBSC colleagues for permission to use data from their reference material studies.

REFERENCES

1. Campbell PJ. International biological standards and reference preparations. II. Procedure used for the production of biological standards and reference preparations. J Biol Stand 1974; 2:255–267.
2. Phillips P. The preparation of international biological standards. Fresenius J Anal Chem 1998; 360:473–475.
3. Matejtschuk P, Andersen M, Phillips P. Freeze drying of biological standards. In: Rey L, May JC, eds. Freeze Drying/Lyophilization of Pharmaceutical and Biological Products. Drugs and the Pharmaceutical Sciences. Vol. 137, 2nd ed. New York: Marcel Dekker, 2004:385–424.
4. Matejtschuk P, Lyophilization of biological standards. Cryo Letters 2005; 26(4): 223–230.
5. WHO. Recommendations for the Preparation, Characterisation and Establishment of International and other Biological Reference Standards. WHO Tech Report Series 932 Annex II, 2006.
6. Ford AW, Dawson PJ. Effect of type of container, storage temperature and humidity on the biological activity of freeze-dried alkaline phosphatase. Biologicals 1994; 22:191–197.
7. Matejtschuk P, Rafiq S, Johnes S, et al. A comparison of vials with ampoules for the storage of biological reference materials. Biologicals 2005; 33:63–70.
8. Phillips PK, Dawson PJ, Delderfield A. The use of DIN glass ampoules to freeze-dry biological materials with a low residual moisture and oxygen content. Biologicals 199119:219–221.
9. May JC, Wheeler RM, Elz N, et al. Measurement of final container residual moisture in freeze dried biological products. Dev Biol Standard 1992; 74:153–164.
10. Matejtschuk P, Malik KP, Duru C, et al. Application of analytical methods and model systems in the lyophilization of complex biomolecules in New Ventures in Freeze Drying. International Institute of Refrigeration, Proceedings no. 2007-3, Strasbourg. ISBN 978-913149-60-1.
11. Hubbard A, Bevan S, Matejtschuk P. Impact of residual moisture and formulation on factor VIII and factor V recovery in lyophilized plasma reference materials. Anal Bioanal Chem 2007; 387(7):2503–2507.
12. Matejtschuk P, Andersen M, Fleck RA, et al. Investigations into anomalies arising in the preparation of biological standards in ampoules and steps for process optimization. Am Pharm Rev 2003; 6(1):28–40.
13. Scott KL, Lecak JP, Acker JP. Biopreservation of red blood cells: past, present, and future. Transfus Med Rev 2005; 19(2):127–142.
14. Bristow AF, Dunn D, Tarelli E. Additives to biological substances IV-Lyophilization conditions in the preparation of International Standards: an analysis by high performance liquid chromatography of the effects of secondary desiccation. J Biol Stand 1988; 16:55–61.
15. Cammack KA, Adams GDJ. Formulation and storage. Animal Cell Biotechnol 1985; 2:252–288.
16. Hsu CC, Ward CA, Pearlman R, et al. Determining the optimum residual moisture in lyophilised protein pharmaceuticals. Dev Biol Standard 1992; 74:255–271.

17. Jiang S, Nail SL. Effect of processing conditions on recovery of protein activity after freezing and freeze drying. Eur J Pharm Biopharm 1998; 45:245–257.
18. Carpenter JF, Chang BS, Garzon-Rodrigues W, et al. Rational design of stable lyophilized product formulation theory and practice. Pharm Res 1997; 14:969–975.
19. Ford AW, Dawson PJ. The Effect of carbohydrate additives in the freeze drying of alkaline phosphatase. J Pharm Pharmacol 1993; 45:86–93.
20. Tarelli E, Mire Sluis A, Tivnann HA, et al. Recombinant human albumin as a stabilizer for biological materials and for the preparation of international reference reagents. Biologicals 1998; 26:331–346.
21. Wang W. Lyophilization and development of solid protein pharmaceuticals. Int J Pharm 2000; 203:1–60.
22. Martini A, Kume S, Rivaldelo M, et al. Use of sub ambient differential scanning calorimetry to monitor the frozen state behaviour of blends of excipients for freeze drying. PDA J Pharm Sci Technol 1997; 51:62–67.
23. Kett V. Modulated temperature differential scanning calorimetry and its application to freeze drying. Eur J Parenter Sci 2001; 6:95–99.
24. Meister E, Giesler H. A significant comparison between collapse and glass transition temperatures. Eur Pharm Rev 2008; 13(5):73–79.
25. Jameel F, Amsberry KC, Pikal MJ. Freeze dried preparations of some oligonucleotides. Pharm Dev Technol 2001; 65:151–157.
26. Jennings TA. Lyophilization an Introduction to Basic Principles. Englewood; CO: Interpharm Press, 1999.
27. Lachman L, Deluca PP, Withnell R. Lyophilisation of pharmaceuticals. II: High sensitivity resistance bridge for low conductance measurements at eutectic temperatures. J Pharm Sci 1965; 54:1342–1347.
28. Mackenzie AP. Physicochemical basis for the freeze drying process. Dev Biol Stand 1977; 36:51–67.
29. Nail SL, Mer LM, Proffitt C, et al. An improved microscope stage for direct observation of freezing and freeze drying. Pharm Res 1994; 11:1098–1100.
30. Matejtschuk P, Fleming J, Fleck RA, Use of a range of analytical techniques in support of the freeze drying of biologicals. Am Pharm Rev 2004; 7(3):45–48.
31. Savage M, Torres J, Franks L et al. Determination of adequate moisture content for efficient dry-heat viral inactivation in lyophilised factor VIII by loss on drying and by near infra red spectroscopy. Biologicals 1998; 26(2):119–124.
32. Lin TP, Hsu CC. Determination of residual moisture in lyophilized protein pharmaceuticals using a rapid and non-invasive method: near infrared spectroscopy. PDA J Pharm Sci Technol 2002; 56(4):196–205.
33. Kamat MS, Lodder RA, DeLuca PP. Near infra red spectroscopic determination of residual moisture in lyophilized sucrose through intact glass vials. Pharm Res 1989; 6(11):961–965.
34. Matejtschuk P. Preparing reference materials for biogenerics. Pharma Technol (Cavendish Group) 2009; (19):28–32.
35. Matejtschuk P, Phillips PK. Product stability and accelerated degradation studies. In: Stacey G, Davis J, eds. Biological Medicines from Cell Culture. Chichester, United Kingdom: Wiley Press, 2007:503–522.
36. Kirkwood TBL. Predicted stability of biological standards and products. Biometrics 1977; 33:736–742.
37. Kirkwood TBL. Design and analysis of accelerated degradation tests for the stability of biological standards III. Principles of design. J Biol Stand 1984; 12:215–224.
38. Gray E, Hawkins JR, Morrison M, et al. Establishment of the 1st international genetic reference panel for factor V leiden (G1691A), human gDNA. Thromb Haemost 2006; 96:215.
39. Saldanha J, Silvy M, Beaufils N, et al. Characterisation of a reference material for BCR-ABL (M-BCR) mRNA quantitation by real-time amplification assays: towards new standards for gene expression measurement. Leukemia 2007; 21:1481–1487.

Molecular Mobility of Freeze-Dried Formulations as Determined by NMR Relaxation Times, and Its Effect on Storage Stability

Sumie Yoshioka
University of Connecticut, Storrs, Connecticut, U.S.A.

Yukio Aso
National Institute of Health Sciences, Tokyo, Japan

INTRODUCTION

Freeze-drying is a useful method for preparing dosage forms of thermally unstable pharmaceuticals without the deleterious effect of heat. The method can also provide a dry product of pharmaceuticals with longer shelf life than solutions or suspensions. Although glassy-state formulations obtained by freeze-drying generally exhibit sufficient storage stability for pharmaceuticals, degradation during storage has been observed in various freeze-dried formulations.

Many studies have demonstrated that storage stability of freeze-dried formulations is related to molecular mobility (1–15). Chemical and physical degradation of small molecules and proteins is enhanced by an increase in molecular mobility associated with moisture sorption. Additives that decrease the molecular mobility of formulations are often effective for the stabilization of the formulation.

This chapter describes molecular mobility of freeze-dried formulations as determined by NMR relaxation times and discusses the relationship between storage stability and NMR-determined molecular mobility.

MOLECULAR MOBILITY AS DETERMINED BY NMR RELAXATION TIMES

NMR has been used to determine molecular mobility of freeze-dried formulations (16–20), along with other techniques like calorimetry, dielectric relaxation spectrometry, and dynamic mechanical measurement (21–25). NMR can determine the mobility of atoms in pharmaceutical molecules such as ^{1}H, ^{2}H, ^{13}C, ^{15}N, ^{17}O, and ^{19}F. To determine the mobility of a specific site in the molecule, high-resolution solid-state NMR with high sensitivity is necessary. Especially, high sensitivity is inevitable for ^{13}C and ^{15}N, which have low natural abundance. In contrast, low-frequency solid-state NMR, which is easier to operate than high-resolution NMR, can be used to determine the mobility of ^{1}H and ^{19}F, which have high natural abundance. This section addresses the molecular mobility of ^{1}H, ^{13}C, and ^{19}F measured by each of low-frequency solid-state NMR and high-resolution solid-state NMR.

Molecular Mobility as Determined by Low-Frequency NMR
Spin-Spin Relaxation Time of Proton
Spin-spin relaxation time (T_2) of protons present in freeze-dried formulations can be determined from free induction decay (FID). Figure 1 shows the FID of

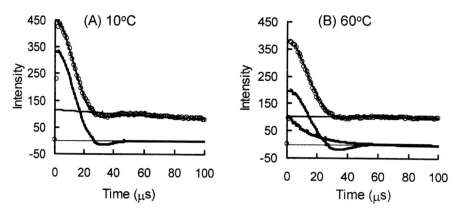

FIGURE 1 Free induction decay of proton in freeze-dried γ-globulin formulation containing dextran at 10°C (**A**) and 60°C (**B**) at 60% RH. *Abbreviation*: RH, relative humidity.

proton in freeze-dried formulation containing γ-globulin as a model protein drug and dextran (molecular weight of 10 kDa) as a polymer excipient, measured by a low-frequency NMR using "solid echo" in the detection stage (26). The FID shows two relaxation processes at 10°C and 60% relative humidity (RH) (Fig. 1A); a slower decay described by the Lorentzian equation (equation 1) and a faster decay described by a Gaussian-type equation (the Abragam equation, equation 2 with a constant c of 0.12). This slower decay is attributed to protons with higher mobility, that is, water protons, and the faster decay is attributed to protons with lower mobility, that is, protons of γ-globulin and dextran. The contribution of protein protons to the FID is not significant because the content of protein was 50 times less than that of dextran. Therefore, the Abragam decay can be considered to be due to dextran protons. The observed FID is describable by an equation representing the sum of the Abragam and Lorentzian equations (equation 3). The T_2 of water protons can be calculated from the FID signals at the latter stage. Subsequently, the T_2 of dextran proton with lower mobility can be calculated from the FID signals at the former stage by inserting the calculated T_2 of water proton into equation (3).

$$F(t) = A \exp\left(-\frac{t}{T_{2(\mathrm{hm})}}\right) \tag{1}$$

$$F(t) = A \exp\left(-\frac{t^2}{2 T_{2(\mathrm{lm})}^{\,2}}\right) \frac{\sin(ct)}{ct} \tag{2}$$

$$F(t) = (1 - P_{\mathrm{hm}}) \exp\left(-\frac{t^2}{2 T_{2(\mathrm{lm})}^{\,2}}\right) \frac{\sin(ct)}{ct} + P_{\mathrm{hm}} \exp\left(-\frac{t}{T_{2(\mathrm{hm})}}\right) \tag{3}$$

where $T_{2(\mathrm{hm})}$ and $T_{2(\mathrm{lm})}$ are the spin-spin relaxation times of protons with higher mobility and lower mobility, respectively. P_{hm} is the proportion of protons with higher mobility.

As shown in Figure 1B, the decay due to dextran protons at 60°C cannot be described by a single Abragam equation, and therefore requires further solving by the Lorentzian equation. This indicates that at 60°C, the dextran protons in the freeze-dried formulation exhibit a slower relaxation process due to higher mobility in addition to a faster relaxation process due to lower mobility. In other words, dextran protons having higher mobility exist in the formulation at 60°C, in addition to solid-like dextran protons with lower mobility. Thus, the FID at 60°C is described by an equation representing the sum of the Abragam and Lorentzian equations for dextran protons as well as the Lorentzian equation for water protons. The proportion of dextran protons having higher mobility can be calculated by fitting FID signals into equation (3) after subtracting signals due to water protons.

The temperature at which the spin-spin relaxation of proton begins to involve the Lorentzian relaxation process due to polymer protons having higher mobility in addition to the Gaussian-type relaxation process due to polymer protons having low mobility is considered to be a glass/rubber transition temperature. Basically, this is a critical temperature of molecular mobility as determined by NMR relaxation measurements and is analogous to glass transition temperature (T_g) determined by differential scanning calorimetry (DSC). This critical mobility temperature is referred to as T_{mc}. The T_{mc} of formulations containing polymer excipients increases as the molecular weight of the polymers increases. The T_{mc} of a formulation containing dextran with a molecular weight of 510 kDa is 5°C higher than that for dextran with a molecular weight of 40 kDa. Similarly, the T_{mc} of molecular weight 120 kDa poly(vinyl alcohol) (PVA) formulation is approximately 5°C higher than that of molecular weight 18 kDa PVA formulation (27). In contrast, the T_2 of water proton calculated by the Lorentzian equation is not significantly affected by the molecular weight of dextran (26). This indicates that the mobility of water molecules in the formulation is determined by the interaction between the glucose unit and water.

Figure 2 shows the effect of water content on the T_{mc} and T_g of freeze-dried γ-globulin formulations containing dextran, polyvinylpyrrolidone (PVP) and α,β-poly(N-hydroxyethyl)-L-aspartamide (PHEA) (28). T_{mc} shifts to a lower temperature as water content increases, indicating that the molecular

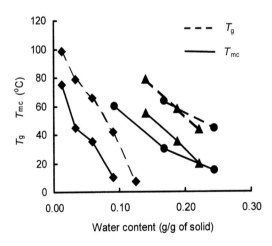

FIGURE 2 T_{mc} and glass transition temperature of freeze-dried γ-globulin formulations containing PHEA (■), dextran (▲), and PVP (●). *Abbreviations*: PHEA, α,β-poly(N-hydroxyethyl)-L-aspartamide; PVP, polyvinylpyrrolidone.

mobility of polymer excipients in the formulation is increased by the plasticizing effect of water. Decrease in T_{mc} with increasing water content is also observed for other freeze-dried formulations containing PVA, methylcellulose (MC), hydroxypropylmethylcellulose, and carboxymethylcellulose sodium salt.

The T_{mc} of freeze-dried formulations containing polymer excipients is generally observed at a temperature of 20°C to 30°C lower than the T_g determined by DSC (Fig. 2). This indicates that these formulations have highly mobile protons even at temperatures below the T_g. T_{mc} can be considered to be the temperature at which a certain region of the molecule, such as terminal units of polymer chains, begins to have greater mobility. T_{mc} is a glass/rubber transition temperature determined by spin-spin relaxation measurements, which can detect local changes in molecular mobility more sensitively than T_g determined by DSC. The T_g of freeze-dried formulations containing polymer excipients with moisture is often difficult to determine because a change in heat capacity at T_g may be overlapped by the peaks of water evaporation and accompanying relaxation processes. Furthermore, certain formulations, especially freeze-dried protein formulations, reveal unclear changes in heat capacity, causing a difficulty in determination of T_g. In such cases, T_{mc} determined by spin-spin relaxation measurement can be a useful measure of T_g.

Laboratory and Rotating Frame Spin-Lattice
Relaxation Times of Proton
Along with T_2, spin-lattice relaxation times in laboratory and rotating frames (T_1 and $T_{1\rho}$) can also be used to measure the molecular mobility of freeze-dried formulations. The T_1 and $T_{1\rho}$ of proton reflect the correlation time (τ_c) of the rotational motion of proton. The relationship between τ_c and T_1 can be described as follows:

$$\frac{1}{T_1} = \frac{3}{10}\gamma^4 \left(\frac{h}{2\pi}\right)^2 r^{-6} \left(\frac{\tau_c}{1+\omega_0^2\tau_c^2} + \frac{4\tau_c}{1+4\omega_0^2\tau_c^2}\right) \tag{4}$$

where γ is the gyromagnetic ratio of ^1H, h is Planck's constant, ω_0 is the ^1H resonance frequencies, and r is the H-H distance. In contrast, $T_{1\rho}$ can be related to τ_c according to equation (5).

$$\frac{1}{T_{1\rho}} = \frac{A\tau_c}{1+4\omega_1^2\tau_c^2} \tag{5}$$

where ω_1 is the frequency of precession generated by the spin locking field and A is a constant.

Figure 3 shows the relationship between τ_c and T_1 (or $T_{1\rho}$) of proton. When τ_c exhibits an Arrhenius behavior, T_1 (or $T_{1\rho}$) exhibits a similar V-shaped pattern with a minimum as a function of temperature. In the temperature range below the minimum (slow motional regime), T_1 (or $T_{1\rho}$) increases in a linear fashion with decreasing temperature (i.e., with decreasing mobility). In the temperature range above the minimum (fast motional regime), in contrast, T_1 (or $T_{1\rho}$) decreases with decreasing mobility associated with decreasing temperature. T_1 minimum is observed at a higher temperature than $T_{1\rho}$ minimum, such that T_1 sensitively reflects faster motion than $T_{1\rho}$ does.

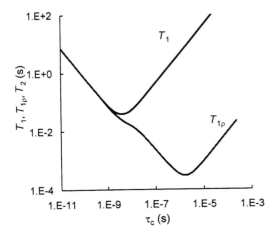

FIGURE 3 Relationship between NMR relaxation times and correlation time.

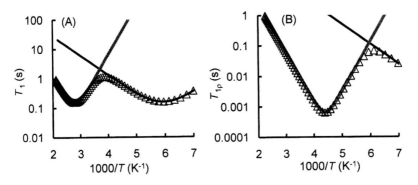

FIGURE 4 Temperature dependence of T_1 (**A**) and $T_{1\rho}$ (**B**) of proton with two correlation times.

When there are multiple protons having different τ_c values in the molecule, spin diffusion occurs between protons located within a short distance, and gives a single T_1 (or $T_{1\rho}$) value. Therefore, the value of T_1 (or $T_{1\rho}$) is determined mainly by a proton that shows the shortest relaxation time. As described by equation (6), relaxation rate (the reciprocal of T_1 and $T_{1\rho}$) can be calculated as the sum of the relaxation rates attributed to each τ_c, when there are two protons having different τ_c values (τ_{c1} and τ_{c2}) in the system. Thus, the observed values of T_1 and $T_{1\rho}$ closely approximate the smaller relaxation times of the two loci, as shown in Figure 4.

$$\frac{1}{T_{1(obs)}} = \frac{P_1}{T_{1(\tau_{c1})}} + \frac{(1 - P_1)}{T_{1(\tau_{c2})}} \tag{6}$$

where P_1 is the fraction of proton having τ_{c1}.

Figure 5 shows the temperature dependence of $T_{1\rho}$ observed for freeze-dried γ-globulin formulations containing dextran, prepared using D_2O (29). Since the ratio of γ-globulin to dextran is 1:50, the calculated $T_{1\rho}$ represents the $T_{1\rho}$ of unexchangeable protons of dextran (5 methine protons and 2 methylene protons in a repeating unit). The temperature dependence of $T_{1\rho}$ exhibits a

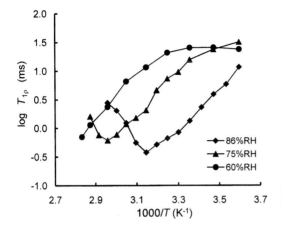

FIGURE 5 Spin-lattice relaxation time of dextran proton in freeze-dried γ-globulin containing dextran.

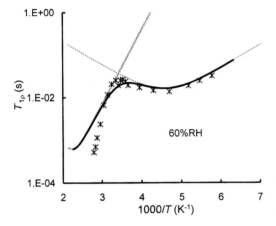

FIGURE 6 Spin-lattice relaxation time of dextran proton in freeze-dried γ-globulin containing dextran in a wider temperature range.

minimum at relatively high humidities (75% and 86% RH). The temperature of the $T_{1\rho}$ minimum shifts to higher temperature as humidity decreases. At 60% RH, minimum is not observed in the temperature range up to 80°C (it may be observed around 90°C), but another minimum is observed at approximately −60°C, as shown in Figure 6. This minimum shifts to approximately 90°C in the dry state. These findings indicate that proton has two different correlation times due to different motions.

The temperature dependence for the $T_{1\rho}$ of proton observed at 60% RH (Fig. 6) can be described by two correlation times (τ_{c1} and τ_{c2}) with an activation energy of 8.0 and 2.5 kcal/mol, and with a pre-exponential factor of 2×10^{-10} and 5×10^{-9} seconds, respectively, at temperatures lower than 35°C ($1000/T$ of 3). The motion represented by τ_{c1} and τ_{c2} may be attributed to methine and methylene protons, respectively, on the basis of the values of activation energy. An activation energy of the same order as the calculated value of τ_{c2} has been reported for the methylene group of amorphous polyethylene (3.72 kcal/mol) (30). $T_{1\rho}$ reflects the motion of methine groups at temperatures between 35°C and 10°C ($1000/T$ of 3 and 3.5), but reflects the motion of methylene groups at temperatures lower than 10°C. The observed $T_{1\rho}$ of methine groups diverges from the values calculated from τ_{c1} at temperatures above 35°C, indicating that the motion of methine groups has greater activation energy at temperature above 35°C. The temperature at which a break is observed in the temperature dependence is coincident with T_{mc}, described in the section of proton T_2.

Molecular Mobility as Determined by High-Resolution NMR
Laboratory and Rotating Frame Spin-Lattice Relaxation Times of Carbon
Figure 7 shows the typical spectra of freeze-dried γ-globulin formulation containing dextran, freeze-dried γ-globulin, and freeze-dried dextran, measured by high-resolution ^{13}C solid-state NMR (31). Peaks at 70 and 180 ppm are assigned to the dextran methine carbon and γ-globulin carbonyl carbon, respectively. The T_1 of each carbon, calculated from the signal decay, decreases with increasing temperature, indicating that relaxation occurs in the slow motional regime. The τ_c of dextran methine carbon then can be calculated from the observed T_1 according to equation (7), if the dipole-dipole interaction between carbon and proton is predominant in the relaxation process, and if the relaxation time can be expressed by a single τ_c.

$$\frac{1}{T_1} = \frac{1}{10}\gamma_C^2\gamma_H^2\left(\frac{h}{2\pi}\right)^2 r_{C-H}^{-6}$$

$$\times \left[\frac{\tau_c}{1 + (\omega_c - \omega_H)^2\tau_c^2} + \frac{3\tau_c}{1 + \omega_c^2\tau_c^2} + \frac{6\tau_c}{1 + (\omega_C + \omega_H)^2\tau_c^2}\right] \quad (7)$$

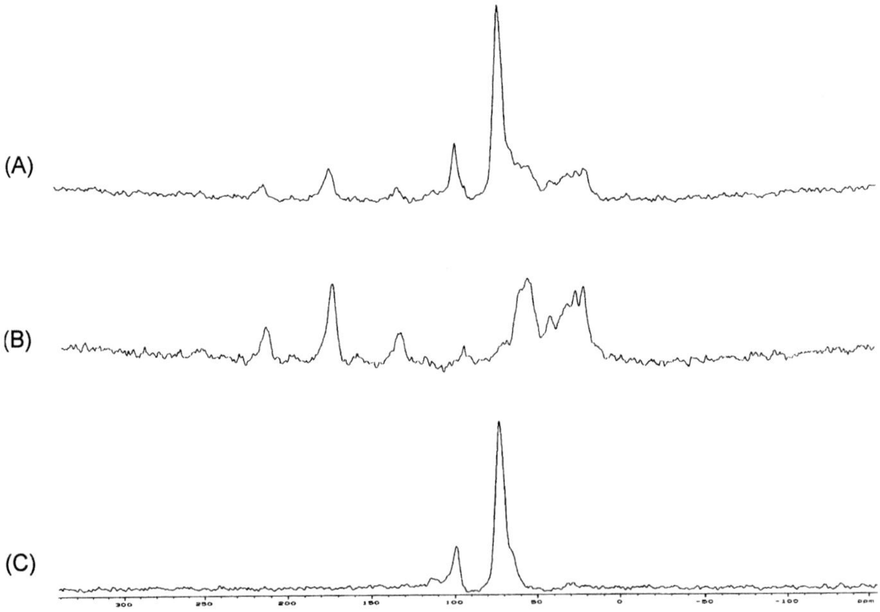

FIGURE 7 ^{13}C-NMR spectra of freeze-dried γ-globulin formulations containing dextran **(A)**, freeze-dried γ-globulin **(B)**, and freeze-dried dextran **(C)**.

where γ_C and γ_H are the gyromagnetic ratios of ^{13}C and ^1H, respectively, h is the Planck's constant, and ω_C and ω_H are the ^{13}C and ^1H resonance frequencies, respectively. r_{C-H} is the C-H distance and a value of 1.2Å was used for the calculation.

In contrast, the τ_c of γ-globulin carbonyl carbon can be calculated from the observed T_1 using equation (8) if the relaxation due to chemical shift anisotropy is predominant, and if the relaxation time in the slow motional regime can be expressed by a single τ_c.

$$\frac{1}{T_1} = \frac{6}{40}\gamma_C^2 B_0^2 \delta_Z^2 \left(1 + \frac{\eta^2}{3}\right)\left(\frac{2\tau_c}{1 + \omega_0^2\tau_c^2}\right) \tag{8}$$

where B_0, δ_Z, and η are the static field, the chemical shift anisotropy, and the asymmetric parameter, respectively. δ_Z and η are defined in terms of three principal components (δ_{11}, δ_{22}, and δ_{33}).

$$\delta_Z = \delta_{11} - \delta_0, \quad \eta = \frac{\delta_{22} - \delta_{33}}{\delta_{11} - \delta_0} \quad \text{when } |\delta_{11} - \delta_0| \geq |\delta_{33} - \delta_0|$$

$$\delta_Z = \delta_{33} - \delta_0, \quad \eta = \frac{\delta_{22} - \delta_{11}}{\delta_{33} - \delta_0} \quad \text{when } |\delta_{11} - \delta_0| < |\delta_{33} - \delta_0| \tag{9}$$

$$\text{where} \quad \delta_0 = \frac{\delta_{11} + \delta_{22} + \delta_{33}}{3}$$

Figure 8 shows the temperature dependence of τ_c determined for dextran methine carbon in freeze-dried dextran and freeze-dried γ-globulin/dextran formulation. For both systems, the τ_c of dextran methine carbon exhibits a significant change in the temperature dependence around the T_{mc} (35°C), the critical temperature of molecular mobility as determined by the spin-spin relaxation of proton. The greater decrease in the τ_c of dextran methine carbon at temperatures above the T_{mc} indicates that the motion of methine groups is significantly enhanced by increased global motion in addition to local segmental

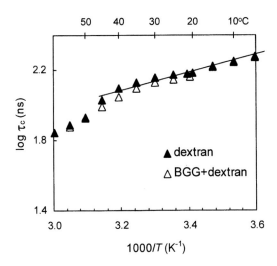

FIGURE 8 Temperature dependence of correlation time for dextran methine carbon in freeze-dried γ-globulin containing dextran 40 kDa.

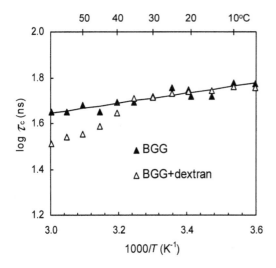

FIGURE 9 Temperature dependence of correlation time for γ-globulin carbonyl carbon in freeze-dried γ-globulin containing dextran 40 kDa.

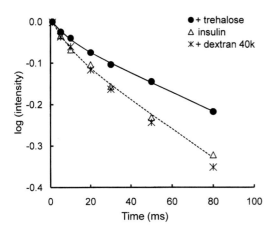

FIGURE 10 Time course of spin-lattice relaxation for insulin carbonyl carbon in freeze-dried insulin, insulin-dextran, and insulin-trehalose systems at 25°C and 12% RH.

motion. This interpretation is supported by the greater decrease in the τ_c of dextran methine proton at temperatures above the T_{mc}, as described in the previous section on proton $T_{1\rho}$ (Fig. 6).

The τ_c of the carbonyl carbon of freeze-dried γ-globulin exhibits linear Arrhenius-like temperature dependence as shown in Figure 9. In contrast, the τ_c of the carbonyl carbon of γ-globulin freeze-dried with dextran exhibits a change in the temperature dependence around 35°C, similar to that observed for the τ_c of dextran methine carbon. This indicates that at temperatures above T_{mc}, the molecular motion of γ-globulin is coupled with that of dextran, even though dextran is well known to cause phase separation with proteins.

Along with the T_1 of carbon, the $T_{1\rho}$ of carbon is useful as a measure of molecular mobility. Figure 10 shows the time courses of spin-lattice relaxation determined for the carbonyl carbon of insulin freeze-dried with dextran or trehalose, compared with that for insulin alone (32). Spin-lattice relaxation is not affected by dextran, but it is significantly retarded by trehalose. The $T_{1\rho}$ of insulin carbonyl carbon in the insulin-trehalose system is longer than in the insulin-dextran system, indicating that the molecular mobility of insulin is decreased by trehalose, since longer $T_{1\rho}$ indicates slower motion in the slow motional regime (Fig. 3).

Retardation of spin-lattice relaxation of protein carbonyl carbon brought about by the addition of sugars is also observed for the carbonyl carbon of β-galactosidase freeze-dried with sucrose, trehalose, or stachyose, as shown in Figure 11 (33). The molecular mobility of the protein is most effectively decreased by sucrose.

The T_1 and $T_{1\rho}$ of protein carbonyl carbon described above represent the average of T_1 and $T_{1\rho}$ for multiple carbonyl carbons present in the protein molecule. More detailed site-specific analysis becomes possible by the ^{13}C-labeling of an amino acid at a specific site of interest.

Laboratory and Rotating Frame Spin-Lattice Relaxation Times of Fluorine

^{19}F-NMR has high sensitivity and specificity and has been used to determine the molecular mobility of ^{19}F-labeled proteins (34) as well as small molecules containing ^{19}F (35). Figure 12 shows the temperature dependence of T_1 and $T_{1\rho}$ of

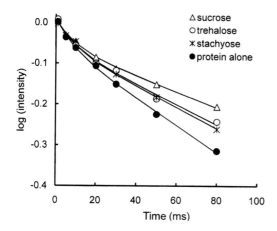

FIGURE 11 Time course of spin-lattice relaxation for carbonyl carbon of β-galactosidase freeze-dried with sucrose, trehalose, or stachyose at 25°C and 12% RH.

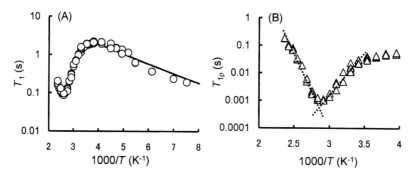

FIGURE 12 Temperature dependence of T_1 (**A**) and $T_{1\rho}$ (**B**) of flufenamic acid ^{19}F in solid dispersions with PVP. *Abbreviation:* PVP, polyvinylpyrrolidone.

^{19}F in amorphous flufenamic acid dispersed with PVP (drug-PVP, 7:3). The T_1 of ^{19}F shows a minimum at approximately 110°C as well as a maximum at approximately −10°C, as shown in Figure 12A. The temperature dependence can be described by assuming that the ^{19}F atom has two Arrhenius-type motions with an equivalent contribution to the process of T_1 but with different activation energies of 50 and 5 kJ/mol, as shown by the solid line in Figure 12A. The motion with greater activation energy may be attributed to β-relaxation and the other one to the rotation of the trifluoromethyl group, which is faster than β-relaxation.

The $T_{1\rho}$ of ^{19}F in flufenamic acid shows a minimum at approximately 60°C, as shown in Figure 12B. The temperature coefficient of $T_{1\rho}$ is about 50 kJ/mol at temperatures below 60°C, suggesting that $T_{1\rho}$ is determined by β-relaxation in this temperature range. In contrast, a greater temperature coefficient of $T_{1\rho}$ is observed at temperatures above 60°C, suggesting that $T_{1\rho}$ is determined by a larger-scale motion than β-relaxation. Thus, motion reflected by the T_1 and $T_{1\rho}$ of ^{19}F varies with temperature.

RELATIONSHIP BETWEEN STORAGE STABILITY AND MOLECULAR MOBILITY AS DETERMINED BY NMR RELAXATION TIMES

The storage stability of pharmaceuticals in the solid state is largely affected by molecular mobility. Changes in the molecular mobility of amorphous pharmaceuticals at T_g bring about changes in the temperature dependence of chemical and physical degradation rates. Coupling between chemical degradation and molecular mobility has been reported for several drugs of small molecular weight in freeze-dried formulation (1–5); hydrolysis of aspirin in freeze-dried hydroxypropyl-β-cyclodextrin/aspirin complex (1), hydrolysis of peptides in freeze-dried formulations containing cross-linked sucrose polymer (3), and deamidation of peptide in freeze-dried formulations containing poly(vinyl-pyrrolidone) (4,5). In addition to chemical instability, physical instability of pharmaceuticals, such as crystallization of amorphous compounds, is related to molecular mobility (36–40). Crystallization of freeze-dried sucrose is inhibited in the presence of PVP at a level as low as 10% due to the decreased molecular mobility of sucrose as indicated by the decreased enthalpy relaxation of the mixtures relative to sucrose alone (41).

Coupling between degradation and molecular mobility has also been reported for degradation of protein pharmaceuticals (6–15). An excellent correlation has been demonstrated between T_g and chemical degradation of freeze-dried antibody-vinca conjugate (8).

This section discusses the relationship between the storage stability of freeze-dried formulations and the molecular mobility as determined by NMR relaxation times, described in the previous section. Focus is placed on the degradation of small molecular weight drugs via bimolecular reaction and protein aggregation in the freeze-dried formulations.

Correlations Between Storage Stability and Structural Relaxation as Reflected by NMR Relaxation Times

NMR relaxation times are useful to determine fast dynamics of freeze-dried formulations, whereas structural relaxation of freeze-dried formulations, which

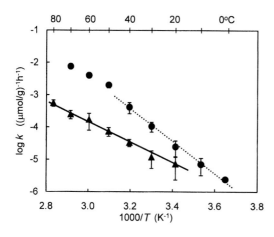

FIGURE 13 Temperature dependence of acetyl transfer between aspirin and sulfadiazine in freeze-dried formulations containing dextran at 12% (▲) and 60% RH (●).

is global mobility rapidly enhanced at temperatures above T_g, is also reflected by NMR relaxation times. Correlations between storage stability and structural relaxation as reflected by NMR relaxation times have been demonstrated for various freeze-dried formulations. Figure 13 shows the temperature dependence of the rate constant for acetyl transfer between aspirin and sulfadiazine in freeze-dried formulations containing dextran. Acetyl transfer is a bimolecular reaction in which the translational diffusion of reactant molecules becomes rate determining when molecular mobility is limited in the solid state (42). The rate constant of acetyl transfer (k_T) and the pseudo rate constant of hydrolysis ($k_{H,pseudo}$) that occurs in parallel with acetyl transfer in the presence of water are described by following equations.

$$\frac{d[SD]}{dt} = -k_T[SD][ASA] \tag{10}$$

$$\frac{d[ASA]}{dt} = -k_T[SD][ASA] - k_{H,pseudo}[ASA] \tag{11}$$

The temperature dependence of acetyl transfer at 60% RH exhibits a distinct break at approximately 40°C, although it is linear at 12% RH. The temperature of this distinct break observed at 60% RH is coincident with the T_{mc} as determined by the spin-spin relaxation measurements described in section "Molecular Mobility as Determined by NMR Relaxation Times." This indicates that the rate of acetyl transfer is affected by a change in the translational mobility of aspirin and sulfadiazine molecules at T_{mc}, resulting in a change in temperature dependence. The temperature dependence of acetyl transfer at 12% RH does not show any break because T_{mc} at 12% RH is higher than the highest temperature for the measurement. Compared with acetyl transfer, hydrolysis of aspirin occurring in parallel with acetyl transfer does not show such a distinct break at T_{mc}, even though hydrolysis is also a bimolecular reaction.

Similarly, no distinct break was observed in the temperature dependence of hydrolysis of cephalothin in freeze-dried formulations containing dextran, as shown in Figure 14. The hydrolysis rate of cephalothin increased with increasing humidity because the rate-limiting step involves water as a reactant. The

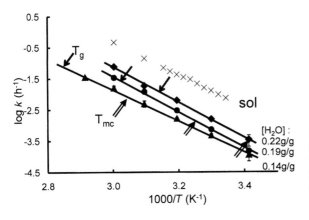

FIGURE 14 Temperature dependence of cephalothin hydrolysis in freeze-dried formulation containing dextran at 23% (▲), 60% (●), and 75% RH (◆).

temperature dependence of the apparent first-order rate constant is linear at all humidities in a manner similar to that of hydrolysis in aqueous solution, regardless of their T_{mc} indicated by arrows in the figure. The temperature dependence is also unaffected by the T_g of the formulations that are approximately 20°C higher than the T_{mc}. Since the translational mobility of drug and water molecules in freeze-dried formulations is affected by T_g and/or T_{mc}, the hydrolysis rate should be affected by T_g and/or T_{mc} if the translational diffusion of the drug and/or water molecules is rate limiting. The absence of a break in temperature dependence around T_g and T_{mc} suggests that the translational diffusion is not rate limiting. Since the translational diffusion of water can be considered to be much faster than that of the larger cephalothin molecule, the diffusion barrier of water molecules may be smaller than the chemical activational barrier. This interpretation is supported by the finding that the activation energy for the hydrolysis of cephalothin in the freeze-dried formulations containing dextran (between 23 and 26 kcal/mol) is close to the apparent activation energy obtained for hydrolysis in solution (24 kcal/mol). Because of the small diffusion barrier of water in freeze-dried formulations compared to the activational barrier, the hydrolysis rate of cephalothin is not affected by T_g and/or T_{mc}, even if the translational mobility of water molecules changes around T_g and/or T_{mc}.

Correlations Between Storage Stability and Fast Dynamics as Determined by NMR Relaxation Times

Protein aggregation, one of the most common degradation pathways of freeze-dried protein formulations, also appears to be closely related to the structural relaxation as reflected by NMR relaxation times. Figure 15 shows the temperature dependence of the time required for 10% protein aggregation (t_{90}) in freeze-dried β-galactosidase formulation containing methylcellulose (43). At 60% RH, the slope changes around the T_g measured by NMR (T_{mc}). No change in temperature dependence of t_{90} is observed at 12% RH, at which the T_{mc} is higher than the highest temperature for the measurement.

Apparent correlation between protein aggregation rate and structural relaxation is also observed for β-galactosidase freeze-dried with sugars. As shown in Figure 16, the slope of $t_{90} - T_g/T$ plots changes at around T_g, suggesting that aggregation rate is related to structural relaxation (33). However,

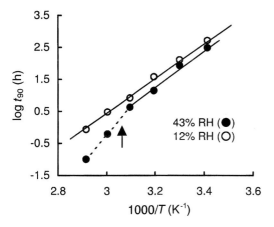

FIGURE 15 Temperature dependence of t_{90} for β-galactosidase aggregation in freeze-dried formulation containing methylcellulose.

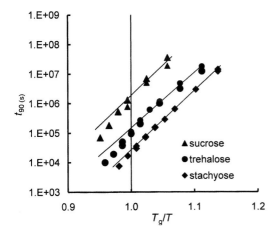

FIGURE 16 t_{90} for β-galactosidase aggregation in freeze-dried formulation containing sucrose, trehalose, or stachyose plotted against T_g/T.

structural relaxation cannot be considered to be the most relevant motion to the protein degradation, because sucrose, which has the shortest structural relaxation time, stabilizes the protein most effectively. These findings suggest that molecular motion other than structural relaxation contributes to the protein degradation. As shown in Figure 11, the $T_{1\rho}$ process of the protein carbonyl carbon is retarded by the addition of sugars, indicating that the fast dynamics of the protein is decreased by sugars. Sucrose is most effective in decreasing the protein dynamics. This suggests that the aggregation rate of β-galactosidase is correlated to the fast dynamics of the protein as determined by NMR relaxation times.

Figure 17 shows the temperature dependence of t_{90} observed for the degradation of insulin freeze-dried with trehalose or PVP (44). For insulin freeze-dried with trehalose stored at 12% RH, the slope of $t_{90} - T_g/T$ plot changed at around T_g, suggesting that degradation rate of insulin is correlated with structural relaxation. In contrast, for insulin freeze-dried with PVP, $t_{90} - T_g/T$ plots are linear at temperatures around T_g without a change in the slope. The value of t_{90} at T_g varies with humidity, indicating that structural relaxation is not the major factor that determines the degradation rate of insulin. If the

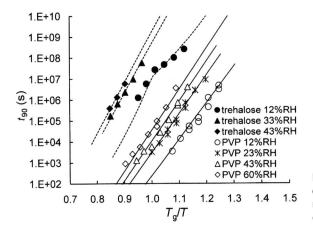

FIGURE 17 t_{90} for insulin degradation in freeze-dried formulation containing trehalose or PVP plotted against T_g/T.

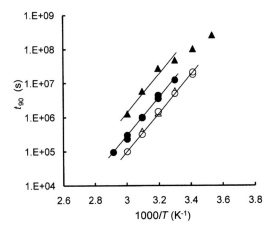

FIGURE 18 Temperature dependence of t_{90} for insulin degradation in freeze-dried formulation containing trehalose or dextran. ●, dextran 12% RH; ○, dextran 43% RH; ▲, trehalose 12% RH; △, trehalose 43% RH.

degradation rate is determined only by structural relaxation, t_{90} at T_g should not vary with humidity, because structural relaxation time is constant (100 seconds) at T_g regardless of humidity. These findings suggest that molecular motion other than structural relaxation contributes the insulin degradation, as suggested for the aggregation of β-galactosidase freeze-dried with sugars.

As shown in Figure 18, the t_{90} for the degradation of insulin freeze-dried with dextran is shorter than that of insulin freeze-dried with trehalose at 12% RH, and this difference in t_{90} is eliminated at 43% RH (32). As shown in Figure 10, the molecular mobility of the carbonyl carbon of insulin freeze-dried with trehalose, as determined by $T_{1\rho}$, is lower than that of insulin freeze-dried with dextran, as described in the section "Molecular Mobility as Determined by NMR Relaxation Times." This difference in mobility is eliminated at 43% RH, similar to the difference in t_{90}. These findings suggest that trehalose decreases the mobility of insulin in low humidity condition, and this decrease in mobility is related to the increase in insulin stability. Thus, the degradation rate of insulin also appears to be correlated to fast dynamics as determined by NMR relaxation times.

REFERENCES

1. Duddu SP, Weller K. Importance of glass transition temperature in accelerated stability testing of amorphous solids: case study using a lyophilized aspirin formulation. J Pharm Sci 1996; 85:345–347.
2. Herman BD, Sinclair BD, Milton N, et al. The effect of bulking agent on the solid-state stability of freeze-dried methylprednisolone sodium succinate. Pharm Res 1994; 11:1467–1473.
3. Streefland L, Auffret AD, Franks F. Bond cleavage reactions in solid aqueous carbohydrate solutions. Pharm Res 1998; 15:843–849.
4. Lai MC, Hageman MJ, Schowen RL, et al. Chemical stability of peptides in polymers. 1: Effect of water on peptide deamidation in poly(vinyl alcohol) and poly(vinyl pyrrolidone) matrixes. J Pharm Sci 1999; 88:1073–1080.
5. Lai MC, Hageman MJ, Schowen RL, et al. Chemical stability of peptides in polymers. 2. Discriminating between solvent and plasticizing effects of water on peptide deamidatin in poly(vinylpyrrolidone). J Pharm Sci 1999; 88:1081–1089.
6. Duddu SP, Zhang G, Dal Monte PR. The relationship between protein aggregation and molecular mobility below the glass transition temperature of lyophilized formulations containing a monoclonal antibody. Pharm Res 1997; 14:596–600.
7. Pikal MJ, Dellerman K, Roy ML. Formulation and stability of freeze-dried proteins: effects of moisture and oxygen on the stability of freeze-dried formulations of human growth hormone. Dev Biol Stand 1991; 74:21–38.
8. Roy ML, Pikal MJ, Rickard EC, et al. The effects of formulation and moisture on the stability of a freeze-dried monoclonal antibody-vinca conjugate: a test of the WLF glass transition theory. Dev Biol Stand 1991; 74:323–340.
9. Hageman M. Water sorption and solid-state stability of proteins. In: Ahern TJ, Manning MC, eds. Stability of Protein Pharmaceuticals. Part A: Chemical and Physical Pathways of Protein Degradation. New York: Plenum Press, 1992.
10. Bell LN, Hageman MJ, Muraoka LM. Thermally induced denaturation of lyophilized bovine somatotropin and lysozyme as impacted by moisture and excipients. J Pharm Sci 1995; 84:707–712.
11. Franks F. Freeze Drying: a combination of physics, chemistry, engineering and economics. Jpn J Freezing Drying 1992; 38:5–16.
12. Pikal M. Freeze-drying of proteins. BioPharm 1990; 3:26–30.
13. Prestrelski SJ, Pikal KA, Arakawa T. Optimization of lyophilization conditions for recombinant human interleukin-2 by dried-state conformational analysis using Fourier-transform infrared spectroscopy. Pharm Res 1995; 12:1250–1259.
14. Duddu SP, Dal Monte PR. Effect of glass transition temperature on the stability of lyophilized formulations containing a chimeric therapeutic monoclonal antibody. Pharm Res 1997; 14:591–595.
15. Costantino HR, Langer R, Klivanov AM. Aggregation of a lyophilized pharmaceutical protein, recombinant human albumin: effect of moisture and stabilization by excipients. Bio/Technology 1995; 13:493–496.
16. Kalichevsky MT, Jaroszkiewicz EM, Ablett S, et al. The glass transition of amylopectin measured by DSC, DMTA and NMR. Carbohydr Polym 1992; 18:77–88.
17. Kalichevsky MT, Jaroszkiewicz EM, Blanshard JMV. A study of the glass transition of amylopectin-sugar mixtures. Polymer 1993; 34:346–358.
18. Oksanen CA, Zografi G. Molecular mobility in mixtrures of absorbed water and solid poly(vinylpyrrolidone). Pharm Res 1993; 10:791–799.
19. Yoshioka S, Aso Y, Otsuka T, et al. Water mobility in poly(ethylene glycol)-, poly(vinylpyrrolidone)-, and gelatin-water systems as indicated by dielectric relaxation time, spin-lattice relaxation time, and water activity. J Pharm Sci 1995; 84:1072–1077.
20. Rubin CA, Wasylyk JM, Baust JG. Investigation of vitrification by nuclear magnetic resonance and differential scanning calorimetry in honey: a model carbohydrate system. J Agric Food Chem 1990; 38:1824–1827.
21. Shamblin SL, Tang X, Chang L, et al. Characterization of the time scales of molecular motion in pharmaceutically important glasses. J Phys Chem 1999; 103:4113–4121.

22. Andronis V, Zografi G. The molecular mobility of supercooled amorphous indomethacin as a function of temperature and relative humidity. Pharm Res 1998; 15:835–842.
23. Liu J, Rigsbee DR, StotzandM C, et al. Dynamics of pharmaceutical amorphous solid: the study of enthalpy relaxation by isothermal microcalorimetry. J Pharm Sci 2002; 91:1853–1862.
24. Hancock BC, Shamblin SL, Zografi G. Molecular mobility of amorphous pharmaceutical solids below their glass transition temperatures. Pharm Res 1995; 12:799–806.
25. Andronis V, Zografi G. Molecular mobility of supercooled amorphous indomethacin, determined by dynamic mechanical analysis. Pharm Res 1997; 14:410–414.
26. Yoshioka S, Aso Y, Kojima S. Dependence of the molecular mobility and protein stability of freeze-dried γ-globulin formulations on the molecular weight of dextran. Pharm Res 1997; 14:736–741.
27. Yoshioka S, Aso Y, Nakai Y, et al. Effect of high molecular mobility of poly(vinyl alcohol) on protein stability of lyophilized γ-globulin formulations. J Pharm Sci 1998; 87:147–151.
28. Yoshioka S, Aso Y, Kojima S. The effect of excipients on the molecular mobility of lyophilized formulations, as measured by glass transition temperature and NMR relaxation-based critical mobility temperature. Pharm Res 1999; 16:135–140.
29. Yoshioka S, Aso Y, Kojima S. Different molecular motions in lyophilized protein formulations as determined by laboratory and rotating frame spin-lattice relaxation times. J Pharm Sci 2002; 91:2203–2210.
30. Chen Q, Yamada T, Kurosu H, et al. Dynamic study of the noncrystalline phase of ^{13}C-laveled polyethylene by variable-temperature ^{13}C CP/MAS NMR spectroscopy. J Polym Sci B 1992; 30:591–601.
31. Yoshioka S, Aso Y, Kojima S, et al. Molecular mobility of protein in lyophilized formulations linked to the molecular mobility of polymer excipients, as determined by high resolution ^{13}C solid-state NMR. Pharm Res 1999; 16:1621–1625.
32. Yoshioka S, Miyazaki T, Aso Y. β-Relaxation of insulin molecule in lyophilized formulations containing trehalose or dextran as a determinant of chemical reactivity. Pharm Res 2006; 23:961–966.
33. Yoshioka S, Miyazaki T, Aso Y, et al. Significance of local mobility in aggregation of β-galactosidase lyophilized with trehalose, sucrose or stachyose. Pharm Res 2006; 23:961–966.
34. Afonin S, Glaser RW, Berditchevskaia M, et al. 4-Fluorophenylglycine as a label for ^{19}F NMR structure analysis of membrane-associated peptides. Chembiochem 2003; 4:1151–1163.
35. Aso Y, Yoshioka S, Miyazaki T, et al. Feasibility of ^{19}F-NMR for assessing the molecular mobility of flufenamic acid in solid dispersions. Chem Pharm Bull (Tokyo) 2009; 57:61–64.
36. Hancock BC, Zografi G. Characteristics and significance of the amorphous state in pharmaceutical systems. J Pharm Sci 1997; 86:1–12.
37. Andronis V, Zografi G. Crystal nucleation and growth of indomethacin polymorphs from the amorphous state. J Non-Cryst Solids 2000; 271:236–248.
38. Aso Y, Yoshioka S, Kojima S. Explanation of the crystallization rate of amorphous nifedipine and phenobarbital from their molecular mobility as measured by ^{13}C NMR relaxation time and the relaxation time obtained from the heating rate dependence of T_g. J Pharm Sci 2001; 89:128–143.
39. Aso Y, Yoshioka S, Kojima S. Molecular mobility-based estimation of the crystallization rates of amorphous nifedipine and phenobarbital in poly(vinylpyrrolidone) solid dispersions. J Pharm Sci 2004; 93:384–391.
40. Korhonen O, Bhugra C, Pikal MJ. Correlation between molecular mobility and crystallization growth of amorphous phenobarbital and phenobaribita with polyvinylpyrrolidone and L-proline. J Pharm Sci 2008; 97:3830–3841.
41. Shamblin SL, Zografi G. Enthalpy relaxation in binary amorphous mixtures containing sucrose. Pharm Res 1998; 15:1828–1834.

42. Yoshioka S, Aso Y, Kojima S. Temperature dependence of bimolecular reactions associated with molecular mobility in lyophilized formulations. Pharm Res 2000; 17:923–927.
43. Yoshioka S, Tajima S, Aso Y, et al. Inactivation and aggregation of β-galactosidase in lyophilized formulation described by the Kohlrausch-Williams-Watts stretched exponential function. Pharm Res 2003; 20:1655–1660.
44. Yoshioka S, Aso Y, Miyazaki T. Negligible contribution of molecular mobility to degradation rate of insulin lyophilized with poly(vinylpyrrolidone). J Pharm Sci 2006; 95:939–943.

14 Scanning Electron Microscopy: A Powerful Tool for Imaging Freeze-Dried Material

Tauno Jalanti
Microscan Service SA, Chavannes-près-Renens, Switzerland

ABSTRACT

Observation of freeze-dried material is a real challenge for both scanning electron microscopy (SEM) and transmission electron microscopy (TEM). The samples are usually very brittle, electrically insulating, highly hygroscopic, and very difficult to embed in resins.

Sample preparation methods are described: cutting, sticking on the SEM stubs, conductive coating with carbon and/or metals like gold or platinum or embedding, and cutting with a microtome.

Examples of structural observations are given including digital image processing.

Over and above SEM observation is the possibility of qualitative elemental analysis by energy dispersive X-ray spectroscopy (EDX) of foreign bodies, which may contaminate the freeze-dried material.

INTRODUCTION

Visualization of microstructures is of the highest importance in material sciences. At present, the tendency for many scientists is to trust in spreadsheets and digital information, thus forgetting the necessary connection to the visible reality of nature. Numeric interpretation of images, so called image analysis is, of course, useful and yields highly important information of structures, but the basis of analysis is the morphological picture of the material, be it at visual size or at very high magnification. The choice of the observational instrument is made according to the desired magnification.

- Macrophotography, magnification $1\times$ to $10\times$.
- Light optical microscopy, magnification $10\times$ to $1000\times$.
- Scanning electron microscopy (SEM):
  - Conventional SEM with a tungsten thermoelectronic gun, magnification $1\times$ to $50,000\times$.
  - Field-emission scanning electron microscopy (FE-SEM) with a field-emission "cold" electron gun, magnification $1\times$ to $500,000\times$.
- Transmission electron microscopy (TEM), magnification $100\times$ to $10^6\times$.

In this chapter, only the contribution of modern SEM to the global subject of this book will be exposed and discussed.

ANALYTICAL SCANNING ELECTRON MICROSCOPY

The SEM is an electron optical device designed to observe and analyse the surfaces of material samples of all nature. An electron beam, focused by electromagnetic lenses, scans the specimen's surface. The interaction of the "primary

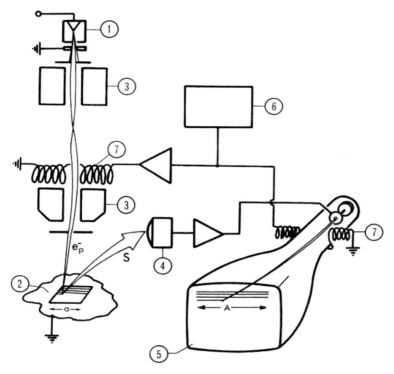

FIGURE 1 Diagram of an SEM. 1, Electron gun; 2, specimen; 3, electromagnetic focusing lenses; 4, signal detector; 5, Screen; 6, Scan generator; 7, scanning coils.

electrons" of that beam with the atoms of the specimen produces various radiations. The resultant image is the transcription on a screen of the intensity of these physical signals scanned at the same rate as the specimen.

The first SEM was developed in 1937 to 1938 by M. von Ardenne (1,2) and the first commercial device was sold in 1965 to DuPont de Nemours by Cambridge Instruments under the name "Stereoscan." These first microscopes had analogical electronics, cathode ray tubes, and rollfilm or Polaroid cameras. Nowadays, the SEMs are fully computerized and digital pictures are stored with full information on mass storage disks that can be directly inserted in reports.

Digital imaging also is an open door for image analysis, computerized 3D presentation, and microrugosimetry.

Basically, an SEM (3,4) is constituted of an electron optical system designed to focalize the electron beam on the sample surface, a scanning system, a signal detector, and amplifier and imaging system (Fig. 1).

Electron optics: The electrons are emitted by a hot tungsten filament or, in modern, top class microscopes, a cold field-emission emitter. The electrons are accelerated by high voltage (typically 5–20 kV) applied between the emitter (cathode) and an anode. The beam is focalized on the sample surface by electromagnetic lenses.

Scanning system: Orthogonal coils produce a crossed magnetic field that deviates electrons in two 90° directions, thus producing a scan on the sample surface. The same scan generator is used for scanning the observation screen. Image magnification is given by the ratio of screen width with the length of a scan line on the sample.

Imaging system: The interaction of the primary electrons (beam) and the atoms of the sample surface yields secondary electrons, backscattered electrons, as well as characteristic X-ray photons. All these signals can be used for imaging.

- The secondary electrons, emitted by ionization of the atoms of the specimen, produce topographic contrast, that is, the information on the geometric structure of the specimen. As these electrons are detected from a very thin surface layer (a few nanometers), they give a very good spatial resolution: superior to 5 nm in the best cases and, in general, 10 to 15 nm. Topographic contrast is the usual imaging method for lyophilized materials.
- The backscattered electrons, primary electrons "bouncing" on the specimen surface, yield information on the atomic number of the atoms encountered, thus giving a basic chemical contrast if the specimen is not coated with too thick a layer of gold (see "specimen preparation"). Backscattered electrons are less sensitive to the electric charging of the specimen and may give better images than secondary electrons on poor conductors.
- X-ray photons, emitted by the ionized atoms, carry accurate information on their chemical nature. The photons are detected by an energy dispersive X-ray spectrometer (EDX). Qualitative and quantitative information is obtained by processing the spectroscopic data. The picture made with intensity of the photons of the specific energy corresponding to an emission ray of an element is a map of the repartition of that element on the sample surface. This method is a more accurate and precise way to localize the element than backscattered electron imaging.
- Other signals, such as absorbed current, induced electromotive force, and cathodoluminescence, are not suitable for the study of lyophilized materials.

Modern SEMs (Fig. 2) are fully computerized with digital imaging and image processing capacities. Typically, the magnification range is from $1\times$ to $50,000\times$. The field-emission gun scanning electron microscopes (FE-SEMs) can magnify up to $500,000\times$ in good conditions. Microscopists prefer to mention the resolution, that is, the size of the smallest detail that can be seen, instead of the magnification. For classical SEM, it is 10 nm. The field depth is very high, producing sharp images even on very rough specimens (Fig. 3).

The analytical SEMs are fitted with EDX equipment for total (morphochemical) material characterization, both morphological and chemical. It should be kept in mind that EDX spectrometry gives information on the nature of the atoms of the specimen from Be to Pu, but not on the chemical bonds. Therefore, it is not suitable for the characterization of organic material.

Specimen Preparation

Specimen preparation is the most important phase of the electron microscope study of any material. It is senseless to own expensive observational equipment if the sample has been modified during preparation.

As the SEM works under vacuum, imperative for the propagation of electrons, only dry specimens can be observed (except in environmental or

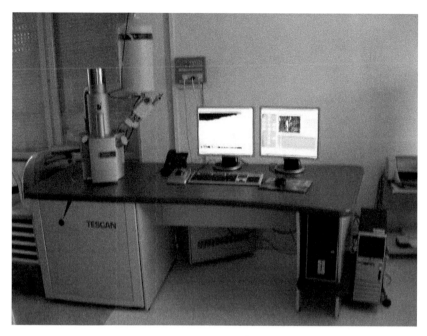

FIGURE 2 A modern computer-based SEM with LCD screens combined with an X-ray spectrometer.

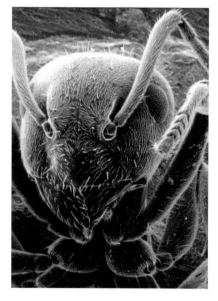

FIGURE 3 Picture of the same specimen of an ant's head taken with a light optical microscope (*left*) and a scanning electron microscope (*right*) showing high depth of field and better resolution.

low-temperature SEM, see under "Other Imaging and Analytical Techniques").
As the SEM is an electric device, the sample has to be an electrical conductor.

Since biological specimens are usually wet and subject to alteration, they
require first a fixation with glutaraldehyde and osmium tetroxide (5), followed
by dehydration in graded acetones and, finally, a critical point drying (6) to
avoid deformations.

The main difficulties in the preparation of lyophilized material for SEM
observation are due to two physical properties:

1. The very brittle nature of the material. Even cutting with ultrasharp blades is
   very difficult and usually alters the structure. Embedding in resins such as
   Epon®, Araldite®, or paraffin is difficult due to poor infiltration. Sticking on
   specimen stubs is also a problem as the material is likely to break easily.
   Very fluid glues and extremely cautious handling are required. Freezing the
   material followed by cryofracture (freeze fracture) can be a good alternative
   to cutting, but poor thermic conductivity of the material can be a problem. A
   carbon-metal replica of cell wall surface can then be made and studied in
   TEM for higher resolution imaging.
2. Very high electric insulation due to their organic nature and their cellular
   structure. This requires a heavy (5–10 nm) sputter coating with gold eventually after carbon coating. This means that high-resolution imaging is not
   realistic and artefacts due to electric charging are very frequent.

CASE EXAMPLE: THE STRUCTURE OF FREEZE-DRIED PLUGS

The samples described are lyophilized mannitol freeze-dried plugs produced by
freeze-drying in glass vials.

Fracture of the vials was initialized with a glazier's diamond and continued by contact with a red-hot iron rod. The plugs were carefully removed
with sharp sickle-shaped tweezers and cut on a hard Teflon® surface with a new
stainless steel scalpel. Pieces, each showing surface, cross section, or bottom,
were glued with fast-setting epoxy glue on aluminum sample stubs.

A carbon layer was first deposited by evaporation under a vacuum of
carbon threads to set a preliminary electric conductivity that facilitates sputter
coating. Approximately 10 nm of gold was then deposited by sputtering
(physical vapor deposition) in a low-pressure argon atmosphere. Conductive
bridges between the sample and aluminum stub were made with silver paint.
Special caution should be taken to prevent absorption of the silver paint by
capillarity in the material.

Observation was made in a Leica S430 SEM, in secondary electron mode,
with an acceleration voltage of 10 or 20 kV.

The pictures shown are a selection from several different sources and,
therefore, are not necessarily taken on the same plug. The magnification figure
has no sense since reproduction factors are not clearly known. However, the
scale bar, usually in microns, is accurate and gives a good idea of the sizes.

The plugs are usually biconcave (Fig. 4) with a membrane-like free surface
(Fig. 5). This "skin" is partially torn revealing the subjacent lamellar structure. In
the medium internal region, the cell structure is rather regular, showing little
perturbation of cell walls (Fig. 6). The region underneath the surface has a
lamellar structure (Fig. 7) that may be related to high pressure effects during

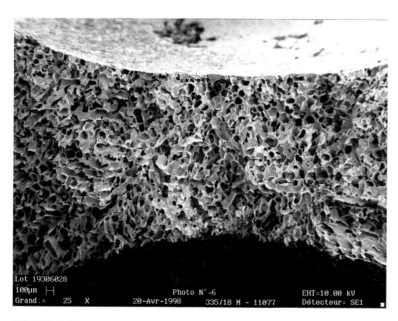

FIGURE 4 Low magnification view of the cross-section of a plug. The top surface is free of contact with glass.

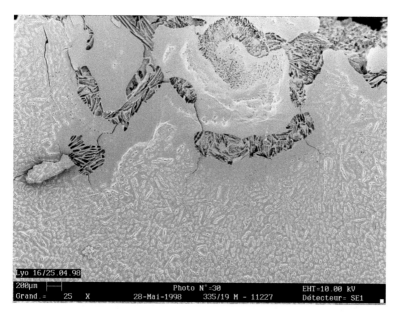

FIGURE 5 Membrane-like top surface of a plug, with the lamellar internal structure partially uncovered.

FIGURE 6 Cellular structure in the internal medium zone.

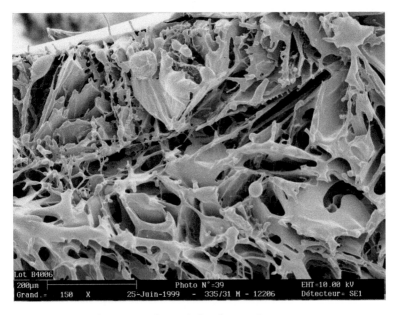

FIGURE 7 Lamellar structure beneath the plugs surface.

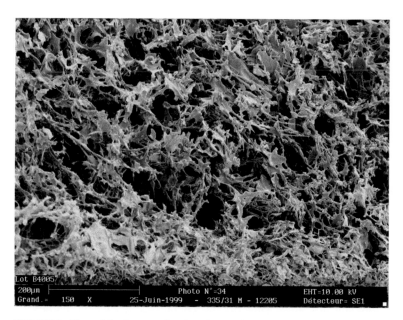

FIGURE 8 View of the bottom of a plug showing high perturbation due to contact with the glass surface.

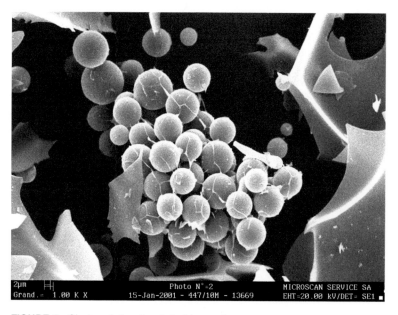

FIGURE 9 Cluster of drug beads inside a cell.

sublimation of the ice. The bottom structure is very much disturbed (Fig. 8): the plug was probably torn away from the glass during the retraction.

Figure 9 represents a good view of the inside of a plug impregnated with a drug, forming clusters of beads.

BEAM DAMAGE

The electron beam, even at low intensity (usually 200–500 pA), can heat the material since cellular structures are poor thermal conductors. This will lead to dilation and fracture of the specimen. The electrons can also penetrate the material, reach the noncoated regions, and burst the sample by electrostatic forces. To avoid these damages, a heavy coating with gold or gold-platinum is necessary, thus reducing the visibility of very small details and making illusory very high-resolution SEM on freeze-dried material. Another solution is observation at very low acceleration voltage (4), for instance 1 kV or less of the primary electrons. This can only be performed in an FEG-SEM.

CONTAMINATION ANALYSIS

Solid particle contaminants are a concern in many high-tech products, especially parenteral pharmaceutical drugs. In lyophilizates, they can be easily isolated by rehydration and further filtering on gold-coated (for electrical conductivity) polycarbonate membrane filters (Fig. 10). Morphochemical characterization is then performed with SEM and X-ray microanalysis (7). The most frequent contaminants are glass (dust from broken vials) as well as metals (for instance stainless steel from production devices) or environmental dust (calcium carbonate, aluminium silicates, etc.). Some organic contaminants such as cellulose or synthetic fibres may be identified by their morphology or by their content in Na, K, Cl, and S, for instance in human skin squames (Fig. 11).

IMAGE ANALYSIS

As the modern SEMs are fully computerized and the images are digital, there is a large potential for developing software for image analysis. This includes the possibilities for quantitative interpretation of cell size, wall thickness, and statistics. Following the taking of stereographic picture pairs (this requires tilting the specimen or the beam), a 3D reconstruction of the specimen structure can be made with specific software as well as quantitative data on the roughness, etc. Adding "false colors" to the topographic images giving a dramatic view is senseless.

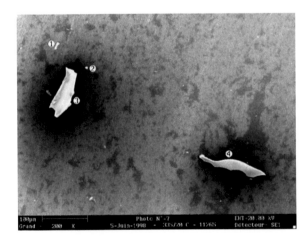

FIGURE 10 Solid contaminant particles found in a lyophilizate.

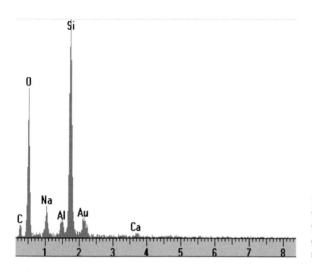

FIGURE 11 The X-ray spectrum shows that all four particles shown on Figure 10 are glass (silicon oxide with sodium).

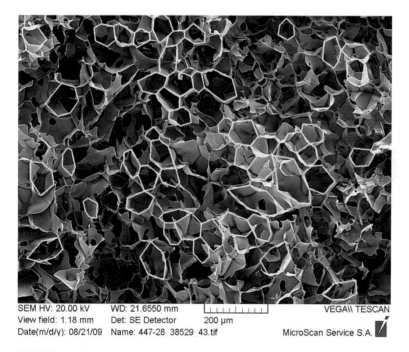

SEM HV: 20.00 kV	WD: 21.6550 mm		VEGA\\ TESCAN
View field: 1.18 mm	Det: SE Detector	200 μm	
Date(m/d/y): 08/21/09	Name: 447-28 38529 43.tif		MicroScan Service S.A.

FIGURE 12 Geometrical analysis of the micrograph of a cross-section showing evidence of hexagonal structures.

Figure 12 demonstrates how image processing can enhance structural features, such as the hexagonal shape of the cells left by the geometrical structure of the ice crystals. Hopefully, this could be done automatically in the future with image processing software.

OTHER IMAGING AND ANALYTICAL TECHNIQUES

Low temperature (cryo) SEM (8): A new orientation in research may be the understanding of the lyophilization process itself. For this, it might be suitable to observe the ice at low temperature and gradually sublimate it in the microscope itself. This requires special skills of observation, avoiding artefacts, and interpretation of results. The "environmental SEM" (9) operating at low vacuum (0.1 atm) is suitable for working with wet specimens. This is usually not the case when studying lyophilizates that are dry and compatible with high vacuum.

Other analytical techniques: Imaging of the cell wall structure with atomic force microscopy (AFM) or scanning tunnelling microscopy (STM) (10) could yield interesting results. However, there remain many unsolved problems during specimen preparation that must be addressed. Secondary ion imaging and spectrometry (SIMS, TOF-SIMS) (11) for chemical analysis and the study of the repartition of the active principles could be very powerful tools, but once again, the main concern is proper specimen preparation.

CONCLUSION

Twenty-first-century analytical microscopy techniques are full of promises for the scientist. However, it should be kept in mind that specimen preparation is the weak but essential link to success. It is senseless to use microscopes or other analysers worth hundreds of thousands euros to work on a nonrepresentative specimen. Proper interpretation of images and analytical results are also essential. In conclusion, it is the observational and analytical skills of the examiner sitting in front of the machine that makes the difference between good and poor results.

REFERENCES

1. von Ardenne M. Das Elektronen-Rastermikroskop. Theoretische Grundlagen (in German). Zeitschrift für Physik 1938; 108(9–10):553–572.
2. von Ardenne M. Das Elektronen-Rastermikroskop. Praktische Ausführung (in German). Zeitschrift für technische Physik 1938; 19:407–416.
3. Goldstein GI, Newbury DE, Echlin P, et al. Scanning Electron Microscopy and X-ray Microanalysis. New York: Plenum Press, 1981.
4. Lyman CE, Newbury JI, Goldstein DB, et al. Scanning Electron Microscopy, X-ray Microanalysis and Analytical Electron Microscopy. New York: Plenum Press, 1990.
5. Karnovsky MJ. A formaldehyde-glutaraldehyde fixative of high osmolality for use in electron microscopy. J Cell Biol 1965; 27:137.
6. Cohen AL. Critical point drying—principles and procedures. Scanning electron microscopy 1979/II 303-24. Microscopy Technique Staining, Ult_SEM.
7. Castaing R. Application des sondes électroniques à une méthode d'analyse ponctuelle chimique et cristallographique. Thèse de doctorat d'état, Université de Paris, 1952, Publication ONERA N. 55, 1951.
8. Jeffree CE, Read ND. Ambient- and Low-temperature scanning electron microscopy. In: Hall JL, Hawes CR, eds. Electron Microscopy of Plant Cells. London: Academic Press, 1991.
9. Danilatos GD, Robinson VNE. Principles of scanning electron microscopy at high specimen pressures. Scanning 1979; 2:72–82.
10. Wiesendanger R. Scanning Probe Microscopy. Cambridge: Cambridge University Press, 1998.
11. Benninghoven A, Rudenauer FG, Werner HW. Secondary Ion Mass Spectrometry: Basic Concepts, Instrumental Aspects, Applications, and Trends. New York: Wiley, 1987.

15

Pharmaceutical Packaging for Lyophilization Applications

Claudia Dietrich
SCHOTT forma vitrum ag, St. Gallen, Switzerland

Florian Maurer and Holger Roehl
SCHOTT AG, Mainz, Germany

Wolfgang Frieß
Ludwig-Maximilians-Universitaet, Munich, Germany

INTRODUCTION

Glass is a material that has been used by mankind for several thousand years, and due to its unique properties there is still a high demand for glass in our everyday life. For instance, glass is used as windows for architectural applications, windshields and windows in vehicles, lenses in optical devices and eyeglasses, containers for food and beverages, as well as for pharmaceutical packaging. Transparency, protection, inertness, and robustness are typical properties of glass. Many of these properties (e.g., transmission and absorption of light) can be tuned in a wide range to meet the requirements of a certain application. A major drawback of glass as a material for packaging is its fragility (see sect. "Mechanical Properties of Glass—Fracture Mechanics").

The properties of glass make it a preferred material for the primary packaging of drugs. New therapeutic research leads to new compositions of agents. Biopharmaceuticals have emerged as a major growth area in the global healthcare market and therefore are a profit-making business. In 2012, 22% of worldwide drug sales will come from biotechnical drugs. In terms of turnover, they will reach 44% of the top 100 products. This growth is driven by therapeutic areas like oncology, antirheumatics, and vaccines, which show a distinct advance in turnover and in market share. These indications require high-potency drugs that are also highly sensitive. Attendant parameters such as temperature, pressure, pH value, organic solvents, or container surface interaction take on greater significance. Simultaneously, the requirements on safety and stability of drugs are also continuously increasing from regulation bodies. Lyophilization is the method of choice to increase the shelf life of sensitive products such as proteins. By removing the water from the formulation and sealing resulting cake in a vial, the drug can be easily stored, shipped, and later reconstituted to its original form for injection.

To maintain the biological activity of the recombinant proteins, a formulation has to be found that preserves the conformational integrity of the molecule and that protects functional groups from degradation. For this purpose, additives such as buffers, salts, amino acids, sugars, and surfactants are used. Typically the most critical parameters for protein stability are the pH value of the formulation and the ionic strength (1).

Thus, special attention has to be taken to minimize the interactions between the drug and the formulation compounds on one side and the glass surface on the other.

Extractable and leachable compounds from the glass can directly influence the formulation of modern active pharmaceutical ingredients (API) based on biomolecules (e.g., recombinant antibodies) whose activity is sensitive to any changes of the storage conditions. Protein adsorption kinetics can be very quick, saturation of the surface taking place in the range of a few hours (2,3). The result is not only a loss of active compound, especially for highly diluted APIs (4–6), but also conformational changes that can occur depending on the stability of the biomolecule (7). These changes can lead to the formation of protein aggregates that might trigger immune responses.

Possible implications are therapy failure and severe diseases (8). Silicone oil can also induce aggregation with the described consequences for the health of a patient (9). Such oils are used to lower the surface energy to minimize the residual volume of highly concentrated drug solutions like antibodies that exhibit a high viscosity (10). Furthermore, in prefilled syringes, a lubricant film is generated to produce a smooth gliding of the piston down the barrel.

Therefore, several demands are made on the usage of glass as a container for (bio)pharmaceuticals:

- Low alkalinity
- Chemical durability
- Low chemical and physical interaction of the container walls with pharmaceuticals ("drug-container interaction")
- Complete removal of dosage (by application of hydrophobic coatings)
- Robustness against sterilization processes (high-energy radiation, ethylene oxide (ETO), water vapor)
- Robustness against mechanical and thermal stresses

If the drug is preserved by freeze-drying (lyophilization) to enhance its lifetime, mechanical requirements on the glass container are also important. For drugs processed by lyophilization, glass is the most frequently used material for vials.

LYOPHILIZATION: REQUIREMENTS ON GLASS VIALS

The three process steps of lyophilization (i.e., freezing, primary, and secondary drying) and the subsequent shipment and storage require a high mechanical strength of a glass vial:

1. Freezing: During freezing, the aqueous solution of protein and excipients is transferred from the liquid to the solid state. While water starts to expand at temperatures below $4°C$ and the volume fraction of ice to water is about 11% (11), the transformation of water molecules to crystals inside a vial, as well as the crystallization of other excipients in the formulation, causes a mechanical load on the walls of the glass vial (12). Some biomolecule formulations that freeze to amorphous solid states lead to glass vial breakage during cooling (13), which is hypothesized to be caused by a sudden detachment of the frozen formulation plug off the walls of the glass vial resulting in a mechanical shock wave through the glass walls (13).

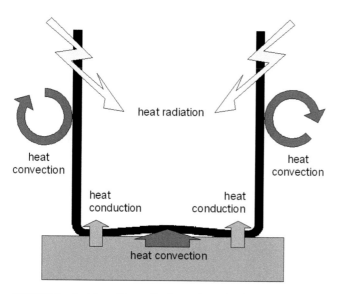

FIGURE 1 Heat transition paths for a glass vial during lyophilization processes.

2. Primary drying: During the primary drying step, frozen water is removed from the suspension by sublimation. To accelerate sublimation, application of a vacuum and gentle heating from the bottom of the vial are common techniques. Heating of the vial from the bottom side can cause glass breakage due to local temperature gradients because of an insufficient heat transition.

3. Secondary drying: In the secondary drying step, remaining nonfrozen water, not sublimated during the primary drying step, is removed. Secondary drying is accomplished at higher temperatures. During the drying processes three mechanisms of heat transition can arise: heat radiation, heat conduction through the bottom of the vial, and heat convection (14) (Fig. 1). The major heat fraction is introduced into the vial by convection, whereas heat radiation and heat conduction are of minor importance.

4. Sterilization and filling of the glass vials can cause damages on the surfaces (glass-to-glass or glass-to-metal contacts, chatter marks, etc.), whereas shipping to and storing at a different site can cause further mechanical loads. Improper handling (e.g., dropping of the vial) can also lead to failure if the impact point is at the site of a defect or residual stress is not removed during the annealing process.

In the pharmaceutical industry, specifications for mechanical stability of a container are very strict because failure of a glass container is unacceptable for safety and economic reasons. Even small flaws on the container can lead to nonsterile conditions inside the product, which can, in the worst case, endanger human health or life. They can also start propagating via subcritical crack growth and finally lead to a time-delayed failure. Therefore, the highest priority for both packaging supplier and pharmaceutical company is to achieve a "zero container-breakage" during production, processing, and usage in combination with low cost and low-cycle times.

Common product specifications (validation procedure) claim failure probabilities of containers used for pharmaceutical applications of less than 1 ppm. To achieve this strict demand, manufacturers of pharmaceutical glass containers spare no effort to continuously improve the material and surface properties and the geometry of their glass products. One simple approach to gain higher vial strengths would be a higher wall thickness of the glass [cf. equations (5) and (6)]. However, thicker glass walls result in a poorer heat transfer that consequently requires longer and more expensive cycle times for freeze-drying. Thus, one improvement for the production of glass vials for freeze-drying applications is an optimal vial design (sufficient mechanical strength) that allows an optimized lyophilization process (sufficient heat transfer). Further improvements of the quality of the vials can be reached by coatings that can have a significant influence on the residual volume or the optical appearance of the lyophilization product.

MECHANICAL PROPERTIES OF GLASS—FRACTURE MECHANICS

Glass is a brittle material, that is, there are no possibilities to reduce mechanical stresses by plastic deformation. If glass is under mechanical (tensile) load, the deformation behavior of the glass is linear elastic until failure. The theoretical strength σ_{theo} of a glass without any flaws (pores, cracks, and other defects) is approximately (15)

$$\sigma_{\text{theo}} \approx \frac{E}{10} \tag{1}$$

where E is the elastic modulus of the glass (E for soda-lime-silica glass $\approx 70\,\text{GPa}$). In reality, the mechanical strength of glass is mainly determined by the quality of its surfaces. Flaws (cracks, inclusions etc.) at the surface reduce the mechanical strength of the glass by concentrating (enhancing) externally applied loads σ_{appl} (stresses). This concentration of stresses of a flaw at a glass surface is expressed by the stress intensity factor K:

$$K = \sigma_{\text{appl}} Y \sqrt{c} \tag{2}$$

Where σ_{appl} is an externally applied load (stress), Y is a geometry factor that takes into account the shape of the defect (deep/low, sharp/blunt), and c is the depth of the defect. Typical values of Y range between 1 and 2 depending on the geometry of the defect (ratio of width over depth). A glass can resist fracture as long as the stress intensity factor K is below the material-specific value K_{IC} ("K-one-c"), also expressed as the fracture toughness of a glass (Table 1).

As can be deduced from equation (2), the critical value of K_{IC} is reached when both an externally applied load σ_{appl} and a critical defect of depth c are present simultaneously. Consequently, failure does not occur if only one

TABLE 1 Values of Fracture Toughness K_{IC} for Brittle Materials

Material	K_{IC} (MPa$\sqrt{\text{m}}$)
Concrete	0.2…1.4
Aluminum oxide	2.8
Soda-lime-silica glass	0.7…0.8

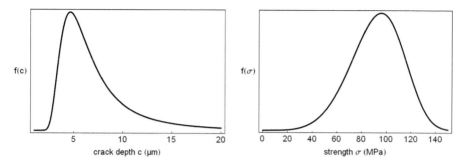

FIGURE 2 Example of the distribution of critical crack depth of a glass sample (*left*) and corresponding distribution of strength (*right*).

criterion is fulfilled (presence of significant flaws but no mechanical loads, or vice versa). Another result from equation (2) is that the scatter of fracture strength σ_{fract} (i.e., the distribution of the value of σ_{appl} at fracture measured from experimental testing) is given by the scatter of the critical crack depth c (Fig. 2):

$$\sigma_{fract} = f(c) = \frac{K_{IC}}{\sqrt{c}Y} \qquad (3)$$

If K_{IC} is reached anywhere in the glass, fracturing will start from this spot immediately (i.e., a glass article will always break at its most severe defect). This behavior is comparable to a chain that will always fail at its weakest link. Strength testing of brittle materials such as glass involves the gathering of the strength data distribution to deduce the distribution of strength-determining surface defects (see sect. "Fracture Statistics"). Of course, the task of glass-container manufacturers is to consequently and continuously improve the quality of a product and to avoid any damage during production, packaging, and shipping of the product.

If the stress intensity factor K is less than K_{IC}, defects extend by subcritical crack growth. The velocity of this extension can range from 10^{-8} m/sec to more than 10^3 m/sec and depends on the value of K and the chemical environment at the crack tip. As the defect depth grows continuously until K reaches the value of K_{IC}, subcritical crack growth is an important topic when considering the strength and reliability of glass products.

STRENGTH TESTING

Strength testing is necessary to determine the reliability of glass vials (i.e., determining its fracture probability for some certain mechanical load). An important requirement for the testing technique used is to simulate the mechanical load under usage conditions. Strength testing of glass is always destructive, thereby the weakest point that could have led to failure if the product was in service is detected for every single specimen. The location of failure (i.e., the origin of fracture) can be determined by means of fractography (see sect. "Fractography"). Knowing the weakest spot of a given geometry, quality improvement can be initiated, for example, by using finite element methods (FEM) as a tool for design optimization (see sect. "Finite Element

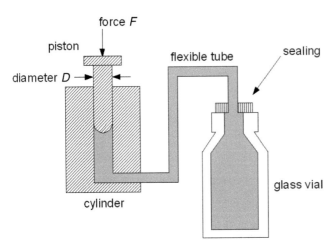

FIGURE 3 Scheme of burst-pressure testing experimental setup.

Methods"). It is recommended as an iterative improvement cycle in the development phase of a new glass-container product to continue strength testing and improving until the container is highly reliable and robust.

Burst-Pressure Testing

Burst-pressure testing is one of the most common methods to investigate the strength of glass containers used for pharmaceutical applications. A liquid medium (water) is used to apply a mechanical load in a single vial via a hydrostatic pressure (Fig. 3).

In most cases the hydrostatic pressure is generated with a constant load rate (e.g., 2 bar/sec) until breakage of the sample. The strength of a vial can then be calculated from the value of the hydrostatic pressure p_{hyd} at failure:

$$p_{hyd} = \frac{4F}{\pi D^2} \tag{4}$$

where F is the force at failure and D is the diameter of the piston to apply the hydrostatic pressure (Fig. 3). The hydrostatic pressure p_{hyd} induces different kinds of mechanical stresses in the glass of a vial (Fig. 4):

- Radial compression stresses σ_{rad} perpendicular to the inner walls
- Circumferential tensile stresses σ_{tan} tangential to the surface(s)
- Longitudinal stresses σ_{ax} parallel to the axis of the vial

The value of the radial compression stresses σ_{rad} equals the hydrostatic pressure p_{hyd} and is present everywhere where the vial is in contact with water:

$$\sigma_{rad} = p_{hyd} \tag{5}$$

As σ_{rad} are compression stresses, they have no influence on the fracture behavior of the vial (in a realistic assumption brittle materials do not fail under compression loads). Nevertheless, whenever radial stresses are present there is also a component of tangential stresses perpendicular to the walls of the vial

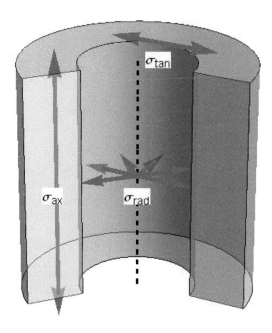

FIGURE 4 Mechanical stresses inside an axially symmetric glass body.

(Fig. 4). In contrast to σ_{rad}, tangential stresses σ_{tan} are tensile and can induce critical stresses on surface defects (15):

$$\sigma_{tan} = \frac{p_{hyd} R}{t} \qquad (6)$$

where p_{hyd} is the hydrostatic pressure, R is the radius, and t is the wall thickness of the glass container. The values of axial stresses σ_{ax} can be estimated by (15)

$$\sigma_{ax} = \frac{p_{hyd} R}{2t} \qquad (7)$$

Equations (6) and (7) are valid for thin-walled containers only. It is obvious from equations (5) to (7) that in a glass body with rotational symmetry the tangential stresses are the most critical stresses determining the fracture behavior. Since σ_{ax} is lower than σ_{tan} by a factor of 2, the direction of crack propagation of a vial at failure normally starts parallel to the axis before bifurcating (Fig. 6).

During the cooling step in the lyophilization process, the formulation freezes and starts expanding. Because of this expansion, radial stresses are applied to the inner wall of the vial, which involve tangential tensile stresses σ_{tan}. The hydrostatic pressure introduced by the burst-testing apparatus simulates the mechanical load of an expanding lyo cake during the cooling step. Thus, burst-pressure testing is a suitable technique to simulate the mechanical loads that appear during the freezing process and to investigate the stress distribution of the vial.

Vial strength (vial breakage) can also be determined qualitatively by simulating the freeze-drying process using sugar alcohols (e.g., mannitol) or protein formulations as a substitute for more expensive pharmaceuticals. With sugar alcohols, a thermally induced crystallization during cooling can be simulated (12,16), whereas protein formulations are appropriate to investigate the thermal expansion behavior of pharmaceuticals freezing in an amorphous state (13). With these kind of experiments, the simulation of the mechanical load during the lyophilization process is more realistic than for burst-pressure testing using water as the pressure medium (Fig. 3). However, the determination of the stresses requires extensive strain-gauge techniques and is less precise.

Thermal Shock Testing

Thermal shock testing is an additional tool to burst-pressure testing to determine the strength of glass containers: A sample is exposed to an instant temperature difference, thus "shocked" either by rapid heating or cooling. This sudden temperature change is accomplished by quenching (e.g., from elevated temperature into an ice bath) or heating up (e.g., from low temperature into a hot liquid bath) the specimen. Because of the rapid temperature change, an enormous temperature gradient is caused within the glass. This temperature gradient can cause strong local stresses due to differing thermal conduction caused by the geometry of the sample (e.g., wall thickness of a glass vial), leading to different thermal expansions of cool and hot regions. Thermal shock testing is an integral testing method, meaning the whole sample is stressed simultaneously at the same time, leading to failure at the weakest point of each specimen.

Fracture Statistics

For strength testing, an intentional mechanical loading in a testing range is required until breakage of the sample. Typically, the strength values for failure during testing (testing range) are much higher than the values that appear to the product in usage (usage range). Nevertheless, strength testing with excess load is necessary to acquire information about the distribution of strength data for every particular product.

The strength data from the tests is plotted in a statistical plot (see Fig. 5 as an example assuming a Weibull distribution of the data): The cumulative failure probability F is plotted versus the burst pressure p. Other strength testing units (on the abscissa/x-axis) can be, for instance, stress σ, temperature difference ΔT, or force F, depending on the technique for strength testing.

The testing range of the experiment is defined by the range of the failure data (in Fig. 5, a testing range between 60 and 125 bar is applied). The dashed lines in Figure 6 enclose the strength data and mark the 95% confidence bounds (i.e., the true distribution of this sample lies within these confidence bounds with a probability of 95%). The axes of the statistical plot in Figure 5 are not linear but intentionally biased. If data can be described by a particular statistical distribution (in this case Weibull distribution), the biasing of the axes leads to a linear arrangement of the data. So plotting strength data in a statistical plot is used to qualitatively check for a particular statistical distribution. To check quantitatively whether strength data can be described by a particular statistical distribution, numerous hypothesis tests ("goodness-of-fit tests") are available (17).

If an appropriate statistical distribution is accepted, the fracture probability F can be calculated for any load by extrapolation. For instance, the fracture

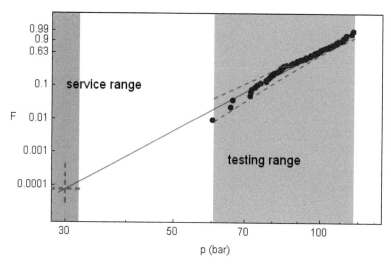

FIGURE 5 Statistical plot of strength (burst-pressure) data assuming a Weibull distribution.

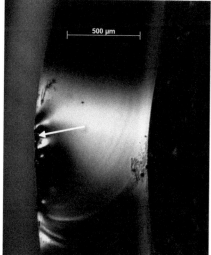

FIGURE 6 Fracture pattern of a glass vial broken by burst-pressure testing (*left*) and location of the fracture origin on the outer surface (*right*).

probability F of the data shown in Figure 5 for a strength of 30 bar is approximately 4×10^{-5}. Of course, the uncertainty of extrapolated values increases with the displacement from the testing range. For lyophilization applications it would be a great benefit to be able to predict the failure probability of a container in combination with a particular pharmaceutical formulation. Glass vial breakage ranges approximately between 1% and 90% (12,13,16), depending on the testing parameters (e.g., filling height, cooling rate, concentration of suspension). To predict the failure probability F for a certain pharmaceutical formulation and process, realistic values of the radial compression stresses σ that are induced by the frozen lyo cake are necessary. From this, the fracture probability can be calculated for stresses applied by the expanded lyo cake and the minimum strength of a glass vial that is required to perform below a particular (given) fracture probability would be accessible. Unfortunately, the real stresses inside a vial caused by the frozen and expanded lyo cake are difficult to determine. However, some authors report on the application of strain gages to measure the strain on the outer surface of a vial (12,13).

Fractography
A fractographic analysis (i.e., the investigation of the fracture pattern and the determination of the fracture origin by means of microscopy) is useful to learn more about the weak points of a glass container. When investigating the crack pattern of a burst vial (Fig. 6), one can locate the fracture origin and thus find the mechanical weak points of the sample. In some cases, significant specific damages introduced during production or processing the vials (e.g., chatter marks or severe scratches) can be detected. Thus, fractography is an essential technique to support product development and improvement.

FINITE ELEMENT METHODS
The FEM is a useful method to calculate mechanical stresses and strains in solid articles. It is possible to investigate complete stress and strain distributions for the surfaces and the bulk of a glass container, even for complex geometries. On the basis of FEM results, a great improvement of a glass vial in terms of mechanical robustness was achieved by optimizing its geometry (11): The flat shape of the bottom part of a standard vial (Fig. 7) was changed to a more

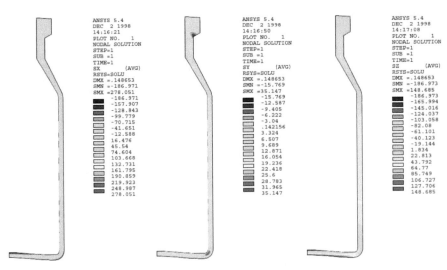

FIGURE 7 Stress distribution for a standard glass vial under hydrostatic pressure of $p = 30$ bar. Left: axial stresses; middle: radial stresses; right: tangential stresses.

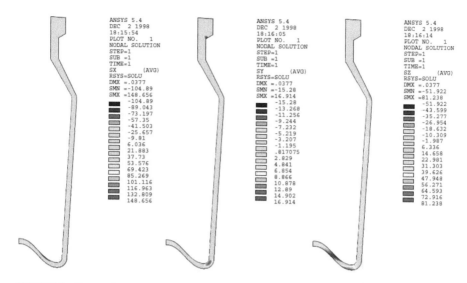

FIGURE 8 Stress distribution for a glass vial with optimized geometry under hydrostatic pressure of *p* = 30 bar. Left: axial stresses; middle: radial stresses; right: tangential stresses.

"champagne-bottle" geometry (Fig. 8). With this geometrical modification, improved heat conduction in combination with enhanced mechanical stability (reduced fracture probability) was realized. Additionally, to further reduce mechanical stresses during the lyophilization process, the shape of the barrel of the vial was changed from cylindrical (Fig. 7) to conical geometry (Fig. 8). With these modifications, the rate of vial breakage was reduced up to 60%. When the frozen lyo cake expands during the cooling step, the radial stresses introduced to the inner walls of the vial increase. At a certain value of radial compression stresses, the ice starts to melt creating a thin film of liquid water between the inner wall of the vial and the frozen lyo cake. This thin film of water acts as a lubricant and supports a detachment of the lyo cake from the vial walls, which results in an instant motion upward where the radius of the vial is greater, and thus the radial stresses are reduced.

SCHOTT TopLyo™ AS AN EXAMPLE FOR OPTIMIZED VIALS FOR LYOPHILIZATION

Reduced risk for glass breakage and improved heat transfer during the freeze-drying process are demands that are met by the improved geometric design and the hydrophobic coating of SCHOTT TopLyo™ vials. This novel and high-quality product was developed for the special requirements of lyophilization.

Apart from the geometric aspect, TopLyo vials exhibit a hydrophobic layer on the inner surface that is applied by SCHOTT PICVD (plasma impulse chemical vapor deposition) coating technique in a validated process (18). A covalent bond is formed between the glass matrix and the layers, which is stable to usual pharmaceutical procedures such as washing, autoclaving, sterilization, and depyrogenization (heat treatment of 300°C), and also against mechanical load. The transparency and the thickness (about 40 nm) of the layer do not have any negative impact on the quality of visual inspection.

Experiments made at Prof Rey and Prof Frieß's laboratories with TopLyo vials further showed the following:

- Less disruption of the lyo cake
- Less dry material pulling up from the edge
- Improved lyo cake aesthetics
- Minimization of residual volume after reconstitution resulting in improved drug removal

All these features aim to reduce reject rates and therefore also to decrease total costs on a market that shows a strong growth.

SUMMARY

Glass vial breakage is one of the most severe failure modes in the lyophilization process and the product lifetime for pharmaceuticals that are preserved by lyophilization. Even small surface flaws can have an influence on the integrity of a glass vial or lead to time-delayed failure due to subcritical crack growth. With strength testing techniques, for example, burst-pressure and thermal shock testing, the failure probability and mechanical weak points of a glass vial can be determined. However, up to now, predictions of failure probabilities for certain pharmaceutical suspensions stored in a glass vial by lyophilization are difficult to acquire. FEM is a suitable tool for optimizing the geometry of a vial, for example, modification of the shape of a glass vial to reduce breakage up to 60%. Moreover, the coating of glass vials can contribute to improvements in the lyophilization process.

REFERENCES

1. McNally EJ, Hastedt JE. Protein Formulation and Delivery. New York: Informa, 2008.
2. Malmsten M, Lassen B. Ellipsometry studies of protein adsorption at hydrophobic surfaces. ACS Symp Ser 1995; 602:228–238.
3. Lubarsky GV, Davidson MR, Bradley RH. Hydration-dehydration of adsorbed protein films studied by AFM and QCM-D. Biosens Bioelectron 2007; 22:1275–1281.
4. Song D, Hsu LF, Au JL. Binding of taxol to plastic and glass containers and protein under in vitro conditions. J Pharm Sci 1996; 85:29–31.
5. Grohganz H, Rischer M, Brandl M. Adsorption of the decapeptide cetrorelix depends both on the composition of dissolution medium and the type of solid surface. Eur J Pharm Sci 2004; 21:191–196.
6. Volkin DB, Middaugh CR. Formulation, Characterization and Stability of Protein Drugs. New York: Plenum Press, 1996:181–217.
7. Norde W. The behavior of proteins at interfaces, with special attention to the role of the structure stability of the protein molecule. Clin Mat 1992; 11:85–91.
8. Schellekens H, Jiskoot W. Erythropoietin-associated PRCA: still an unsolved mystery. J Immunotoxicol 2006; 3:123–130.
9. Jones LS, Kaufmann A, Middaugh CR. Silicone oil induced aggregation of proteins. J Pharm Sci 2005; 94:918–927.
10. Mundry T. *Einbrennsilikonisierung bei pharmazeutischen glaspackmitteln - analytische studien eines produktionsprozesses.* PhD Thesis, Berlin, 1999.
11. Nattermann K, Spallek M, Seibert V, et al. Patent DE000019925766A1: Behälter für medizinische und feinchemische Zwecke, 1999.

12. Jiang G, Akers M, Jain M, et al. Mechanistic studies of glass vial breakage for frozen formulations. I. Vial breakage caused by crystallizable excipient mannitol. PDA J Pharm Sci Technol 2007; 61:441–451.
13. Jiang G, Akers M, Jain M, et al. Mechanistic studies of glass vial breakage for frozen formulations. II. Vial breakage caused by amorphous protein formulations. PDA J Pharm Sci Technol 2007; 61:441–451.
14. Nail SL, Gatlin LA. Freeze-drying: principles and practice. In: Avis K, Lieberman H, Lachmann L, eds. Pharmaceutical Dosage Forms. Vol. 2. New York: Marcel Dekker Inc., 1993.
15. ICG Advanced Course 2006: Strength of Glass—Basics and Test Procedures (2006). Verlag der Deutschen Glastechnischen Gesellschaft; ISBN: 3-921089-47-6.
16. Williams A, Dean T. Vial breakage by frozen mannitol solutions: correlation with thermal characteristics and effects of stereoisomerism, additives, and vial configuration. J Parenter Sci Technol 1991; 45(2):94–100.
17. D'Agostino RB, Stephens M. Goodness-Of-Fit-Techniques. New York: Marcel Dekker Inc., 1986.
18. Walther M. SCHOTT type I plus™ containers for pharmaceutical packaging. In: Bach H, Krause D, eds. Thin Films on Glass. Berlin: Springer, 1997:370–377.

16

Closure and Container Considerations in Lyophilization

Fran L. DeGrazio
West Pharmaceutical Services, Inc., Lionville, Pennsylvania, U.S.A.

INTRODUCTION

Lyophilization is a process that was developed many years ago, yet became more prominent in the pharmaceutical industry with the growth of biopharmaceutical drug products. The purpose of freeze-drying is to extend the shelf life of products that would have minimal life in liquid format and give them the ability to have a standard shelf life that would typically consist of a minimum of two years.

In the last five years (2004–2008), 27 New Molecular Entities (NMEs) or Biologic License Applications (BLAs) were approved by the FDA in lyophilized form (1). Currently there are over 100 lyophilized drug products in use in the United States alone.

OVERVIEW

The container/closure system, which consists of the elastomeric stopper and the glass vial, impacts the effectiveness of the lyophilization process in two fundamental ways:

1. Resistance to heat transfer
2. Reduction in throughput

The container/closure system can impact final product quality and pharmaceutical elegance in an assortment of ways:

1. General chemical compatibility
2. Extractables/leachables
3. Moisture ingress
4. Surface adsorption
5. Silicone contamination
6. Seal integrity
7. Functionality in the field

PROCESS EFFECTS

Resistance to Heat Transfer

The limiting factors relating to heat transfer are the contact between the product and the chamber shelf and their thermal conductivity. The limit to the heat transfer is principally due to the lack of intimate contact between the product and the shelf and, to a lesser degree, to the lower thermal conductivity of the container and the frozen product. These resistances need to be compensated for in the process of getting heat to the sublimation front. The greatest impact on the level of quality for batch uniformity is the efficiency and effectiveness of the heat transfer system; therefore, the container is a critical component of the entire heat transfer system in the lyophilization chamber (2).

Reduction in Throughput

The volume of production over a period of time is defined as throughput. Throughput can be affected by problems with the container/closure system such as vial breakage, incomplete stoppering, extended drying times for closures, or poor sublimation rate due to a poor package design.

In choosing an appropriate container/closure system, it is important to address all of these considerations, in addition to issues such as chemical compatibility and stability.

If all of these needs are identified early in the process of choosing a primary package for lyophilization, a container/closure system that is optimal for all critical factors will be investigated for development.

VIAL CONSIDERATIONS FOR LYOPHILIZATION (2)

The process of lyophilization is affected by the vial's dimensions, physical characteristics, chemical properties, and interactions with both the drug product contained and the chamber shelf on which the vial rests. The text that follows will detail various key characteristics that can be either an advantage or disadvantage to the lyophilization process.

Physical Properties of Vials

In the pharmaceutical and biotechnology industries, type 1 glass tubing vials are typically used as the primary container for lyophilized drugs. One of the key reasons tubing vials are used over molded vials is the dimensional tolerance consistency. The capability of tubing vials to meet tight dimensional tolerance is superior to that of molded vials. This is due to mold-to-mold variation that can occur in the manufacturing process leading to variations in wall weight, outside diameter, and bottom thickness. In addition, the lower the coefficient of expansion for a glass type, the better it is for use in the lyophilization process. Because of all of these reasons, much of the discussion below will focus on glass tubing vials.

Because of recent development with newer plastic materials, plastic vials are also being considered for certain lyophilized applications. The development of new materials, such as cyclic olefins, has addressed some of the weaknesses associated with plastic materials in small-volume parenteral applications. Because of these more recent developments, an overview of these types of plastic vials will follow the glass vial discussion.

Bottom Thickness

The bottom thickness of a tubing vial is not normally the same as the wall thickness, because of the characteristics in its manufacturing process. As a general rule, the bottom thickness is 55% to 65% of the wall thickness. For example, in a tubing vial with a standard wall of 1.20 mm, the minimum bottom thickness would be in the range of 0.66 to 0.78 mm. The uniformity of glass distribution in the bottom is also important to prevent or minimize vial bottom breakage. If the bottom thickness is too thin, "ring-out," or bottom breakage, becomes a problem and lyophilization chamber throughput can be greatly

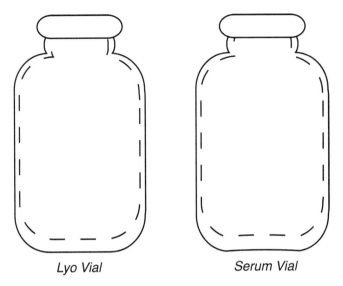

FIGURE 1   Flat bottom vials versus concave bottom vial.

reduced. Conversely, if the bottom becomes too thick, the lyophilization process is affected by an increase in the resistance to heat transfer, again impacting chamber throughput.

Bottom Flatness or Concavity
The relative flatness of the bottom of a tubing vial is essential to limiting the resistance to heat transfer. The greater the concavity of the bottom, the further away the product in the tubing vial is from the shelf, which is the source of the energy for the lyophilization process (Fig. 1). Excessive concavity also allows more air underneath the tubing vial, adding insulation and further increasing resistance to the heat transfer from the shelf to the product. In both tubing and molded vials, bottoms that have a very concave structure can create angles where the wall and bottom meet, which can act as stress points, potentially causing the bottom of the vial to break away from the sides, as well as creating cracks in the wall. Chamber throughput can be severely lessened as a result.

Bottom Inside Radius
This dimension is controlled by the formation of vial bottoms with a minimum of concavity. The greater the concavity, the sharper the angle inside. If this angle is too sharp, the resulting radius may become a stress point. As a result, when the product is frozen, the ice crystals will expand into the area available, pushing on the bottom, inside radii. This force will create a weakness that can lead to the "lens out" of the bottom of the vial.

Bottom Outside Radius
This angle is critical for maximizing the surface area of the tubing vial on the lyophilization chamber shelf. It is also necessary for the elimination of stress points on the outside radius where the vial contacts the chamber shelf. While

independent of the inside radius, the effects of a sharp outside radius are essentially the same. It is important to have the optimal radii that will lead to minimization of stress in the glass, yet lead to a reduction in the resistance to heat transfer. The optimal rounded diameter will help to absorb and alleviate the stress buildup that occurs during the lyophilization cycle.

Amount of Fill in the Container

The amount of internal pressure generated during the lyophilization cycle is directly related to the amount of fill in the vial. Fill levels of product up to a maximum of 35% of the vial's capacity are generally recommended as these proportions seem to create the fewest processing problems.

Plastic Vial Systems

Because of the introduction of cyclic olefin polymers (COP) and copolymers (COC), these types of materials are now being evaluated in various applications especially for biopharmaceutical and biological applications. These materials combine the inherent break-resistant nature of plastic with the clarity and inert surface that is beneficial to sensitive biopharmaceutical applications.

In addition to the benefits mentioned above, the use of a COC or COP vial brings the additional benefits of lower extractables resulting in a lower likelihood of leachables in the drug product. Glass contains free alkali oxides and traces of metals. At higher pH conditions glass can undergo delamination, as discussed earlier within this chapter. All of these things can lead to a negative impact on the stability of the drug product.

Cyclic olefin–based plastic vials have been studied for packaging lyophilized products. It has been found that lyophilization in some of these materials has led to more uniformity within the freeze-dried cake (3).

The major weakness of these materials is that, even though they have decreased moisture vapor transition rates versus other plastics, they still are not equal to glass from a moisture barrier standpoint. It would always be recommended that a secondary packaging barrier, such as an aluminum pouch, be used to assure adequate shelf life.

LEACHING AND DISSOLUTION OF GLASS

Several studies of glass vials have identified the potential of the glass to react in different ways with liquid products stored in the vials over time. Dissolution or leaching reaction between the liquid and the glass is very probable and is dependent on various specifics of the drug product and its processing in the vial. For instance, basic or high pH products can lead to dissolution of the glass surface.

The lyophilization process minimizes the potential for interactions between the glass and the drug product. In some circumstances, a pharmaceutical manufacturer may decide that lyophilization is the best alternative for extended product shelf life because a compatible glass container for the liquid product cannot be found. An example of this may be a strongly alkaline solution. For this reason, this chapter will briefly explain the concepts of leaching and dissolution as they relate to the glass vial; there will be more focus on the potential of the rubber closures to have extractables that may leach into the lyophilized drug product since this process is more likely to occur.

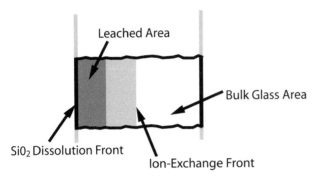

Leached Area

Bulk Glass Area

SiO$_2$ Dissolution Front

Ion-Exchange Front

FIGURE 2 Schematic illustration of the leaching process.

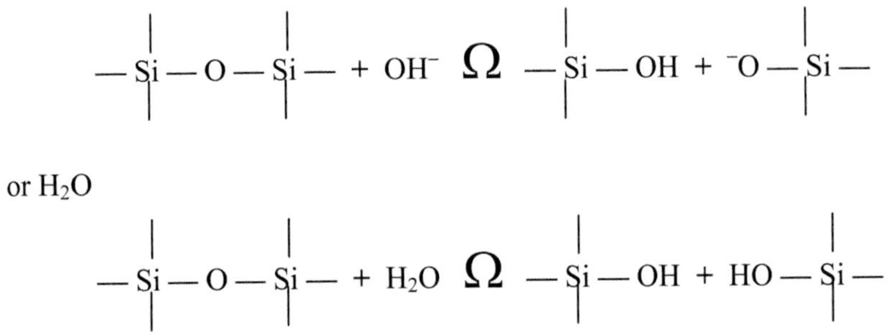

$$-\underset{|}{\overset{|}{Si}} - O - \underset{|}{\overset{|}{Si}} - + OH^- \ \Omega \ -\underset{|}{\overset{|}{Si}} - OH + {}^-O - \underset{|}{\overset{|}{Si}} -$$

or H$_2$O

$$-\underset{|}{\overset{|}{Si}} - O - \underset{|}{\overset{|}{Si}} - + H_2O \ \Omega \ -\underset{|}{\overset{|}{Si}} - OH + HO - \underset{|}{\overset{|}{Si}} -$$

FIGURE 3 Dissolution schematic.

Glass leaching is a selective process. It is primarily an ion exchange process for glass modifiers such as Li$^+$, Na$^+$, K$^+$, Mg^{2+}, Ca^{2+}, and Mg^{3+} (4):

Na$^+$ (glass) + H$^+$(solution) → H$^+$ (Glass) + Na$^+$ (solution) (5)

A schematic view of the general glass leaching process is given in Figure 2.

There is an exchange of metal ions with hydrogen ions from water. This is typical of an acidic solution (6).

The other predominant reaction of water with glass is dissolution. This typically occurs with basic solutions. Glass dissolution reactions result in the release of metal ions and other inorganic materials (5) (Fig. 3).

ELASTOMERIC CLOSURES FOR LYOPHILIZATION
Composition and Manufacturing Process

Elastomeric closures for lyophilization are composed primarily of proprietary rubber formulations based on butyl polymer. Rubber is used because of its capacity to seal well against the surface of a glass vial. Nonhalogenated and halogenated butyl rubbers have been used for years in these types of applications for two primary reasons: low moisture vapor transmission (MVT) rate and low extractability. Each of these characteristics will be covered in detail in this chapter.

There are many considerations in developing and choosing a closure for freeze-drying. These are identified in Table 1.

TABLE 1 Considerations in Developing and Choosing a Closure for Freeze-Drying

Low moisture permeation
Low moisture absorption
Low headspace volatiles
Low extractables
Low absorption and adsorption characteristics
Low oxygen transmission
Low coring/fragmentation
Good reseal
Low surface tackiness during processing
Capability to have good seal integrity with gas or vacuum
Good handling properties during initial and final stoppering

A rubber closure, although predominately rubber polymer, is typically composed of 6 to 12 materials. Each material gives the finished rubber formulation its chemical and functional characteristics. The typical rubber composition is composed of the following ingredients:

Polymer
Filler
Primary curative
Accelerator
Activator
Plasticizer
Pigment
Antioxidants/stabilizers

There may be a combination of inert mineral fillers used in the rubber formulation to reinforce the rubber. Additionally, the primary cure, accelerator, and activator would be considered the "cure system" for the formulation.

The raw materials are weighed according to a formulation card, which is specific for the type and amount of each ingredient. The raw materials are then blended together in a mixer. A final blending step is typically completed on an open mill to assure that adequate dispersion has occurred. The rubber batch is then calendered or extruded in preparation for molding. Typically, compression molding technology is used to produce the stoppers. Compression molding takes sheets of precisely weighed uncured rubber and puts them into a tool that has two molding surfaces, one for the top of the stopper and one for the plug, or drug contact side.

The mold is closed and a specified amount of heat, pressure, and time are applied, forming chemical cross-links within the rubber matrix. This is the phenomenon of curing rubber.

The web of stoppers is removed from the mold and then trimmed into individual closures using an appropriate die. These stoppers are then ready for washing, siliconization, sterilization, and any other processing that is needed.

One important point to remember about traditional thermoset rubber closures, such as those explained above, is that the raw materials undergo a chemical reaction. The application of various types of energy, such as heat, will continue that reaction. At some point, the energy input can begin to break down the cross-links within the rubber matrix, leading to the beginning of degradation of the rubber. This is an important consideration when identifying the drying and sterilization cycles for closures. Too much energy input can lead to issues such as surface rubber degradation. This phenomenon can be evidenced by the tackiness of the closures. Additionally, butyl rubber, by nature, is tacky. The addition of other ingredients and coatings are used to help minimize this tendency.

Physical Properties

Each raw material can have an effect on the various characteristics necessary for an optimal lyophilization closure formulation.

One key physical characteristic of the stopper that relates to its ability to handle well in the stoppering process is the hardness of the vial closure. Typically, a durometer between 45 and 55 (Shore A durometer units) will permit good handling during stopper insertion, yet will not have a negative effect on features such as needle penetration, coring, and reseal. An elastomeric closure that is too hard will have detrimental effects on some of the key functional characteristics that will be important when the final drug product is in the market.

Since a needle will penetrate the lyophilization stopper at least twice, the characteristic of reseal is very important. Reseal is the ability of the rubber to close the hole made by penetration of a needle once it is removed.

The optimal lyophilization formulation will have the correct hardness and reinforcement materials, which will balance the need for good seal integrity (softer, low compression set) and good machinability (harder, higher durometer). The compression set of an elastomeric formulation is important as it will dictate the formulation's ability to keep its rubber characteristics—specifically its ability to bounce back after deformation. The ability of a rubber closure to seal properly is extremely important for several reasons. Typically in a liquid parenteral application, the secondary seal (aluminum crimp) will be applied shortly after stoppering the vial and while it is still in the sterile suite. In the case of lyophilized product, the stoppers are pressed into the vials at the end of the cycle; the stoppered vials are then removed from the chamber. They may then sit for a period of several hours before the aluminum seal is put onto the vial. In some cases, in addition to the concern of microbial ingress, there is also a need to keep a vacuum or gas in the vial. The subject of container/closure system integrity will be discussed in great detail later in the chapter.

Another interesting physical characteristic of the rubber is glass transition temperature. This feature is important because many lyophilized products are stored at temperatures below ambient. The need for the rubber formulation to be able to keep its rubber characteristics under those conditions is extremely important. The glass transition of uncured butyl rubber is typically in the range $-75°C$ to $-67°C$ (7). Finished rubber formulations have typically been found to have glass transition temperatures in the range of $-65°C$ to $-55°C$ (8). Quite often frozen products are shipped under dry ice conditions. These temperatures are below the glass transition temperature for rubber, so special precautions may need to be taken. The glass transition point is the temperature at which rubber becomes more plastic-like. If this occurs, it loses its elastic characteristics.

Closure Configurations Considerations (2)

Drug or biological drying time is critical to the efficiency of the overall lyophilization process. Closure vent design plays a direct factor in allowing the efficient sublimation of water that occurs during the primary drying stages of lyophilization. A theoretical mathematic model for the conductance of water vapor through the stopper vent was developed to aid in the optimization of stopper design. If moisture vapor is considered a gas, the enhancement of its flow depends on flow rate, the properties of the gas, and the geometry of the stopper vent. Two properties can be calculated from this information. These are mean free path, which is the average distance a molecule will travel before colliding with another molecule, and the effective vent diameter, which is the diameter of the largest circle that can be inscribed in the cross-section of the vent. On the basis of these properties, the flow can be characterized as viscous, molecular, or mixed. Through experimentation, the flow of water vapor through a lyophilization stopper vent was determined to be a mixed flow. This means the diameter of the vent is several times greater than the mean free path. The mathematical model for mixed gas flow is shown in Figure 4.

The first term is the viscous flow contribution and the second is the molecular flow contribution.

The important conclusions that can be derived from this model are as follows:

–Conductance is a function of the geometry as well as the area of the vents.
–Conductance is greater for a vent when the width and height are equal.

Overall stopper conductance is computed by calculating the conductance of each individual vent area and summing the results. A small number of large vents is more efficient than a large number of small vents, even if the total vent area is equal. Variations in vent areas are illustrated in Figure 5.

Traditionally there are two styles of lyophilization closure configurations. These are the two-legged style (double vent) and the igloo style (single vent). These styles are illustrated in Figure 6.

It is important when choosing a closure configuration to understand the washing, sterilization and stoppering process, the amount of lubricant on the closure, and the glass vial type and dimensions. A combination of these factors can have an effect on the insertion of the stopper into the vial.

$$C = \frac{ab^3}{1000\mu LK}\overline{P} - 9.71\sqrt{\frac{T}{M}}\ \frac{a^2b^2}{(a+b)L+(8/3)\,ab}$$

a = Width (cm)
b = Height (cm)
μ = Viscosity $\frac{dynes \cdot sec}{cm^2}$
L = Length (cm)
$\underline{K}$ = Constant based on a/b
$\overline{P}$ = Average pressure across vent (dynes cm²)
T = Temp (K)
M = MW

FIGURE 4 Calculation of conductance.

FIGURE 5   Various vent geometries.

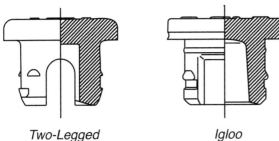

Two-Legged   *Igloo*

FIGURE 6   Two-legged stopper versus Igloo stopper.

Incomplete stoppering can occur as a result of stoppers sticking together or if the stopper-container interference fit is not optimal. This is an important factor in giving the freeze-drying process a good start. When the stopper is inserted into the filled container, it needs to be at the proper position. It cannot be angled or misplaced prior to chamber loading. Second, when the lyophilization chamber shelves are lowered after completion of the lyophilization cycle, the stopper must be able to be pushed into the vial and fit snuggly. The stopper cannot be pulled out because of tackiness nor can it pop out of the vial because of inappropriate interference fit between the stopper and the glass or because of overlubriciousness. A common mistake is applying too much silicone oil to lyophilization closures to assure they would not stick to the chamber shelves. This may cause a pop-out problem during stopper insertion because of the excessive lubricity provided by the silicone oil. Issues with lyophilization and siliconization will be discussed to a greater depth later in this chapter.

Chemical Properties

The chemical properties of any drug primary container/closure system are very important. These properties bear a direct relationship to compatibility, stability, and leachables in the final dosage form.

Chemical properties are an important consideration when choosing a primary closure. Each lyophilization cycle and drug product will have its own characteristics and processes, so understanding the closure formulation that will be used for each independent application is critical.

Extractables/Leachables

In the Guidance for Industry entitled "Container Closure Systems for Packaging Human Drugs and Biologics," released by the Center for Drug Evaluation and Research (CDER) and Center for Biologics Evaluation and Research (CBER) divisions of the U.S. Food and Drug Administration (FDA), there is information to help the manufacturer better prepare for New Drug Application (NDA) submissions. One of the newer areas of interest discussed in the guidance is extractables and leachables. Extractables are species that can extract from the packaging component under stressed conditions with various solvents. Leachables are the container/closure extractables that are found in the drug product. There is concern with liquid drug products that the potential to extract is greater because of direct solvent-vehicle interaction and refluxing of solvent into the component over time. However, the issue of volatile extractables plays a major consideration for lyophilized products. Extractables from a rubber compound are important to consider even with freeze-dried products because they may alter the composition of the reconstituted drug product either directly, by interaction, or indirectly, by changing a formulation parameter such as pH.

There have been several fully documented cases of ingredients from a rubber closure migrating from the closure to the lyophilized drug product, quite often resulting in haze of the reconstituted solution. Pikal and Lang (9) in their study concluded that at the low-pressure characteristics of the environment in the freeze-dryer, vapor phase diffusion is fast enough to allow significant quantities of sulfur and wax, two raw material components of a specific stopper formulation, to transfer from the closure to the product surface where absorption occurs. The lyophilization cycle and the elastomeric closure formulation are two critical variables in this type of occurrence that can lead to unsatisfactory product elegance in the market.

Low molecular weight materials such as oils, waxes, and polymer fragments may become absorbed on the surface of the freeze-dried products, and this may prevent complete dissolution upon reconstitution.

The absorption of other volatiles from the stopper may not lead directly to a solubility issue but are still leachables in the drug. Examples of this are breakdown products of a peroxide-cured rubber formulation. Alcohols and ketones are examples of some of these resultant leachables.

Dependent on the uniqueness of the situation, these extractables may also leave deposits within the lyophilization chamber, another issue to consider.

These examples show the importance of selecting the appropriate leachables testing for the drug products. Volatilization can continue over the drug product's shelf life; therefore, it is important to evaluate leachables over the shelf life of the product to make sure that the closure, process, and drug product are optimal.

MOISTURE

There are at least three sources of moisture in a lyophilized drug. The first is residual moisture in the drug following the lyophilization cycle. The second source is moisture from the environment that may pass through the closure or the seal. The third source is moisture from the stopper itself. If the stopper contains moisture when applied to the vial, the cake has the potential to absorb this moisture.

Moisture Vapor Transmission

The permeation of moisture through the stopper is MVT. Typically lyophilization closure recommendations are for stoppers that contain nonhalogenated butyl or halogenated butyl rubber. The MVT rate of these polymers is low in comparison to alternative polymers. Table 2 lists a comparison of typical MVT rates.

Moisture Absorption

As discussed earlier, a finished rubber formulation is composed of many different materials. These materials bring a certain amount of retained moisture to the rubber batch; additionally, each may have its own implications in the formulation's capacity to absorb water. This can occur directly from the environment, but more often occurs during the sterilization cycle. Most closures are steam sterilized through an autoclave process. This process pushes steam into the matrices of the rubber. The composition of the rubber will have a direct impact on the amount of moisture absorbed and retained.

Under normal conditions, the closures should be put through a drying cycle to remove residual water. This cycle should be designed not only to

TABLE 2 Comparison of Typical Moisture Vapor Transmission Rates

Formulation type	Moisture vapor transmission (g/m² per day)
Natural rubber	9.0
Synthetic isoprene	13.7
Ethylene propylene diene monomer	1.8
Nonhalogenated butyl	0.2
Bromobutyl	0.3
Chlorobutyl	0.1

TABLE 3 Moisture Vapor Transmission Vs. Moisture Absorption Comparison

Stopper formulation	Moisture vapor transmission (g/m^2 per day)	Weight gain (%)
EPDM 1	2.15	0.48
Butyl 4	0.65	0.87
Butyl 5	0.50	1.95
Butyl 2	0.40	1.39
Butyl 1	0.35	0.27
Butyl 3	0.10	1.31

Abbreviation: EPDM, ethylene propylene diene monomer.

remove the water on the surface of the stopper but also to dry the internal moisture.

The final cake weight for biotechnology-derived drugs may be smaller than that of a traditional pharmaceutical product. The same amount of moisture that may have been acceptable with a larger cake weight product may cause problems for biotech products.

Research (10) found that MVT at equilibrium bears no relationship to the moisture absorption capacity of a rubber formulation. This was a critical determination because, previously, it was assumed that MVT rate was the only real characteristic of concern in relationship to a rubber closure and its use for lyophilization. Table 3 lists both MVT and percentage weight gain for several rubber formulations. Weight gain can be used to understand the capacity of an individual stopper to absorb moisture after an autoclave cycle or some other environmental conditioning.

It is also important to understand how a stopper formulation will "dry" or rid itself of moisture. Weight loss methods are not satisfactory for this characteristic (10). These methods will count not only moisture loss but also the loss of other volatiles from the rubber. To quantify moisture in rubber closures accurately, it is recommended that the Karl Fischer coulometric titration be used. A drying oven is used to drive moisture out of closures that have been cut into small pieces to facilitate the drying. Dry nitrogen is used to carry the moisture from the oven to the titrator. Methods have been developed and validated to assure accuracy of measurement. These methods can be used to help develop and validate drying cycles used in production for processing lyophilization stoppers.

Another important fact is that closures must be dried before the lyophilization cycle. The drug lyophilization process does not dry the rubber closure because the stopper does not experience the same cooling and heating cycle as the drug. Typically, a two-step process is used to optimize the drying cycle for a load of closures. Again, it is important to realize that the validation of this drying cycle is specific for the rubber formulation, its configuration, the amount of closures in a container, the type of container holding the closures, and the equipment being used to dry the closures.

The first phase of designing the appropriate drying cycle is to understand, under normal processing conditions, the effect extended drying has on the residual moisture in the closure. This can be understood by plotting the amount of moisture in the closures by the amount of time they were dried. The drying time should be extended until the amount of moisture found in the closures has leveled off. This value can then be used as a target to understand process consistency. Multiple samples should be taken from multiple process runs to validate that the designed drying process does routinely get the closure moisture load to an acceptable level. It is important to take samples from different areas of the container holding the stoppers—for instance, the sterilizable bag or the stainless steel vessel—to assure that sufficient drying has occurred in all areas of the container.

The other consideration in designing extended drying cycles is to understand the implication of temperature and time on the rubber formulation for characteristics other than moisture content. For instance, does surface tackiness of the closures increase or does coring of the closure change because of additional heat input?

Moisture Transfer from Closure to Product

Studies have shown (11) that the amount of moisture in a stopper bears a direct relationship to the amount that may be transferred to the freeze-dried product (Table 4).

In this study, the average percentage of moisture initially after freeze-drying compared with that after three-month storage gives an indication of the amount of moisture gained by sucrose that had been lyophilized using closures containing varying levels of moisture. Initial testing of the sucrose shows that it contains approximately 2% to 2.5% moisture immediately after freeze-drying. If 3% was a maximum specification for stability of the product, only the sucrose packaged with one of the closure formulations would be within specification after three months of storage. Refer to Table 5.

The study demonstrated that the larger the amount of initial water held by a closure, the more the moisture is passed to the dry product. The product,

TABLE 4 Milligrams of Water Absorbed by Sucrose After Lyophilization Compared to Moisture Load in 20-mm Lyo Closure

Stopper formulation	Amount of water in 20-mm closure when vials were sealed[a]	Amount of water gain by sucrose in 3 mo[b]
Butyl Y	10.9	3.6
Butyl Z	9.5	2.5
Butyl X	6.3	1.7
Butyl W	2.8	1.4

Per Karl Fischer testing.
[a]Average milligrams of water per triplicate testing.
[b]Average milligrams of water per quadruplicate testing.

TABLE 5 Percentage of Moisture in Sucrose After Lyophilization Vs. Three-Month Storage

Name	Percentage of water in sucrose immediately after lyophilization[a]	Percentage of water in sucrose after 3 mo storage at room temperature[a]
Butyl Y	2.25	3.91
Butyl Z	2.45	3.62
Butyl X	2.34	3.12
Butyl W	1.95	2.65

Per Karl Fischer testing.
[a]Average percentage of water per triplicate testing.

however, will reach a maximum amount of moisture as equilibrium develops between the closure and the dry product.

SILICONIZATION OF CLOSURES

Another area where problems may be encountered in processing and use of closures for lyophilization is siliconization. Siliconization is the process of applying silicone oil to the surface of a rubber closure to give the surface adequate lubricity. Lubricity aids in processing or machinability. This is critical in relation to several specific areas, such as the application of the stopper to the vial. Typically a stopper moves down a chute, is applied to the mouth of the vial, which is then placed in the lyophilization chamber. The stoppers are seated in a position that allows sublimation of the water to occur. After sublimation, a vacuum or nitrogen may be used to backfill the vial headspace. The shelves of the lyophilization chamber move to press the stopper into the vial in a closed position, sealing the vacuum or inert gas inside the vial. The seal between the rubber and glass vial must hold for a period of time before the secondary aluminum seal is applied. The application of silicone at an optimized level is important for multiple reasons. In applying the stopper to the vial, the closures are typically placed in high-speed sealing equipment. Any hesitation of the rubber closures can cause a problem with this system. Siliconization is critical to the closures when they are in the lyophilization chamber to prevent the closures from sticking to the shelves after they are pressed into the vials so they cannot be pulled out of the vial when the shelving is raised. Additionally, insertion into the vial needs to be smooth to facilitate the entire process; too much or too little silicone may cause problems with friction during insertion and may cause pop-up of the stopper after it is inserted if there is too much lubricity.

Process for Siliconization

Typically silicone oil is applied to closures during one of the final rinses in the wash process prior to sterilization. A mechanical emulsion of the silicone oil (e.g., Dow Corning 360 Medical Fluid) and water should be made. This is then applied to the rinse water in the washer. The wash cycle should be optimized to assure adequate distribution of the silicone throughout all of the closures.

Some companies may use a pre-made emulsion such as Dow Corning 365 Medical Grade emulsion; however, these types of emulsions contain materials other than silicone oil and water to keep them in the emulsion phase.

Problems with Silicone

Excess silicone may cause various problems with injectable products. These range from droplets in solution to hazing of lyophilized product.

A documented study (12) conducted by Preston et al. showed the potential for the lyophilization cycle with an aqueous solvent vehicle create a residue, found to be sodium bicarbonate from the type I vial and silicone oil from the rubber closure. This caused hazing in the vials.

In addition to the silicone oil itself being found as droplets, there are also documented examples of the silicone oil particles acting as a nucleus to attract other materials, such as proteins. An example of this phenomenon is documented in a study of particle generation of siliconized stoppers and tumor necrosis factor formulations (13). One of the conclusions drawn from the study was that oil droplets promoted growth of particles by adsorbing to them, thus allowing sites for additional particle growth.

Testing for Silicone Oil

The presence of silicone oil on closures can be easily identified qualitatively through infrared (IR) spectroscopy. Typically strong bands are found in the 1200 to 1000 cm^{-1} region, near 2900 cm^{-1} with additional sharp bands in the 1270 to 1250 cm^{-1} and 874 to 740 cm^{-1} regions. A quantitative method for understanding the amount of silicone oil applied to the closures is required to determine optimization of silicone oil application. This can be conducted through a quantitative IR method or through atomic absorption methods. In both cases, the silicone oil is removed from the stoppers with an organic solvent. The amount of silicone oil is then quantified in the solution by comparing the samples to predetermined standards and calibration curves.

These methods can be used to assist in silicone level and application optimization.

Methods have also been developed to quantify the amount of silicone oil in some drug products or placebos. Typically atomic absorption is used following a liquid/liquid extraction of the drug solution. These methods are valuable in helping to analyze the cause of a hazing or leaching problem.

Alternatives to Silicone

For many years silicone oil was the only material added to closures for lubricity purposes. Closure manufacturers have developed several alternatives to the traditional silicone oil that can be used in a similar function. The alternatives range from a cured silicone that is retained onto the closure surface better than traditional silicone oil, to films that are based on fluoroelastomer polymers. Each of these alternatives can be beneficial. Their use helps avoid the need to optimize silicone application by the drug manufacturer and, in some cases, such as with the film coatings, additional benefits in areas such as reduction of extractables is typical.

CONTAINER CLOSURE SYSTEMS

One extremely important feature of the entire package is its need to keep a seal after the lyophilization process. Many features have to come together to achieve this in a satisfactory manner. This is more of a challenge for a lyophilized product than a liquid injectable because the seal must be maintained for several hours without the application of the aluminum overseal. The elastomeric formulation, the closure shape, any closure coating, and the glass vial all must fit together perfectly to achieve the required sealing characteristics. The fit between the rubber and the glass is critical. The use of, for instance, a rubber formulation that is too hard may cause a problem because harder formulations become less elastic. Closures with coatings must be evaluated carefully. The addition of a coating may increase surface hardness or may, depending on its area of application, become a barrier between the glass and the rubber. This barrier or increase in hardness prevents the rubber from sealing adequately, especially where there may be nicks or lines in the surface of the glass vial finish. If the rubber does not seal these areas, they create routes for ingress of contamination or egress of gases.

CONTAINER CLOSURE INTEGRITY METHODS

Many methods have been used by industry to evaluate seal integrity. These range from very insensitive methods such as bubble testers to extremely sensitive methods such as helium leak detection.

Methylene blue testing is a common method used to assure seal integrity. Methylene blue, however, tests to assure that the product meets sealing requirements for liquid transfer or product loss. In many cases, lyophilized products are actually holding an inert gas or vacuum. Methylene blue methods or other more traditional procedures are not sensitive enough to evaluate a package for loss of these gases. Neither are they sensitive enough to optimize or validate a sealing process for a package that is designed to hold a gas. The best alternative in this case is a method that uses a gas such as helium. Additionally, a method such as helium leak detection can give quantitative data on the package and its leak rate, thus giving better information that allows for improved problem solving and optimization.

ALUMINUM CRIMP ALTERNATIVES

Injectable products, including lyophilized products, are typically sealed with aluminum seals with plastic buttons on top that must be removed to access the medicament. These plastic buttons are typically made of polypropylene material. In some cases, certain lyophilized products may be held at extreme conditions during shipment or storage. These plastic buttons may not be able to withstand these conditions because of their composition. In these cases, specially designed buttons manufactured from materials that can withstand extreme temperature and shipping conditions are used.

CONCLUSION

Many facets of the container/closure system should be taken into consideration to expedite the development, scale-up, and production processes for lyophilized materials. It is important to remember that just as every drug product is unique, so is its packaging requirements. A thorough evaluation early in the development process will help to avoid problems later in the cycle.

REFERENCES

1. The Pink Sheet; FDAs CBER and CDER web sites (April 7, 2009). Excludes diagnostic and blood products.
2. The West LyoTec® System Manual. Technology to Optimize Your Cost Effective Lyophilization Process. Lionville, Pennsylvania: West Pharmaceutical Services Inc. (Formerly The West Company, Inc.), 1990.
3. West Pharmaceutical Services (formerly The West Company) technical study. Vial-to-Vial Temperature Mapping of Groups of Vials During Freezing and Freeze Drying. November 30, 1995.
4. John WY, Chien Yie W. Sterile Pharmaceutical Packaging: Compatibility and Stability. Parenteral Drug Association, Inc., Technical Report No. 5, 1984.
5. Borchert SJ, Ryan MM, Davison RL, et al. Accelerated extractable studies of borosilicate glass containers. J Parenter Sci Technol 1989; 43(2):67–79.

6. Doremus RH. Chemical durability of glass. In: Tomozawa M, Doremus RH, eds. Treatise on Materials Science and Technology, Glass II. Vol. 17 New York: Academic Press1979:41–69.

7. Roff WJ, Scott JR, Pacitti J. Handbook of Common Polymers. Fibres, Films, Plastics and Rubber. Cleveland, OH: CRC Press, 1971.

8. Alexander B. West Pharmaceutical Services Inc. (formerly The West Company, Inc.), Inter-Office Correspondence Reference Project # WUTS950821TS, August 21, 1995.

9. Pikal MJ, Lang JE. Rubber closures as a source of haze in freeze dried parenterals: test methodology for closure evaluation. J Parenterl Sci Technol 1978; 32(4):162–173.

10. DeGrazio FL, Smith EJ. Moisture Content of Lyophilization Stoppers. Parenteral Drug Association Summer Meeting, Rosemont, IL, June 7–8, 1990.

11. DeGrazio FL, Flynn K. Lyophilization closures for protein based drugs. J Parenter Sci Technol 1992; 46(2):54–61.

12. Preston WA, Baldoni J, Haeger BE, et al. Residue in vials after lyophilization J Parenter Sci Technol 1987; 41(1):40–41.

13. Hora MS, Rana RK, Smith FW. Lyophilized formulations of recombinant tumor necrosis factor. Pharma Res 1992; 9:33–36.

17

Extractables and Leachables as Container Closure Considerations in Lyophilization

Diane Paskiet and Fran L. DeGrazio
West Pharmaceutical Services, Inc., Lionville, Pennsylvania, U.S.A.

INTRODUCTION

The suitability of the container closure systems (CCS) for lyophilized pharmaceutical and biological products can be enhanced using the principles of quality by design (QbD) and employing risk management tools. Leachables, chemical constituents that migrate from packaging materials into the pharmaceutical product under normal conditions, can be minimized or eliminated when components are considered early in the pharmaceutical development process. A set of experiments can be designed to facilitate material selection by profiling materials for extractables using exaggerated conditions and assessing the impact to the pharmaceutical product and ultimately patient safety.

The classification of a CCS is not limited to the final package with respect to extractables and leachables in lyophilized pharmaceutical products. Materials that directly or indirectly come into contact with the pharmaceutical product at any time during the manufacturing, cleaning process, storage, or administration are candidates to be considered for extractables/leachables evaluation. In addition, secondary and ancillary or associated components may have an impact on patient safety and should be considered in the overall design space.

Sources of potential hazards can be traced back to the chemistries of the CCS materials of construction. Functional parameters for CCS materials used with a lyophilized pharmaceutical can be identified from a set of experiments designed to lead to an in-depth understanding of the chemistries of the critical components. Knowledge gained from these studies will enable science-based decisions for identifying critical quality attributes (CQAs) and provide a proactive approach to achieving quality. This chapter illustrates the selection and qualification of critical CCS components intended to be in contact with a lyophilized pharmaceutical or biological product and strategies for acquiring the appropriate data to contribute to the overall drug product quality.

OVERVIEW

The principles of QbD can be qualitatively applied to selection and control of materials used in the manufacture, storage, and administration of lyophilized pharmaceuticals. The following concepts are considered with respect to CCS:

1. CQAs for a CCS
2. Identification of critical CCS components for extractable/leachable—assessment using risk management tools
3. Strategies for CCS qualification and control

The design of experiments for qualifying CCS for lyophilized pharmaceutical products is centered on extractable studies, incorporating factors associated to the manufacturing process and variables indicating worst-case exposure conditions. Recommendations for a practical leachable assessment model are given based on

1. Qualitative and quantitative extraction studies
2. Assessment of CCS analytical profile
3. Evaluation of lyophilized pharmaceutical products for leachables
4. Correlation of leachables to extractables and lifecycle management

CRITICAL QUALITY ATTRIBUTES FOR CCS

Pharmaceutical product CQAs are managed through the critical process parameters (CPP) established during the validation of the manufacturing process. A subprocess to the pharmaceutical product manufacture is that of the CCS in which CQAs are established to ensure suitability for use, along with employing good manufacturing practices (GMP).

> Device containers should not be reactive, additive or absorptive as to alter the safety, identity, strength, quality or purity of the drug beyond the official or established requirements for the drug product. Standards or specifications, methods of testing, and where indicated methods of cleaning, sterilizing and processing to remove pyrogenic properties shall be written and followed for drug product container and closures (1).

CCS for lyophilized products and their associated components have a direct impact on the pharmaceutical product CQA. Suitability of CCS is based on the four major building blocks shown in Figure 1 (2). The sum of these characteristics define the target CCS and establish CQAs.

Safety is paramount to package suitability and the factors that define patient safety are linked to the chemistries of the materials of construction and the potential to contaminate the pharmaceutical product. A CQA of a CCS is the absence or control of substances that may leach into the pharmaceutical product from the CCS in conjunction with aspects of performance, protection, and compatibility. Building quality into the manufacturing process requires an enhanced understanding of the component manufacture, properties, and chemistries to identify CPP. Chemical entities that are needed to enhance the properties of the materials for superior performance can influence the overall compatibility as well as the functionality and container closer integrity. There are unique approaches for evaluating the key aspects indicating that the CCS is suitable for the intended purpose; however, there is a universal approach to address the safety aspect. A chemical evaluation of the critical materials of construction for extractables, followed by a leachable study, will indicate the risk for pharmaceutical product contamination and patient safety contributing to a scientifically informed decision for material selection and qualification of the system components.

FIGURE 1 Key aspects of CCS suitability.

IDENTIFICATION OF CRITICAL CCS COMPONENTS FOR EXTRACTABLE/LEACHABLE ASSESSMENT USING RISK MANAGEMENT TOOLS

Components used in a multistage manufacturing process, along with the final package and drug administration system, encompass many material variables that can be classified as part of the CCS.

> The choice and rationale for selection of the container closure system for the commercial product should be discussed. Consideration should be given to the intended use of the drug product and the suitability of the container closure system for storage and transportation (shipping), including the storage and shipping container for bulk drug product, where appropriate (3).

Identification of the components likely to contaminate or interact with the pharmaceutical product is crucial to the pharmaceutical product CQA. These defined CCS materials can be exposed to conditions of low pressures, high and low temperatures, and solvents for extended periods of time. Exposure to harsh conditions has the potential to cause constituents from the CCS materials to volatilize, solubilize, bloom, or degrade, impacting the purity of the pharmaceutical product. Exploring functional relationships between a lyophilized pharmaceutical product and the CCS will enable specifications to be defined, building CQAs into the design space. The chemistries of the CCS materials are integral to the manufacture and performance of the components, and detailed

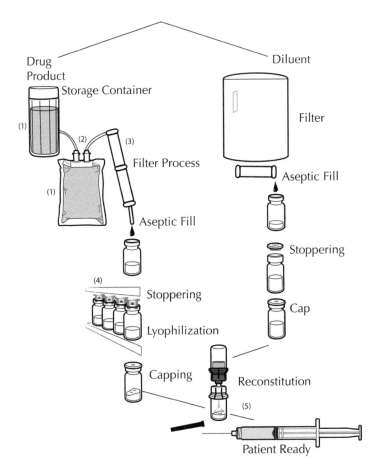

FIGURE 2 Progression of the manufacture and administration of a lyophilized pharmaceutical.

understanding of the component formulations is necessary to indicate suitability. The complexity and the wide range of materials associated with the CCS for a lyophilized pharmaceutical product is made evident from Figure 2 and Table 1. There are various manufacturing processes and combination of material options; this list is a representative sampling to provide an understanding of possible materials.

As described previously, the CCS for a lyophilized drug cannot be limited to the final container and there are wide ranges of components that may have an impact on pharmaceutical product quality. As an example, a model CCS used for administration of a lyophilized pharmaceutical is shown in Figure 3.

The material formulations should be optimized to produce an acceptable CCS product low in extractables; however, this may not be possible when certain performance requirements are to be attained. The impact to the materials used in lyophilization processing, cleaning, storage, reconstitution, and administration can be assessed for the likelihood of drug product contamination based on a leachable and extractable study. The initial challenge is to rationalize the selection of the critical components to be assessed in studies for a lyophilized

TABLE 1 Typical CCS Components Used in Lyophilization and Materials of Construction

Component	Materials of construction
Chambers, storage containers, tank liners, single-use bags (1)	Stainless steel, polycarbonate, polyvinyl chloride, thermoplastic elastomers, cyclic polyolefin
Tubing (2)	Polyvinyl chloride, silicone, polyurethane, latex
Gaslets, seals (2)	Styrenebutadiene rubber, ethylene proplyene diene monomer, thermoplastic elastomers
Filters, connectors (3)	Polyesters, polypropylene, nylon, polytetrafluoroethylene, polyethersulfone
Elastomeric closures for lyophilization and reconstitution (4)	Butyl, chlorobutyl, bromobutyl
Vials for lyophilization and reconstitution (4)	Glass, cyclic olefins
Reconstitution, transfer, and administration devices (syringe, plunger, tip caps, needle) (5)	Cyclic olefins, polyethylene, butyl rubbers, polycarbonate, tungsten pins, stainless steel

Abbreviation: CCS, container closure systems.

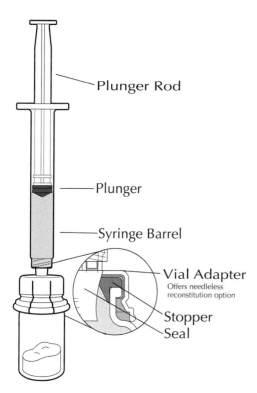

Plunger Rod

Plunger

Syringe Barrel

Vial Adapter
Offers needleless
reconstitution option

Stopper
Seal

FIGURE 3 Diluent in prefilled syringe and transfer device on lyophilized product.

pharmaceutical product. An adaptation of a preliminary hazard analysis (PHA) model to indicate the critical components for leachable and extractable evaluation is shown in Table 2 (4). The model assesses each potential hazard for the likelihood of leachables and consequence is based on the highest risk score of 120 (severity, 0–5; frequency, 1–3). The model highlights the key considerations for CCS properties relative to pharmaceutical lyophilization and a system for

TABLE 2 Example PHA for Materials Vs. Hazards for Potential Leachables

Potential for leachables	Investigate hazard	S	F	Imp $S \times F$
Component	Significant contact area	5	3	15
	Long duration	5	3	15
Primary material	Solubility in media	5	3	15
	Exposure to high temperatures	5	3	15
	Sublimation of chemical constituents	5	3	15
	Conditions above T_g	5	3	15
	Previous data unknown	5	3	15
	Known constituents of concern	5	3	15

Abbreviations: PHA, preliminary hazard analysis; S, severity; F, frequency.
(3) ICH Q9 Quality Risk Management

ranking. A standard method for identification of hazards or rating of those hazards does not exist for CCS materials used with pharmaceutical products as it depends on the particular system being analyzed and the target profile. Material rankings are defined relative to the individual system and are subjective; typically a conservative approach is taken without prior experience or regulatory direction. The PHA of the existing system will enable proper components to be selected and prioritized for further studies early in the development process when little information on the materials of construction is known.

QUALIFICATION AND CONTROL OF CCS

The studies required to qualify as CCS are extensive, incorporating both chemical and toxicological evaluations. The testing strategy can build depending on the pharmaceutical phase of development. Initially, CCS features such as performance attributes, compendial tests, and Food and Drug Administration (FDA) sanctioned (GRAS, generally recognized as safe) materials and/or designated as medical grade should be verified. Further testing can be planned in advance of the next stage to allow for appropriate protocols and sampling plans to be established. The work can be initiated relative to risk-benefit factors. Advanced planning for CCS studies can be guided by the pharmaceutical development phases; Figure 4 illustrates activities to be considered.

The depth of data to acquire becomes increasingly important as phase 2 approaches. Knowledge drawn from initial studies will facilitate a proper experimental design. During the preclinical phase, critical CCS components used for manufacturing, storage, and administration should be investigated so that in phase 1 the material selection can be narrowed down. Compatibility studies should be planned for the following indications:

- Loss of active/denaturing/binding
- Detection of interaction and degradation products
- Migration of CCS chemical constituents
- Formation of precipitates
- Detection of particles
- Discoloration of product/package
- Drug product pH change
- Brittleness of package component
- Change in CCS physical properties
- Inadequate CCS protection/performance

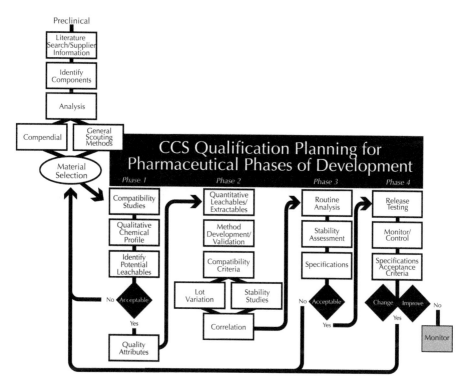

FIGURE 4   CCS Qualification planning for pharmaceutical phases of development.

Quality attributes can be recognized from compatibility evaluations for all CCS, which include materials from primary, secondary, ancillary, intermediate, cleaning, filling, sterilization, and lyophilization processes. Components introduced at scale-up or during upstream processes also have the potential to impact the leachables profile. An extractable profile using aggressive solvents can be developed and the data compare to that of the pharmaceutical product after exposure to CCS. As a first pass, the potential for constituents to leach can be targeted for further studies or could trigger investigation for an alternate material.

In phase 2, reliable quantitative measurements should be considered for critical extractable components and asymptotic levels of extractables generated to correlate to leachables. The methods should be optimized to achieve limits that can be associated to patient safety. Criteria for the overall compatibility can then lead to the CCS CQA and target methods to validate for analysis of multiple CCS lots and leachable stability studies.

Routine testing can be initiated in phase 3, which will contribute to acquiring a database on component extractables for the purpose of setting specifications and developing acceptance criteria. Indications for process improvements may be recognized at this point and requalification of a new component may be triggered to achieve the desired CQA. When leachables are to be controlled, routine testing and/or verification of CCS for CQA will continue into phase 4 as needed to ensure the reliability of the CCS supply chain.

QUALITATIVE AND QUANTITATIVE EXTRACTION STUDIES

The goal of an extraction study is to provide information on the chemistry of the CCS materials of construction. The major objectives for acquiring extractable data are to (*i*) screen potential CCS components under exaggerated conditions to learn the chemistry and identify possible "bad actors", (*ii*) select and qualify the components under worst-case conditions, (*iii*) gain knowledge to design and conduct a leachable study, and (*iv*) provide data for quality control and lifecycle management.

> Chemistry manufacturing and control information is expected for every material used in the manufacture of packaging components. The complete chemical composition and results of extractable/toxicological evaluations for pharmaceutical products that are likely to interact with packaging component and introduce extracted substances into patients should be submitted in the original application 2. Furthermore, the Common Technical Document format CTD section 3.2.S.6 indicates the suitability should be discussed with respect to choice of materials, protection from moisture and light, compatibility of the materials of construction with the drug substance, including sorption to container and leaching, and/or safety of materials of construction (5).

It is the regulatory expectation that extracted substances from CCS that have migrated into a pharmaceutical product (leachables) should be evaluated to identify, eliminate, and/or control the leachable substances. Identification of potential leachables is accomplished through comprehensive extractables studies. Migration of minor constituents from component materials into a solvent is dependent on the migrant's transport and thermodynamic properties in both the component and solvent. The potential for chemicals to migrate from the component(s) is ascertained through extraction studies and linking data to leachables, thereby helping to assure patient safety.

The presence or absence of leachables is substantiated using aggressive extraction conditions. Voluminous extractable data can be generated; however, all extractables detected are not necessarily leachables but all drug product leachables need to be traced back to extractables. Extractables may have a direct correlation to substances found in the pharmaceutical product that may have originated from the CCS formulation ingredients or indirect relationship resulting from an extractable degradation or interaction product. The extractable experimental design should be developed for broad application. The first phase of the study would screen to discover potential leachables using aggressive solvents of varying polarity and one with similar propensity to the pharmaceutical product. The CCS components would be exposed to exaggerated conditions and analyzed using multiple analytical techniques. The extractable data would then benchmark potential substances, which may contaminate the drug product and cause harm to a patient. Evidence of CCS performance and compatibility can also be derived from the extractable studies. Performance additives can be monitored and measured, when necessary, to indicate stability and proper function of the components. CQAs for protection, performance, compatibility, and safety developed for candidate CCS components will need to be managed holistically, and the understanding of the component chemistries is essential to show that a CCS is suitable for its intended purpose.

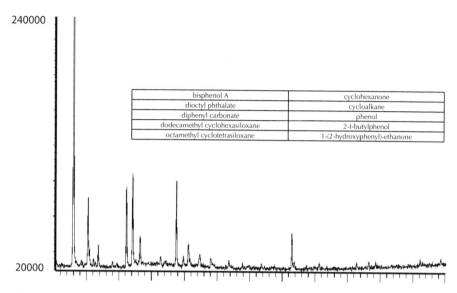

bisphenol A	cyclohexanone
dioctyl phthalate	cycloalkane
diphenyl carbonate	phenol
dodecamethyl cyclohexasiloxane	2-t-butylphenol
octamethyl cyclotetrasiloxane	1-(2-hydroxyphenyl)-ethanone

FIGURE 5 Chromatogram and data of a GC/MS screen.

Discerning attributes that impact patient safety is a complex task, encompassing the discovery, identification, and measurement of extractables. Information from the product description, literature searches, and required compendial tests will only provide basic information. Screening the actual materials using a harsh solvent and conditions will provide data that would signal potential substances of concern. General scouting techniques such as pH, ultraviolet/visible spectroscopy, conductivity, nonvolatile residue, and infrared spectroscopy can be used to obtain corroborative data. Gas chromatographic (GC) and liquid chromatographic (LC) separation techniques with mass spectrometric (MS) detection methods would provide explicit extractable data indicating chemical entities that may require a toxicological evaluation. An example of a typical extractables profile for polycarbonate material refluxed in 2-propanol and analyzed by GC/MS is shown in Figure 5 along with identified chemical constituents.

The potential of identified extractables to contaminate the pharmaceutical product in relation to other aspects of compatibility, performance, and protection will enable an informed decision for CCS selection. Once the selection of the appropriate CCS materials is made, further testing is done to build knowledge for acceptance criteria and to associate CPP. The critical extractables associated with a negative impact to patient safety should be identified, monitored, and controlled.

ASSESSMENT OF CCS ANALYTICAL PROFILE

The extractable profile is tied to patient safety based on the probability of the chemical entity to migrate and capacity to cause harm. Extractable migration is determined from a properly conducted leachable study followed by a toxicological evaluation of leachables to indicate patient safety. A toxicological evaluation of all extractables may not be practical and a PHA can be employed during this stage to assess individual extractables for the potential to migrate under normal conditions into the pharmaceutical product matrix. All extractable data should be judged for the purpose of selecting the analytes to target in a leachable study and for information that may be useful to control component performance characteristics, both capable of affecting patient safety. The capacity of an extractable to become a leachable and negatively impact a patient can be predicted based on (*i*) total amount of the extractable per part(s), (*ii*) potential for the extractable to migrate from the material and diffuse into the pharmaceutical product matrix based on properties of solubility and volatility, (*iii*) duration of pharmaceutical product contact relative to temperature and pressure, (*iv*) patient contact with the materials, and (*v*) toxic effects of the chemical entity. Leachables are linked to patient safety and depend on predicted total daily intake, duration, route of administration, and patient population. Resources should be managed to provide the necessary data to indicate possible mutagenic, carcinogenetic, and reproductive effects as well as possible sensitization and acute and chronic systemic effects. Careful selection of the extractable species to target for leachable evaluation and subsequent toxicological study is a major consideration when building quality into the CCS design space.

EVALUATION OF PHARMACEUTICAL PRODUCTS FOR LEACHABLES

The awareness of the level of risk for leachables begins with extractable data but an understanding of the leachable trends over the shelf life of the product is needed to assess patient safety and to assure an understanding of the maximum value expected over its shelf life. Sources of leachables can be traced to formulation ingredients or processing aids (i.e., lubricants, plasticizers, monomers, antioxidants, and curatives) that are used in the component manufacturing and identified from the extraction data. Leachables may also result from unexpected interactions with the pharmaceutical product or by-products occurring from component extractables. The extractable data may have been predictive of leachables, but detection of interaction products can only be identified from examination of the actual pharmaceutical product in contact with the packaging under representative conditions. The experimental variables to consider for a leachable study are based on the product configuration, temperature, pressure, and time. Appropriate sampling would take into consideration exposure to the environmental extremes during manufacture, filling, shipping, and storage. The extractable profile of the migrants from multiple component lots, reaching thermodynamic equilibrium, can be used to establish trends and be correlated to leachables.

The extractable analytical procedures can be used as a starting point for leachable screening but the methods usually require optimization to achieve the appropriate separation and sensitivity. Measurement of the leachable species can be challenging in certain matrices; interferences from the pharmaceutical product are commonly encountered and concentration of the sample may be necessary to achieve the target sensitivity. Accurate and precise leachable measurements will allow a correlation to the extractable data. To achieve accurate measurements of the specific compounds, the methods require validation. It is expected that validation data will be acquired to demonstrate the proper recovery, precision, linearity, and robustness for the established limits of detection and quantitation. The FDA and International Conference on

Harmonization (ICH) have published guidelines describing the process for method validation (6).

The CCS components used in the final packaging should be included in the stability protocol and evaluated for signs of instability as well as correlated to drug product leachables. Measurement of leachables at an accelerated stability point can be used as an initial worst-case indicator, but there is a risk of generating anomalous data leading to inconclusive or inaccurate results. It is recommended to evaluate the stability of the CCS and the pharmaceutical product for CCS interaction products throughout the entire stability program on several lots and multiple sample configurations. "The ultimate proof of stability of the container closure system and packaging process is established by full shelf life stability" (2). Any observed changes in the components should signal an investigation and corrective actions; any observed leachables should be trended and evaluated for safety. In the end, correlation of the leachable data to that of extractable data should show higher levels of extractables. Accurate measurements are required to establish specifications and support a comprehensive safety evaluation of leachables, which will provide the necessary data for acceptance criteria.

CORRELATION OF LEACHABLES TO EXTRACTABLES AND LIFECYCLE MANAGEMENT

Correlation of the leachable data to that of extractables will provide the basis for control strategies and associate CQAs to components. An adequate relationship established between leachables and the type and amount of extractables may justify control of leachables through extractable specifications. Development of extractables acceptance criteria can benefit pharmaceutical product CQAs by monitoring the profiles for control of specific leachables and detection of unexpected material changes if an adequate correlation has been made. Once a CCS is qualified for a particular lyophilized product, it cannot be presumed to be adequate for all products. The suitability of the CCS for every pharmaceutical product needs to be demonstrated. Detection of a change in an extractable profile would trigger an investigation that may solve a problem in advance rather than potentially putting a patient at risk. In the event of a planned change to the pharmaceutical product, the components' manufacturing process or the components' materials, a comparison of an established extractable profile would link back to leachables and the need for additional safety assessments. Continual monitoring of extractables will assure CCS safety resulting in reduced risk of exposing a patient to a contaminated drug product.

ACKNOWLEDGMENTS

Andrea Straka, Amy Miller, and Adrienne Williams for technical support and Curtis Hoover for illustrations; West Pharmaceutical Services, 101 Gordon Drive, Lionville, Pennsylvania, 19341.

REFERENCES

1. Code of Federal Regulation Title 21 Food and Drugs Part 211, Current Good Manufacturing Practice for Finished Pharmaceuticals. Food and Drug Administration, Department of Health and Human Services.

2. Container Closure Systems for Packaging Human Drugs and Biologics, Chemistry Manufacturing and Controls Documentation. Food and Drug Administration, CDER/CBER, Guidance for Industry, 1999.

3. Q8 Pharmaceutical Development, Office of Communications, Training and Manufacturing. Food and Drug Administration, CDER/CBER, Guidance for Industry, International Conference on Harmonization (ICH), 2006.

4. Q9 Quality Risk Management, Office of Communications, Training and Manufacturing. Food and Drug Administration, CDER/CBER, Guidance for Industry, International Conference on Harmonization (ICH), 2006.

5. M4Q-Quality, Office of Communications, Training and Manufacturing. Food and Drug Administration, CDER. Industry, ICH, 2001.

6. Q2B Validation of Analytical Procedures: Methodology, Office of Communications, Training and Manufacturing. Food and Drug Administration, CDER/CBER, Guidance for Industry, International Conference on Harmonization (ICH), 1996.

18 Optimizing the Throughput of Freeze-Dryers Within a Constrained Design Space

Jim A. Searles

Aktiv-Dry LLC, Boulder, Colorado, U.S.A.

INTRODUCTION

The objective of freeze-drying process development is to deliver to manufacturing a cycle that achieves all of the following:

1. Acceptable product quality, consistent within a batch and from batch to batch
2. Operation within the capabilities of the equipment, with appropriate safety margins to ensure robustness
3. Efficient plant utilization via the shortest possible cycle time and full loading of the lyophilizer

   This chapter highlights key aspects of how to achieve all three at the same time with a methodology for optimizing the throughput of a group of freeze-dryers. It includes an actual case study in which most of the dryer time was spent on a single product, and the vast majority of processing time was taken up with primary drying. The case study therefore focused on maximizing primary drying rate while avoiding product failure. The reader will also find sections further exploring heat transfer phenomena within the freeze-dryer's shelves, and advice for specifying, procuring, testing, qualifying, and validating freeze-dryers.

   This chapter is aimed at those supporting freeze-dryers in manufacturing, as well as scientists and engineers in manufacturing technology, technical services, and process development groups. The latter will find this information very useful for ensuring that the freeze-drying cycles that they develop on laboratory-scale freeze-dryers will be transferrable to production-scale units.

   This chapter employs symbols and nomenclature that engineers are taught in heat transfer, fluid mechanics, and transport phenomena classes. For the reader's convenience, the symbols are consistent with those defined and used in Ref. 1, which is an excellent, well-respected heat transfer textbook that can be downloaded free on the Internet (1). The terms "drying" and "sublimation" are used interchangeably.

ANALYZING FREEZE-DRYER TIME

The overall productivity of a freeze-dryer can be defined in terms of annual throughput, or number of vials of product produced per year. Manufacturing organizations should always have an eye toward maximizing throughput of quality products because it increases the return on investment for the facilities, equipment, personnel, and supporting infrastructure. Freeze-dryer time should

---

This chapter was completed while the author was affiliated with Eli Lilly & Company, Inc., Manufacturing Science, and Technology, Indianapolis, Indiana, U.S.A.

be routinely tracked and analyzed. It is useful to categorize time for each product and dryer in the following manner:

- Manufacturing (matrix of the following for each product)
    - Loading
    - Equilibration
    - Freezing
    - Annealing (if applicable)
    - Primary drying
    - Secondary drying
    - Re-cooling (if applicable)
    - Stoppering
    - Unloading
- Turnaround
    - Defrosting
    - Cleaning
    - Sterilization
    - Vacuum testing
- Maintenance
    - Planned/routine/preventive
    - Unplanned
- Idle time

Any variability in time required for any of the above steps can also have a negative impact on overall throughput. Reduction in variability allows planners to cut planning "cushions," driving higher throughput by scheduling more batches and reducing idle time.

A key aspect of this analysis is to include a metric for time spent on failed batches or maintenance that was not effective. For a failed production batch, the impact extends to the additional time spent on turnaround for that batch as well. This will help justify the costs of improving expertise within the organization and the quality control improvements.

Another impact on throughput is the extent of freeze-dryer loading. If each freeze-dryer is not being fully loaded with each product, it may be worth examining the reasons and increasing to a full batch size. If the limitation is due to freeze-dryer capabilities, note that this chapter's case study also considers partial versus full loading of two of the freeze-dryers.

PRIMARY DRYING DESIGN SPACE

The "design space" is defined by International Conference on Harmonization (ICH) Q8 as "The multidimensional combination and interaction of input variables (e.g., material attributes) and process parameters that have been demonstrated to provide assurance of quality" (2).

For primary drying, the process parameters constituting the design space are chamber pressure and shelf fluid inlet temperature.

Outputs are the resulting product temperature, product drying rate, and all product quality attributes. One important *caveat* is that other input variables that can affect performance during primary drying are the choice of vial and stopper, choice of formulation, and processing conditions for freezing and annealing (3). Particulate levels present in the vials during freezing can also be a factor (4).

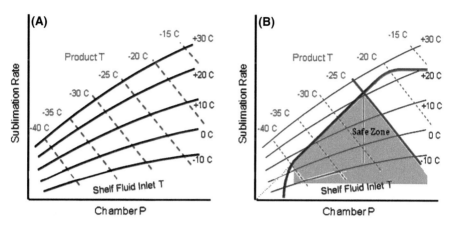

FIGURE 1   Example of primary drying design space.

Another important aspect is the extent of product characterization and identification of all relevant product quality attributes. In considering changes to an established product's manufacturing conditions, one must often take another look at product attributes that are not part of routine release or stability testing—especially product crystallinity by powder X-ray diffraction. Acceptance ranges for routine tests such as moisture and reconstitution time may need to be adjusted once sufficient data are accumulated.

   An excellent example of a primary drying design space can be found in a paper by Chang and Fischer (5). Figure 1A is a sketch representing the nature of their findings. Process conditions are the chamber pressure on the *x*-axis and shelf fluid inlet temperature shown as a family of curves. Outputs are product temperature (family of dashed curves) and sublimation (drying) rate on the *y*-axis. This way of presenting the design space for primary drying, which will look slightly different for each product, is that it gives prominence to the drying (sublimation) rate by putting it on the *y*-axis. Observations that can be made from this figure are

1. Increasing chamber pressure (up to a point) leads to higher sublimation rate but also higher product temperature.
2. Increasing shelf fluid inlet temperature also leads to higher sublimation rate and product temperature.
3. At a given product temperature, the highest sublimation rate can be found at combinations of lower chamber pressure and higher shelf fluid inlet temperature.

   Minimizing primary drying time means maximizing the sublimation rate during primary drying. Chang and Fischer measured primary drying rate by removing vials through the cycle and weighing them. The third observation above is a cornerstone of this chapter—the drying rate can be maximized "for a given product temperature" by using combinations of lower chamber pressure and higher shelf fluid inlet temperature. It has long been known that there is a maximum product temperature for primary drying, above which the product will collapse. This maximum allowable temperature constitutes a "boundary" on the design space. Figure 1B shows a dashed black line over the −25°C curve to illustrate that boundary for a product that would have that as the maximum allowable temperature during primary drying. The case study below (and other chapters of this book) covers how to determine this temperature.

One would assume that the freeze-dryer itself has inherent limitations in the product drying rate that it can support. However, it was only recently discovered that the maximum drying rate that a given dryer could support is strongly influenced by the chamber pressure (6). The phenomenon responsible, choked flow, is explained in the next section. Briefly, exceeding the maximum allowable drying rate causes the chamber pressure and thereby product temperature to increase. Choked flow represents a second boundary on the design space depicted in Figure 1B as a bold black line.

The area within these boundaries constitutes a "safe zone" (Fig. 1B) within which the freeze-dryer will be able to support the product drying rate, and the product will remain below its collapse temperature. Note that the safe zone becomes more constrained as one moves toward higher drying rates. Therefore, one must be fully aware of appropriate boundaries within the design space to achieve both objectives of producing quality product while remaining within capabilities of the equipment.

CHOKED FLOW

It has recently been recognized that, given the design and operating conditions of pharmaceutical freeze-dryers, the water vapor velocity within them could be reaching the speed of sound (6). This is important because the flow of water vapor between the chamber and condenser can become "choked" if the water vapor reaches the speed of sound at any point between them. Choking can lead to the chamber pressure climbing above its set point (6).

The speed of sound is known as "sonic" velocity, and depends on the medium (e.g., water, air, water vapor) and is weakly dependent on temperature and molecular mass. For ideal gasses the speed of sound is

$$V_s = \sqrt{\frac{\gamma RT}{M}} \tag{1}$$

where γ is the ratio of specific heats (C_P/C_V) for the gas (1.3 for water vapor), R is the ideal gas constant, T is absolute temperature, and M is the molecular weight. Note that it is independent of pressure. The speed of sound of water vapor at 0°C is about 400 m/sec. One is in danger of choking the duct between the shelf chamber and condenser chamber if the velocity approaches this value in the duct between the two chambers (6). The limitation is on the gas velocity and not the mass flow of gas. At twice the chamber pressure, the same velocity of water vapor through the duct translates to twice the mass flow rate. Therefore for choked flow, higher chamber pressures also mean higher "carrying capacity" of water vapor.

There are two things one can do to prevent choked flow in freeze-dryers. One is to reduce the drying rate by reducing the shelf fluid inlet temperature, and the other is to operate at a similar (or even higher) drying rate, but at a higher chamber pressure where choking will occur at a higher mass flow rate of water vapor.

There are other phenomena that will limit the maximum possible sublimation rate. At very high sublimation rates (irrespective of pressure) one should be concerned about limitations imposed by

- available heating capacity for shelf heat transfer fluid,
- the heat transfer coefficient from the shelf heat transfer fluid to the top of each shelf,
- available condenser coil (or plate) surface area,[a]
- the heat transfer coefficients within the condenser coils, and
- available cooling capacity for the condenser coils.

Regardless of the nature or cause of a given limitation, one should test lyophilizers to determine their maximum drying rate capability over the range of expected operating pressures.

MEASURING FREEZE-DRYER DRYING RATE CAPABILITIES

A freeze-dryer is first and foremost a dryer. Its principal function is to move water out of the product and onto the coils of its condenser. The rate at which the freeze-dryer can do this should be measured periodically. For modern freeze-dryers with separate product and condenser chambers, the maximum drying rate that the dryer can support increases substantially with chamber pressure. This section outlines procedures for measuring the maximum possible drying rate over a range of chamber pressures.

Freeze-dryer drying rate capability studies are often referred to as "ice slab sublimation studies" because they use slabs of ice on the trays instead of product-filled vials or trays. The ice slabs are easily formed by freezing 2.5-cm depth of water on each shelf, held in place by shelf border rings into which plastic film has been installed. One note of caution—do not attempt to reuse ice slabs or use a given set of them for more than about six hours of sublimation because after that period of time the ice tends to lose contact with the top of the plastic film, having "burned off" those areas first. In addition, each freeze-dryer must be tested separately. Even nominally identical dryers may have differences that can only be detected by these tests.

The tests are conducted in two parts. In part A, a characteristic data trend of chamber pressure versus shelf fluid inlet temperature is generated. These data are used to decide shelf fluid inlet temperatures to use for separate subsequent executions of part B in which the (maximum supportable) drying rate is measured for each selected chamber pressure.

The rationale behind the conduct of part A bears some explanation. Figure 2 shows data from three separate experiments in the same laboratory-scale freeze-dryer. After the water was frozen on the shelves, vacuum was pulled and the shelf fluid inlet temperature was increased at 0.5°C/min. This figure shows the resulting trend of chamber pressure for three cases: chamber pressure set to 0, 100, and 200 mTorr (0, 13.3, and 26.6 Pa).

For the latter two cases, at a point during the ramp, the chamber pressure began to climb above set point. This is because at that point, the rate of ice

---

[a]A rule of thumb is that the condenser coil area should be greater than the shelf area used for product, but the effect of condenser coil surface area on condenser performance has yet to be explored and published.

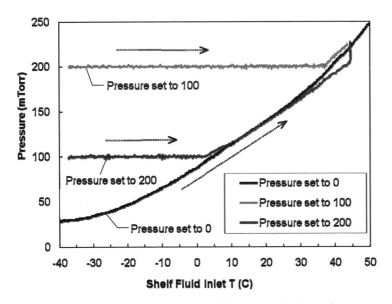

FIGURE 2   Ice slab sublimation data for a laboratory-scale freeze-dryer.

sublimation exceeded the ability of the freeze-dryer to transport it to the condenser chamber and to condense it as ice on the condenser coils. Note that for the 100-mTorr set-point case, the shelf fluid inlet temperature at which this occurred is about the same as for the case in which the chamber pressure was set to 0. And for the 200-mTorr set-point case, the shelf fluid inlet temperature at which that set point was exceeded was approximately the same as for when the 0 mTorr and 100 mTorr cases reached 200 mTorr. The key to understanding this is that the point of departure from the pressure set-point is the time when the flow rate of purge gas has also decreased to zero. The condenser pressure reached 20 mTorr, suggesting choked flow, and the Pirani pressure gauge reading decreased to equal that of the capacitance manometer, confirming that nitrogen was no longer being injected into the chamber as a purge gas.

One can observe from the figure that it is unnecessary to make *separate* measurements of the shelf fluid inlet temperature required to achieve chamber pressure failure at a range of separate pressures (e.g., 100 and 200 mTorr in this case). Rather, a continuous shelf fluid inlet temperature ramp with the chamber pressure set to zero will provide a continuum of shelf fluid inlet temperatures and their corresponding pressures. Therefore, routine testing can be streamlined by conducting part A with the purge gas turned off (chamber pressure set to zero).

Part A of Ice Slab Studies

1. Tape a thermocouple to the top surface of each shelf, at least 0.5 m from the tray ring edges.
2. Load a tray ring with attached plastic sheeting onto each shelf. The sheeting (or film) should be as thin as possible (0.002 in. or less). The tray rings should cover as much as the shelves as possible.

3. Pump a 2.5-cm depth of water into each shelf's lined tray ring. Record the mass of water pumped onto each shelf, ensuring consistency between shelves and with that used for subsequent tests.
4. Close the door of the freeze-dryer and freeze the water by reducing the shelf fluid inlet temperature to $-45°C$.
5. Hold the shelf fluid inlet temperature at $-45°C$ or below for at least 30 minutes to equilibrate.
6. Pull vacuum to 100 mTorr (13.3 Pa).
7. Set the shelf fluid inlet temperature to increase at $0.5°C/min$ up to $+45°C$.
8. End the run by breaking vacuum and returning the shelf fluid inlet temperature to 20°C, allowing time for all of the remaining ice to melt. For part A it is not necessary to weigh the mass of water remaining on each shelf.

It is important to understand the meaning of the graph of chamber pressure versus shelf fluid inlet temperature that results from part A. The curve is characteristic of that dryer only, with that thickness of plastic film, surface area, and depth of ice. It does not give any drying rate data directly—it is only used to inform what conditions to use for the part B experiment(s). However, that freeze-dryer should always exhibit a similar pressure versus shelf fluid inlet trend with such an ice loading etc. Deviations from this trend mean that some aspect of performance has changed.

Part B of Ice Slab Studies

The part B procedure can be carried out as many times as desired—usually once for each chamber pressure to be tested. Analysis of a single part B result yields the maximum drying rate (in kg/hr) supportable by that freeze-dryer at the chamber pressure that was measured for that test. The chamber pressure is not directly controlled, rather the shelf fluid inlet temperature is set having been chosen to correspond with the desired pressure according to the results of part A.

1. Same as for part A.
2. Same as for part A.
3. Pump about 2.5-cm depth of water into each shelf's lined tray ring. Record the mass of water pumped onto each shelf, ensuring consistency between shelves and with that used for previous tests.
4. Same as for part A.
5. Same as for part A.
6. Set chamber pressure set point to zero and/or disable purge gas flow.
7. Increase the shelf fluid inlet temperature, at the maximum possible rate, directly to that chosen for this test. Choose the shelf fluid inlet that gave the desired chamber pressure in part A.
8. Hold for six hours.
9. Break vacuum and ramp the shelf fluid inlet temperature to $+20°C$.
10. Pump the remaining water from each shelf into a container on a scale and record the mass of water remaining on each shelf.

Calculate the average sublimation rate for each shelf by dividing the mass sublimed by the hold time (e.g., 6 hours). Compare the results from each shelf. Significant differences may mean that there is a shelf heat transfer fluid flow problem or air bubbles in affected shelves. The total sublimation rate of all of the individual shelf results represents the maximum supportable sublimation rate at that chamber pressure (the average chamber pressure over the six-hour hold).

Run part B over the desired chamber pressure range—usually 100, 200, and 300 mTorr. Plot each freeze-dryer's results as a different series, with the measured sublimation rate on the *y*-axis, and average chamber pressure for the six-hour hold on the *x*-axis. This will enable comparison of the capabilities of all freeze-dryers tested.

CALCULATING SHELF HEAT TRANSFER COEFFICIENTS

In the freeze-drying of pharmaceutical and biological products, an aqueous solution is filled into vials, or sometimes trays, for drying. For freeze-dryer capacity testing discussed above, tray rings lined with plastic film are used to create a 2.5-cm deep ice slab on the top of each shelf. The heat and mass transfer paths for these are shown in Figure 3. Comparing panels A and B of the figure,

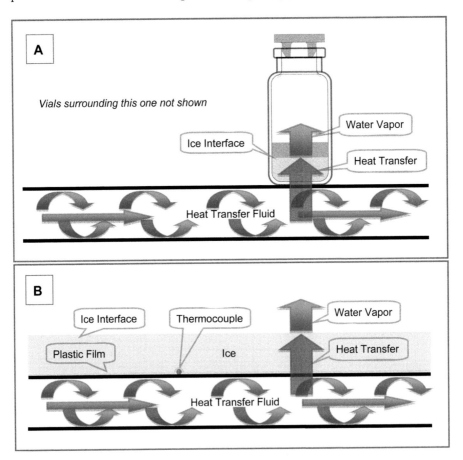

FIGURE 3 Heat transfer pathways from the bulk heat transfer fluid in a shelf to the ice interface (**A**) for product loaded in vials and (**B**) for ice slab studies. The heat transfer pathway from the bulk heat transfer fluid in a shelf to the top of the shelf surface is the same for both cases, therefore the heat transfer coefficient calculated from ice slab studies using data from a thermocouple under the plastic film applies regardless of trays or vials or ice slabs.

one can see that regardless of the type of load on the shelf (vials vs. ice slabs vs. trays), the heat transfer path from the bulk heat transfer fluid in a shelf to the top surface of the shelf is the same. Therefore, heat transfer coefficients calculated from ice slab studies (dryer capacity testing) are applicable regardless of the type of freeze-dryer load (vials vs. trays vs. ice slabs).

Heat transfer fluid is pumped through the freeze-dryer shelves to remove or add heat to the product loaded onto them. The flow of this liquid is driven, usually by a single pump, through all shelves in parallel. Unfortunately the flow rate of this fluid is almost never monitored or explicitly controlled. The fluid temperature is controlled at the inlet to the shelves; however, it is almost universally referred to as the "shelf temperature." However, we use a more exact name—*shelf fluid inlet temperature*, T_s. The shelf fluid inlet temperature equals that of the shelf's top surface *only* when no heat transfer is taking place, which is almost never.

Heat transfer is driven by a difference in temperature. The magnitude of heat transfer between two points can be described by a simple equation introducing a heat transfer coefficient.

$$q = h(T_f - T_s) \tag{2}$$

where q is the rate of heat transfer per unit area of shelf surface (W/m^2), h is the heat transfer coefficient representing the efficiency of the heat transfer ($W/m^2 \cdot K$), T_f is the temperature of the heat transfer fluid (K), and T_s is the temperature at the shelf surface (K). To analyze the results of ice slab studies and to calculate a heat transfer coefficient for each shelf we use equation (2) to calculate h, with known terms being the other three variables in the equation. The term q, which is in units of W/m^2, is the energy supplied over the six-hour hold period to drive sublimation divided by the surface area of shelf that was actually covered by the ice slab as in equation (3).

$$q = \frac{Q_{subl}}{A_{ice}} \tag{3}$$

in which Q_{subl} is the energy delivered for sublimation (W) and A_{ice} is the area of shelf covered by ice (m^2).

We are fortunate that we are carrying out sublimation during this process because the observed sublimation rate can readily be converted to the rate of heat flow by using the heat of sublimation, which for water is 2.97 J/kg. The equation for this relationship is

$$Q_{subl} = h_{sg} \cdot m' \tag{4}$$

where h_{sg} is the latent heat of sublimation (2.97 J/kg) and m' is the average rate of sublimation over the dwell period (kg/hr).

Combining Equations 2, 3, and 4 yields a simple equation for finding the heat transfer coefficient for each shelf using the results from ice slab studies.

$$h = \frac{h_{sg} \cdot m'}{A_{ice}(T_f - T_s)} \tag{5}$$

As defined, the heat transfer coefficient is a lumped, fit parameter that encompasses many phenomena. Table 1 summarizes these. The energy is transported to the sublimation interface from the bulk fluid to the interior shelf

TABLE 1 Heat Transfer Path and Mechanisms

From	To	Mechanism(s)
Bulk heat transfer fluid	Inner shelf surfaces	Convection (highly subject to the fluid type, shelf thickness, flow patterns, flow rate, temperature, viscosity), and conduction via the heat transfer fluid
Inner shelf surface	Outer shelf surface	Conduction
Outer shelf surface	Lower film surface	Gas-phase conduction (pressure dependent) and radiation
Upper film surface	Ice interface	Conduction through ice

surfaces by convection and conduction via the heat transfer fluid. Convection is extremely difficult to characterize analytically unless the flow paths are simple and fluid properties are well defined (and in the present case they are not well defined). Furthermore, the fluid properties are heavily dependent on temperature, which will vary widely over the course of sublimation tests. One may even encounter a turbulent to laminar flow transition in the heat transfer fluid as temperature is reduced, resulting in a step change in heat transfer properties. Heat transfer through the thickness of the shelf plate itself will be by pure conduction.

The case study below includes calculation of heat transfer coefficients from ice slab studies.

CASE STUDY

This case study covers the investigation and resolution of a projected capacity shortfall of freeze-dryer capacity for a particular product across 12 production freeze-dryers across three facilities on two continents. One of the facilities was in start-up mode with their three units.

Freeze-Dryer Time Analysis

The freeze-drying cycles run for this single product accounted for 58% of all freeze-drying time. The next largest block of time was spent on turnaround. A full 84% of this product's freeze-drying cycle was taken up with primary drying. In addition, in one facility there was significantly greater variability in primary drying time from batch-to-batch than in the other. This was later resolved by harmonizing thermocouple placement techniques and procedures for assessing the end of the step. Savings from this alone was estimated at $3.9 M/yr. Because almost half of all freeze-dryer time was being spent on primary drying for this single product, it became the focus of continued investigations described below.

Product Quality Attributes

The collapse temperature during primary drying is a critical parameter that must be firmly established. It is well-known that for a given formulation, the thermal history prior to primary drying can have dramatic impacts on how it behaves during primary drying. The product under consideration has a significant mannitol content in its formulation, so freezing and annealing studies were carried out to determine how these steps should be carried out across all of the freeze-dryers. A key result was that an annealing step was added to the process to ensure consistent, complete crystallization of the mannitol prior to initiating primary drying. A characteristic powder X-ray diffraction pattern was found for this product and archived in technical reports for future reference.

After analyzing the temperature trends of this product through primary drying, it was decided to use as a consistent metric the initial temperature plateau early in primary drying as that to which we correlated the failure modes. Microcalorimetry as well as laboratory- and pilot-scale freeze-drying studies were used to determine that, for a properly frozen and annealed product, at $-10°C$ this product suffered from microcollapse due to mannitol crystal melting, at $-6°C$ some vials (particularly those at the edge of a shelf or tray ring) collapsed at the edge of the cake, and temperatures at or above $-3°C$ yielded product with slow reconstitution. Photographs were taken of acceptable product as well as those that should be classified with various types of defects. The images were included in manufacturing procedures for product inspection to ensure continuity over the ensuing years.

Primary Drying Optimization

The failure modes testing of the product defined the maximum temperature during primary drying (specifically the initial plateau early in primary drying) to be $-10°C$. This is a relatively high value, which is due to the presence of a significant quantity of crystallized mannitol. The next step was to minimize the duration of primary drying with new set points for shelf fluid inlet temperature and chamber pressure. A range of these parameters was tested, and product quality attributes as well as overall duration of primary drying and maximum sublimation rate were measured. It was found that with the cycle currently in place, the product dried at a maximum rate of 0.45 kg/hr·(m^2 of product loading). The drying rate is per square meter of product loading—this is with a hexagonal packing. It was found that the primary drying duration could be reduced from an average of 60 hours to just 34 hours if the primary drying shelf fluid inlet temperature was changed from $-5°C$ to $+20°C$, and the chamber pressure was increased from 110 to 200 mTorr. The new primary drying conditions (with product frozen and annealed properly) were found to yield excellent product quality, however, the product dried at a higher maximum rate of 1.0 kg/hr·(m^2 of product loading). The next section shows results of freeze-dryer capability measurements to determine if the dryers could reliably handle this higher sublimation rate.

Freeze-Dryer Capabilities Assessment

Figure 4 shows the results of the combined primary drying optimization and the freeze-dryer capability assessments. The y-axis of both panels is the maximum sublimation rate during primary drying, normalized by square meter of shelf space loaded with this product. The green squares and diamonds show the drying rate generated by product for the old and new processes respectively. Panel A shows the results for dryers C through J fully loaded with product. Panel B shows the results for the current partial load (108 trays) as well as full load (144 trays) in freeze-dryers A and B.

From Figure 4A, the interpolated line for dryers C and D shows that they would be able to accommodate a rate of about 1 kg/hr (m^2 of product loading) at the current pressure of 110 mTorr, but none of the others would. In particular, dryer J had always shown stress in the form of increasing condenser coil

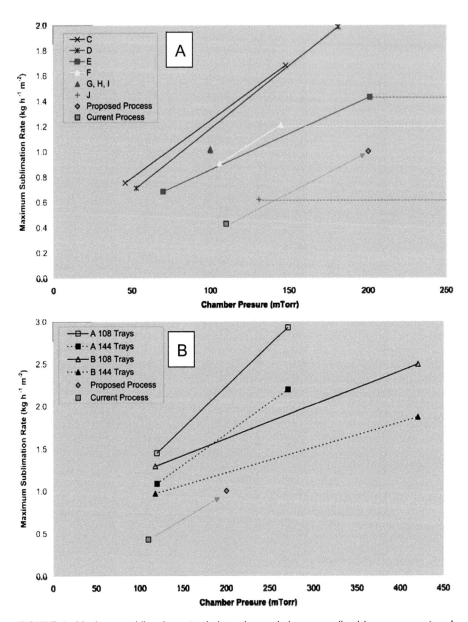

FIGURE 4 Maximum sublimation rate during primary drying, normalized by square meter of shelf space loaded with this product. Combined results from primary drying optimization (*bold black squares and diamonds*) compared to freeze-dryer capabilities. Panel A: Full load in dryers C through J. Panel B: Current partial load (108 trays) as well as full load (144 trays) in freeze-dryers A and B.

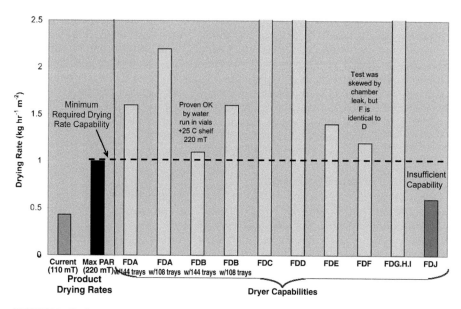

FIGURE 5 Drying rate capabilities (*bold black and dashed black bars*) compared to the required drying rate for the new process (*black bar*) and the current process (*gray bar*).

temperatures early in primary drying, and the single ice slab study result confirmed that it is indeed very close to being unable to hold pressure. Two other dryers, E and F, showed a similar sign of stress at their higher-pressure tests, so dotted lines were also extended to the right of those points. Three identical dryers that were still in start-up mode, dryers G, H, and I, gave identical results. They were able to run at a set point of 200 mTorr, with fresh ice slabs and shelf fluid inlet temperature of +70°C, so they have no data points shown for that pressure.

Overall, Figure 4 reveals that if primary drying were to remain at 110-mTorr pressure, there would be no room for an increased drying rate because the current process is close to the limit measured and/or interpolated for a number of the dryers. However, at the higher pressure of 200 mTorr, we can achieve a significantly higher drying rate of product while having sufficient additional drying rate capability in all of the machines except unit J.

Figure 5 shows the results in a different format. All dryers except J are projected to be able to support the higher drying rate of the new cycle with a full load at 200 mTorr. Dryer F's test had been successful except a small leak into the dryer, so its results were not as close as the bar height would indicate. In addition, dryer B was projected to be close to failure but able to handle the new cycle fully loaded, so a water run was conducted to verify the projection. It was known that a load of vials filled with only water would initially dry faster than product. To make the test even more conservative the test was run at slightly higher shelf fluid inlet and chamber pressure set points than those for the new cycle. The water run was successful with pressure set point holding and no other signs of stress to that dryer was also approved for the new cycle.

Other aspects of freeze-dryer capability were assessed by calculation. The surface area of condenser coils/plates was compared to the shelf surface area of each dryer. A condenser coil surface area as low as 0.6 m² was able to accommodate a drying rate of 1 kg/hr. Most of the production dryers had a ratio of condenser to shelf surface areas of just over 1.0, which appears to be sufficient. The cross-sectional areas of the ducts connecting the product and condenser chambers were also analyzed. Of course other aspects of the water vapor flow path can be critical as well—the presence of any obstructions or bends and the length of the connection are also important. At least as far as the cross-sectional area is concerned, a ratio of duct cross-sectional area to shelf area for product of 0.03 should be sufficient.

Shelf Heat Transfer Coefficients

Shelf heat transfer coefficients were calculated per equation (5). In Figure 6 one can see the dramatic range of shelf heat transfer coefficients over the range of shelf fluid temperatures, for a laboratory-scale dryer as well as for a production freeze-dryer. It is not unexpected that there should be such variation with temperature, because the viscosity of the heat transfer fluid will increase with decreasing temperature, reducing the turbulence and therefore the efficiency of heat transfer from the fluid to the inner surfaces of the shelf. The viscosity change will also affect the flow rate of the heat transfer fluid. It was surprising to learn that the flow rate of this critical part of the system was not monitored either on the laboratory-scale unit or on those in production. It is somewhat understandable given that the rate of heat transfer from the fluid to the top of the shelf is rarely a limiting factor. However, in terms of process and product consistency and protection against known failure modes, one would surely find the low cost of a flow meter a small price to pay.

Another surprise was that the heat transfer coefficients for the production freeze-dryer would be so much lower than that for the small dryer. This is another reason to be cautious about transferring a process from a laboratory-scale dryer into one of production size.

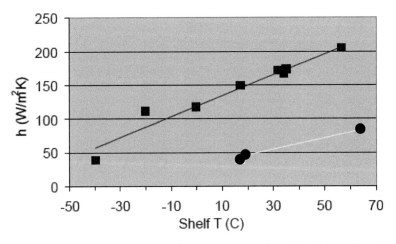

FIGURE 6 Shelf heat transfer coefficients measured during ice slab studies calculated using Equation 5. ■: Laboratory scale dryer. ●: Production scale dryer.

Case Study Impact

We used the finding of choked flow during the primary drying of freeze-drying to further define the primary drying design space. By elucidating the failure limits of each freeze-dryer and showing how sensitive they were to chamber pressure, we were able to successfully optimize the primary drying conditions for a key product. This led to a 29% cycle time reduction, projected to increase freeze-drying productivity by $20 M/yr. In addition, it made available 23% more dryer time for other products of more of the same product.

A six-sigma project resulted in rapid implementation of new product thermocouple holders and improved decision making for the end of freezing, primary drying, and secondary drying. The benefit of this project was projected to be $3.9 M/yr.

SPECIFYING A NEW FREEZE-DRYER

When spending millions to purchase a new freeze-dryer it is important to ensure that it will have the highest drying rate capability possible. Information in this chapter is intended to help engineers accomplish this. First and foremost, be aware of how to accurately measure the drying rate capabilities of a freeze-dryer (as described in this chapter), and put these procedures into freeze-dryer specification and procurement documents. Test freeze-dryers to ensure they meet your requirements for drying rate capability before they leave the factory. A minimum guidance value of 1.0 kg/hr·(m^2 of shelf surface for product), when run at 100-mTorr chamber pressure, is recommended. The freeze-dryer should be able to handle twice that rate at 200 mTorr. In addition, the heat transfer coefficient for each shelf should be at least 40 W/m^2·K at a shelf fluid inlet temperature of +20°C. Condenser coil/plate surface area should be greater than that for product on the shelves. Shelf fluid flow rate should be continuously monitored and alarmed. Ice slab studies should be repeated every year after major maintenance to ensure that each shelf is performing to specification and that the entire system is functioning properly.

SUMMARY OF RECOMMENDATIONS

This chapter provides definition of critical freeze-dryer performance attributes and, just as important, how to measure them. This will help ensure the procurement of quality equipment, ensuring that equipment capabilities are maintained over the service life, and also provide for surprise-free transfers of new products and processes from development into full-scale manufacturing and from one production freeze-dryer into others. Specific recommendations are

- Use ice slab studies to measure the drying rate capability over the desired range of chamber pressures (before taking delivery, after taking delivery, during qualification, and periodically through the service life).
- Thoroughly characterize and document each product's failure modes.
- Ensure that freeze-dryers are equipped with flow meters or switches on the shelf heat transfer fluid, and that they notify operators when flow falls below an established limit.
- Remember that what is measured by the thermocouple on the shelf heat transfer fluid is just that—it is not the actual shelf surface temperature when

the dryer is under load. Heat flows only where there is a pressure difference; therefore, when supporting a high load such as freezing or early primary drying there must necessarily be a significant difference between the shelf fluid inlet temperature and the shelf surface temperature just under the product.

- Track and analyze step times, product quality attributes, and process measurements for each product.
- Know the drying rate that each product will be generating when it is at its maximum—early in primary drying. These can be measured in small-scale freeze-dryers.
- Understand the "design space" for your process (2). This chapter defines it for significant parts of freeze-drying. For primary drying one will find that the design space gets smaller at higher drying rates.

ACKNOWLEDGMENTS
I wish to acknowledge and thank Professor Ted Randolph at the University of Colorado, and the following people from Eli Lilly & Co. for their contributions to this work: David Lowndes, Steve Nail, Gordon Livesey, Adam Scott, Ted Tharp, Keith Glad, Rebecca Elliott, Robert Bell, Lisa Rathburn, Rick Owens, Carmel Egan, Ralph Riggin, Sandrine Favre, and Bob Cole.

REFERENCES
1. Lienhard JC IV, Lienhard JC V. A heat transfer textbook. 3rd ed. Cambridge, MA: Phlogiston Press, 2008. Available at: http://web.mit.edu/lienhard/www/ahtt.html.
2. International Conference on Harmonization of Technical Requirements for Registration of Pharmaceuticals for Human Use. Pharmaceutical development q8(r1). ICH Harmonised Tripartite Guideline 13 November 2008.
3. Searles J, Carpenter J, Randolph T. Annealing to optimize the primary drying rate, reduce freezing-induced drying rate heterogeneity, and determine tg′ in pharmaceutical lyophilization. J Pharm Sci 2001; 90(7):872–887.
4. Searles J, Carpenter J, Randolph T. The ice nucleation temperature determines the primary drying rate of lyophilization for samples frozen on a temperature-controlled shelf. J Pharm Sci 2001; 90(7):860–871.
5. Chang BS, Fischer NL. Development of an efficient single-step freeze-drying cycle for protein formulations. Pharm Res 2001; 12(6):831–837.
6. Searles JA. Observation and implications of sonic water vapor flow during freeze-drying. Am Pharm Rev 2004:58–69. Available at: http://americanpharmaceuticalreview.com/ViewArticle.aspx?ContentID=27.

19

Monitoring and Control of Industrial Freeze-Drying Operations: The Challenge of Implementing Quality-by-Design (QbD)

Miquel Galan
Telstar Technologies, S.L., Terrassa, Spain

INTRODUCTION

Lyophilization (freeze-drying) is a well known drying technology especially suitable for many pharmaceuticals mainly because of the very low temperatures involved in the process, making it almost the unique available alternative for long-term stabilization of heat-sensitive products.

During a freeze-drying process, the initial liquid solution is first frozen, and then, water (as the most common solvent) is separated by exposing it to a very low solvent partial pressure in the drying chamber, promoting the direct sublimation of the ice to vapor, and leaving a porous dry structure with an apparent volume equal than the original frozen solution. After a well-performed cycle, the product exhibits long-term stability, and, at the time of use, can be reconstituted with a very fast rehydration.

The drawbacks of this extremely gentle process are the high energy consumption and the long processing cycle time, due to the very low temperatures and low pressures involved. The ability of running this process under sterile conditions helps, however, to make it especially attractive to the pharmaceutical industry.

PROCESS CHALLENGES

Primary drying (sublimation) is the step that concentrates most of the risks of the process. It is when there may happen different failures that can ruin a batch: shrinkage, collapse, and melting, etc. To better understand the risks associated in primary drying, it is a must to have a good knowledge of what happens during the previous step of freezing:

As the solution is cooled down and the nucleation temperature is reached (with more or less supercooling), crystals of pure ice are separated while the remaining solution becomes more concentrated. Very often, the products involved in pharmaceutical lyophilization do not crystallize, but this interstitial solution becomes more and more viscous until a glassy structure appears, with an apparent solid behavior.

But over many vials, during the freezing step, the scatter in supercooling may be shown as a probability distribution, so there can be a large degree of heterogeneity in the nucleation temperature within a batch, affecting directly the size of the ice crystals. If there is a large supercooling, the ice crystal sizes are significantly smaller than in the case of experiencing small supercooling (1–3).

During primary drying, where these ice crystals sublimate, it has to be avoided that the already dried phase is exposed to a temperature above its glass transition temperature, so the entire structure might collapse losing its morphology and elegance, which in turn would lead to loss of activity and difficulty of reconstitution. So, primary drying must be performed maintaining the sublimation interface temperature below the formulation specific upper limit: the collapse temperature.

During the drying process, a mass and heat transfer balance can be established; the heat to sublimate is supplied by the combined effects of conduction, convection, and radiation from the heated shelves. The sublimation process, being endothermic, produces a cooling effect over the product. The balance of both heat input and removal dictates the equilibrium temperature of the sublimation interface.

There can be two types of sublimation processes, one where the limiting factor lays in the heat supply or the opposite where the bottleneck is the transport of the sublimated vapors, directly affecting the heat removal.

The strategies to run a cycle in each case are very different and antagonist. It is not infrequent having cycles with heat transfer, as the limiting factor, at the beginning of the primary drying, experiencing mass transfer limitation at the end.

It is in this way that one can understand why, in some cases, there are processes that may experience collapse at the end of primary drying: as the sublimation proceeds, the thickness of the already dry layer increases and, if the crystals are small, their conductance for the generated vapors is decreasing, reducing the sublimation mass flow and thus its associated cooling effect. So, if the heat input remains constant, the sublimation interface temperature rises significantly. If the process is performed at a product temperature close to its maximum temperature, this can lead to the collapse.

To define the lyophilization process, a "recipe" is specified, which is a set of shelf temperatures and chamber pressures versus time steps, but this does not guarantee repeatable conditions for freezing and also does not guarantee that the variable of interest, that is, the sublimation interface temperature is perfectly defined.

It is not uncommon listening in lyophilization courses that "in this process there are two independent variables, shelf temperature and chamber pressure, and when they are fixed, then the dependent variable, product temperature, becomes also fixed." The problem is that the product temperature is not only a function of these two mentioned variables, but at least two others: the heat transfer resistance (how difficult is the heat supply from the shelf to the sublimation interface) and the vapor flow resistance in the dry structure (how difficult it is for the sublimated vapors to flow through the dry layer) that increases as the drying proceeds.

Heat transfer resistance depends on the actual barriers between the sublimation interface and the fluid circulating within the shelves. And as there are three different mechanisms of heat transfer, each one will involve different parameters. Conduction will depend directly on solid layers, materials, and contact between these layers (shelf metal thickness, the eventual use of trays, vial bottom shape and thickness, and product height, etc.). Convection will strongly be influenced by the dimensions of the interstitial space between the product container (usually glass) and the tray or the shelf, and the chamber residual gas density (directly related to pressure). Radiation can induce batch heterogeneity depending on the exposure of each vial to its direct sight of the surrounding surfaces, so the edge vials will be influenced by the chamber walls in a different way than the central ones (Fig. 1).

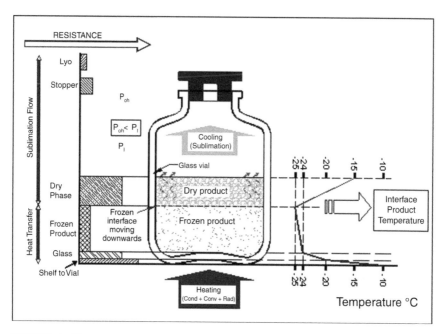

FIGURE 1 Heat and mass transfer during primary drying as well as one example of temperature profile in a vial.

Mass transfer depends basically on the morphology of the dry cake, so nucleation temperature, product crystallinity, etc., are the key governing factors.

From the process perspective there are two really important variables that show a great interest to be measured in-line: sublimation interface temperature (to avoid the loss of morphology, that in turn would lead to loss of activity and difficulty of reconstitution, the primary drying must be performed maintaining the sublimation interface temperature below the formulation specific upper limit: the collapse temperature) and sublimation mass flow to minimize the process time in an attempt to optimize productivity.

Besides the temperature, also the residual amount of frozen water has to be monitored to detect the end point of the primary drying. In fact, if secondary drying is started before the end of the previous step, the product temperature may exceed the maximum allowed value, thus causing melting or collapse, while if secondary drying is delayed, the cycle is not optimized and the operational cost increases.

Finally, the residual water content at the end of secondary drying has to be monitored; for most products the target level of residual water is very low, usually from less than 1% to 3%, even if for certain products it has been demonstrated that a too low level of residual water should be avoided and the final residual moisture must be in a definite range (4).

With the actual technology, it is impossible to obtain a direct measure of the sublimation front interface temperature without interfering with the process dynamics or impairing the sterile conditions needed by some products. An example of commonly used, but invasive, monitoring device is the insertion of a thin thermocouple or—a significantly more bulky—RTD (resistance temperature detector) inside the vial. This method alters the elementary phenomena of nucleation and ice crystal growth, as well as the heat transfer to the product. As a consequence, the drying kinetics is faster in the monitored vial and the results are not representative of the whole system; nevertheless, this method is vastly used to monitor the primary drying and to detect the end point of the primary drying stage. It is also impossible to place the temperature probe in a way that can monitor the ice temperature up to its complete disappearance, so it is common knowledge that "temperature probes only tell the truth during freezing, at the early phase of primary drying, and then one can trust them again during secondary drying," but, where the product is most at risk, its information can be totally nonmeaningful. Last, but not least, the probe insertion itself compromises the sterility of the product and it is also impossible to place them in a process with automatic loading and unloading.

Sublimation mass flow can only be measured quantitatively, nowadays, by the TDLAS (tunable diode laser absorption spectroscopy) technique (5), based on the simultaneous measurement of the water vapor concentration (selective infrared absorption) and the vapor velocity in the duct connecting the chamber and the ice condenser (wavelength shift by Doppler effect). This is a physically nonintrusive sensor, but its main disadvantages are its reported accuracy (typically about 90%), the significant extra size needed in the duct connecting the chamber and the condenser, its cost and impossibility of retrofitting it in an existing unit.

There are other several sensors claimed to be useful for monitoring the primary drying. In fact most of them are based only in the partial pressure measurement of the solvent (water), but their real utility is limited to the detection of the end of primary drying and cannot give any information on the kinetics of the process.

The state of the art of the techniques available to monitor primary drying has been recently reviewed and discussed by Barresi et al. (6). Some points to be stressed are that information on the whole batch can be available, and the end of primary drying can be effectively determined through different techniques, allowing data reconciliation, and resulting in a very robust system.

QUALITY BY DESIGN: A CHANGE IN MENTALITY FROM TRADITIONAL VALIDATION

Automated lyophilization plants dominate the industry. Excipients and active ingredients are mixed, the solution is prepared and filtered, and the vials are washed, sterilized, and depyrogenated. The solution is dosed into the vials and loaded into the freeze-dryer. After the cycle, they are unloaded, capped, and are tracked throughout the packaging and delivery processes.

Problems in these automated steps can result in large quantities of vials, ampoules, etc. that must be quarantined, retested, rejected, reprocessed, or destroyed, all at significant expense.

Of course, the worst case scenario would be that defective manufactured products were not detected but were inappropriately shipped for use by patients who at best would receive ineffective medications and potentially could receive toxic or harmful products.

In 1987, the U.S. Food and Drug Administration (FDA) issued its Guideline on General Principles of Process Validation (7). It defined validation as "establishing documented evidence which provides a high degree of assurance that a specific process will consistently produce a product meeting its predetermined specifications and quality attributes."

It seems that over time, validation activities have become centered mostly on documentation instead of on ensuring quality. A nonnegligible amount of activities have grown up around process validation with a proliferation of validation protocols, validation reports, and validation documentation, but there are still processes that work poorly, and lyophilization development is often seen by many professionals more like an art than a process that can be scientifically approached. It has been lost the goal which is that before trying to demonstrate that the process reliably does what it is supposed to do, the process "must be known" in depth.

The traditional approach presupposes that if nothing is changed from the validation batches, everything will remain the same. But this assumption is false because neither ingredients nor processing conditions can remain perfectly invariant. There will be changes from batch-to-batch, especially due to stochastic supercooling leading to different nucleation temperatures. There may be further changes over time, that operators can introduce, or the equipment will age or will be moved from one site to another. There will be a new supplier for a certain material, and this new material may be within specifications. It was never real that everything could be kept the same.

In September 2004 the FDA released its Guidance for Industry: PAT—A Framework for Innovative Pharmaceutical Development, Manufacturing, and Quality Assurance (8). This guidance was intended to describe a regulatory framework (Process Analytical Technology, PAT) that should encourage the voluntary development and implementation of innovative pharmaceutical development, manufacturing, and quality assurance. Citing the FDA:

> the Agency considers PAT to be a system for designing, analyzing, and controlling manufacturing through timely measurements (i.e., during processing) of critical quality and performance attributes of raw and in-process materials and processes, with the goal of ensuring final product quality. It is important to note that the term analytical in PAT is viewed broadly to include chemical, physical, microbiological, mathematical, and risk analysis conducted in an integrated manner. The goal of PAT is to enhance understanding and control the manufacturing process, which is consistent with the current drug quality system: quality cannot be tested into products; it should be built-in or should be by design.

The focus of PAT is understanding and controlling the manufacturing process. A process is well understood when all critical sources of variability are identified and explained, variability is managed by the process, and product quality attributes can be accurately and reliably predicted. The strategy should accommodate the ability and reliability of process analyzers to measure critical attributes and the achievement of process end points to ensure consistent quality. These end points have to be understood as the achievement of the desired material attribute, not just the process time. This process understanding is perceived as inversely proportional to risk.

Very soon after the publication of this guideline, first case studies were presented in specialized forums. A high percentage of them reported the use of near-infrared (NIR) spectroscopy sensors matched with multivariate analysis software to monitor classical blending and drying processes. The easy part of all these examples was that all of them were agitated processes where a single sensor was able to monitor representatively the whole batch. As already mentioned before, lyophilization has been a process up to now still dominated by non-PAT approaches due to the lack of available sensors.

This guidance, nevertheless, has to be understood as the trigger enabling the regulated pharmaceutical industries to embrace new technologies that can monitor the real process parameters of interest, which in turn should allow manufacturing better products, increasing their quality, maximizing the production, and reducing the rejections.

Soon after the release of the PAT guideline, the first contradictions appeared. If a process has to be flexible to adapt itself to the inherent cycle variations to produce a constant output, then it conflicts with the existing paradigm of validation. The first attempt to smooth it was with the introduction of the term design space. FDA defines design space in its Guidance for Industry Q8: Pharmaceutical Development (9), as

> The multidimensional combination and interaction of input variables (e.g., material attributes) and process parameters that have been demonstrated to provide assurance of quality. Working within the design space is not considered as a change. Movement out of the design space is considered to be a change and would normally initiate a regulatory postapproval change process. Design space is proposed by the applicant and is subject to regulatory assessment and approval.

So, in the development of a lyophilization cycle, different sets of variables, that is, shelf temperatures and chamber pressures, could be tested to produce an acceptable product, but unfortunately any change on these variables have a direct impact on the mass flow and thus the drying speed. If there is no way to monitor the flow, the longest cycle resulting from all these combinations should be defined in the step durations. Unfortunately, this has a severe impact on productivity, just to cover the unlikely event of having a batch with deviations but still within specifications (i.e., within the design space) to be considered fully compliant.

A new concept emerged: quality by design (QbD) as the opposite of the classical approach of quality by testing (QbT). QbD consists of three key elements: the use of design space to establish elastic quality standards, the use of risk assessment to define the boundaries of those standards, and the implementation of PAT to monitor and adjust to those standards. The resulting cost controls and regulatory streamlining should significantly increase the efficiency of the industry and the oversight process.

The acceptance of a risk-based approach to regulation, emerging within the FDA since 2001, made possible the monitoring of process variables without forcing the costs and complexity of trying to measure every conceivable variable in an ever expanding potential situation. Without this acceptance, the application of statistical tools and regulatory guidelines would create a self-defeating situation jammed in expensive and unnecessary data.

It is clear that two of the three elements needed for the QbD concept can be readily available: design space definition and risk-based approach, but it is strongly needed that the PAT tools should allow monitoring the variables of interest in the lyophilization process.

DYNAMIC PARAMETERS ESTIMATOR: THE ALTERNATIVE

A noninvasive monitoring technique, useful for estimating the average state of the whole batch, is proposed as a valuable alternative to the lack of sensors previously mentioned. Barresi et al. have described this method in detail in chapter 20. In many engineering applications it is desirable to have estimates of hard-to-measure or even nonmeasurable quantities. A dynamic model sensor combines a priori knowledge about the physical system (mathematical model) with experimental data (in-line measurements) to provide an in-line estimation of the sought quantities.

There are two types of models: statistical (empirical) and mechanistic. A mechanistic model is derived from the knowledge about the underlying physics of the unit operation. If there is not any knowledge about the mechanism, there is only the option of traditional statistical design of experiments (DoE). In advance of an explanatory theory, these statistical models have the ability to identify the key variables and their relationships, providing focus for monitoring and, ultimately, providing the data that theory building and proving requires.

In this case, nevertheless, the dynamic model sensor is the mechanistic model describing the heat and mass transfer both in the frozen phase and the already dry phase of a product being lyophilized. The equations are transient state, so they are valid even if the process is not in steady state.

However, the actual parameters of these equations are not known because they depend on the product properties and the specific conditions at which the process is run.

The way to overcome it is introducing a small perturbation in the process and acquiring the system response to this perturbation. The set of equations are then solved to interpret the experimentally acquired pressure rise curve used as the perturbation, in order that the system response can be reproduced (Fig. 2).

The easiest perturbation sensitive to both heat and mass transfer is closing for a few seconds the valve between the chamber and condenser. The system to analyze is, then, a closed system (unless the unit has a significant leak) and the

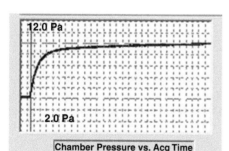
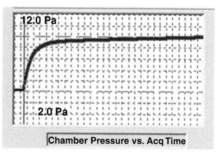

FIGURE 2 Experimental pressure rise acquisition (*left*) and calculated one after solving the equations. Test parameters: vials filled with 1 mL 10% sucrose solution; product height: 7.21 mm; vial internal diameter: 14.25 mm; no. of vials: 609; shelf temperature: −32.6°C; elapsed drying time: 30 minutes; chamber volume: 201 L; and acq. time: 30 seconds.

model is less complicated. This pressure rise technique has been widely used by many analyzing systems as a perturbation. It was first cited by Neümann in 1956 (10) and since then also cited by different authors (11,12).

Several algorithms have been proposed in the past to interpret the pressure rise test (PRT), namely the barometric temperature measurement (BTM) (13–15), the manometric temperature measurement (MTM) (16–19), the dynamic pressure rise (20), and the pressure rise analysis (21–23). What differentiates these methods is the type and the details of the mathematical model used and the parameters that are estimated. Some of the previous approaches, in fact, are based on the sum of elementary mechanisms or rely on simplifications.

The actual novelty is the model and the way of interlacing with the lyophilizer.

The dynamic parameters estimation (DPE) algorithm is an advanced tool proposed by Velardi et al. to interpret the results of the PRT in a more reliable way. It implements an unsteady-state model for mass transfer in the drying chamber and heat transfer in the product, given by a set of partial differential equations describing

- Conduction and accumulation of heat in a frozen layer of the product
- Mass accumulation in the drying chamber during the PRT
- Time evolution of product thickness.

The details of the model can be found in Barresi et al. (6) and in Velardi et al. (24). The discretized system of equations is integrated in time in the internal loop of a curvilinear regression analysis. The cost function to minimize in a least square sense is the difference between the simulated values of the pressure in the drying chamber and the actual values measured during the PRT.

The solution of the model supplies a full-state estimation of the system at the beginning of the PRT: sublimation interface temperature and position, temperature profile over the frozen layer thickness at any axial position (and, thus, at the product bottom as well), mass transfer coefficient, heat transfer coefficient, sublimation mass flow, and an estimation of the required remaining time to complete the primary drying.

These parameters can be considered very robust explaining the process at the moment of the analysis, but there is no guarantee of their accuracy after a certain time horizon. For this reason, this process is repeated in a timely manner along the whole primary drying step.

In Figure 3 an example of the results obtained with DPE algorithm monitoring a freeze-drying cycle is given. It is possible to observe that in the first part of the cycle the temperature at the bottom of the vials estimated by means of DPE algorithm is very close to the value measured by the thermocouples placed in some vials. The earlier increase of the temperature measured by the thermocouples is due to the fact that in the monitored vials the sublimation rate is generally higher with respect to the other vials of the batch and, thus, the primary drying is completed early. The ratio of the signals of a capacitance manometer and of thermal conductivity gauge, like the Pirani gauge, is used to assess the end of the primary drying. At this point the concentration of water into the drying chamber becomes very low, so the pressure measured by Pirani (that is calibrated for air as default) approaches the one measured by the capacitive gauge. This method has been employed in this case to verify independently the end of primary drying, even if

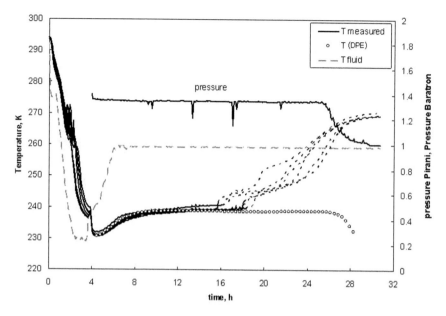

FIGURE 3 Comparison between experimental and estimated temperatures at the vial bottom; cycle at 5 Pa with 609 vials. Pirani to Baratron pressure ratio is shown.

the Pirani gauge is not generally employed in production plants because they generally are not robust enough to withstand repetitive steam sterilization cycles.

In Figure 4 an example of the results obtained with two different batches are shown.

The estimated product temperature decreases nearby the end point. Similar results have been obtained also with the other algorithms based on the PRT. One of the causes that has been proposed to explain this is that when the primary drying is approaching the end, a fraction of vials (mainly the edge vials, because of radiation from the chamber walls) has already finished sublimating, while DPE continues interpreting pressure rise curves assuming batch uniformity or rather a constant number of sublimating vials. A decrease in pressure rise, corresponding to a lower sublimation rate, may thus be interpreted by the algorithm as a reduction of the front temperature. As a consequence, the monitoring methods based on the PRT approach cannot be used in the last part of the primary drying. Nevertheless, the estimations of DPE, which uses a much more complex model to describe the pressure rise occurring during a PRT, are consistent for a large fraction of the primary drying. Some mathematical artifacts can even magnify this problem (see extra explanations in chap. 20). In any case, care must be paid when information obtained using monitoring methods based on the PRT are used to control the process. In fact, all these algorithms estimate the mean value of the product temperature, but in some vials (e.g., those placed at the edges of the tray) product temperature can be higher.

One important aspect of the cycle operation is switching from primary drying to secondary drying at the right time. If the action is taken before the

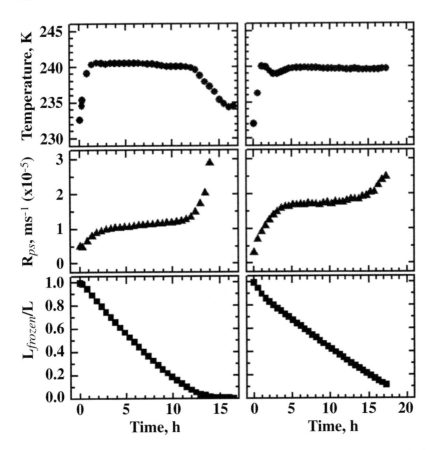

FIGURE 4 Example of DPE estimations obtained in two cycles run with vials of different dimensions. Left: type A vials, placed on a medium size rectangular tray and not shielded (PC = 10 Pa, T_{shelf} = 263 K; total primary drying time 16 hours 35 minutes). Right: type B vials, placed on a smaller circular tray and shielded by empty vials (PC = 10 Pa, T_{shelf} = 253 K; total primary drying time 17 hours 17 minutes). (●) Moving front temperature; (▲) global mass transfer resistance; (■) ice thickness estimated by the DPE solver. *Abbreviation*: DPE, dynamic parameters estimation.

completion of primary drying, then as there is still some ice present and generally the temperature is raised significantly, fatal melting or at least important shrinkage at the bottom of the vial may occur. In the opposite case, productivity will be affected.

Traditional approaches are based on a direct measurement of process variables: product temperature response (thermocouples), thermal conductivity (Pirani) and capacitance manometer pressure ratio, residual gas analyzer (quadrupole mass spectrometer), mass evolution (microbalance), and simple empirical pressure rise technique, etc.

A method combining the sublimation mass flow evolution with the mathematical model can give an accurate determination of the end of primary drying.

DPE can also be used to perform in a single cycle a complete determination of the design space of the manufactured product. A cycle with a collection of steps with different chamber pressures and shelf temperatures can be launched. The DPE will provide for each step condition that are the corresponding product temperature and sublimation mass flow, allowing to generate the map where the acceptable conditions of the process are possible as well as the impact in the process time. Also exact boundary conditions where unacceptable product conditions occur can be generated from these data.

SECONDARY DRYING

The secondary drying phase aims to remove unfrozen water molecules, since the ice was removed in the primary drying phase. This part of the freeze-drying process is governed by the material's sorption isotherms. In this phase, the temperature is raised higher than in the primary drying phase, typically above $0°C$, to break any physicochemical interactions that have formed between the water molecules and the "dry" material. Usually the pressure is also lowered in this stage to encourage desorption (typically in the range of microbars or fractions of a pascal); however, sometimes the same pressure than primary drying is maintained.

At the end of the freeze-drying operation, the final residual water content in the product is very low. In general, for small molecules, the drier, the better. At the end of the operation, the final residual water content in the product is around 0.5% to 4%, and typically, a goal of less than 2% residual moisture in the final lyophilized product is obtained. But for large biopharmaceutical molecules it is not infrequent that the optimal moisture content is within a definite range and overdrying is as harmful as leaving the product with excess humidity. Even in the lyophilized state, peptides and proteins depend on small quantities of water to keep spatial structure, so they need to remain within a range of residual moisture in the final product, for example, 2% to 4%.

In secondary drying, if sufficient time allows equilibrium to be reached, product residual moisture content will depend only on the nature of solid, the product temperature (in this step the product temperature is practically equivalent to the shelf temperature), and the relative humidity in the lyophilizer (a direct function of the water partial pressure at the ice condenser temperature).

Up to now, the typical procedure followed when there is an attempt of process optimization is as follows:

First, rising the shelf temperature up to a value that will not jeopardize product properties (avoiding that the glass transition temperature fall below the process temperature when the mobility of the molecules has increased so much that crystallization of the excipients may occur expelling the absorbed water from the crystal lattice allowing unwanted reactions with the drug).

Then, maintaining the conditions for several hours to reach the expected final moisture level.

And, at the best, during the development stage, taking periodic samples (i.e., with a "sample thief") and measuring off-line their moisture content progression by means of Karl Fischer titration, thermal gravimetric analysis, NIR spectroscopy, etc.

There have been very few rationalization attempts for the secondary drying, basically because the main process concerns have been focused in the primary drying where most challenges exist, and secondary drying has been perceived as a significantly less risky step.

Nevertheless, in the literature appear some studies. One of them (25), applying some heuristics according to the type of product. In another case (26), a statistic modeling with the help of DoE and a set of experiments having shelf temperature, time, and excipient concentration as independent variables is proposed. For a specific product with a specific concentration, a nonlinear equation can be experimentally derived to predict the final moisture as a function of these process parameters. This technique requires a certain number of experiments and is valid only for that specific studied formulation.

One elegant attempt, proposed by Oetjen (27), takes advantage of the quasilinear variation of the logarithm of the desorption rate versus the process time. After some measurements of the desorption rate (with the help of the well-known PRT), the time at which the desorption rate will be negligible (i.e., practically zero) can be extrapolated, so the water content can be integrated at any time during the secondary drying. The only drawback of this process is the need of knowing the residual water content at the equilibrium (when zero desorption has been reached) to know the real total moisture content remaining in the product. Nevertheless, if the formulation and the process conditions are not changed from the first experimental work, there is no need to repeat this experimental residual water determination.

Barresi et al. (28,29) have proposed a mechanistic model-based method to monitor the secondary drying without the help of performing any analysis. The aim of the proposed approach is to monitor the secondary drying stage to determine the residual moisture content of the dried product and to give a reliable estimation of the time that is necessary to complete this stage according to the specified target (residual water content and/or desorption rate of water). The method requires periodical PRT, closing for a short time interval (typically from 30 seconds to few minutes) the chamber-to-condenser valve and recording the time evolution of the total pressure in the chamber. From the slope of the curve at the beginning of the test, the actual water desorption rate (DR, mol/sec) is calculated. The PRT is then repeated at prespecified time intervals to know the time evolution of the desorption rate. A mathematical model of the secondary drying coupled with the measure of desorption rate obtained from PRT allow the in-line moisture content calculation.

The proposed method requires modeling the dependence of the desorption rate and the residual moisture content in the product. Various mathematical equations can be used to this purpose, the easiest one assuming that the desorption rate depends on the residual moisture content. Integrating the resulting differential equation and iterating the process after some desorption rate determinations, the equations can be solved to know both the initial moisture content at the beginning of the secondary drying and the actual product moisture content.

Figure 5 shows a secondary drying step with the comparison between the experimental (sampling and analyzing) values and those predicted by the proposed algorithm of the desorption rate and of the residual water content (original solution was 10% sucrose. The equipment was pilot freeze-dryer LyoBeta 25 with a vial sampling thief from Telstar (Terrasa, Spain).

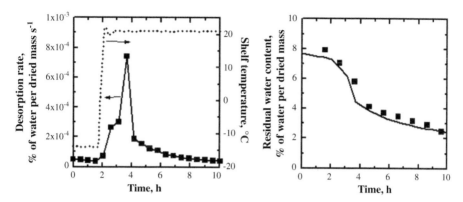

FIGURE 5 Secondary drying. Comparison between the experimental (sampling and analyzing) values (symbols) and those predicted by the proposed algorithm (*solid line*) of the desorption rate (*left*) and of the residual water content (*right*). The time evolution of the shelf temperature is also shown (*dotted line*). Product: sucrose. Equipment: pilot freeze-dryer LyoBeta 25.

DYNAMIC MODEL SENSOR MONITORING: THE ADVANTAGES

This system allows monitoring any production cycle. Its main advantages are as follows:

- Detailed tracking of the whole primary drying process allow further improvements to maximize productivity without impairing product quality.
- The dynamic model sensor provides also information of the completion of primary drying.
- The model is valid for both bulk and lyophilization in vials.
- Definition of the cycle design space coupling the product, its container, and the lyophilizer capabilities extremely simplified.
- It allows plant characterization, so scaling-up or simply transferring the production from one lyophilizer to another one can be straightforward, helping to generate the documentation for full support of the transfer process.
- Much more robust process understanding has an inverse relationship with the risk of producing a poor quality product. Significantly less restrictive regulatory approaches and scrutiny should be expected.

CLOSING THE LOOP

Availability of in-line monitoring of the cycle variables of interest (sublimation interface temperature and sublimation speed) plus the ability to detect the end of the primary drying make the perfect scenario impelling the development of a control system targeted for industrial machines that can close the loop, enabling an in-line control with a predictive action. Detailed description (30) by Barresi et al. can also be found in chapter 20 of this book.

    The main goal of this control software is to minimize the drying time using an optimal heating shelf temperature control strategy that continuously adjusts the heating fluid temperature throughout the main drying phase. In such a way it is possible to maintain the product at its maximum temperature without

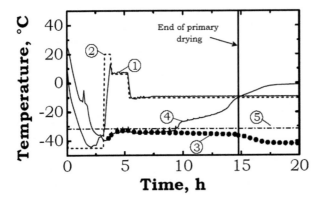

FIGURE 6 Example of an optimal freeze-drying cycle obtained using LyoDriver controller to set the optimal fluid temperature for the primary drying stage. Sucrose 10% solution. Pressure: 10 Pa. Legend: (1) Heating fluid temperature, °C; (2) set-point heating fluid temperature, °C; (3) bottom product temperature estimated through DPE, °C; (4) product temperature measured by a thermocouple in a vial, °C; (5) target maximum temperature, K. *Abbreviation*: DPE, dynamic parameters estimation.

impairing its integrity and, at the same time, minimizing the primary drying time. Finally, it must be pointed out that the control action is determined also taking into account the actual thermal dynamics of the freeze-dryer.

Results obtained in a small industrial-size prototype apparatus manufactured by Telstar (Fig. 6) equipped with this LyoDriver control software, which takes advantage of the DPE monitoring system, are presented in this section. In this case the only necessary extra input is the maximum allowable product temperature from which the software calculates the shelf temperature required to reach it as fast as possible without overshoots.

Figure 6 shows one typical example of experimental freeze-drying cycle run using LyoDriver to control the heating fluid temperature. The cycle is shortened, without risk for the product, because since the beginning the heating up is set at the maximum value allowed, but the future temperature of the product is predicted and overshoot is avoided taking also into account the cooling dynamics of the apparatus. A 10% by weight sucrose solution was freeze-dried at constant pressure (10 Pa), and the maximum temperature was set at −32°C, corresponding to the collapse temperature measured by a freeze-drying microscope.

It can be noticed that the product temperature detected through thermocouples at the bottom never overcomes the limit temperature, not even during the DPE tests when the temperature slightly increases. In this case it is the heat transfer from the shelf that controls the sublimation rate, and the fluid temperature is set at −10°C in the second part of the primary drying step; this value, significantly higher than the product temperature, is the one that assures maximum sublimation rate, while pressure set point (providing it is high enough to guarantee significant heat transmission by convection) is not very influent.

It can be pointed out that in both cases DPE gives good results up to the end of the primary drying and the product temperature estimated agrees with thermocouple measurements, at least until the monitored vials are representative of the entire batch. The system can also determine the end of primary drying, switching to process conditions selected for secondary drying.

Some already commercially available systems use the PRT as sensing device, for example, Oetjen and coworkers (13,14) proposed to use the results of the BTM and a set of heuristics for the calculation of the control actions. The use of the MTM in a control algorithm that manipulates the shelf temperature and the chamber pressure has also been proposed and demonstrated to be a useful tool for a lyophilization cycle development using, presumably, a single freeze-drying run, but it has no predictive capacity and it cannot be used to perform a true process optimization (25,31,32).

In spite of these, LyoDriver is the only alternative that can be proposed as a control tool to be used in an industrial lyophilizer.

CLOSED LOOP CONTROL: MANUFACTURING ADVANTAGES

-   Process development: The most immediate application is using it as a recipe developer. It has several advantages with respect to the few existing tools (as described above), especially because the system at any moment can compute the optimum conditions, so it delivers the optimum primary drying steps since the early moments up to the end.
-   Process transfer and scale-up: The recipe, as provided by LyoDriver, is the optimum recipe that can be run in the final production lyophilizer. Of course it might need slight adjustments to match the slight differences of the industrial lyophilizer with the pilot one (different heat transfer due to shelf geometry and industrial unit limitations with respect to the development one, such as cooling power, instant condensing capacity, etc.). The previous statement would be in case that the production unit would be a standard one. In case that the production unit would also be equipped with the LyoDriver, then it would automatically correct any difference without any human intervention.
-   Production: A lyophilizer with just the DPE would be extremely useful because of the window opened into the process monitoring (one would be aware, for example, of a change of vial supplier with apparent equivalent vials but with a noticeable impact in the heat transfer process), but in case that it is equipped with the LyoDriver then the process would cope perfectly with the QbD paradigm: the process would be self-adjustable to give a constant output, adapting any change in product properties (this would include batch-to-batch changes in nucleation during freezing) or process variables, and it would be run close to the optimum limit.

PLANT REQUIREMENTS

These soft sensors require a modest investment in the plant hardware; a capacitance vacuum gauge is a must, but it is generally a standard component in production units. Either the monitoring or control software described above need to be resident in a computer, but it can share the same computer that is used for the SCADA (supervisory control and data acquisition) software commonly used in any industrial or laboratory plant. The SCADA application has to be adapted as the input/output interface for the required input parameters needed at the beginning of the process (i.e., number of containers, product area in each container, filling volume, amount of solids, etc.) and logging the calculated soft variables (i.e., sublimation interface temperature, sublimation mass flux, diffusivity of the dry phase, heat transfer coefficient, etc) and events (i.e., end of primary drying, mass transfer control situation, etc.).

Interaction with the freeze-dryer is also needed: periodical closing/opening the valve between chamber and condenser at prespecified intervals, actuating (closing) the leak valve used for pressure control during the PRT, acquiring the PRT data at high speed (generally 10 Hz is sufficient), and the shelf fluid temperature at the beginning of each PRT.

SCADA software must communicate in real time with the DPE and/or the control software (namely LyoDriver) without compromising 21CFR Part11 requirements about electronic records and electronic signatures (33).

THE UNKNOWN: REGULATORY APPROACH

No matter the advantages and importance of the perceived benefits of all these presented technological tools, there is, nowadays, still a lot of uncertainty about the regulatory approach to implement tools like them.

In November 2008 the FDA issued a new document "FDA Draft Guidance—Process Validation General Principles and Practices" (34). In the chapter IV—B "Specific Stages and Activities of Process Validation in the Product Lifecycle" specifically states:

> More advanced strategies, such as process analytical technology (PAT), use timely analysis and control loops to adjust the processing conditions so that the output remains constant. Manufacturing systems of this type can provide a higher degree of process control. In the case of PAT strategy, the approach to process qualification will be different from that for other process designs.

Each pharmaceutical industry planning to implement these tools need, therefore, to start since the early beginning a joint work with the Regulatory Authorities to establish the roadmap of the specific process validation.

CONCLUSIONS

It is possible to design a process with a consistent output, despite a very variable input, and with a mechanistic model; it is possible to perform powerful analysis of the correlation between the process parameters and the process output. These simulations allow identifying key parameters and spend the limited resources where it is gained the most.

The associated complexity and hardware associated to the implementation of these mechanistic models are much more modest than other accepted practices when implementing typical PAT tools (as for example, modern process analyzers or process analytical chemistry tools and multivariate data acquisition and analysis software, generally requiring high processing power and high investment).

The pharmaceutical industry should benefit from the implementation of QbD as it

- ensures better design of products with an expectation of fewer problems in manufacturing;

- reduces the number of extra filing supplements for postapproval changes—relying on process and risk understanding with commensurate risk mitigation;
- allows implementation of new technology to improve manufacturing without extraordinary regulatory scrutiny;
- reduces overall costs of manufacturing, less waste;
- may improve interaction with the Regulatory Authorities allowing industry to deal with them on a science level instead of on a process level, resulting in reduced deficiencies and quicker approvals; and
- continuously improves products and manufacturing processes, which are viable and significant outcomes of QbD.

The FDA has also reported (35) the benefits of implementing QbD for them as it

- enhances scientific foundation for review;
- provides better coordination across review, compliance, and inspection;
- improves information in regulatory submissions;
- provides better consistency along with improvements in quality of review;
- gives more flexibility in decision making that is beneficial to both the industry and the FDA;
- ensures decisions will be made on science and not on merely empirical information; and
- involves various disciplines in decision making and uses resources to address higher risks.

Further improvements can be proposed in the monitoring and control algorithm to take into account heterogeneity within the batch that can be caused by different reasons: radiation from chamber walls, shelves, and door; by the fluid dynamics of the sublimated water vapor in the drying chamber; by non-uniform temperature of the heating shelf; and by nonuniform initial filling height.

The mechanistic model described above could be coupled with the fluid dynamics model of the chamber and condenser. A two-scale model could be obtained, and the effect of the local pressure distribution in the chamber could help to assess the batch variance.

ACKNOWLEDGMENTS

The Lyolab research team of the Department of Materials Science and Chemical Engineering of the Politecnico di Torino in Italy.

REFERENCES

1. Pikal MJ, Rambhatla S, Ramot R. The impact of the freezing stage in lyophilization: effects of the ice nucleation temperature on process design and product quality. Am Pharm Rev 2002; 5(3):48–53.
2. Searles JA, Carpenter JF, Randolph TW. The ice nucleation temperature determines the primary drying rate of lyophilization for samples frozen on a temperature-controlled shelf. J Pharm Sci 2001; 90:860–871.
3. Galan M. Abstracts of papers: understanding and controlling the nucleation step in lyophilization. Barnett International's Annual Conference on Freeze-Drying. Washington, DC, May 18–20, 2005.

4. Hsu CC, Ward CA, Pearlman R, et al. Determining the optimum residual moisture in lyophilized protein pharmaceuticals. Dev Biol Stand 1992; 74:255–271.
5. Gieseler H, Kessler W, Finson M, et al. Evaluation of tunable diode laser absorption spectroscopy for in-process water vapor mass flux measurements during freeze drying. J Pharm Sci 2007; 96(7):1776–1793.
6. Barresi A, Pisano R, Fissore D, et al. Monitoring of the primary drying of a lyophilization process in vials. Chem Eng Process 2009; 48:408–423.
7. FDA. Guideline on General Principles of Process Validation. Available at: http://www.fda.gov/Drugs/GuidanceComplianceRegulatoryInformation/Guidances/ucm124720.htm. Accessed July 2009.
8. FDA. Guidance for Industry: PAT—A Framework for Innovative Pharmaceutical Development, Manufacturing, and Quality Assurance. Available at: http://www.fda.gov/downloads/Drugs/GuidanceComplianceRegulatoryInformation/Guidances/UCM070305.pdf. Accessed July 2009.
9. FDA. Q8 Pharmaceutical Development. International Conference on Harmonization (ICH):Guidance for Industry—Q8 Pharmaceutical Development. Available at: http://www.fda.gov/RegulatoryInformation/Guidances/ucm128028.htm. Accessed July 2009.
10. Neümann K. Freeze drying apparatus. US patent 2,994,132. 1961.
11. Nail SL, Johnson W. Methodology for in-process determination of residual water in freeze-dried products. Dev Biol Stand 1991; 74:137–151.
12. Willemer H. Measurement of temperature, ice evaporation rates and residual moisture contents in freeze-drying. Dev Biol Stand 1991; 74:123–136.
13. Oetjen GW. Drying. In: Oetjen GW, ed. Freeze-Drying. Weinheim: Wiley-VCH, 1999:58–109.
14. Oetjen GW, Haseley P. Process automation. In: Oetjen GW, Haseley P, eds. Freeze-Drying. 2nd ed. Weinheim: Wiley-VHC, 2004:268–291.
15. Oetjen GW, Haseley P, Klutsch H, et al. Method for controlling a freeze-drying process. US patent 6,163,979. 2000.
16. Milton N, Pikal MJ, Roy ML, et al. Evaluation of manometric temperature measurement as a method of monitoring product temperature during lyophilization. PDA J Pharm Sci Technol 1997; 5:7–16.
17. Tang XC, Nail SL, Pikal MJ. Evaluation of manometric temperature measurement, a Process Analytical Technology tool for freeze-drying: part I, product temperature measurement. AAPS PharmSciTech 2006; 7(1):E95–E103.
18. Tang XC, Nail SL, Pikal MJ. Evaluation of manometric temperature measurement, a Process Analytical Technology tool for freeze-drying: part II, measurement of dry layer resistance. AAPS Pharm Sci Tech 2006; 7(4):E77–E84.
19. Tang XC, Nail SL, Pikal MJ. Evaluation of manometric temperature measurement (MTM), a Process Analytical Technology tool in freeze drying, part III: heat and mass transfer measurement. AAPS Pharm Sci Tech 2006; 7(4):E105–E111.
20. Liapis AI, Sadikoglu H. Dynamic pressure rise in the drying chamber as a remote sensing method for monitoring the temperature of the product during the primary drying stage of freeze-drying. Drying Technol 1998; 16:1153–1171.
21. Chouvenc P, Vessot S, Andrieu J, et al. Optimization of the freeze-drying cycle: a new model for pressure rise analysis. Drying Technol 2004; 22:1577–1601.
22. Chouvenc P, Vessot S, Andrieu J, et al. Optimization of the freeze-drying cycle: adaptation of the Pressure Rise Analysis to non-instantaneous isolation valves. PDA J Pharm Sci Technol 2005; 5:298–309.
23. Hottot A, Vessot S, Andrieu J. Determination of mass and heat transfer parameters during freeze-drying cycles of pharmaceutical products. PDA J Pharm Sci Technol 2005; 59:138–153.
24. Velardi S, Rasetto V, Barresi A. dynamic parameters estimation method: advanced manometric temperature measurement approach for freeze-drying monitoring of pharmaceutical solutions. Ind Eng Chem Res 2008; 47(21):8445–8457.
25. Pikal MJ, Tang X. Automated process control using manometric temperature measurement. US patent 6,971,187. 2005.

26. Mayeresse Y. Abstracts of Papers: Moisture content in freeze dried product. Cambridge HealthTech Institute, PepTalk Conference: Lyophilization. San Diego, CA, January 9–11, 2008.
27. Oetjen GW. Method of determining residual moisture content during secondary drying in a freeze-drying process. US patent 6,176,121 (2001).
28. Barresi A. Abstracts of Papers: Monitoring and control of freeze drying using innovative software devices. The Freeze Drying of Pharmaceuticals and Biologicals Conference Breckenridge, CO, August 6–9, 2008.
29. Galan M. Abstracts of Papers: Secondary drying. Cambridge HealthTech Institute, PepTalk Conference: Lyophilization. San Diego, CA, January 12–14, 2009.
30. Fissore D, Pisano R, Velardi, et al. PAT tools for the optimization of the freeze-drying process. Pharm Eng 2009; 29(5):58–70.
31. Tang XC, Nail SL, Pikal MJ. Freeze-drying process design by manometric temperature measurement: design of a smart freeze-dryer. Pharm Res 2005; 22:685–700.
32. Gieseler H, Kramer T, Pikal MJ. Use of manometric temperature measurement (MTM) and SMART (TM) freeze dryer technology for development of an optimized freeze-drying cycle. J Pharm Sci 2007; 96:3402–3418.
33. FDA. Part 11, Electronic Records; Electronic Signatures—Scope and Application. Available at: http://www.fda.gov/RegulatoryInformation/Guidances/ucm125067.htm. Accessed July 2009.
34. FDA. Guidance for Industry: Process Validation—General Principles and Practices. Available at: http://www.fda.gov/downloads/Drugs/GuidanceComplianceRegulatoryInformation/Guidances/UCM070336.pdf. Accessed July 2009.
35. Winkle HN. Abstracts of Papers: Implementing Quality by Design. PDA/FDA Joint Regulatory Conference. Washington, DC, September 24–28, 2007.

20

Process Analytical Technology in Industrial Freeze-Drying

Antonello A. Barresi, Davide Fissore, and Daniele L. Marchisio
Dipartimento di Scienza dei Materiali e Ingegneria Chimica, Politecnico di Torino, Torino, Italy

INTRODUCTION

Process analytical technologies (PAT) are intended to be systems for designing, analyzing, and controlling manufacturing through timely measurements (i.e., during processing) of critical quality and performance attributes of raw and in-process materials and processes, with the goal of ensuring final product quality. PAT tools are thus required to implement a true QbD (quality-by-design) manufacturing principle, rather than the classical QbT (quality-by-testing) approach: quality is no longer tested into products, but it is built-in or is by design. This is the framework described by the Guidance for Industry PAT—A Framework for Innovative Pharmaceutical Development, Manufacturing, and Quality Assurance issued by U.S. FDA in September 2004. This guidance encourages the design and implementation of innovative pharmaceutical development, manufacturing, and quality assurance to support innovation and efficiency, with the goal to have safe, effective, and affordable medicines. In fact, besides a proper design of the formulation containing the active pharmaceutical ingredient (API), a comprehensive understanding and design of the manu-facturing process is required to guarantee product quality.

Pharmaceuticals freeze-drying is a process where significant improve-ments can be obtained if PAT tools are used. In fact, although regarded as a soft dehydration process, due to the low operating temperatures, the API can be damaged as a series of stresses are applied to the molecules of the product, which can be rather labile, both during freezing, due to the large variation of solute concentration, ionic strength, and, eventually, pH, and during drying (1). Moreover, in case of pharmaceutical products also the final appearance and the reconstitution time, which can be strongly affected by processing, must be taken into account (2). To this purpose, during the process, product temperature has to be carefully maintained below a limit value that is a characteristic of the product. In case of solutes that crystallize during freezing, the limit value cor-responds to the eutectic point to avoid formation of a liquid phase and suc-cessive boiling due to low pressure. In case of solutes that do not crystallize during freezing, the maximum allowed product temperature is close to the glass transition temperature to avoid the collapse of dried cake. This collapse as well as the shrinkage (that is generally due to limited and localized collapse phe-nomena) can be responsible for a higher residual water content in the final product, a higher reconstitution time, and the loss of activity of the pharma-ceutical principle; besides this, a collapsed product is often rejected because of the unattractive physical appearance. Finally, the residual water content at the end of secondary drying has to match a target value, which is dependent on the product considered.

Process monitoring systems are required to guarantee product quality. This is a challenging task due to the impossibility of measuring in-line the variables of interest, product temperature and residual water content, without interfering with the dynamics of the system and/or impairing the sterile conditions generally required when pharmaceuticals are manufactured. Wiggenhorn et al. (3,4) discussed the use of near-infrared spectroscopy methods, microbalances, and mass spectrometry to monitor the evolution of the residual water content during the drying. While the first two techniques can be used only in pilot-scale apparatus, for research purposes and recipe development, as they cannot be sterilized, the mass spectrometer can be used also in commercial-scale production setups, where full aseptic conditions are mandatory, if an aseptic sterile filter is placed between drying chamber and the spectrometer. An analysis of the technology available and of its possible use either in research and development units or in industrial production apparatus was later on presented by Mayeresse et al. (5), who tested a device based on inductive coupled plasma/optical emission spectroscopy ("cold plasma" probe) showing its potentiality for industrial applications.

Recently, the use of Raman spectroscopy has been proposed for product monitoring (6). A complete review of the techniques available to monitor freeze-drying both in pilot-scale equipment and in large-scale units can be found in Barresi et al. (7) and in Barresi and Fissore (8). They divided various sensors into two groups. The first group comprises temperature probes (thermocouples and resistance thermal detectors), conductivity probes, balances, spectroscopy-based methods, and off-line analysis after sampling via the sampling theft. These devices allow monitoring of single vials or a group of vials. The second group is formed by all those methods that provide information about the status of the whole batch: temperature measurements using the pressure rise test (PRT), windmill devices, mass spectrometer, pressure gauges, cold plasma sensor, tunable diode laser absorption spectroscopy (TDLAS) sensor, and moisture sensors have to be named here.

Besides process monitoring, a proper design of the recipe, that is, of the values of shelf temperature and chamber pressure versus time, is required. A recipe can be obtained in a pilot-scale equipment using a trial-and-error procedure. As an alternative, it is possible to use a monitoring and control system. The controller receives the information about the status of the product from the monitoring system and calculates the values of shelf temperature and chamber pressure according to a performance requirement, like the minimization of the drying time without violation of the constraint on the maximum product temperature (9–13). After this, the recipe has to be scaled-up to the industrial-size apparatus. Generally, a trial-and-error procedure is used to this purpose, even if it is expensive and does not guarantee acceptable and reproducible product quality and performance throughout a product's shelf life. Also for this step the use either of control tools that assure the respect of the constraint on product temperature (instead that on the equipment set points), if available in the industrial freeze-dryer, or of software tools for reliable process transfer would be extremely useful. It must be stressed that a proper design of the apparatus is essential not only to minimize the variability of the conditions in different parts of the equipment, and thus the intra-batch variance, but also to assure the highest possible sensitivity and reliability to the monitoring sensors employed.

Process design, control, and qualification are the focus of the Guidance for Industry Process Validation: General Principles and Practices issued by U.S. FDA in November 2008. According to this guidance, each step of a manufacturing process has to be controlled to ensure that the finished product meets all design characteristics and quality attributes including specifications.

This chapter is intended to address these issues, and, to this purpose, it is structured as follows: The first section points out how equipment design can affect product quality and batch heterogeneity during operation (mainly during primary drying), through pressure gradients in the drying chamber and temperature gradients over the various shelves, besides the well-known effect of radiation. The use of computational fluid dynamics (CFD) can be very effective, especially if dual-scale modeling approaches are adopted, to improve the design not only of the drying chamber but also of the condenser, and for a selection of duct size and valves that assure the required performances. Recipe development in pilot units is discussed in the second section, showing that the recipe can be obtained *in-line*, by using some tools to monitor the process, and then manipulating the operating variables to achieve the desired goals, or can be designed *off-line*, by using mathematical models to build the *design space*, and then validating it. The third section discusses how process transfer or scale-up can be made easy and robust either using specific software tools after a process identification step or using a monitoring and control system, if available in the large-scale equipment. Finally, future perspectives concerning further improvement in the performance of monitoring systems and the possibility of using more sophisticated control systems, which can take into account batch heterogeneity, is discussed; a short account about the use of model predictive control (MPC) in freeze-drying is also given.

EQUIPMENT DESIGN

When designing a freeze-dryer, a number of critical issues must be properly addressed. These can be grouped into two categories. The first category is related to the problem of batch heterogeneity; as it is well known, especially at the industrial scale, it is challenging to design an equipment capable of guaranteeing the very same operating conditions for the entire batch. It is very common, in fact, that the product contained in vials positioned in different points of the chamber experience different temperature and pressure histories, resulting, for example, in significant variations for the drying time and the final residual water content (as well as other important characteristic properties). Typical design and operating parameters affecting the final batch heterogeneity are chamber geometry, clearance between shelves and number of shelves, position of the duct leading to the condenser, number and position of the inert gas injection nozzles, as well as temperature gradients of the heating fluid circulating through the shelves (14–16). An example of the effect of these factors on the drying of a pharmaceutical product in vials is shown in Figure 1.

The second category of critical issues concerns the capability of the freeze-drier to evacuate the requested water vapor flow rate to operate under the desired operating conditions (i.e., specific sublimation rates and thus batch drying times and chamber pressure). Typical design parameters that affect the overall performance of the freeze-drier with respect to these issues are mainly the geometry of the condenser and, for freeze-driers with condensers separated

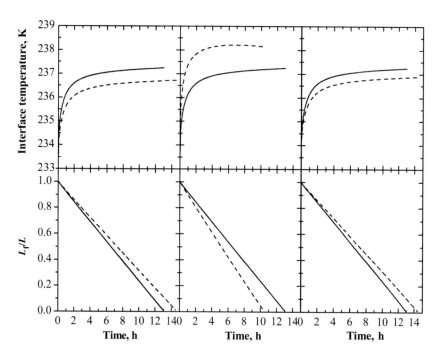

FIGURE 1 Comparison between the time evolution of the temperature (*upper graphs*) and of the frozen layer thickness (*lower graphs*) in a freeze-drying cycle of a 10% wt/wt sucrose solution (operating pressure = 10 Pa, $d_v = 14 \cdot 10^{-3}$ m, $L = 7.2 \times 10^{-3}$ m). Left-hand graphs: Dynamics in a vial placed over a shelf with $T_{shelf} = -15°C$ (*dashed line*) and in a vial placed over a shelf with $T_{shelf} = -12°C$ (*solid line*). The vials are considered perfectly shielded from radiation. Central graphs: Dynamics in a vial perfectly shielded for radiation (*solid line*) and in a vial at the border of the shelf, where radiation is relevant (*dashed line*). $T_{shelf} = -15°C$. Right-hand graphs: Dynamics in a vial placed at the border of the shelf, but shielded from radiation, (*solid line*) and in a vial placed in the central part of the shelf (*dashed line*), where pressure is higher. $T_{shelf} = -15°C$.

from the freeze-drying chamber (which is the most common case in industrial apparatuses), the size of the duct and of the isolating valve.

Since the experimental investigation of these issues is costly and time consuming, it is very interesting to explore them by using a computational approach, capable of giving meaningful predictions yet significantly reducing the extent of the experimental campaign. Such an approach is represented by CFD that, through a finite-volume numerical scheme (17), solves the continuity and Navier–Stokes equations along with other relevant governing equations (e.g., enthalpy balance) not only delivering reliable predictions, which can replace long and expensive experiments, but also providing engineers and scientists involved in equipment design with precious and very detailed insights. The main limitation of this technique stands in its description of the sublimating gas as a continuum (18), being therefore valid only when the Knudsen number (Kn) is smaller than 0.01. Although at the low pressures typical of freeze-drying operations the characteristic mean free path of the involved gas molecules can be quite large, under usual operating conditions (both for chamber and condenser) and for the most common industrial geometries this condition is almost always satisfied. Nevertheless, in the transitional regime ($0.01 < Kn < 0.1$) CFD can still be used with specific boundary conditions imposing non-null gas velocity at the wall (i.e., partial-slip boundary condition). In the so-called molecular regime ($Kn > 1$), the flow is not dominated anymore by collisions between gas molecules, and nonequilibrium effects start being important, resulting in velocity distributions far away from the Maxwellian one (19). In these cases it is necessary to resort to alternative simulation frameworks, such as the Lattice–Boltzmann scheme.

CFD can be used to simulate pseudostationary conditions at a certain time or the entire freeze-drying cycle and the analysis can be limited to specific parts of the equipment (i.e., chamber, condenser, duct and valve, etc.), or can include the entire apparatus. As an example, in Figure 2 the computational geometry used to investigate a pilot-scale freeze-dryer (LyoBeta^{TM} by Telstar, Terrassa, Spain) and some results are shown. In this case, the grid describes the entire equipment from the sublimating sources, i.e., the product contained inside vials positioned on the shelves or directly displayed in trays (on the shelves), to the condenser where the solvent (typically water) is sublimated and the inert gases are extracted with a vacuum pump. The possibility of working with unstructured codes (such as Fluent or openFoam) combined with the possibility of using very irregular grids (with hexahedral and tetrahedral elements) allow to resolve most of the relevant geometrical details with adequate accuracy.

From Figure 2 (upper right) it is possible to see that water vapor to flow from the chamber to the condenser has to overcome some pressure drop, and this increases with increasing the sublimation rate. This factor plays a crucial role since the local pressure values and thus the product temperatures (as highlighted in Fig. 1) change from point to point in the chamber, potentially resulting in batch heterogeneity. CFD can be profitably used to design chambers that minimize this effect and can also allow estimating the minimum pressure obtainable in the chamber, and the solvent and inert gas partial pressure, as a function of the operating conditions.

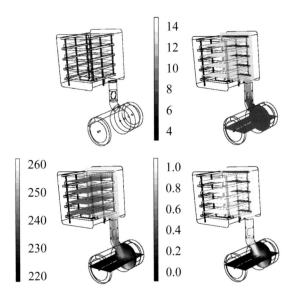

FIGURE 2 Simulation of a small pilot-scale equipment (LyoBeta^{TM} by Telstar) using CFD; the grid was created with Gambit and consisted of about 300,000 cells, and a finite rate mechanism has been included as submodel for ice formation in the condenser. From left to right and top to bottom: Geometry of apparatus and contour plots for absolute pressure (Pa), static temperature (K), and water mass fraction (dimensionless) for an average sublimation rate on each single shelf equal to 0.4 kg/h m^2. By permission of Telstar, Terrassa, Spain.

For example, CFD can be used to test different design solutions concerning the position of the duct connecting the chamber to the condenser, whose effect depends also on the distance between shelves. Figure 3 shows the pressure distribution over the shelves for different duct positions and shelf clearances (compare cases A and B, with the same number of shelves in the pilot scale, and cases A and C with different shelf size but same clearance). It can be seen that when the clearance between the shelves is large, and the shelf has a small size,

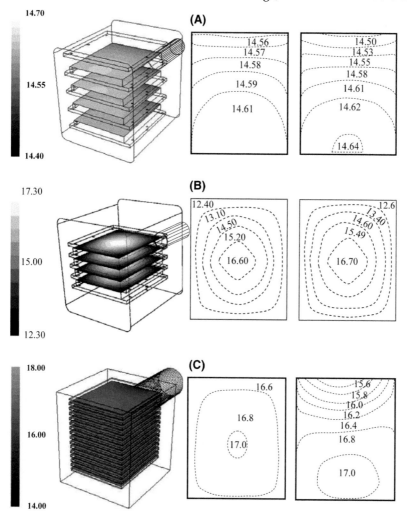

FIGURE 3 Global three-dimensional representation of pressure contour plots computed for all the plates (*left*), for the shelf at the bottom of the chamber (*center*), and for the shelf close to the duct (*right*) **(A)**: small-scale apparatus, with clearance between the shelves equal to 100 mm and duct positioned on the right side of the back wall, at middle height; **(B)**: small-scale apparatus, with clearance between the shelves equal to 60 mm and duct positioned in the center of the back wall; **(C)**: large-scale apparatus (LyoMega™ 400 by Telstar) with clearance between the shelves equal to 100 mm and duct off-centered, in the upper-right quadrant. By permission of Telstar, Terrassa, Spain.

the pressure difference between the center and the edges of the shelf is quite small; in this case it is more influenced by the position of the duct (not shown). On the contrary, when the shelf-to-shelf distance is quite small, or the size of the shelf increases, as in the industrial apparatus, the pressure difference between different points of the chamber becomes significant, but it is weakly affected by the position of the duct (as proven by the fact that pressure profiles are in this case slightly asymmetric). As it has been already reported, this significantly impacts the sublimation rate, and we will come back to this aspect later on.

Another important factor to consider when designing freeze-drying equipments is the presence of composition gradients in the vapor inside the chamber, due to spatially heterogeneous partial pressure of solvent (water or organic). In fact, according to the geometry of the chamber and condenser, and depending on the operating conditions, not only the total absolute pressure in every point of the chamber can vary, but also the composition in different points may be substantially different, and both these effects potentially lead to batch heterogeneities. This latter one is strongly determined by the number and position of the inert gas nozzles, very often used to manipulate and control the overall chamber pressure. These have to be positioned to guarantee as much as possible uniform inert gas distributions inside the chamber.

Vapor and inert gas distribution and vapor pathways can strongly influence efficiency and performance of the condenser; the inert mass fraction increases in the condenser approaching the outlet. Of course, optimal design of the condenser should guarantee the best performances in terms of both heat and mass transfer. An example of how CFD allows to estimate the local variation of water mass fraction is shown in Figure 2 (lower right).

A proper design is important in pilot-scale equipments where the geometry is generally much simpler, but most of these issues become even more important when considering an industrial-scale apparatus. In particular, the design of the condenser can largely benefit from the knowledge of the real fluid dynamics inside the condenser itself and from the evaluation of the ice deposition rate on coils and surfaces: computations can become extremely heavy in this case, especially in the complex geometry of an industrial apparatus, due to the necessity of modeling the vapor disappearance (and the ice formation) with a realistic mechanism that takes into account the proper kinetics.

Regarding the fluid dynamics of the vapor in the drying chamber, in Figure 3 the typical absolute pressure profiles encountered on the shelves of a laboratory-scale and on an industrial-scale apparatus are reported and compared.

As it is seen, under similar operating conditions the pressure values and the pressure differences (in different points of the same shelf or in different shelves) may significantly differ in the two apparatuses. As already reported, this has a huge impact on the final drying history of each vial, since the local water vapor pressure determines the interface equilibrium temperature of the drying product and, as a consequence, the sublimation rate. This can be estimated by resorting to a two-scale approach, where the evolution of the product in each vial is linked to the mathematical description of the entire chamber. Different strategies can be used and a very brief discussion is reported in what follows; however, readers interested in the details are remanded to the work of Rasetto (20). In many practical cases it is very interesting to construct the two-scale model within an online framework, by resorting for example to user-defined scalars and subroutines available in many commercial and open source

CFD codes. In fact, the sublimation rate can be easily correlated to the operating conditions (shelf temperature and chamber pressure) and to the vial and product characteristics by means of a simple mass balance equation, coupled with an equation stating that the heat flow from the shelf to the product is used to sublimate the ice. This simple model (for the vial) results in the prediction of the sublimation flux value in each point of the computational domain of the CFD simulation (at the chamber scale) and in each instant of time of the drying process, depending on the local vapor pressure, shelf temperature, and thickness of frozen layer. Results obtained with this approach are described below whereas results obtained by using an off-line approach, more suitable for process transfer and scale-up, are presented in the next section.

In Figure 4 (upper line), an example of the typical pieces of information obtained from the dual-scale simulation is shown; in particular, the contour plots for the absolute pressure (Fig. 4A), the interface temperature (Fig. 4B), and the resulting sublimation rate (Fig. 4C) are reported, quantifying the effect of pressure gradients on vials history. As it is possible to see, a difference of about 2 to 3 Pa in the local water vapor pressure results in a variation of about 1 to 2 K in the equilibrium interface temperature, resulting in turn in a variation of 3% to 4% in the sublimation rate. This shows how the CFD two-scale model can be used to design the freeze-dryer. By fixing the maximum allowed sublimation rate difference, it is possible to calculate the maximum pressure difference across a single shelf or in different shelves.

Exploiting even further the two-scale model, it is possible to calculate the maximum tolerated gradients in the heating fluid circulating inside the shelf that result in a given variation of sublimation rate. As an example, in Figure 4 (lower line) the profiles across a shelf of the industrial-scale apparatus for the resulting sublimation rate with a constant heating fluid temperature (Fig. 4C) and with a constant gradient of the heating fluid temperature (Fig. 4D) are reported.

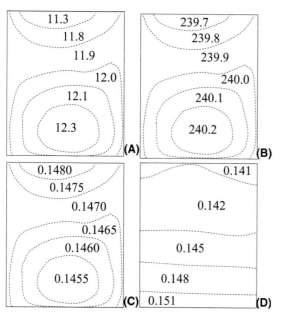

FIGURE 4 Contour plots on the 12th shelf of a LyoMega^{TM} 400 (by Telstar) of absolute pressure, Pa (**A**), interface temperature, K (**B**) and sublimating flux, kg/m^2 h, in the case of heating fluid with constant temperature (**C**) and undergoing a linear variation of 2 K from one end to another (**D**). By permission of Telstar, Terrassa, Spain.

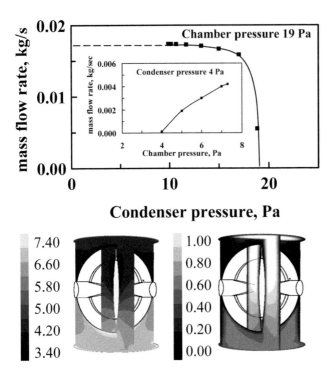

FIGURE 5 Mass flow rate versus condenser and chamber pressures for a typical valve/duct (DN 700) used in industrial freeze-dryers (*top*); contour plots of absolute pressure (Pa) and Mach number (dimensionless) under critical conditions (*bottom*). By permission of Telstar, Terrassa, Spain.

Great care must be paid to the possibility of choking flow in the duct connecting chamber and condenser, mainly due to the presence of the valve. In fact, due to the very low pressure values (and therefore very high water vapor velocities), critical sonic flow conditions may be encountered. The diameter and length of the duct, as well as the geometry of the isolating valve, must be properly designed to guarantee under a wide range of operating conditions that the desired sublimation rate is evacuated. Figure 5 reports the mass flow rate as a function of condenser and chamber pressures, for a specific case. As the pressure difference increases critical flow conditions are reached, resulting in a maximum flow rate, known as critical flow rate; it must be noted anyway that the flow rate in critical conditions increases with pressure in the chamber because this affects the static density of the fluid. This critical flow rate depends only on the chemical nature of the vapor and the length-to-diameter ratio of the duct or the equivalent length-to-diameter ratio of the isolating valve. Typical values for water vapor are reported in the figure.

RECIPE DEVELOPMENT
A freeze-drying cycle for a given formulation containing an API is generally specified as a sequence of values of shelf temperature and chamber pressure that allows obtaining the product with acceptable quality. The design of such recipe

is generally carried out by means of a trial-and-error approach in a small-scale apparatus. According to the new paradigm of the Guidance for Industry PAT, the QbT approach has to be replaced by a QbD manufacturing principle. This result can be obtained in the following two different ways:

1. The recipe can be obtained *in-line*, by using some tools to monitor the process, and then manipulating the operating variables to achieve the desired goals.
2. The recipe can be designed *off-line*, by using mathematical models to build the *design space*, that is, the multidimensional space that encompasses combinations of product design and processing variables that preserve product quality, and then validating it.

Both approaches are discussed in the following.

In-line Recipe Design

A recipe for a freeze-drying process can be obtained in-line by means of a monitoring system that provides information about the state of the system (i.e., product temperature and sublimating interface position), and of a control system that calculates the values of the operating conditions (shelf temperature and chamber pressure) on the basis of the status of the product and of the goal (and of the constraints) of the process.

Many tools are available to monitor a freeze-drying process, and they have been recently reviewed and compared by Barresi and Fissore (8). Nevertheless, only few of them provide information about the status of the product and can thus be used for recipe development. The salient feature of these tools is that they make use of experimental measures (of temperature, vapor flux, or pressure) and of a mathematical model to estimate the temperature of the product, the residual ice content, and the parameters required by the control system.

The measurement of the temperature of the product (or of the vial wall) is used by the "smart vial." It consists of a soft sensor (or observer), that is, a device that uses the temperature measure and a mathematical model of the process to provide a real-time estimation of some parameters and state variables (7,11,13,21–23) and can be coupled with a wireless transmission system to transmit the measured values to a PC (12,24). Both active or passive transponder can be used, as shown in Figure 6A. The passive device is able to measure temperature into several vials and to communicate the results to an external receiver through a central unit. The low-frequency RF-ID technology is employed to avoid the use of batteries (13,25). As an alternative, a small battery can be completely embedded into the vials. This device can transmit the measurements within a range of several meters by means of a 2.4-GHz radio (24). The synthesis of the observer is a complex task and a lot of algorithms have been proposed in the literature. Velardi et al. (26,27) proposed to use a simplified pseudostationary model (28) and the measurement of the product temperature at the bottom of the vial to design an observer using either the extended Kalman filter or the high-gain technique. Both observers are able to estimate product temperature and the position of the moving interface, as well as product resistance to vapor flow in the dried layer and the heat transfer coefficient between the shelf and the vial. The two algorithms have been validated by means of numerical simulations using a detailed one-dimensional model as a

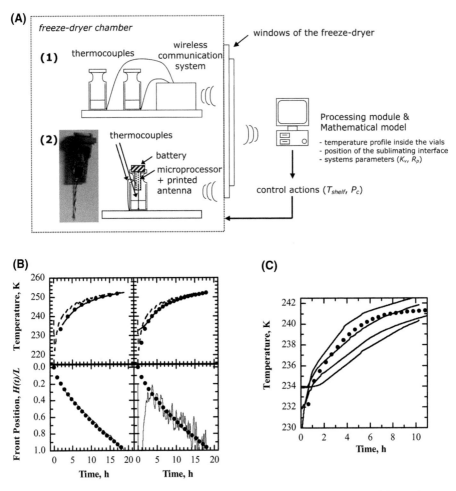

FIGURE 6 (**A**) Sketch of the "smart vial" using either a passive transponder (1) or a battery hidden in the stopper (2). The thermocouple can measure the temperature of the glass vial or that of the product. (**B**) Comparison between high-gain (*left-hand graphs*) and extended Kalman filter (*right-hand graphs*) predictions when a noisy measure is used. Upper graphs: comparison between true and estimated temperatures. Solid line: T_i estimated by the observer; solid circles: detailed model; dashed line: T_B noisy. Lower graphs: moving front position estimated by the observers (*solid line*) and calculated using the detailed model (*circles*). (**C**) Estimations of the front temperature in four vials placed in different positions on the shelf (*solid lines*); the average value estimated using DPE (•) is shown for comparison.

source of experimental data (28). The performances of both observers are roughly the same in terms of quality of the estimations, but the computational effort required by the Kalman filter is higher and its tuning is quite tricky while the estimations obtained using the high-gain observer are provided faster and the tuning is simpler. Moreover, the high-gain observer exhibits less sensitivity toward noisy measurements as it can be seen in Figure 6B, where the performance of both observers is compared in case of a noised measurement, obtained adding to the values of T_B a white noise with a 1 K standard deviation. The interface temperature is accurately estimated by both algorithms while the position of the sublimation front estimated using the extended Kalman filter suffers for the noise introduced with the measure that is amplified making the estimation less reliable. The main drawback of the high-gain algorithm is that it is based on a change of variables that, due to the number of parameters to be estimated, requires the computation of high-order derivatives of the functions involved. To this purpose, Velardi et al. (27) proposed to estimate the vapor diffusivity in the dried layer by using an identification algorithm within a small interval at the beginning of the sublimation, and then designed the high-gain observer to estimate the other variables, thus reducing the order of the problem.

Since the insertion of a probe, although extremely tiny, in contact with the product should be avoided as it can interfere with the evolution of the product, a different observer has been designed, using the measurement of the external temperature at the bottom of the vial (11,13,22) and a mathematical model that takes into account also the heat transfer in the glass wall (28). This is the noninvasive version of the "smart vial" that has been recently patented by Barresi et al. (21).

Other devices that can monitor the process use the direct measure of the vapor flow. The measurement of vapor flux by using a windmill sensor placed in the spool connecting the drying chamber to the condenser has been proposed by Tenedini and Bart (29) to monitor the batch. More recently, the use of TDLAS has been proposed. This device measures water vapor concentration and gas flow velocity in the duct connecting the freeze-drying chamber and the condenser using Doppler-shifted near-infrared absorption spectroscopy (30,31). The concentration and gas velocity measurements are combined with the knowledge of the duct cross-sectional area to determine the vapor mass flow rate. Such devices allow for process monitoring as the sublimation flux can be integrated in time to determine the total water removed throughout the process. TDLAS seems particularly promising since it can be installed in laboratory-scale as well as in production-scale equipment, and it can be easily placed in a sterile process environment, even if it could be difficult to retrofit existing units as it should be located in the freeze-dryer spool. The main drawbacks are the cost and the difficulty of calibration; fluid flow modeling can be used to provide acceptable density and velocity determinations (32). Product temperature can be estimated from the vapor mass flow rate if the heat transfer coefficient between the shelf and the product in the vial (K_v) is known (33,34):

$$T_B = T_{shelf} - \frac{j_w \Delta H_s}{A_v K_v} \tag{1}$$

Thus, the use of equation (1) requires a preliminary estimation of the heat transfer coefficient K_v in the same equipment and using exactly the same vial. To this purpose, it is possible to run a test where the value of T_B is measured by means of a thermocouple inserted in the vial and in close contact with the bottom, and calculate K_v using the following equation:

$$K_v = \frac{j_w \Delta H_s}{A_v (T_{shelf} - T_B)} \tag{2}$$

The simultaneous estimation of sublimation flux and product temperature, as well as K_v and product resistance to vapor flow (R_p), can be obtained using the PRT, that is, the measurement of the pressure rise due to the closure of the valve in the spool connecting the drying chamber to the condenser. Both graphical methods and mathematical models have been proposed to relate the pressure rise to the sublimating interface temperature. Moreover, the sublimation flux can be estimated, as it is directly related to the slope of the pressure rise curve at the beginning of the test. Thus, it can potentially provide information about the end of the primary drying, characterized by the strong decrease of the vapor flux flowing from the vials into the chamber.

In early works (35–38), the pressure rise measurement was proposed as a method for determining the end of primary drying and for estimating the product temperature on the basis of the saturating vapor pressure of the ice. Oetjen proposed and patented a method, the barometric temperature measurement (BTM), to estimate the temperature of the sublimating interface using the value of the pressure at which the first derivative of the pressure rise curve has a maximum (39–41). Milton et al. (42), Liapis and Sadikoglu (43), Obert (44), Chouvenc et al. (45), and Velardi et al. (46) proposed to use mathematical models to describe the pressure rise in the drying chamber during the PRT, and an optimization algorithm to estimate the product temperature, the residual water content, K_v, and R_p by looking for the best fit between the pressure measurement and the values obtained by means of mathematical simulation. What differentiates one algorithm with respect to the others is the mathematical model used and the parameters estimated.

The manometric temperature measurement (MTM) algorithm, originally proposed by Milton et al. (42), is based on an equation that describes the pressure rise as a consequence of the sum of various mechanisms and provides estimations of the temperature at the interface of sublimation, of a parameter that is a function of R_p, and of K_v. A slightly different algorithm was proposed by Pikal et al. (47) and by Tang et al. (48–51). It estimates R_p directly, while K_v is calculated by equating the heat consumed by ice sublimation and the heat flux from the shelf to vials. A modification of the MTM algorithm was proposed by Obert (44). It takes into account the heat capacity of the whole glass wall of the vial and the contribution of the desorption of the bound water to the pressure rise.

The pressure rise analysis (PRA) proposed by Chouvenc et al. (45) is based on a simple heat balance for the frozen product, taking into account the desorption of bound water and the thermal capacity of the portion of the vial glass in contact with the frozen product. The product temperature at the interface of sublimation, R_p, and the desorption rate of bound water are estimated, even if the desorption rate is generally considered to be negligible.

The dynamic parameters estimation (DPE) algorithm proposed by Velardi and Barresi (46, 52) is based on an unsteady-state model for mass transfer in the drying chamber and on the energy balance for the frozen layer, thus taking into account the different dynamics of the temperature at the interface and the bottom of the vial.

It must be remarked that generally the estimated product temperature decreases nearby the end point of the primary drying, even if the shelf temperature remains constant. This drop may be an artifact because a fraction of vials, the edge-vials, has already finished sublimating while DPE, MTM, and PRA continue interpreting pressure rise curves assuming batch uniformity, that is, a constant number of sublimating vials. Thus, a lower pressure rise, corresponding to a lower sublimation rate, may be interpreted by these algorithms as a reduction of the front temperature. Recent investigations by Fissore et al. (53) evidenced that the decrease of the estimated temperature in the last part of the main drying can be due to the ill-conditioning of the model-based algorithms used to interpret the PRT, beside batch heterogeneity. Thus, they proposed an improved version of DPE algorithm, called DPE+, to cope with this problem. The salient feature of DPE+ algorithm is that problem dimensionality is reduced, as R_p is calculated directly from the slope of the curve of pressure rise at the beginning of the test, and only the interface temperature is estimated through the usual least-square minimization. Besides, Fissore et al. (53) investigated the problem of the optimal selection of the duration of the test, evidencing that the optimal duration is equal to the time constant of the process:

$$\tau = \frac{V_c M_w R_p}{N_v A_v R T_i} \tag{3}$$

If the duration of the test is lower than τ, then is not possible to "capture" the complete dynamics of the process. Moreover, it must be taken into account that very low values cannot be feasible from a practical point of view. On the contrary, if the duration of the test is too high, then bad estimations are obtained as a consequence of the ill-conditioning of the problem. To this purpose, the DPE+ algorithm, where the ill-conditioning problem is partly solved, appears to be more robust than other algorithms, as it is less sensitive to the duration of the test.

A comparison among the estimations obtained by means of MTM, PRA, DPE, and DPE+ algorithms in case of a vials freeze-drying process is shown in Figure 7. While the original DPE algorithm estimates a (wrong) decreasing value for T_i after eight hours since the beginning of the primary drying, similarly to the results obtained in this case by other PRT-based algorithms, the DPE+ algorithm allows obtaining correct estimations up to 13 hours since the beginning of the primary drying. The estimated T_i starts decreasing only when the ratio of the pressure signals given by a capacitive gauge (Baratron) and by a Pirani sensor, generally used to assess the end of the primary drying, also starts decreasing. With respect to the estimations of K_v and R_p, the values obtained by means of the various algorithms are very close, apart from MTM that estimates, in the case-study investigated, a higher product resistance and thus a lower sublimation flux and a lower coefficient K_v (which is calculated from the sublimation flux).

Information obtained from PRT-based algorithms, as well as from soft sensors, can be used to control the system, that is, to calculate in-line the best values of T_{shelf} and drying chamber pressure (P_c) according to the target required. A control system is required to manage the process: possible application to industrial processes have been discussed in chapter 19 of this book. Here, it is shown how the same system can be used to automatically find the optimal recipe for a given product. When using a soft sensor as measuring device only a single vial is monitored, while when one of the algorithms based on the PRT is used, a "mean" value of the status of the system is obtained. This issue has to be taken into account to guarantee product quality in the whole batch, and it will be extensively discussed in the following.

Fissore et al. (9) showed the results that can be obtained when using a simple feedback proportional controller that calculates the value of T_{shelf} as a

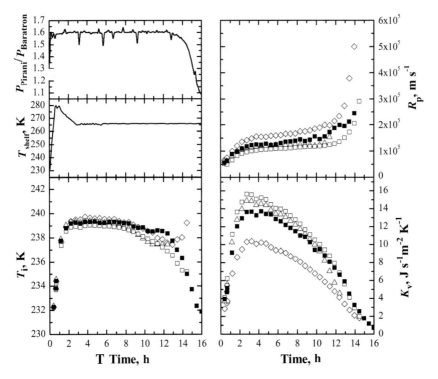

FIGURE 7 Comparison between the estimated values (◇: MTM, △: PRA, □: DPE, ■: DPE+) of the temperature at the interface of sublimation (*lower graph, left-hand side*), of the resistance of the dried product to vapor flow (*upper graph, right-hand side*), and of the heat transfer coefficient between the shelf and the bottom of the vial (*lower graph, right-hand side*) during the freeze-drying of a 10% by weight sucrose solution (P_c = 10 Pa, N_v = 175, d_v = 14.25 × 10^{-3} m, V_c = 0.2 m^3). Vials have been shielded to minimize radiation effects. The duration of each PRT is 30 seconds; the time interval between two PRT is 30 minutes. The ratio between the signals of the Pirani and of the Baratron sensors (*upper graph, left-hand side*) and the shelf temperature (*middle graph, left-hand side*) are also shown. Experiments carried out in LyoBeta™ 25 freeze-dryer by Telstar, with LyoDriver automatic control.

function of the difference between product temperature, estimated by using the smart vial, and the limit value. The gain of the controller is calculated looking for the value that minimizes this difference along the drying time. A mathematical model is thus required to calculate the evolution of product temperature as a function of the control actions. Various observers can be used to monitor the product in several vials placed in different positions in the drying chamber, and the highest product temperature can be used by the controller to calculate the control actions. In this case the control action would be much more conservative, but it would guarantee the fulfillment of the constraint on maximum product temperature in all the vials of the batch.

Tenedini and Bart (29) patented a method for controlling both the primary and the secondary drying by using the measurements provided by a windmill sensor and some "rules of thumb."

A simple control system that calculates T_{shelf} from the BTM output was first proposed by Neumann (35) and described by Oetjen et al. (54), even if in these earlier works the BTM was generally used for monitoring the batch and the shelf temperature was set manually. Neumann (55) later on proposed a simple control system where chamber pressure is manipulated by means of a throttling valve placed in the spool connecting the chamber to the condenser (or acting on condenser temperature) to maintain the product at the desired temperature. More recently Willemer (56) showed a control system where T_{shelf} and P_c were manipulated according to BTM estimations. The Thermodynamic Lyophilization Control (TLC) system was later on proposed by Oetjen (39) and by Oetjen and Haseley (41). It is based on a set of heuristics for the calculation of the control actions using the results of the BTM algorithm and was mainly used for recipe development.

Recently Tang et al. (48) and Pikal et al. (47) proposed and patented an expert system, named SMART^{TM} Freeze-Dryer, for manipulating T_{shelf} and P_c using the results obtained by means of the MTM algorithm and some empirical and good practice rules. Gieseler et al. (57) validated experimentally the SMART Freeze-Dryer with different excipients and formulations. Results confirm that the algorithm can be a useful tool for development of a lyophilization cycle during a single freeze-drying run. Nevertheless, it has no predictive capacity as it does not take into account the evolution of the product as a consequence of the actions taken and, thus, in our opinion a wide margin for optimization may exist.

Barresi et al. (7,11–13) proposed to use DPE outcomes in a control loop where T_{shelf} is the manipulated variable. The salient feature of this tool, named "LyoDriver," is the use of a mathematical model of the process to calculate the value of T_{shelf} that maintains the product temperature close to the maximum allowed value, thus minimizing the duration of the primary drying and avoiding any temperature overshoot. The use of LyoDriver does not require any previous knowledge about the product and the process, except obviously the value of the maximum allowed temperature of the product, and the truly optimal operating conditions are calculated in-line by the software. Moreover, such control system takes into account the dynamics of the heating and cooling systems of the equipment, as well as the temperature rise that occurs during a PRT. The key parameters of LyoDriver are the *prediction horizon*, which is the time throughout which the algorithm estimates the evolution of the product temperature and calculates a proper heating policy, and the *control interval*, which is the time between two successive PRTs. LyoDriver calculates a sequence of set points for the temperature of the fluid used to heat/cool the product ($T_{fluid,sp}$), one or more for each control interval through all the prediction horizon, in such a way that the product temperature is as close as possible to its target. Either a constant $T_{fluid,sp}$ can be considered for each control interval or a multiple sequence of set points in each control interval can be calculated. At the beginning of the main drying, when the temperature of the product is well below the upper limit, the heating fluid temperature is raised at the maximum rate that can be obtained in the equipment; a maximum temperature for the fluid can be set to avoid overheating, even at this early stage. Then, a PRT is performed and LyoDriver calculates the optimal heating policy according to the current system state. This is regularly repeated after each DPE run so that potential mismatches between the model used by LyoDriver to make the calculations and the actual process behavior can be corrected.

Two controllers, based on a feedback and on a model-based algorithm, have been designed and compared. The former is a simple proportional controller that calculates the optimal gain, and thus $T_{\text{fluid,sp}}$, minimizing the difference between the maximum product temperature, that is, the temperature of the product at the bottom of the vial (T_B), and the target value along the prediction horizon:

$$
\begin{aligned}
t_0 \leq t < t_1 &\rightarrow T_{\text{fluid,sp,1}} = T_{\text{fluid}}(t_0) + K_P \left[T_B(t_0) - T_{\text{target}} \right] \\
t_1 \leq t < t_2 &\rightarrow T_{\text{fluid,sp,2}} = T_{\text{fluid}}(t_1) + K_P \left[T_B(t_1) - T_{\text{target}} \right] \\
&\vdots \\
t_{N-1} \leq t < t_N &\rightarrow T_{\text{fluid,sp,N}} = T_{\text{fluid}}(t_{N-1}) + K_P \left[T_B(t_{N-1}) - T_{\text{target}} \right]
\end{aligned}
\tag{4}
$$

The value of the gain of the controller has to be calculated using some performance criteria, for example, the minimization of the integral square error (ISE) between the product temperature and the set-point value from the current time (t_0) up to the prediction horizon (t_N):

$$
\min_{K_P} (\text{ISE}) = \min_{K_P} \int_{t_0}^{t_N} \left(T_{B,\text{predicted}}(t) - T_{\text{target}} \right)^2 dt
\tag{5}
$$

The mathematical model used to estimate the time evolution of the product temperature required by equation (5) has to be a compromise between accuracy, that is, the capability of describing reliably product evolution during the process, and simplicity, that is, it has to be formulated on the basis of few parameters that can be either measured or estimated in-line with good accuracy. The simplified model of Velardi and Barresi (28) has been used for this purpose.

If a model-based approach is implemented, the optimal sequence of $T_{\text{fluid,sp}}$ throughout the prediction horizon is calculated as a piecewise-linear function in such a way that the bottom product temperature is equal to the target value. Again, the simplified mathematical model of Velardi and Barresi (28) has been used:

$$
\begin{aligned}
t_0 \leq t < t_1 &\rightarrow T_{\text{fluid,sp,1}} = T_{\text{target}} + \left[T_{\text{target}} - T_i(t_0) \right] \left[K_v \left(\frac{1}{K_v} + \frac{L_f(t_0)}{k_f} \right) - 1 \right]^{-1} \\
t_1 \leq t < t_2 &\rightarrow T_{\text{fluid,sp,2}} = T_{\text{target}} + \left[T_{\text{target}} - T_i(t_1) \right] \left[K_v \left(\frac{1}{K_v} + \frac{L_f(t_1)}{k_f} \right) - 1 \right]^{-1} \\
&\vdots \\
t_{N-1} \leq t < t_N &\rightarrow T_{\text{fluid,sp,N}} = T_{\text{target}} + \left[T_{\text{target}} - T_i(t_{N-1}) \right] \left[K_v \left(\frac{1}{K_v} + \frac{L_f(t_{N-1})}{k_f} \right) - 1 \right]^{-1}
\end{aligned}
\tag{6}
$$

When the model-based controller is used the drying time is slightly higher than that obtained through the feedback algorithm, but the simpler mathematical formulation, since no optimization is involved in the calculation, and the smaller computation time make the model-based approach more suitable for in-line control (58).

Figure 8 shows two examples of a recipe developed in-line using LyoDriver. The ice temperature at the bottom of the vial estimated by means of

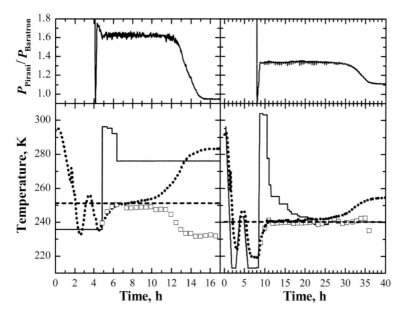

FIGURE 8 Examples of freeze-drying cycles obtained using LyoDriver (model-based algorithm) to set the fluid temperature for the primary drying stage. Left-hand side: 5% wt/wt aqueous solution containing mannitol (2.5%) and sucrose (2.5%). Operating conditions: $L = 6.6 \times 10^{-3}$ m, $d_v = 14.70 \times 10^{-3}$ m, $P_c = 10$ Pa, $T_{max} = -22°C$. Right-hand side: 5.2% wt/wt aqueous solution containing mannitol (4%), sucrose (1%), and other excipients. Operating conditions: $L = 5 \times 10^{-3}$ m, $d_v = 21.0 \times 10^{-3}$ m, $P_c = 6$ Pa, $T_{max} = -33°C$. Upper graphs: Ratio between the signals of the Pirani and of the Baratron gauges. Lower graphs: Set point of the fluid temperature (*solid line*), product temperature at bottom of the vial estimated by DPE (□) and measured through a thermocouple (*dotted line*), in a single vial (that shows faster drying than the average of the batch). The maximum allowed product temperature is shown (*dashed line*). Experiments have been carried out in Aérial (Prof. L. Rey's group) using a LyoBeta™ 25 freeze-dryer by Telstar.

DPE algorithm is shown, as well as the value of the product temperature measured by a thermocouple placed in close contact with the bottom of the vial. It can be observed that the estimated maximum product temperature never overcomes the limit value and, thus, the maximum allowable heating rate is obtained throughout primary drying, thus minimizing the duration of this step (59). As said before, it can be noted that when the primary drying is approaching the end, the product temperature estimated by means of DPE algorithm decreases. This behavior must be taken into account when setting the parameters of the control algorithm, and in the last part of the cycle it is advisable to use the sequence of control actions calculated in the previous step, without updating on the basis of the new (misleading) measurements. The same problem was evidenced also by Gieseler et al. (57) when using the SMART™ Freeze-Dryer. In the example of Figure 8 (left-hand graph), the heat transfer from the shelf controlled the sublimation rate, and the fluid temperature was usually maintained almost constant in the second part of the drying. The value of the shelf temperature assured the maximum sublimation rate since it was significantly greater than the product temperature. In some cases the vapor transport through the solid matrix controls the drying rate: transition to mass transfer control may occur as a consequence of the increase of dry cake thickness, of the formation of a crust, or of the increase of the cake resistance during the process due to a change in its structure. In this case there is a strong risk of failure because if the heat flux is not reduced, the product temperature increases. LyoDriver is still effective in this case. Figure 8 (right-hand graph) shows the results obtained during the freeze-drying of a complex formulation containing mannitol and sucrose, with other minor components (excipients and buffer). The freeze-drying of this solution is quite difficult because collapse starts at −45°C, but after an annealing cycle the collapse temperature rises to −26°C. The optimal cycle determined by LyoDriver, with T_{max} = −33°C, is shown (chamber pressure is constant and set at 6 Pa for primary drying). In this case the set point for the fluid temperature calculated by the controller approaches the product temperature, and the driving force for the heat transfer becomes very small because the constraint on the maximum temperature of the product is effective. Of course, an in-line change of the chamber pressure might be useful to reduce the drying time, restoring heat transfer control; the most effective ways to manage this type of situation would be using a MPC approach, as discussed in a next section.

It can be pointed out that in both cases shown in Figure 8 DPE gives good results up to the end of the primary drying and that the estimated product temperature agrees with thermocouple measurements, at least until the monitored vials are representative of the entire batch. Finally, the system can determine the end of primary drying, switching to process conditions selected for secondary drying.

Off-line Recipe Design

As it has been previously stated, a freeze-drying cycle for a given product is generally developed by identifying, through trial and error, a set of process conditions (shelf temperature, chamber pressure, and drying time) that give a product with acceptable quality. An alternative approach to the use of a control system to develop in-line the optimal recipe is presented here. It can be particularly interesting when not only the best cycle is of interest but also the whole design space, that is, the multidimensional space that encompasses combinations of product design and processing variables that guarantee product quality, is desired. A lot of time can be saved if the design space is built off-line, using a mathematical model of the process, and then validating it.

Various models are available in the literature and can be used for this purpose. They are based on mass and heat balance equations for the frozen and dried product in each vial (9,28,60–63). The use of these models requires to know the values of the physical properties of the product (e.g., density, specific heat, and thermal conductivity), of the heat transfer coefficient between the shelf and the product in the vial (K_v), and of the resistance of the dried product to vapor flow (R_p). In very few cases the values of K_v and of R_p can be found in the literature, and, thus, experimental investigation is required to estimate these parameters. Some of the techniques previously described to monitor the process can be used to this purpose, for example, the smart vial and the techniques based on the PRT.

When the model parameters are known, the following steps are required to build the design space:

1. Mathematical simulation of the evolution of product temperature and of sublimating interface position, given the values of shelf temperature and chamber pressure. Generally, the higher is the value of T_{shelf}, the higher is

the temperature of the product and the sublimation rate. The effect of chamber pressure is more complex as it affects both heat and mass transfer. Higher values of P_c decrease the driving force for water vapor transport from the sublimating interface to the chamber and increase the rate of heat transfer as the thermal conductance of the gas in the gap between the shelf and the bottom of the vial is increased. Figure 9 (graph A) shows an example of the results obtained using the previously quoted model (28).

2. From the results of the mathematical simulation of the process it is possible to know the time required to complete the main drying and to identify the ranges of T_{shelf} and P_c that allow maintaining product temperature below the limit value.

3. Constraints imposed by the equipment used to carry out the operation have to be taken into account. In particular, there is an upper limit to the vapor

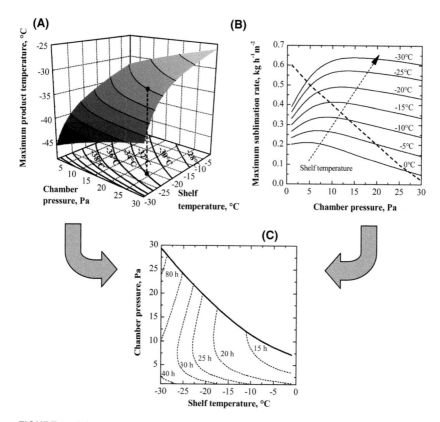

FIGURE 9 (**A**) Maximum product temperature versus shelf temperature and chamber pressure; various limit temperatures are indicated by solid lines. (**B**) Maximum sublimation rate versus shelf temperature for various values of chamber pressure; the operating limit imposed by the equipment characteristics is indicated by the dashed line. (**C**) Design space obtained assuming a limit temperature of −32°C. Dashed lines identify values of shelf temperature and chamber pressure resulting in the same drying time. Data refers to a 10% by weight aqueous sucrose solution in tubing vials ($d_v = 14.2 \cdot 10^{-3}$ m) filled with 1 mL of solution.

flow rate that can be condensed, and a further limitation is due to the fluid dynamics of the vapor in the duct connecting the chamber to the condenser, as the velocity of the vapor cannot become higher than the speed of sound in water vapor (Fig. 9B).

The result of these calculations is a diagram like graph C in Figure 9. The space of T_{shelf} and P_c is divided into two parts by a line that is a function of the limit temperature. The points below this line correspond to values of T_{shelf} and P_c that guarantee that product temperature is below the limit value and that sublimation rate is compatible with equipment characteristics. Besides, the drying time has to be equal or higher than the value calculated by means of mathematical simulation of the process.

What has been discussed above is just an example to demonstrate the concept, as it refers to a very simple case, where T_{shelf} and P_c are maintained constant throughout the drying, and a simplified model is adopted for simulation. It is possible, in fact, to realize a tool that takes into account variations of shelf temperature and chamber pressure with time.

Finally, the design space has to be validated. This requires to perform cycles under increasingly aggressive conditions until the product is unacceptable. The limiting operating conditions can be compared with that in the previously calculated values.

It is important to highlight that such design space is valid for a well-defined product and for a well-defined equipment. Moreover, results are affected by the values of the parameters used for the mathematical simulations of the process. These values are affected by uncertainty that has to be taken into account when building the design space. A possible approach consists of assuming that the values of the parameters are distributed around a mean value according to a certain distribution (e.g., a Gaussian function), and then calculating the probability of avoiding temperature overshoot as a function of this uncertainty (e.g., the variance of the distribution), as shown in Figure 10. By this way we move from a "deterministic" to a "probabilistic" design space. Given the desired probability of avoiding temperature overshoot (e.g., 99%), it is possible to calculate the values of T_{shelf} and P_c that allow to get this result.

A similar approach could be used to take into account that various vials in a batch do not have the same behavior because of nonuniform filling height, radiation from chamber walls, vapor fluid dynamics in the drying chamber, and uneven distribution of shelf temperature (14–16). Also in this case a distribution of the values of the various parameters among the vials of the batch can be assumed (or can be measured experimentally) and, thus, can be taken into account to calculate the probability of avoiding temperature overshoot versus the operating conditions.

Finally, it has to be remarked that if the operating conditions, for any reason, change during the process, product can be damaged or not, but, in any case, the duration of the main drying would be different. Thus, in case of failures or deviations that result in changes of T_{shelf} and/or P_c that maintain product temperature below the maximum allowed value, that is within the design space, the mathematical simulation of the effect of the disturbance is required to calculate the new duration of the process; as an alternative a monitoring system that automatically detects the end of primary drying (7,64) can be employed.

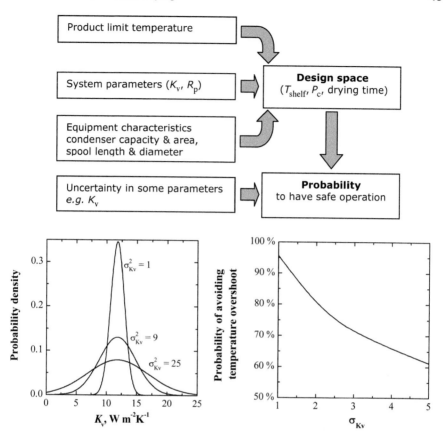

FIGURE 10 Sketch of the calculation procedure required to take into account parameters uncertainty when building the design space. The characteristics of the system are the same as used to get the results shown in Figure 9. The operating conditions used for the example are $P_c = 10$ Pa, $T_{shelf} = -8°C$, $T_{max} = -32°C$.

Of course, if the effect of the disturbance is the increase of product temperature above the limit value, the operation can be stopped.

ROBUST PROCEDURES FOR RECIPE SCALE-UP AND PROCESS TRANSFER

Previous sections have evidenced that the local values of the process variables, which significantly influence the drying of the single vials, not only depend on the selected set points of the operating variables, but are strongly influenced by apparatus geometry and design, equipment size and loading. This explains why even serious problems may be experienced with sensitive products when transferring from one equipment to another, or sometimes even if the loading in the same equipment is changed. Much more severe may be the problem in case of process scale-up from a laboratory or pilot equipment to the industrial one.

Modeling can be a very efficient tool also in this case in order to speed up the procedure strongly, reducing the development costs. In the section about equipment design we discussed of the use of a dual-scale model to investigate the effect of the characteristics of the equipment on product dynamics and to choose the best design. In that case a model for the vapor source (the ice sublimation) was directly implemented in the CFD code; as only relatively simple models can be implemented in a CFD code, thus resulting in less accurate predictions for the product evolution, this approach is preferable when a certain degree of uncertainty on the predictions of product evolution is acceptable, while the highest accuracy is desired for the chamber hydrodynamics. When the evolution of the product has to be estimated with the best possible accuracy, as is the case considered here, where for example the effect of a reduced loading on the shelves, or of a smaller clearance between the shelves to increase productivity, has to be investigated, it is better to adopt a detailed model to simulate the dynamics of the product in the vials. A simple and efficient iterative algorithm that couples the detailed simulation of the dynamics in each vial with the results obtained using CFD to model the fluid dynamics in the chamber can be used.

An alternative algorithm is based on the extraction from the results of the CFD simulations of correlations linking the pressure in several points of the shelves with the sublimation flow rate. Much faster calculations are obtained when the detailed model of the vial is coupled to these correlations, which can be provided as a sort of look-up table (65). This algorithm can be used to track the dynamics in single vials placed in a particular position (e.g., where the radiation effects are more important or where the pressure is higher), as well as that of the whole batch, thus calculating the mean value of the product temperature and of the residual water content, as well as the standard deviation around this mean value.

An example of the calculations that can be done is given in Figure 11, which shows the distribution of the drying time and of the maximum product temperature among the various vials of a batch. The effect of radiation from chamber walls, of nonuniform shelf temperature, and of pressure gradients has been considered when simulating the evolution of the product in the various vials of the batch. Values obtained in a laboratory-scale and in the industrial-scale equipment are shown, thus evidencing that in a small-scale equipment the batch has a higher degree of heterogeneity, that is, the variance of the maximum product temperature and of the drying time is higher than those obtained in an industrial-scale apparatus. In fact, in the laboratory-scale equipment pressure is very uniform but radiation effect is more relevant. It must be evidenced that multimodal distribution is mainly due to radiation effect (the temperature peaks correspond to the vials at the corners or at the edges of the shelves); the width of the main peak is mainly due to the range of variation of the shelf temperature, that, in this case, has been considered equal in the two pieces of equipment but can be related to equipment size and loading.

It is clear that the previous approach can be used only in case the characteristics of the equipment are completely known: a step of process identification on both pieces of equipment may be necessary, but once this has been done, a process transfer tool can be realized that allows a robust and reliable process transfer. Of course, when possible, it would be preferable to develop the recipe in the pilot freeze-dryer limiting the range of process variables to those that will be possible to use in the industrial one. Obviously, once the operating characteristics of the industrial equipment are known, using the same procedure described above, it will be possible to draw the design space for the industrial process.

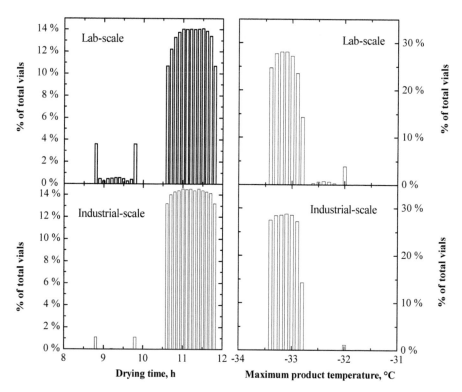

FIGURE 11 Distribution of the drying time (*left-hand side*) and of the maximum product temperature (*right-hand side*) among the various vials of a batch in a laboratory-scale (*upper graphs*) and in an industrial-scale (*lower graphs*) equipment. In both devices tubing vials with a diameter of 14.25 mm, arranged in an hexagonal array and filled with 1 mL of a 10% by weight sucrose solution, have been considered. In the industrial-scale equipment there are 16 shelves (with a size of 1500 mm × 1800 mm), with a clearance of 50.5 mm; in the laboratory-scale equipment there are 4 shelves (with a size of 450 mm × 455 mm), with a clearance of 40 mm. Shelf temperature is uniformly varying between −12°C and −15°C (operating pressure = 10 Pa).

It must be kept in mind that also the capacity of the duct, the characteristics of the valve and the condenser efficiency can affect the final result, strongly influencing the quality of the pressure control, the propagation of disturbances, and the atmosphere composition in the chamber. A dynamic model of the whole apparatus (where a lumped model of the various components of the equipment can be obtained through CFD or detailed simulation) may be extremely useful to this purpose.

Finally, it can be presented a completely new way to develop a cycle for the industrial apparatus. If it is equipped with one of the monitoring and control tools described in the previous section, it may be sufficient that a cycle is launched imposing the proper restrictions on product temperature. The optimal cycle will be automatically obtained, taking eventually into account the constraints that have been added. Previous cycles on a pilot scale may be advisable to verify the quality of the product using a reduced quantity of the product, but, after this experiment, the cycle that assures that the same operating conditions are really experienced by the product is easily obtained.

It must be remarked once again that when using the PRT as sensing device to obtain a full-state estimation of the system, that is, the temperature and position of the moving front and the mass and heat transfer coefficients, the average values for the batch are obtained, as the status of the product is supposed to be the same in all the vials. The existence of a distribution of the values of product temperature and of drying time around a mean value has to be taken into account when monitoring and controlling the process, both in the step of recipe development and then during normal operation (14,16,66).

FUTURE DEVELOPMENTS AND PERSPECTIVES

Previous sections of this chapter have shown how equipment design and process development can be improved by use of modeling, monitoring, and control. Some of the latest developments in this field either already commercially available or that will be available soon have been discussed; the focus was on the technology applicable in pilot-scale apparatus and in the R&D stage. In chapter 19 the possible use of most of these recent developments in industrial production has been discussed.

To conclude the analysis of the PAT in freeze-drying, the topics that require further investigation and new solutions and the latest developments currently subject of active research will be briefly discussed here.

One of the principal needs is the availability of product probes that are non-intrusive, reliable and possibly cheap, allow monitoring both the average and the variance of the product temperature, supplying the value of the process parameters of interest, are steam sterilizable, can be loaded and recovered through the automatic loading and unloading system, and can eventually be used inside a control loop. Unfortunately the devices currently available, and in particular the different types of wireless temperature sensors, some of which were already discussed, have only a few of the desired characteristics, even if they already represent a significant improvement and are very promising.

Recently, a new generation of wireless probes was developed ("passive transponders") that generate the energy required for transmission from an electromagnetic field. The Temperature Remote Interrogation System (TEMPRIS) sensors, originally developed for automotive applications, have been recently proposed by Schneid and Gieseler (67). This system allows for real-time temperature measurement at the bottom center of a vial (if placed correctly) and can be beneficial for scale-up as the same sensors can be used in laboratory and manufacturing scale. The main drawbacks are the dimension of the sensor that must be inserted in the product, and is thus very invasive, even if it has been proposed to set it in the upper empty part of the vial. In addition, the operating principle poses a limit to the number of devices that can be used in parallel, and only the temperature is measured. The smart vials described before are more sophisticated devices because they are also able to evaluate other process variables (interface position, sublimation flux, heat transfer coefficients, and product resistance) from the temperature measurements. They are compatible with different wireless technologies and can employ the low-frequency RF-ID to communicate the results from the passive measuring device to an external receiver or make use of a small battery that can be completely embedded into the vials, and it is able to transmit the measurements within a range of several meters by means of a 2.4-GHz radio. To increase life and performances of all these systems, the characteristics of the battery and the electronics must be improved to cope with the very low temperature of the process, and the data communication must be also improved to simplify the calibration procedure and make it independent on the geometric characteristics of the freeze-dryer.

The most important aspect is the possibility to obtain an evaluation of the distribution of product temperature in the batch. In fact, several vials of the batch can be easily monitored; Figure 6 showed the evolution of the interface temperature estimated in different vials, compared with the batch average front temperature determined through the DPE method previously described. The figure shows that the results of the two methods are consistent and highlights at the same time the difference in the type of information that can be obtained. The use of several observers can allow the estimation of the variance of the batch; in fact, ice nucleation is a stochastic process and thus different ice structures and different sublimation rates can result. In addition, the effect of temperature differences in the shelf and eventually a different contribution from wall radiation can result in different thermal histories for vials in different positions.

Concerning the pressure rise methods, in the previous part it has been shown that they can be successfully used to determine the average interface temperature of the product undergoing primary drying. It must be noted, however, that all the approaches based on MTM can only provide scattered data, since they do not involve a continuous measurement during the drying phase, and they should be used with care when significant heterogeneities in the batch are present or when radiating heating is important, especially with a small lot of vials (14,16). The limits of applicability of the PRT methods are currently under investigation; there are concerns about their use at very low temperature (typical cases are those of most vaccines productions) or in the second part of the primary drying step, but it has been shown that the reliability of the method can be largely improved if the algorithm is based on a proper model, the structure of the algorithm is made more robust and the conditions are optimized. The PRT approach is not invasive as it does not interfere with freezing and drying; nevertheless, it introduces a small perturbation because during the pressure rise step the temperature of the product slightly increases: it is therefore of the outmost importance to minimize the duration of the test to minimize this disturbance.

Notwithstanding their limits, the observers and the pressure rise method, and in particular the DPE, are very powerful; to get an even better device, which gets together the best of both systems, partly eliminating their weaknesses, can be obtained coupling the two devices. The value of the interface temperature, the heat transfer coefficient, and the mass transfer resistance calculated with the PRT can be used to initialize the observer equations after each pressure rise test, as already discussed. In this way the observer dynamics is corrected at any DPE run, strongly improving its convergence. This procedure can be repeated in the central part of the dying phase where the pressure rise method is known to give best product temperature predictions. Conversely, at the end of primary drying, temperature estimation can be only provided by the nonlinear observer.

It has been observed that the method of the PRT is effective and can also be carried out safely in large apparatus, and Chouvenc et al. (68) showed how it can be adapted also to slow closing valves; nevertheless, in industrial applications, people are usually not very comfortable in periodically closing and opening very large valves, and thus "valveless" methods could be very much appreciated. The TDLAS is one of them, but as discussed is still not very reliable and is very difficult to calibrate. It is not suitable for retrofitting and is quite expensive. Alternative solutions that can be easily implemented also in existing devices are currently under investigation.

The availability of reliable sensors allows realizing advanced systems for process control, as shown before, based on state estimates of the system and the use of a simple mathematical model of the drying process to calculate the optimal values of the shelf temperature that minimize the drying time and guarantee product quality; as they can take into account the actual heating/cooling capacity of the freeze-dryer, predicting potentially damaging temperature overshoots and anticipating accordingly the control actions, it is possible to avoid process failure, even when the process becomes mass transfer controlled or when the pressure in the chamber is varied. An example of a control system of this type, called LyoDriver, developed at Politecnico di Torino, has been presented.

Further improvements can be done in the monitoring and control algorithm. In fact control actions calculated by LyoDriver are a function of some parameters (e.g., the cake resistance and the overall heat transfer coefficient between the heating/cooling fluid and the product) that can be variables among the vials of the batch. If an aggressive control policy based on the estimation of the mean state of the batch is used, a fraction of the product that experiences higher temperatures can be damaged.

Even if not an easy task, it can be possible to improve the control system to manage heterogeneous batches; two strategies can be followed.

The first one consists in monitoring in-line not only the average state of the batch but also that of the edge vials characterized by a higher product temperature, choosing the best heating strategy according to their state. A hybrid system can be realized, taking advantage of the characteristics of the pressure rise method and of the observer, as already discussed. The DPE algorithm can be coupled with a system that estimates the variance of the batch using sensors that monitor the single vials, and the control system can calculate the control actions on the basis of the mean product temperature and the temperature of the radiated vials. Otherwise an advanced DPE algorithm would be requested to monitor radiated vials; in particular, the DPE algorithm should be modified in such a way that it is able to interpret the pressure rise curve simulating the batch not as a whole but as the sum of several vial types (both vials placed in the core and on the side of the batch); this is possible in principle, but requires to be tuned for each apparatus geometry and condition, and appears to be less robust than the previous solution.

A different strategy can employ a tool that couples the results of the fluid dynamics of the water vapor inside the drying chamber obtained through CFD simulations and a detailed one-dimensional model of the drying that takes also into account radiation from the chamber sidewall, similarly to the dual-scale models previously described (20,65,69). By this way, it is possible to track the dynamics of vials placed in different positions of the batch and, thus, to estimate the batch variance for certain operating conditions. Then, this information could be used to set the tuning parameters of the controller in such a way that it chooses the best control strategy to meet the product specifications of all the vials.

Another way to improve the control is to pass from single loop to multi-variable controller, as it would be very interesting to manipulate simultaneously both chamber pressure and shelf temperature. The chamber pressure is an important parameter. It influences the product temperature, and in case of mass transfer control conditions, a reduction of P_c during primary drying phase can speed up ice sublimation even if a low value of P_c might decrease too much the overall heat transfer coefficient and, thus, lower the heat transferred to the product. In addition, pressure manipulation is interesting because the response is very fast. Process control based on pressure manipulation has been proposed since the early 1960s, as mentioned above, and there is a large patent literature (see Ref. 7 for a review), but current practice is on the manipulation of shelf temperature because this approach is considered safer, even if the "optimal pressure" (generally kept constant during the process) can be calculated off-line (9,28,48).

A mathematical simulation of the consequences of P_c manipulation is required to determine the optimal operating conditions all along the process. Control schemes so far proposed for controlling the sublimation phase of a freeze-drying cycle (i.e., LyoDriver controller) cannot be applied to multiple-input multiple-output systems, even if it is possible to switch to a new pressure value when the system is moving to mass-transfer control (70). A more sophisticated control approach, named MPC, has been proposed. MPC is a descriptive name for a class of control algorithms that uses a process model to explicitly predict the future response of a plant or a piece of equipment. At each control interval the MPC algorithm attempts to drive the predicted output (i.e., product temperature and/or sublimation rate) as close as possible to a desired target value (i.e., maximum value allowable for the controlled variable) by computing a proper sequence of future manipulated variable adjustments (T_{shelf} and P_c). Unlike traditional control schemes, MPC can be easily used for multi-variable systems, handle process interactions with ease and take into account the uncertainties of model predictions. An application of MPC algorithm has been recently proposed by Daraoui et al. (71) with the goal of minimizing the drying time, even if in this case the only manipulated variable was the shelf temperature. A detailed mathematical model is used (72) in an algorithm that combines the internal model control (IMC) structure and the MPC framework. The purpose was to correct the modeling errors introduced in the model-based optimization, taking advantage of the fact that constraints on the manipulated and on the controlled variables can be easily taken into account in the optimization algorithm.

Preliminary results on MPC control systems manipulating simultaneously shelf temperature and chamber pressure have shown the potential advantages of the approach (73).

CONCLUSIONS

The use of a mathematical model of the drying process, and of some experimental measures, is at the basis of a number of tools that can be used successfully to improve freeze-dryer design and freeze-drying operations to save time when looking for the best recipe for a certain product and to preserve product quality.

It must be evidenced that the model best suited for each application must be chosen balancing the required accuracy, the computation cost and time, the complexity of the problem, and the robustness required, as convergence problems may limit the utilization of complex models, computation time and memory required may become limiting when very different scales are involved, and models very detailed from the point of view of the physical description of the phenomena involved may give practically no advantage if the many parameters added cannot be estimated a priori with the required accuracy. It must be remarked that even if CFD simulations are expensive in terms of computing time and the results are strictly valid only for the configuration considered, they can be very useful in cases where direct measurements are extremely difficult and/or very expensive due to difficulty or impossibility to measure the quantity of interest and due to constrain of the industrial environment. Freeze-drying in industrial apparatus is a very representative case. But using CFD it is also possible to obtain correlations that may be generalizable and easily coupled to the detailed model of the vials. The result is a tool that can supply the information required with very short computing time (minutes) on a personal computer, and can thus be used either for process or equipment design. Moreover, process modeling can be used also for troubleshooting and for estimating the consequences of process troubles (such as uncontrolled pressure or temperature variations in the drying chamber, reduction of condenser efficiency) on product quality. This can save a significant amount of time as it is possible to estimate if temperature overshoot occurred, and thus whether it is worthwhile to continue the operation (12).

Moreover, it must be highlighted that detailed modeling of the apparatus, resulting in an improved design of the freeze-dryer, can help to improve the reliability of the monitoring sensors (model-based monitoring), and thus of the control system, taking into account batch unevenness.

Finally, it has been shown how a PAT framework can be used to develop and implement effective and efficient innovative approaches in lyophilization development and manufacturing. In depth analysis, matching a mathematical model with some physical measurements helps to gain valuable process knowledge assessing important lyophilization parameters. Examples of monitoring tools based either on (continuous) temperature measurement or on (discontinuous) pressure measurement have been discussed, as well as their use in a closed loop control system.

A predictive control system provides a way of process optimization from the initial R&D stage. The R&D stage of recipe development can be tremendously simplified (time saving) as the optimum process can be obtained in one single run.

But tools based on PRT show high potential to be used to monitor industrial processes, where a collection of process parameters along the cycle can be obtained without the need of physically inserting any probe. In the same way, in-line closed loop control appears as a realistic possible process control approach.

ACKNOWLEDGMENTS

The authors would like to acknowledge their PhD students and postdoctoral fellows Salvatore Velardi, Roberto Pisano, Valeria Rasetto, and Miriam Petitti for their contribution to this work. Telstar-Industrial (Spain) is gratefully acknowledged for its permission to publish some confidential results, and for its financial support.

LIST OF SYMBOLS

A_v	cross surface of the vial, m^2
d_v	outer vial diameter, m
H	moving front position
ΔH_s	heat of sublimation, J/kg
j_w	water vapor flow, kg/s
K_p	gain of the proportional controller
K_v	heat transfer coefficient between the shelf and the vial, W/m^2K
k_f	thermal conductivity of the frozen product, W/m K
L	product thickness after freezing, m
L_f	frozen layer thickness, m
N_v	number of vials
P_c	chamber pressure, Pa
R_p	resistance of the dried product to vapor flow, m/s
T	temperature, °C
T_{fluid}	temperature of the fluid used to heat/cool the product, °C
$T_{fluid,sp}$	set point of the temperature of the fluid used to heat/cool the product, °C
T_i	temperature of the product at the interface of sublimation, °C
T_{max}	maximum temperature allowed by the product, °C
$T_{predicted}$	product temperature predicted by means of mathematical model, °C
T_{shelf}	shelf temperature, °C
T_{target}	target value for product temperature, °C
t	time, s
V_c	volume of the drying chamber, m^3

Greeks

σ	variance of a parameter
τ	time constant of the PRT, s

Subscripts and superscripts

0	value at $t = 0$, initial time for the interval considered
B	value at the bottom of the vial
N	value at the end of the horizon time

Abbreviations

API	Active pharmaceutical ingredient
BTM	Barometric temperature measurement
CFD	Computational fluid dynamics
DPE	Dynamic parameters estimation
ISE	Integral square error
IMC	Internal model control
MTM	Manometric temperature measurement
PAT	Process analytical technology
PRA	Pressure rise analysis
PRT	Pressure rise test
QbD	Quality by design
QbT	Quality by testing
TLC	Thermodynamic lyophilization control

REFERENCES

1. Franks F. Freeze-drying of Pharmaceuticals and Biopharmaceuticals. Cambridge: Royal Society of Chemistry, 2007.
2. Barresi AA, Ghio S, Fissore D, et al. Freeze-drying of pharmaceutical excipients close to collapse temperature: influence of the process conditions on process time and product quality. Drying Technol 2009; 27:805–816.
3. Wiggenhorn M, Winter G, Presser I. The current state of PAT in freeze-drying. Am Pharm Rev 2005; 8:38–44.
4. Wiggenhorn M, Winter G, Presser I. The current state of PAT in freeze-drying. Eur Pharm Rev 2005; 10:87–92.
5. Mayeresse Y, Veillon R, Sibille PH, et al. Freeze-drying process monitoring using a cold plasma ionization device. PDA J Pharm Sci Technol 2007; 61:160–174.
6. Wiggenhorn M, De Beer T, Geier F, et al. Evaluation of advanced approaches to monitor product temperature during lyophilization and the implementation of process analyzers (PAT). Proceedings of Freeze-drying of Pharmaceuticals and Biologicals Conference, Breckenridge, CO, 2008, August 7–9.
7. Barresi AA, Pisano R, Fissore D, et al. Monitoring of the primary drying of a lyophilization process in vials. Chem Eng Process 2009; 48:408–423.
8. Barresi AA, Fissore D. Product quality control in freeze drying of pharmaceuticals. In: Tsotsas E, Mujumdar AS, eds. Modern Drying Technology. Vol. 3. Chapter 4. Weinheim: Wiley-VCH (in press).
9. Fissore D, Velardi SA, Barresi AA. In-line control of a freeze-drying process in vials. Drying Technol 2008; 26:685–694.
10. Fissore D, Pisano R, Rasetto V, et al. Applying process analytical technology (PAT) to lyophilization processes. Chim Oggi/Chem Today 2009; 27(suppl 2):VII–XI.
11. Fissore D, Pisano R, Velardi SA, et al. PAT tools for the optimization of the freeze-drying process. Pharm Eng 2009; 29(5):58–70.
12. Barresi AA, Fissore D, Pisano R. Freeze-drying techniques. New ways to enhance process control and recipe development in pharmaceuticals freeze-drying. Pharm Manuf Packing Sourcer 2009; (May):36–42.
13. Barresi AA, Velardi SA, Pisano R, et al. In-line control of the lyophilization process. A gentle PAT approach using software sensors. Int J Refrig 2009; 32:1003–1014.
14. Barresi AA, Pisano R, Rasetto V, et al. Model-based monitoring and controlling of industrial freeze-drying processes. In: Thorat BN ed. Drying 2008—Proceedings of the 16th International Drying Symposium (IDS2008), Hyderabad, India, November 9–12, 2008; B:746–754.
15. Rasetto V, Pisano R, Barresi AA, et al. Modelling and experimental investigation of radiation effects in a freeze-drying process. In: Scura F, Liberti M, Barbieri G, et al. eds. Proceedings of the 5th Chemical Engineering Conference for Collaborative Research in Eastern Mediterranean Countries (EMCC5), Cetraro, Italy, May 24–29, 2008:394–397.
16. Barresi AA, Pisano R, Rasetto V, et al. Model-based monitoring and control of industrial freeze-drying processes: effect of batch non-uniformity. Drying Technol (in press).
17. Ferziger JH, Peric M. Computational Methods for Fluid Dynamics. Berlin: Springer, 2002.
18. Batchelor GK. An Introduction to Fluid Dynamics. Cambridge: Cambridge University Press, 1965.
19. Chapman S, Cowling TG. The Mathematical Theory of Non-uniform Gases. Cambridge: Cambridge University Press, 1939.
20. Rasetto V. Use of mathematical models in the freeze-drying field: process understanding and optimal equipment design (PhD dissertation). Italy: Politecnico di Torino, 2009.
21. Barresi AA, Baldi G, Parvis M, et al. Optimization and control of the freeze-drying process of pharmaceutical products. US patent application 20090276179 A1, 2009.
22. Barresi AA, Velardi S, Fissore D, et al. Monitoring and controlling processes with complex dynamics using soft sensors. In: Ferrarini L, Veber C, eds. Modeling,

Control, Simulation and Diagnosis of Complex Industrial and Energy Systems. Ottawa: ISA Series on Industrial Automation, 2009:139–162.

23. Velardi S. Frreze-drying: modelling, monitoring and control (PhD dissertation). Italy: Politecnico di Torino, 2004.

24. Corbellini S, Parvis M, Vallan A. A low-invasive system for local temperature mapping in large freeze dryers. Proceedings of International Instrumentation and Measurement Technology Conference—I2MTC, Singapore, May 5–7, 2009; 80–85.

25. Vallan A, Corbellini S, Parvis M. A Plug&Play architecture for low-power measurement systems. Proceedings of Instrumentation and Measurement Technology Conference—IMTC 2005, Ottawa, Canada, May 16–19, 2005; 1:565–569.

26. Velardi SA, Hammouri H, Barresi AA. In-line monitoring of the primary drying phase of the freeze-drying process in vial by means of a Kalman filter based observer. Chem Eng Res Des 2008; 87:1409–1419.

27. Velardi SA, Hammouri H, Barresi AA. Development of an High Gain observer for in-line monitoring of sublimation in vial freeze-drying. Drying Technol 2010, 28(2):256–268.

28. Velardi SA, Barresi AA. Development of simplified models for the freeze-drying process and investigation of the optimal operating conditions. Chem Eng Res Des 2008; 86:9–22.

29. Tenedini KJ, Bart SS Jr. Freeze drying methods employing vapor flow monitoring and/or vacuum pressure control. US patent 6226997 B1. 2001.

30. Kessler WJ, Davis SJ, Mulhall PA, et al. Lyophilizer monitoring using tunable laser absorption spectroscopy. Proceedings of 18th International Forum Process Analytical Chemistry, Arlington, VA, 2004, January 12–15.

31. Gieseler H, Kessler WJ, Finson M, et al. Evaluation of Tunable Diode Laser Absorption Spectroscopy for in-process water vapor mass flux measurement during freeze drying. J Pharm Sci 2007; 96:1776–1793.

32. Kessler WJ, Caledonia G, Finson M, et al. TDLAS-based mass flux measurements: critical analysis issues and product temperature measurements. Proceedings of Freeze-drying of Pharmaceuticals and Biologicals Conference, Breckenridge, CO, 2008, August 7–9.

33. Kuu WY, Nail SL, Sacha G. Rapid determination of vial heat transfer parameters using tunable diode laser absorption spectroscopy (TDLAS) in response to step-changes in pressure set-point during freeze-drying. J Pharm Sci 2009; 98:1136–1154.

34. Schneid S, Gieseler H, Kessler WJ, et al. Non-invasive product temperature determination during primary drying using tunable diode laser absorption spectroscopy. J Pharm Sci 2009; 98:3406–3418.

35. Neumann KH. Freeze-drying apparatus. US patent 2994132, 1961.

36. Neumann KH. Determining temperature of ice. In: Noyes R, ed. Freeze-Drying of Foods and Biologicals. Park Ridge: Noyes Development Corporation, 1968.

37. Nail SL, Johnson W. Methodology for in-process determination of residual water in freeze-dried products. Dev Biol Stand 1991; 74:137–151.

38. Willemer H. Measurement of temperature, ice evaporation rates and residual moisture contents in freeze-drying. Dev Biol Stand 1991; 74:123–136.

39. Oetjen GW. Freeze-Drying. Weinheim: Wiley-VCH, 1999.

40. Oetjen GW, Haseley P, Klutsch H, et al. Method for controlling a freeze-drying process. US patent 6163979. 2000.

41. Oetjen GW, Haseley P. Freeze-Drying. 2nd ed. Weinheim: Wiely-VHC, 2004.

42. Milton N, Pikal MJ, Roy ML, et al. Evaluation of manometric temperature measurement as a method of monitoring product temperature during lyophilization. PDA J Pharm Sci Technol 1997; 51:7–16.

43. Liapis AI, Sadikoglu H. Dynamic pressure rise in the drying chamber as a remote sensing method for monitoring the temperature of the product during the primary drying stage of freeze-drying. Drying Technol 1998; 16:1153–1171.

44. Obert JP. Modélisation, optimisation et suivi en ligne du procédé de lyophilisation. Application à l'amélioration de la productivité et de la qualité de bacteries lactiques lyophilisées (PhD dissertation). INRA Paris-Grignon, 2001.

45. Chouvenc P, Vessot S, Andrieu J, et al. Optimization of the freeze-drying cycle: a new model for pressure rise analysis. Drying Technol 2004; 22:1577–1601.

46. Velardi SA, Rasetto V, Barresi AA. Dynamic Parameters Estimation method: advanced manometric temperature measurement approach for freeze-drying monitoring of pharmaceutical. Ind Eng Chem Res 2008; 47:8445–8457.

47. Pikal MJ, Tang X, Nail SL. Automated process control using manometric temperature measurement. US patent 6971187 B1. 2005.

48. Tang XC, Nail SL, Pikal MJ. Freeze-drying process design by manometric temperature measurement: design of a smart freeze-dryer. Pharm Res 2005; 22:685–700.

49. Tang XC, Nail SL, Pikal MJ. Evaluation of Manometric Temperature Measurement, a Process Analytical Technology tool for freeze-drying: part I, product temperature measurement. AAPS PharmSciTech 2006; 7(1): Article 14.

50. Tang XC, Nail SL, Pikal MJ. Evaluation of Manometric Temperature Measurement, a Process Analytical Technology tool for freeze-drying: part II, measurement of dry layer resistance. AAPS PharmSciTech 2006; 7(4): Article 93.

51. Tang XC, Nail SL, Pikal MJ. Evaluation of Manometric Temperature Measurement (MTM), a Process Analytical Technology tool in freeze drying, part III: heat and mass transfer measurement. AAPS PharmSciTech 2006; 7(4): Article 97.

52. Velardi SA, Barresi AA. Method and system for controlling a freeze drying process. US patent application 12/441752.

53. Fissore D, Pisano R, Barresi AA. On the methods based on the Pressure Rise Test for monitoring a freeze-drying process. Drying Technol (in press).

54. Oetjen GW, Ehlers H, Hackenberg U, et al. Temperature-measurement and control of freeze-drying processes. In: Fisher FR, ed. Freeze-drying of foods. Washington, D.C.: National Academy of Sciences—National Research Council, 1962:178–190.

55. Neumann KH. Temperature responsive freeze drying method and apparatus. US patent 3077036. 1963.

56. Willemer H. Additional independent process control by process sampling for sensitive biomedical products. Proceedings of 17ème Congrès International du Froid. Wien, Austria, 1987, August 24–29; C:146–152.

57. Gieseler H, Kramer T, Pikal MJ. Use of Manometric Temperature Measurement (MTM) and SMART^{TM} Freeze Dryer technology for development of an optimized freeze-drying cycle. J Pharm Sci 2007; 96:3402–3418.

58. Pisano R, Fissore D, Velardi SA, et al. In-line optimization and control of an industrial freeze-drying process for pharmaceuticals. J Pharm Sci (in press).

59. Rasetto V, Pisano R, Barresi AA, et al. Fast development of freeze-drying cycles for temperature and moisture sensitive products. Proceedings of the European Drying Conference AFSIA 2009, Lyon, France, May 14–15, 2009. Cahier de l'AFSIA Nr. 23. AFSIA-ESCPE, Villeurbanne, France, 2009:112–113.

60. Liapis AI, Litchfield RJ. Optimal control of a freeze dryer—I. Theoretical development and quasi steady-state analysis. Chem Eng Sci 1979:975–981.

61. Lombraña JI, Diaz JM. Heat programming to improve efficiency in a batch freeze-dryer. Chem Eng J 1987; 35:B23–B30.

62. Lombraña JI, Diaz JM. Coupled vacuum and heating power control for freeze-drying time reduction of solutions in phials. Vacuum 1987; 37:473–476.

63. Sadikoglu H, Ozdemir M, Seker M. Optimal control of the primary drying stage of freeze drying of solutions in vials using variational calculus. Drying Technol 2003; 21:1307–1331.

64. Pisano R, Guler SB, Barresi AA. In-line detection of Endpoint of sublimation in a freeze-drying process. Proceedings of the European Drying Conference AFSIA 2009, Lyon, France, May 14–15, 2009. Cahier de l'AFSIA Nr. 23. AFSIA-ESCPE, Villeurbanne, France, 2009:110–111.

65. Rasetto V, Marchisio DL, Fissore D, et al. On the use of a dual-scale model to improve understanding of a pharmaceutical freeze-drying process. J Pharm Sci (in press). DOI: 10.1002/jps.22127.

66. Rasetto V, Marchisio DL, Fissore D, et al. Model-based monitoring of a non-uniform batch in a freeze-drying process. In: Braunschweig B, Joulia X, eds. Computer-Aided Chemical Engineering—Proceedings of 18th European Symposium on Computer

Aided Process Engineering (ESCAPE18), Lyon, France, June 1–4 2008. Vol. 25, CD Edition. Paper FP_00210. Amsterdam (The Netherlands): Elsevier Science.

67. Schneid S, Gieseler H. Evaluation of a new wireless temperature remote interrogation system (TEMPRIS) to measure product temperature during freeze drying. AAPS PharmSciTech 2008; 9:729–739.

68. Chouvenc P, Vessot S, Andrieu J, et al. Optimization of the freeze-drying cycle: adaptation of the Pressure Rise Analysis to non-instantaneous isolation valves. PDA J Pharm Sci Technol 2005; 5:298–309.

69. Barresi AA, Rasetto V, Pisano R, et al. Multiscale modelling of freeze-drying for optimisation and quality control of pharmaceutical products. In: Scura F, Liberti M, Barbieri G, Drioli E, eds. Proceedings of the 5th Chemical Engineering Conference for Collaborative Research in Eastern Mediterranean Countries (EMCC5), Cetraro, Italy, May 24–29, 2008:390–393.

70. Fissore D, Pisano R, Barresi AA. On the design of an in-line control system for a vial freeze-drying process: the role of chamber pressure. Chemical Product and Process Modeling 2009; 4: Article 9.

71. Daraoui N, Dufour P, Hammouri H, et al. Optimal operation of sublimation time of the freeze drying process by predictive control: application of the MPC@CB software. In: Braunschweig B, Joulia X, eds. Computer-Aided Chemical Engineering—Proceedings of 18th European Symposium on Computer Aided Process Engineering (ESCAPE18)), Lyon, France, June 1–4 2008; 25:453–458.

72. Sadikoglu H, Liapis, AI. Mathematical modelling of the primary and secondary stages of bulk solution freeze-drying in trays: parameter estimation and model discrimination by comparison of theoretical results with experimental data. Drying Technol 1997; 13:43–72.

73. Pisano R. Monitoring and control of a freeze-drying process of pharmaceutical products in vials (PhD dissertation). Italy: Politecnico di Torino, 2009.

21

Practical Considerations for Freeze-Drying in Dual Chamber Package Systems

Dirk L. Teagarden, Stanley M. Speaker, and Susan W. H. Martin
Pfizer Biotherapeutics Pharmaceutical Sciences, Pfizer, Inc., Chesterfield, Missouri, U.S.A.

Thomas Österberg
Pfizer Product and Process Development, Global Manufacturing, Pfizer Health AB, Strängnäs, Sweden

INTRODUCTION

The market for injectable drugs is increasing due in large part to the development of biotherapeutics for previously untreatable ailments, such as rheumatoid arthritis and multiple sclerosis. Many biotherapeutics are administered parenterally and are lyophilized, due to low-intestinal absorption and poor physicochemical stability, respectively. An increased use of lyophilized dosage forms and the demand for more doses per unit have created a requirement for new package forms (1). Additionally, the growth of the home healthcare market has created an increased need for self-injection of subcutaneously delivered parenterals (2). Thus, patient convenience in drug delivery is important when developing self-administered parenteral drug products. A common mode to ensure ease-of-use by the patient is delivery of subcutaneous drugs in prefilled syringes/cartridges (3). These syringe/cartridge systems are available either "as is" or with a delivery device such as an autoinjector or pen injector (4). Recent evaluations of the parenteral drug market expect an 11% to 20% growth for prefilled syringes/cartridges (2,3,5,6). Prefilled syringes/cartridges enhance patient convenience by simplifying dosage preparation and administration. Other advantages of prefilled syringe packaging include increased patient safety due to decreased number of dose preparation steps, decreased administration errors (including inadvertent needle sticks), decreased costs due to smaller overfill, lack of need for a diluent vial and transfer syringe, less packaging, the extension of product exclusivity (i.e., life cycle management), and a unique branding opportunity for a product. These packaging systems also enable easier recruitment of patients for clinical trials in a crowded therapeutic area where the standard-of-care includes prefilled syringes and/or devices (e.g., diabetes and growth hormone–deficiency markets).

Many parenterals, especially biotherapeutics, are unstable in aqueous-based formulations for long periods of time and are, therefore, lyophilized. Lyophilized drug products must be reconstituted with sterile diluent prior to use. The reconstitution operation is an additional step the patient or caregiver must perform correctly to ensure safety and appropriate dosing to the patient. Lyophilizing in dual chamber packages [e.g., dual chamber vial (DCV), dual chamber syringe (DCS) syringe, or dual chamber cartridge (DCC)] eliminates

(A)

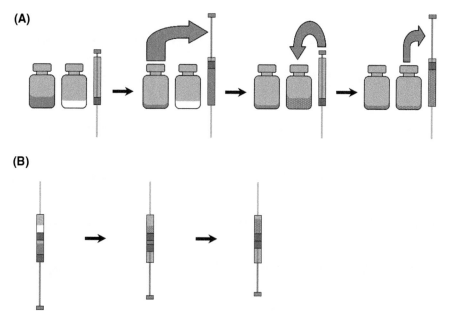

(B)

FIGURE 1 Simplified comparison of reconstitution of a lyophile in vial (**A**) versus dual chamber syringe (**B**).

the manual reconstitution step and greatly improves patient convenience and safety. A comparison of the reconstitution steps between traditional vials and a DCS is illustrated in Figure 1. Genotropin® [recombinant human growth hormone (rhGH)] was the first lyophilized protein product that transitioned in 1987 from a standard vial to DCC packaging. Presently, several rhGH products are marketed in lyophilized dual chamber package systems [e.g., three pharmaceutical companies market lyophilized rhGH products in DCC with multidose pen devices (Eli Lilly, Pfizer, and Merck Serono) and one markets it as a single use DCS product (Pfizer)].

However, lyophilizing a drug in a dual chamber system can be challenging compared to lyophilization in a standard vial configuration because of differences in heat transfer during lyophilization and the presence of siliconized glass surfaces. The purpose of this chapter is to review practical considerations for lyophilizing drugs in dual chamber systems and outline strategies to address these challenges.

DUAL CHAMBERED PRIMARY PACKAGING SYSTEMS

There are several dual chamber package systems that contain lyophilized products. These systems include dual chamber vials, dual chamber cartridges, and dual chamber syringes. One of the main benefits of dual chamber packaging is that the reconstitution step is built into the package and thereby improves patient convenience and safety. Each of these dual chamber package systems is briefly discussed below. Examples of marketed lyophilized drug products in dual chamber packages are listed in Table 1.

TABLE 1 Examples of Marketed Products in Dual Chamber Packages

Drug product	Indication	Company	Primary package	Device information and/or notes
Cardizem Lyo-Ject (diltiazem hydrochloride)	Atrial fibrillation	Biovail Pharmaceuticals	DCS, 1 mL	No device; Lyo-Ject DCS from Vetter; removed from market in 2007
Caverject (alprostadil)	Erectile dysfunction	Pfizer	DCC, 0.6 mL	CAVERJECT IMPULSE; disposable, variable dose pen injector
Edex (alprostadil)	Erectile dysfunction	Schwarz Pharma	DCC, 0.6 mL	Edex Injection Device; reusable, variable dose pen injector
Genotropin (recombinant human growth hormone)	Growth hormone deficiency, Prader–Willi syndrome, Turner syndrome, short for gestational age	Pfizer	DCC, 1 mL / DCS, 0.25 mL	Genotropin Pen; reusable, variable dose pen injector; Ypsomed / No device; Miniquick DCS from Pfizer
Humatrope (recombinant human growth hormone)	Growth hormone deficiency, Turner syndrome, idiopathic short stature, short stature homeobox gene	Eli Lilly	DCC, 3 mL	Hematro Pen or Hematro Pen 3; reusable, variable dose pen injectors; Ypsomed
Lupron depot (leuprolide acetate)	Prostate cancer	TAP Pharmaceuticals, Inc	DCC	No device; Atrix technology for controlled release
NeoRecormon (Epoetin Beta)	Anemia	Roche	DCC	Reco-Pen; reusable, variable dose pen injector; Ypsomed
Peg-Intron (pegylated interferon α2)	Hepatitis C	Schering	DCC	RediPen (based on BD Liquid Dry Injector Pen); disposable, single use pen injector
Preotact (parathyroid hormone)	Osteoporosis	Nycomed, NPS Pharmaceutical	DCC, 3 mL	Preotact Pen; reusable, fixed dose pen injector; Ypsomed
Saizen (recombinant human growth hormone)	Growth hormone deficiency	Merck Serono	DCC	One click Pen; reusable, variable dose pen; Haselmeier
Solu-Cortef (hydrocortisone sodium succinate)	Acute corticosteroid therapy	Pfizer	DCV (Act-O-Vial)	No device
Solu-Medrol (methylprednisolone sodium succinate)	Acute corticosteroid therapy	Pfizer	DCV (Act-O-Vial)	No device

Abbreviations: DCC, dual chamber cartridge; DCS, dual chamber syringe; DCV, dual chamber vial.

Traditional Vials

Traditional vials for lyophilization range in size from 2 to 100 mL and may be made from tubular or molded glass. However, tubular glass is usually preferred due to better contact with the lyophilizer shelf and, hence, better heat transfer. The elastomeric closures used for a standard freeze-dried vial have openings to allow the passage of water vapor during lyophilization. The elastomeric closures used for lyophilized drug products are generally 13 or 20 mm in diameter. The closure formulation should preferably have a low propensity to absorb moisture during sterilization/storage and exhibit a low-extractability/leachability profile. Additionally, the elastomers may be coated or laminated with a fluorinated polymer to minimize leachables. At the conclusion of the lyophilization cycle, the elastomeric closures are fully seated and a crimp cap is placed over the closure.

Dosing a lyophilized drug product in a vial to a patient requires many steps prior to dosing the patient. These steps include determination of the diluent amount, disinfection of the diluent stopper, loading the diluent into a syringe, disinfection of the lyophilization stopper, adding the diluent into the lyophilization vial, reconstitution of the lyophilized cake, determination of the amount of reconstituted drug product to withdraw, disinfection of the lyophilization stopper, and loading of the dosing syringe. Additional steps may be required if the drug is to be administered via intravenous injection. These multiple steps are not only cumbersome for the health professional or patient but also increase the opportunity for error in dosing or aseptic technique.

Dual Chamber Vials

A DCV is a two-compartment vial where the lyophilized drug product is in the lower compartment and diluent is in the upper compartment. The two compartments are separated by a middle elastomeric closure. An illustration of this package system is shown in Figure 2. The user simply presses a plastic activator to force the middle stopper and diluent to drop into the lower compartment to reconstitute the lyophilized powder. The user is now free to aseptically withdraw the appropriate dose from the vial with a separate syringe. Thus, at time of use, there are six steps prior to dosing the patient. These include activation of the DCV, reconstitution of the lyophilized cake, removal of the cap on the upper chamber, disinfection of the lyophilization stopper, determination of the amount of drug to withdraw, and loading the dosing syringe. Additional steps may be

FIGURE 2   Lyophilized cake in dual chamber vial (Act-O-Vial).

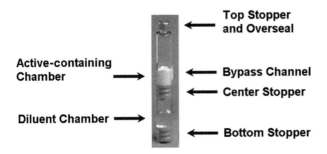

FIGURE 3 Dual chamber cartridge with external bypass (Genotropin).

required if the drug is to be administered via intravenous injection. Two corti-costeroids are marketed in the Act-O-Vial (Table 1), which is a DCV available in the following vial sizes: 1, 2, 4, 6, and 10 mL. A recent publication reviews the rational selection of glass and elastomeric components for a DCV package (7).

Dual Chamber Cartridge

A DCC consists of a syringe barrel with two chambers that are separated by a middle plunger. The front chamber contains a lyophilized drug and is capped by an elastomeric closure part that is thin enough to be punctured by a double-sided needle (i.e., a pen needle). The rear chamber contains the diluent and is closed at the bottom with an elastomeric end plunger. A bypass exists between the two chambers to allow for liquid to flow from the rear chamber to the front chamber. The bypass may consist of either an external channel or internal grooves, where the grooves are cut into the inside of the glass. Figure 3 shows an example of DCC with an external bypass feature. DCCs on the market are available in 1-mL and 3-mL standard formats.

A DCC is typically used in conjunction with a pen injector for recon-stitution and delivery to the patient. Pen injectors are available in disposable and reusable formats. The DCC is preloaded in the pen injector for the disposable ready-to-use format. The user/patient must load the DCC into the pen injector for the reusable format. The first time the pen is utilized the user has to disinfect the rubber septum, attach a pen needle, reconstitute the cartridge, prime the pen to remove air, and set the dose. Additional uses for this system only require the user to disinfect the rubber septum, attach a pen needle, set the dose, and inject. The needle is removed after each injection. Thus, at time of use there are five steps prior to dosing. Several examples of DCCs are illustrated in Table 1.

Dual Chamber Syringe

A DCS is similar to a DCC except that the DCS is designed to be used without a device at time of use. Therefore, a DCS becomes equivalent to a prefilled syringe after reconstitution. A typical DCS, which is illustrated in Figure 4, includes a plunger rod and flange. Prior to reconstitution, the user affixes a needle to the DCS. This may be a pen needle or luer needle depending on the type of front closure. The plunger rod pushes liquid from the rear chamber through the bypass into the front chamber in a controlled fashion. This is typically

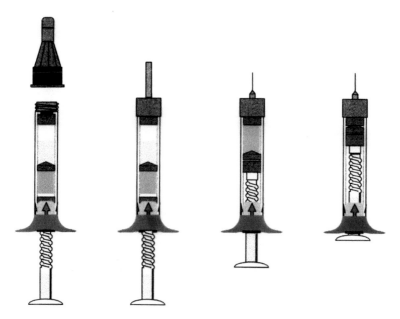

FIGURE 4 Use of dual chamber syringe (Genotropin MiniQuick).

performed by turning the plunger rod through a thread mechanism in the flange. There are approximately four steps prior to dosing the patient for the DCS (Fig. 4). These steps include removing the tip cap or disinfection of the rubber disk, attaching the luer or pen needle, attaching the plunger rod, and pushing/turning the plunger rod to move diluent from the rear to the front chamber to reconstitute the lyophilized cake. There are two DCS products that have been marketed: Cardizem, which is no longer on the market, and Genotropin Miniquick (Table 1).

FORMULATION DEVELOPMENT FOR THE ACTIVE-CONTAINING CHAMBER
It is critical to evaluate and gain an understanding of the fundamental relationships between the formulation, the process, and the package since all must work in unison for a successful product to be developed. Although formulation development activities for the lyophilized powder in the dual chamber package are similar to those for standard vial systems, there are some activities that are specific to package and process compatibility. Foremost, the formulator must consider the impact of moisture on product quality and ensure the cake mass is sufficient to withstand moisture ingress from the elastomer and diluent chamber. The formulator must also ensure the formulation is compatible with silicone and the selected elastomer. If the formulation is not compatible with silicone, either the formulation or the siliconization process needs to be optimized. Novel primary packaging, for example, silicone-free plastic, should be investigated in

cases of extreme silicone incompatibility. The main concern for compatibility of the formulation with the elastomer is leachable release from the elastomer. This can be controlled with fluoropolymer coating of the elastomer or optimization of the formulation. Finally, the formulator should attempt to maximize product concentration in the formulation to minimize cake height since this significantly impacts overall lyophilization time.

Product Characterization

It is crucial to understand the physicochemical characteristics of the formulation bulk solution that are critical in determining the optimal conditions for primary drying. Primary drying (ice sublimation) must be carried out by maintaining the product temperature below the glass transition temperature (T_g') or collapse temperature (T_c) for amorphous formulations or below the eutectic temperature for totally crystalline formulations. The glass transition temperature is a valuable guide, indicating the temperature at which the product viscosity reaches a point at which the solid matrix can begin to "flow." Differential scanning calorimetry (DSC) is typically used to measure the T_g' of the frozen drug product bulk solution. While it is a "critical" temperature for cycle design, most amorphous products will produce an acceptable cake structure at product temperatures slightly above the T_g'. It is only at the collapse temperature where a product will exhibit collapsed cakes. A more valuable tool in determining the optimum primary drying temperature is the T_c of the product. The T_c, as determined by one of several methods such as freeze-dry microscopy, is the temperature at which the frozen matrix physically collapses. Depending on the product, the T_c can be considerably higher than the T_g' (8). Collapse is a phenomenon common to frozen amorphous matrices. Collapse can occur when the product temperature during primary drying exceeds the T_g' to the extent that the frozen concentrate no longer possesses sufficient structural strength to support its own mass once the pure ice matrix is removed by sublimation. Eutectic melting, though a distinctly different physical phenomenon than collapse, is nevertheless dependent on the same process variables (shelf temperature, chamber pressure, and resultant product temperature) and results in similar catastrophic failure to the product. Although collapse and melting describe different phenomena, they are from a practical perspective similar and the corrective measures taken from a lyophilization cycle design standpoint are identical. The terms "collapse or melt-back" are regularly used to describe such a failure during primary drying.

FORMULATION DEVELOPMENT FOR THE DILUENT-CONTAINING CHAMBER

A key component of the dual chamber freeze-dried product is the selection of the appropriate diluent. The diluent chosen should enable rapid dissolution of the freeze-dried cake on transfer to the dry powder chamber, be compatible with the freeze-dried powder, and not negatively impact the chemical or physical stability of the reconstituted product. Typical diluents range from simple systems, such as water for injection, to tonicity modifying solutions, to water containing a preservative. Several of the dual chamber pen products have a preservative added to enable the reconstituted solution to be used in a multidose pen injector. Example preservatives used in several marketed dual chamber package systems include benzyl alcohol and *m*-cresol. If a multidose product is desired then appropriate compendial antimicrobial effectiveness testing (AET) must be preformed to assure that the correct preservative and concentration are chosen for the product. It should be noted that, depending on where the product is to be marketed, there may be different AET requirements. It is also important that the stability of the preservative is monitored over the product shelf life to assure that sufficient preservative is present to meet compendial AET. Losses of the preservative due to degradation or sorption to the package are two potential routes that should be evaluated during development studies and formal International Conference on Harmonization (ICH) stability studies to support product registration.

FILL VOLUME DETERMINATION FOR EACH CHAMBER

The goal for any freeze-dried product is that it can be reconstituted to produce a specific drug concentration on addition of the reconstitution vehicle and that the desired dose can be properly delivered to the patient. Therefore, it is important for the formulator to understand all of the parameters that must be determined to enable the label dose in these dual chamber packages to be properly administered. The equations below summarize the relationships for those parameters that need to be experimentally measured to determine the appropriate fill for each chamber to manufacture a product that can administer the labeled dose.

Equation 1: Label dose = (deliverable volume) × (reconstituted drug concentration)

Equation 2: Deliverable volume = (total reconstituted volume) − (hold-up volume for reconstituted package)

Equation 3: Total reconstituted volume = (volume transferred from diluent chamber) + (volume displacement factor created by the reconstituted freeze-dried cake)

Equation 4: Volume transferred from diluent chamber = (volume in diluent chamber) − (hold-up volume from the diluent chamber)

Equation 5: Reconstituted drug concentration = (amount of active ingredient)/(total reconstituted volume)

Therefore, the key parameters that need to be experimentally determined include the following:

a. The volume displacement factor created by the reconstituted freeze-dried cake
b. The hold-up volume of the diluent chamber
c. The hold-up volume for the reconstituted package
d. The deliverable volume

Once these parameters have been determined it is simple to calculate the appropriate fill volume for each chamber using the mathematical equations listed above. Note that other corrections to the fill would be required if there are any significant adsorption losses of the active ingredient to the package surfaces (i.e., to either the glass or the rubber components). It is also important to realize that many of the dual chamber freeze-dried products (especially syringes) utilize low-fill volumes. These low-fill volumes can be difficult to accurately control. Therefore, it is paramount that careful control of the fill volume for both the active containing and diluents chambers be maintained to satisfy product release specifications.

PACKAGING SELECTION PROCESS

A key component to development of a sterile freeze-dried product is the understanding that the product is a sum of the formulation, process, and package. The interrelationships of all three must be understood to develop a product of acceptable quality that will meet its specifications over its intended shelf life. It should be noted that the container closure system design can impact the lyophilization process with respect to resistance to heat transfer and processing throughput. Additionally, the container closure system can impact the product quality in the following ways: chemical compatibility, extractables/leachables, moisture ingress, sorption, particulate matter (e.g., via silicone oil, etc.), seal integrity, stability, and functionality. When working with new drug candidates it is not unusual for formulators to be presented simultaneously with problems of drug substance impurities, chemical instability, and packaging-derived instability. Carefully designed experiments should be integrated into the initial formulation design exercise to ensure that the impact of packaging components be identified. Therefore, one or more of the product components (i.e., the formulation, the package, or the process) may need to be optimized to meet the desired characteristics for a specific product within available manufacturing capability and at an acceptable cost of goods.

Elastomer Selection

An example of a compatibility issue that can be a concern includes those that impact the physical/chemical stability of the active ingredient or cause sorptive loss of a key excipient. Several of the marketed DCS/DCC/DCV lyophilized products contain a preservative in the diluent chamber. It is well-known that many of the preservatives utilized (e.g., benzyl alcohol or parabens) can be lost from solution due to sorption to rubber closures (9,10). Typically, a butyl rubber is selected as the closure material for most dual chamber packages because this rubber is generally inert, has low moisture/gas permeability, and has a low sorption potential. Despite this low sorption potential, it must be considered that the diluent-containing chamber exposes a relatively high surface area of rubber to the diluent due to the exposure of liquid to two rubber surfaces for most DCS package systems. The high surface area of exposed rubber compared to the low diluent volume present can contribute to a significant preservative loss to the butyl rubber.

Rowles et al. (9) reported on studies evaluating the sorptive loss of benzyl alcohol to rubber closures. The sorptive loss data fit a passive diffusion model whereby the amount of benzyl alcohol lost was proportional to the square root of time. According to this model, the benzyl alcohol loss to the package was relatively rapid but slows with time. Data for the sorptive loss of the preservative, benzyl alcohol, for the DCS product (CAVERJECT IMPULSE®) is summarized in Figure 5. Fitting this data to the passive diffusion model whereby the amount of benzyl alcohol lost is plotted versus the square root of time is illustrated in Figure 6. The temperature dependence of this sorptive process is

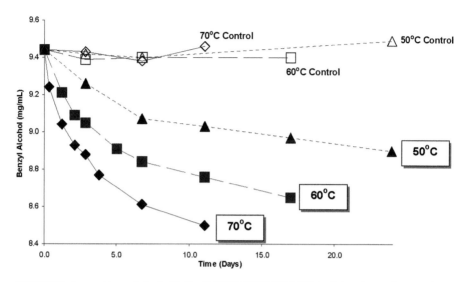

FIGURE 5 Loss of benzyl alcohol to the CAVERJECT IMPULSE dual chamber syringe package at accelerated temperatures.

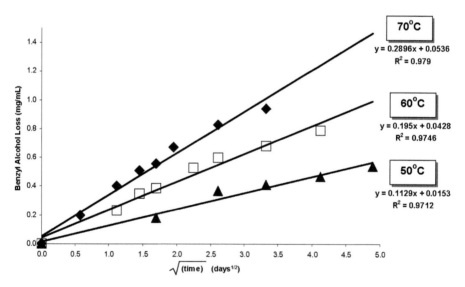

FIGURE 6 Sorption profile for benzyl alcohol to the CAVERJECT IMPULSE dual chamber syringe package (data fitted to passive diffusion model).

described by an Arrhenius treatment whereby the logarithm of the rate constant (obtained from slope of the benzyl alcohol loss vs. the square root of time) is proportional to 1/Temperature (K). The fit of the data to the Arrhenius equation is illustrated in Figure 7. An excellent fit to the model is demonstrated and enables one to accurately predict the sorptive loss of benzyl alcohol to this package system as a function of time at a given storage temperature.

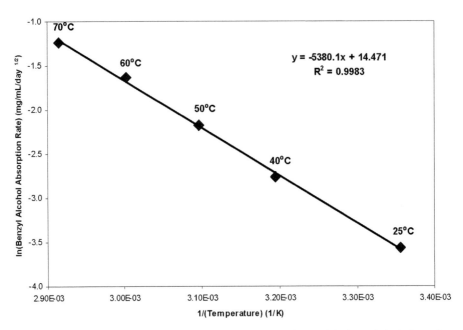

FIGURE 7 Arrhenius plot illustrating the temperature dependence of the benzyl alcohol sorption to the CAVERJECT IMPULSE dual chamber syringe package.

Siliconization of Components

Syringe barrels and elastomer components are siliconized for machineability, functionality, reduced sorptive loss to the glass walls, and to assure maximal volume delivery. Medical-grade silicone oil (polydimethylsiloxane) is applied to the barrel by wiping, spraying, or spraying followed by baking (i.e., spray-baking). The application method impacts the amount of residual silicone left on the barrel that may be extracted/leached into the drug product solution. The wiping application method usually leaves the highest level of residual silicone, followed by spraying and finally spray-baking. Because of the need to produce lower levels of residual silicone, siliconization of most syringes is achieved by spraying or spray-baking. The spray nozzle used during silicone application may be either static or moving (i.e., the nozzle actually enters the barrel and sprays a uniform layer of silicone as it is removed at a controlled rate). A moving spray nozzle normally delivers a more uniform layer of silicone on the entire length of the barrel than a static nozzle.

Spray-baking is specifically used for dual chamber packaging because it gives a lower total content of silicone in the barrel and a more homogenous/reproducible layer of silicone on the glass surface compared to the other siliconization methods. Another key advantage with spray-baking is that the break-loose and glide force of the plungers have less variability and are more stable during long-term storage. Therefore, spray-baking is the most reliable option for DCCs in pen injectors. This siliconization technique involves spraying a diluted medical-grade silicone oil emulsion (e.g., a 1–3% solution of Dow 365™ medical-grade silicone emulsion) into the barrel of the syringe and then heat-curing the silicone to the glass surface at approximately 300°C. Heat-curing leaves two layers of silicone on the glass as illustrated in Figure 8. The first layer is a

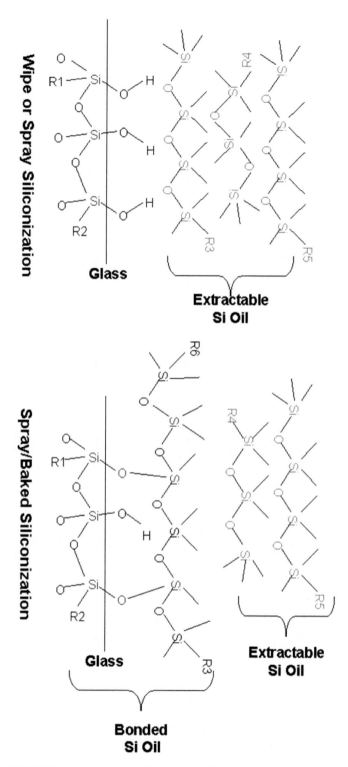

FIGURE 8 Comparison of wipe or spray siliconization versus spray/baked siliconization.

monolayer that is chemically bonded via the hydroxyl groups to the glass. Therefore, it is not available for extraction or leaching into the drug product solution. The second layer is noncovalently associated with the first layer and is enriched in higher molecular weight branched silicone polymers plus a reduced fraction of lower molecular weight silicone. The viscosity of the silicone used for application, the temperature, and the length of curing directly impact the amounts of high and low molecular weight silicones in this second layer (11,12). Heat curing also drives off the stabilizers and water present in the original emulsion. Importantly, it is the second layer of noncovalently bound silicone that can be extracted or leached into the drug product. Additionally, the lower molecular weight silicones have lower vapor pressures and may be volatized during the freeze-drying process (12) and deposited in the frozen solution. If silicone does enter the drug product via leaching and/or deposition of volatile silicones during lyophilization, it may lead to opalescence after reconstitution due to silicone globules refracting light. An additional concern is that exposure of certain protein drugs to silicone has led to protein-related aggregates (13–15). Protein aggregates are considered a potential risk factor for the safety and efficacy of the product (16,17).

Siliconization of the dual chamber package should be optimized for functionality, for example, break-loose and glide force (18). Both break-loose and glide force should be monitored over the lifetime of the product since silicone disposition over the barrel may shift with time and potentially adversely impact compatibility with the device. Wen et al. (19) described spectroscopic and optical methods that can be utilized to monitor silicone disposition in prefilled syringes. These analytical tools could be adapted to monitor silicone disposition after spray-baking in dual chamber package systems. Also, as previously noted, it is important to optimize the siliconization process in terms of compatibility with the formulation.

Extractables/Leachables

Because of the intimate contact of the active and diluent formulations with the syringe barrel and plunger, extractable/leachable (E/L) profiling is important for product development in dual chamber syringes and cartridges. It is especially important to investigate the E/L profile early because low-fill volumes in the syringe or cartridge may lead to high E/L concentrations compared to the same formulation in vial. Notable extractables and leachables may come from the glass, the silicone coating, or the elastomer.

Needle Type

The needle is not affixed to the syringe/cartridge barrel for lyophilized products in dual chamber packaging systems, rather it is attached at the time of use. Lyophilization in a container with a staked needle would not be feasible for a number of reasons, including increased vapor flow resistance during sublimation due to the narrow bore of the needle and the impracticality of sealing the syringe within the freeze-dryer. Additionally, the preferred spray-baked siliconization process for preparation of the syringe barrel would be impossible due to the fact that the adhesive used to affix the needle would decompose during the heat-curing process. Therefore, a pen needle is generally used for a DCC in pen injectors. However, for DCS systems one can utilize either pen needles or luer tipped needles depending on closure utilized for the selected package system.

FREEZE-DRY CYCLE DEVELOPMENT

This section provides a brief overview of lyophilization cycle development, with particular attention to considerations for lyophilization in dual chamber packaging systems due to salient differences from lyophilization in vial formats. The process flow diagram shown in Figure 9 outlines the various unit operations

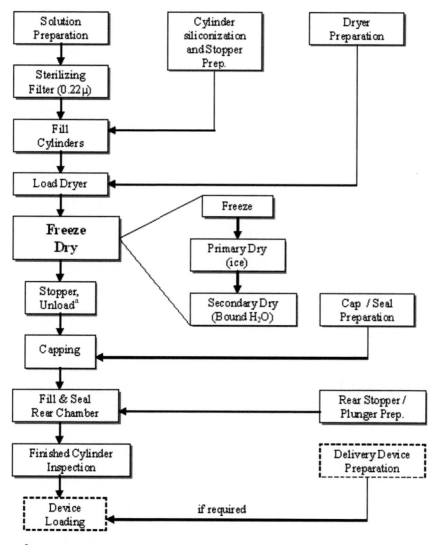

[a] Some systems require external stoppering (after unloading)

FIGURE 9 Process flow diagram for manufacture of a lyophilized product in a dual chamber package system.

involved in the manufacture of a freeze-dried product in a dual chamber package.

Freezing

The objective of freezing is to solidify most or all of the water in the system by formation of ice and to convert solutes into crystalline or glassy solids. Proper freezing will solidify the product in all containers to a temperature below the critical temperature required for the product (e.g., glass transition temperature, eutectic temperature, or collapse temperature). As a sample is cooled, the system may remain liquid well below the equilibrium freezing point due to super-cooling. The degree of supercooling may significantly affect the time to sublime ice since higher supercooling results in smaller ice crystals and a higher product surface area. As the surface area increases, the cake resistance to sublimating vapors increases and results in slower primary drying. Less supercooling (hence larger ice crystals) is desired to minimize primary drying time. Dual chamber package systems (particularly syringes/cartridges) typically have the drug product in the front chamber, which is not in direct contact with the lyophilizer shelf. This leads to significant challenges to development of the freezing phase as well as other aspects of lyophilization cycle development. Figure 10 illustrates these differences and highlights some of the considerations for dual chamber package systems of this type.

The heterogeneity between samples during freezing will generally be magnified in dual chamber packages. Differences in freezing rate will occur for both containers in different locations within a lyophilizer and across a single lyophilizer shelf. These differences will be more pronounced with increasing distance between the drug product compartment and the lyophilizer shelf and also with increasing product fill volume. Uniformity of freezing across a batch can be improved by selection of a loading system that provides appropriate spacing between syringes and enables greater heat transfer contact area between the syringe barrel and the conductive surface of the holder. A similar concern is typically noted as the manufacturing scale changes. Generally a smaller scale (e.g., laboratory) lyophilizer will be poorly insulated compared with a com-mercial unit. The susceptibility of the product to radiative heat from the

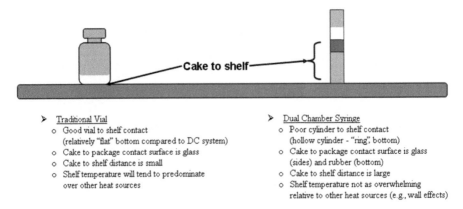

> Traditional Vial
  o Good vial to shelf contact
    (relatively "flat" bottom compared to DC system)
  o Cake to package contact surface is glass
  o Cake to shelf distance is small
  o Shelf temperature will tend to predominate
    over other heat sources

> Dual Chamber Syringe
  o Poor cylinder to shelf contact
    (hollow cylinder - "ring" bottom)
  o Cake to package contact surface is glass
    (sides) and rubber (bottom)
  o Cake to shelf distance is large
  o Shelf temperature not as overwhelming
    relative to other heat sources (e.g., wall effects)

FIGURE 10   Traditional vial versus dual chamber syringe comparison.

chamber walls will be more significant as the distance between the product compartment and the shelf increases. It should also be noted that proper evaluation of the freezing stage can only be properly assessed for samples under clean conditions since particulate matter can impact the nucleation of ice crystals. Understanding the behavior of the active drug substance itself during freezing is also of critical importance. This is especially crucial for biological products such as proteins. Slower freezing of samples can lead to denaturation of proteins since a longer residence time for the active substance in the freeze-concentrated liquid may favor degradation routes such as aggregation. Annealing has been shown to reduce heterogeneity in ice crystal structure across a batch during freeze-drying in syringes (20). Annealing above the glass transition temperature causes growth of ice crystals, which may decrease product resistance during sublimation of ice in the primary drying phase.

Primary Drying

Primary drying consists of the removal of frozen water (ice) by means of sublimation. Following the freezing phase, the drug product bulk solution is present as a frozen matrix of amorphous and/or crystalline material plus pure ice. During primary drying, the pressure is reduced to a suitable level and the shelf temperature is increased to deliver energy to the system to promote sublimation. Determination of optimum conditions for primary drying is generally the most crucial aspect of lyophilization cycle design. The primary drying phase is also typically the longest step in the lyophilization process. Therefore, optimization of primary drying conditions presents the best opportunity for increased efficiency, reduced cycle time, and decreased process cost. However, during the sublimation process, the product is also most vulnerable to failure. During primary drying it is crucial to stay below the identified critical temperature (i.e., T_g' or T_c) to prevent collapse of the product. It is also important that the sublimation of ice be complete before the end of the primary drying phase. If ice remains when the secondary drying phase begins, the product will likely experience melt-back and result in failure of some or all of the batch.

Dual chamber packages present a unique challenge to primary drying phase development. As noted for the freezing phase, package systems that employ a product chamber separated from the lyophilizer shelf behave differently from traditional vial systems. The radiative heating effects from the chamber walls are much more significant and the ability of the shelf temperature to control these effects is greatly reduced. Therefore, it is often difficult to maintain a uniform product temperature and, thus, uniform drying rate across an entire shelf and/or load of DCCs. One must be cognizant of the fact that product located near the center of the lyophilizer shelves can experience temperatures on the order of 5°C to 15°C lower than their counterparts on the edge of the load near the chamber walls. This product temperature difference will result in significant differences in drying rate. Figure 11 illustrates the higher product temperatures (i.e., higher drying rates) for the perimeter samples as compared to the lower product temperatures (i.e., lower drying rates) for the insulated inside samples. Because of this fact, it is critical when interpreting product temperature data to ensure that both hotter (edge) and colder (center) regions for each shelf within the dryer are identified and evaluated for the impact on the product. One should also note the difference in equilibrium

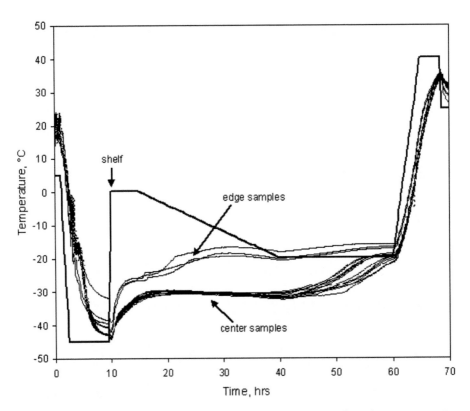

FIGURE 11 Comparison of product temperature and drying rate for edge versus center cylinders.

product temperature between inner and outer samples after sublimation has been completed. When the sublimation phase is completed, the inner samples equilibrate to the shelf temperature in a manner similar to that observed for typical vial products. However, the outer (edge) samples are observed to equilibrate at a significantly higher temperature (as high as 5°C above the shelf temperature). This illustrates the degree to which the radiative heat from the chamber walls is influencing the product temperature of the edge samples due to the distance between the cake and the lyophilizer shelf. This heterogeneity in product temperature can differ significantly depending on the syringe holder type, the dryer design, and the size of the freezer-dryer. It has been observed that syringe racks providing greater spacing between cartridges can result in sublimation behavior in syringes (across the batch load) similar to freeze-drying in glass vials (21). Properly designed development experiments are the key to appropriately guide the freeze-dry cycle development and subsequent scaling up of the process.

Secondary Drying
The secondary drying phase of the freeze-dry cycle consists of the removal of unfrozen water by desorption. Once the primary drying phase has been

completed the drug product is present as a bulk solid matrix devoid of frozen water (ice) with the remaining unfrozen water dispersed in the solid matrix. The lyophilizer chamber pressure is maintained at a low level during secondary drying, often not changing from the primary drying chamber pressure. The shelf temperature is usually increased to deliver additional energy to the system to drive the desorption of water from the product matrix. Secondary drying is a relatively high-temperature process, often ranging from $+25°C$ to $+40°C$. The maximum secondary drying temperature is dependent on the glass transition temperature (T_g) of the dry product. This temperature is typically determined by DSC of the dry cake. Note that the dry cake glass transition temperature is inversely dependent on the moisture level in the lyophilized powder. The secondary drying temperature selected should allow a significant margin for the critical temperature in case powder of slightly higher moisture is encountered during secondary drying. As a general rule, the ramp to secondary drying should be relatively rapid since the risk should be minimal and goal is typically to reduce total cycle time. However, in those cases for products which have a relatively low critical temperature (or one which is very sensitive to moisture content), a slower ramp may be preferable to allow desorption to occur during the ramp time and provide for a slightly lower moisture content when the final secondary drying temperature is reached. Secondary drying time allows for some flexibility in lyophilization cycle design. Although removal of moisture is still critical, total desorption of all unfrozen water is not a requirement of the end of secondary drying as total sublimation is for all frozen water (i.e., ice) is to the end of primary drying. Though many dual chamber packages can be stoppered within the lyophilizer similar to vial products, there are still a number of dual chamber products on the market that are stoppered externally after being unloaded from the lyophilizer. This means that the cake will have a residence time between lyophilizer unloading and final stoppering where the cake is exposed to its environment. This exposure should be under controlled humidity conditions, and the exposure time should be minimized to prevent additional moisture uptake.

Loading Configuration

A unique consideration for dual chamber packaging is the configuration of the syringes in the dryer. Syringes cannot be loaded into a lyophilizer by conventional tray or trayless systems since some variety of cartridge holder is required. Loading systems for syringes can vary significantly; however, there are four broad categories of loading systems. When developing a lyophilization cycle for product in a dual chamber package it is critical to understand the potential impact of the loading configuration on the drying cycle (e.g., cartridge holder design, contact surface area for heat transfer, and cartridge spacing). A loading system that is representative or identical to the commercial process is essential for proper cycle development. This is true of traditional (vial) cycle development and of even greater importance for dual chamber packaging. The basic categories of syringe holders are discussed below.

Packed Cassette

A diagram of cartridges loaded in a packed cassette configuration is shown in Figure 12. The packed cassette or "shoe box" configuration consists of cartridges tightly packed against one another in a "box" or cassette (usually stainless steel).

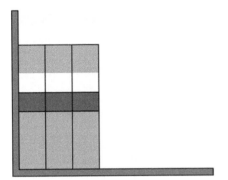

➤ Cylinders tightly packed
➤ Poor glass-to-metal contact
➤ Nonuniform across load

FIGURE 12 Packed cassette loading configuration.

This is perhaps the least efficient loading system and most prone to cycle development problems. The difficulty arises from a combination of poor contact (heat transfer) as well as heterogeneity within a given cassette. As discussed previously, syringes generally do not receive heat from the shelf well due to the poor contact area. As most cartridges are effectively hollow glass cylinders, the bottom surface for heat transfer usually consists of a thin ring of glass. The packed cassette-loading configuration exacerbates the already problematic heat transfer issues of syringes by introducing significant heterogeneity. Syringes packed against the edges of the cassette that are in contact with the stainless steel walls will receive relatively good heat transfer due to direct contact with the metal walls (as the cassette bottom is in contact with the lyophilizer shelf). Syringes closer to the center of the cassette find themselves not only in poor contact with the cassette/shelf but also are tightly packed with neighboring (frozen) cylinders. This means that cylinders near the cassette wall will dry relatively rapidly, while those near the center of the cassette will remain insulated due to the tight packing with their neighbors. Cylinders loaded in the packed cassette configuration tend to have the greatest disparity between "edge" and "center" cylinder product temperature.

Post/Peg Support

A diagram of syringes loaded in a post support configuration is shown in Figure 13. This loading system involves the use of stainless steel or aluminum pegs to support the cylinders. The diameter of the posts is slightly less than that of the hollow cylinder of the syringe barrel. Thus, the syringes can fit over the pegs providing stability during loading and unloading as well as improved homogeneity. Post loading systems provide improved heat transfer characteristics over cassette systems due to the increased contact surface area between the peg surface and the interior wall of the syringe. Post systems also provide spacing between cylinders, so that they are not tightly packed against one another. This significantly increases the homogeneity within the batch.

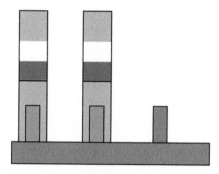

➤ Increased cylinder spacing
➤ Improved glass-to-metal contact
➤ Increased uniformity across load

FIGURE 13 Post/peg cassette loading configuration.

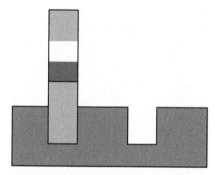

➤ Increased cylinder spacing
➤ Improved glass-to-metal contact
➤ Increased uniformity across load

FIGURE 14 Sleeve cassette loading configuration.

Sleeve Support

A diagram of syringes loaded in a post support configuration is shown in Figure 14. This loading system involves the use of stainless steel or aluminum blocks with holes cut slightly larger than the outer diameter of the syringe cylinders. Syringes are then loaded into the holes, or "sleeves," which support the cylinders. The syringes fit securely into the sleeves, providing stability during loading and unloading as well as improved homogeneity. Sleeve loading systems provide improved heat transfer characteristics over cassette systems (and generally postsystems as well) due to the increased contact surface area between the outer wall of the syringe cylinder and the surface of the sleeve. Sleeve systems also provide spacing between cylinders (similar to a post loading configuration), so that cylinders are not tightly packed against one another. This significantly increases the homogeneity within the batch.

Sleeve-and-Post Support

A diagram of syringes loaded in a post support configuration is shown in Figure 15. This loading system combines elements of both the post and sleeve loading

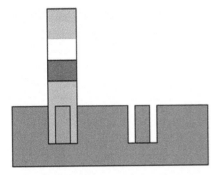

➤ Increased cylinder spacing
➤ Best glass-to-metal contact
➤ Increased uniformity across load

FIGURE 15 Sleeve/post cassette loading configuration.

configurations. The posts on which the cylinders rest are located within the holes (sleeves) cut into a stainless steel or aluminum block. Cartridges are loaded into the holes that are in contact with both the sleeves as well as the support post. This provides stability during loading and unloading as well as improved homogeneity. Sleeve-and-post loading systems provide improved heat transfer characteristics over the other systems discussed due to greater heat transfer contact surface area between the syringe cylinder walls and the surface of the sleeve and post. Sleeve-and-post systems provide spacing between cylinders so that cylinders are not tightly packed against one another. This significantly increases the homogeneity within the batch.

Proper characterization of the drug product and an understanding of the fundamental stages of lyophilization are critical for the development of a suitable freeze-dry cycle for any product. DCS systems introduce a variety of unique considerations that can challenge even a veteran of lyophilization process development. Poor heat transfer due to low shelf-to-package contact surface area as well as distance between the shelf and the drug product are just a few significant factors influencing the drying of material in dual chamber packaging. The loading configuration is critical to the development of the process and should be identified and implemented in development studies as early as possible.

SCALE-UP AND OPTIMIZATION

Process scale-up and optimization are challenges for any lyophilization cycle design since numerous factors discussed previously can have a significant and sometimes difficult to predict impact on technology transfer between freeze-dry units. Cycle scale-up for dual chamber package systems can be particularly problematic since the impact of the contrasts between vials and syringes can vary significantly and unpredictably between small and large-scale lyophilizers. One should not expect laboratory or even pilot-scale cycle data to transfer well to commercial scale lyophilizers. Therefore, it is often necessary to move to large-scale development studies much earlier in the development process for dual chamber products than for traditional vial products. Typically a significant number of commercial scale "engineering" or "development" batches must be produced to optimize the freeze-dry cycle in the final production unit.

Regardless of development efforts to alleviate issues of scale, one should be prepared for scale-up issues to be encountered. Lyophilization is a process inherently dependent on and influenced by the unit in which the process is run and the scale of that unit. The total shelf area, refrigeration/compressor capacity, condenser area, chamber wall thickness, ability for the wall temperature to be maintained etc. have a tremendous influence on the product temperatures encountered. As a lyophilization process is transferred to a larger scale unit (whether from a development lyophilizer to a pilot or clinical dryer or from a lab or pilot dryer to production) one should expect to see differences and plan engineering/qualification runs in advance to address these concerns. The issue that will most commonly be encountered is that product temperature as a function of shelf temperature and chamber pressure may be lower in a production unit. Therefore, as a result of the lower-than-expected product temperatures, the total time required for primary drying might be longer than expected. The implication of this is that when scaling up the freeze-dry cycle to larger freeze-dry units one must make sure that primary drying is complete before secondary drying is started or significant product melt-back/collapse may occur.

Other factors to consider when determining optimum primary drying conditions include the shelf-loading pattern. When dealing with syringe and dual chamber syringe products, the configuration of the shelf load (i.e., the positioning of the containers relative to each other in the load) can have a significant impact on the primary drying temperature and, in particular, on the susceptibility of the load to be influenced by factors such as radiative and convective heat within the chamber. Also note that for such products, extreme differences between product temperatures for containers on the "outside" and the "inside" of the shelf or tray-load can result. This can be an even greater challenge when dealing with products with relatively low T_c and, hence, low primary drying product temperatures. Conservative cycle design for such systems can often result in extremely long drying cycles. The low shelf temperatures required to maintain the outside samples below the T_c result in extremely low product temperatures for the inside containers and, thus, slow drying. Optimization of this type of configuration for a reduced total drying time can be achieved by operating in the region of "microcollapse." Microcollapse is a phenomenon where a small but acceptable degree of collapse occurs (at the microscopic level), to an extent that allows for greatly enhanced vapor flow through the cake and thus rapid drying. If a set of freeze-dry conditions can be achieved where microcollapse is achieved for the outside samples, this will also provide the maximum allowable drying conditions for the inside, rate determining, samples. It is important to ensure the reconstitution time and product stability remain acceptable. Also, this type of cycle development may be extremely sensitive to scale-up issues, and final cycle design will typically require development cycles to be performed in the production units.

INSPECTION CRITERIA

DCS/DCC products present an additional challenge in the establishment of inspection criteria for the final lyophilized cake. Residual moisture level in the lyophilized cake is considered a key stability impacting parameter for freeze-dried

Slight to No Collapse

External Bypass   Internal Bypass

Severe Collapse

External Bypass   Internal Bypass   **FIGURE 16** Examples of varying degrees of collapse by inspection.

products. Often, problems with the physical appearance of the lyophilized cake due to partial collapse can be indicative of variations in moisture level in the cake. As discussed previously, moisture content can vary dramatically between syringe samples in the same batch due to factors such as product temperature variations across the load leading to heterogeneity in freezing and drying plus moisture uptake from the environment postlyophilization for externally stoppered products. Variations in cake appearance may or may not be linked to critical quality problems. Inspection of syringe samples can be further complicated in some cases by the presence of the internal bypass (for transfer of diluent from the rear to the front chamber), which can in some cases obscure the lyophilized cake. During process development, it is valuable to isolate and evaluate samples showing different degrees of collapse by physical inspection. Inspection criteria can be more easily established after developing a correlation between appearance, moisture content, and quality attributes (including stability).

   The example shown in Figure 16 shows freeze-dried cakes from a representative DCC product, which was freeze-dried in both an internal and external bypass DCC package system. The presence of the internal diluent bypass partially obstructs the cakes and made inspection more difficult for this package system. Product quality was assessed for samples representing each "degree of collapse" by inspection. It was found for this example that only the "severely collapsed" samples were unacceptable from a product quality perspective. However, this correlation would need to be evaluated on a case-by-case basis to establish appropriate inspection criteria for other dual chamber freeze-dried products.

TERMINAL STERILIZATION FEASIBILITY

Traditionally, freeze-dried products are processed using established sterilizing filtration and aseptic filling methods. Recently, however, European regulatory guidance requires the evaluation of terminal sterilization for freeze-dried products (22). Undoubtedly, terminal sterilization is a reasonable request from a microbiological safety standpoint. However, there are major technical and scientific challenges associated with the application of terminal sterilization to freeze-dried products. Dry powder heat sterilization, which is the first technique required for evaluation, is not expected to be compatible with freeze-dried protein products because of the high temperatures involved (usually $\geq 160°C$). Sterilization by ionizing irradiation may be more feasible from a technical standpoint; however, there are major challenges associated with this technology as well. Proteins are known to be sensitive to ionizing radiation (23,24), although the extent of degradation might be lower than that associated with the high-temperature treatment during dry heat sterilization. Additionally, lyophilization stoppers are often incompatible with irradiation whereas radiation-resistant stoppers may not be compatible with lyophilized products that require special rubber formulations with low water retention (25).

IMPACT AND CONTROL OF MOISTURE

A critical aspect of most freeze-dried products is an evaluation of the impact of residual moisture on product quality. Since the freeze-drying process typically produces amorphous systems, residual moisture can have a profound effect on the glass transition temperature of freeze-dried products. The presence of residual moisture can plasticize the dried cake and greatly increase the molecular mobility of the solid and water (26). This is especially a concern if the dried product is stored near or above its T_g. The presence of moisture can promote a variety of "solid-state" interactions and lead to instability of the final product (27). Therefore, the impact of variable moisture levels on chemical and/or physical stability is paramount to enabling the appropriate moisture specification to be set. Consequently, it is important to establish a rationale for the appropriate residual moisture specification for each freeze-dried product based on critical product characteristics such as chemical or physical stability. This moisture specification rationale should be included as part of the regulatory file for new drug applications. Typical moisture specification limits for most freeze-dried products range from 1% to 5%; however, there may be exceptions that fall outside this range. It is also noteworthy that there may be both a time of release moisture specification and a separate shelf life moisture specification. Data should be generated to justify both sets of moisture specifications.

Proper control of moisture for the product throughout its shelf life is best met by understanding the sources of moisture for the dried material. The secondary drying conditions for the freeze-dry cycle usually determine the residual moisture level at the end of drying (assuming no cake collapse occurred during the primary drying phase of the freeze-dry cycle). Use of lyoclosures with packages that allow internal stoppering of the package enable tight control of residual moisture after drying is completed. It should be noted that some dual chamber container closure systems allow internal stoppering within the freeze-dryer; however, others do not have this capability. Therefore, those products that are sealed outside of the dryer (e.g., externally stoppered vials, some DCCs, or DCVs, etc.) may absorb atmospheric moisture before they are properly sealed. The rate of moisture uptake

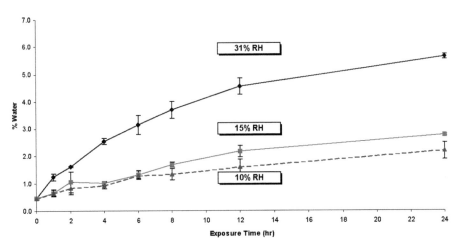

FIGURE 17 Moisture adsorption isotherm (22°C) for CAVERJECT IMPULSE dual chamber syringe prior to sealing.

as a function of various potential postdrying exposure relative humidities should be evaluated to determine acceptable storage conditions (i.e., time and relative humidity) prior to sealing the container. An example data set generated for the CAVERJECT IMPULSE dual chamber product is illustrated in Figure 17. The data reveal that, as expected, the greater the relative humidity the greater the rate and extent of moisture absorption. The product moisture level increases as function of time for all humidity conditions studied. It does not appear that a product equilibrium moisture value has been reached for any of the humidity conditions studied since the moisture levels appear to still be increasing for the last time interval studied (i.e., 24 hours). The ability to adequately control the moisture level of the final product will be influenced by how dry the product is at the end of drying, how much moisture is absorbed prior to capping, and how much moisture increases during its product shelf life due to moisture absorption from packaging components. The data from this study adequately demonstrate that the time of exposure at 15% RH (i.e., the approximate relative humidity of most clean room unloading/capping environments for externally stoppered lyophilized products) will need to be controlled to minimize moisture uptake. Alternatively, the data reveal that storage of the uncapped cartridges in a lower humidity condition (e.g., in some type of transfer cabinet with desiccant present to lower the relative humidity in the cabinet to less than 10% RH) should decrease the rate of moisture uptake.

   All sources of moisture for the product should be identified and be controlled throughout its shelf life. It should be noted that the dried product may absorb moisture from the package during its shelf life. This additional moisture can come from either the rubber closure or the diluent chamber. Selection of the appropriate plunger and its processing conditions is critical to minimize moisture absorption from the package. It is noteworthy that how the stoppers/plungers are washed, autoclaved, dried, and stored after processing can have a

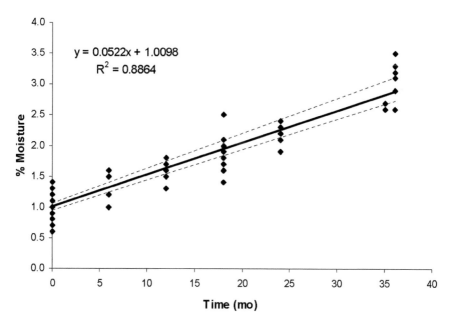

FIGURE 18 Moisture permeation for CAVERJECT IMPULSE stored at 25°C.

profound impact on the occluded or trapped moisture that can diffuse from the closure to the product during storage (28). Additionally, it is common that the plungers used in the dual chamber packaging are rather large in comparison to the dried cake mass and that there may be multiple plungers exposed to the freeze-dried cake. This is especially the case for most DCS/DCC products. Therefore, adsorption of moisture from the plunger should be evaluated. The other potential source of moisture change in the dried product during its shelf life is the permeation of moisture from the liquid chamber through the rubber plunger. Data illustrating the change in moisture content for the CAVERJECT IMPULSE product when stored at 25°C is shown in Figure 18. It is difficult to determine from this data how much of the moisture change is due to release of occluded moisture from the plunger and how much is from transfer of water from the diluent chamber. However, studies of moisture changes for CAVERJECT IMPULSE when stored at 60°C to 80°C (with and without diluent in the diluent chamber) revealed that the relative amount of moisture change was approximately 35% to 55% from transfer from the diluent chamber and 45% to 65% from the plunger itself. The calculated moisture vapor permeability coefficient generated from this data was consistent with the literature values reported for the moisture vapor permeability coefficient for butyl rubbers (10). Others have reported that the predominant moisture change is from the plunger and little movement of water occurs between the chambers (29). No matter what the source of moisture change, it is important to assess the potential moisture changes that may occur for dual chamber package systems, which one might utilize. Once this data has been generated one should be able to predict the moisture change for known freeze-dried cake mass as a function of time and temperature.

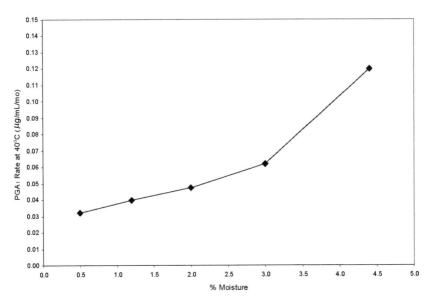

FIGURE 19 The impact of various moisture levels on the degradation rate constants of CAVERJECT IMPULSE (as measured by degradation product, PGA_1, formation at 40°C).

It is also possible for the dried product to absorb atmospheric moisture for container closure systems that leak. Therefore, one must assure that appropriate container closure integrity testing (CCIT) data has been generated for the package system chosen to minimize this risk.

The most important parameter to assess for establishing an appropriate moisture specification is the impact of moisture on product quality. Typically this is evaluated by determining the effect of various moisture levels on product stability. Samples of product in a DCS/DCC package can be prepared at different moisture contents and the resulting stability evaluated. An example of the measured degradation rate constants at 40°C for CAVERJECT IMPULSE determined as a function of moisture content is illustrated in Figure 19. Data such as this can be used to justify an appropriate moisture specification. Note that, as stated earlier, it may be necessary to establish a moisture specification for product at the time of quality control release and a second moisture specification for the product shelf life.

Therefore, the moisture specifications should be established using the following criteria:

1. Evaluate what the process can achieve (at the end of the freeze-dry cycle and through final sealing).
2. Perform moisture absorption isotherms for externally stoppered products to estimate additional moisture uptake.
3. Determine moisture absorption from the package with time and storage condition (i.e., from the closure or diluent chamber).
4. Determine the effect of variable moisture contents on product quality (e.g., stability).

SPECIFICATIONS FOR DUAL CHAMBER LYOPHILIZED PRODUCT PACKAGE SYSTEMS

The product specifications describe tests that are utilized to demonstrate that each batch of product has met appropriate quality criteria. However, it should be noted that these may differ depending on which countries the product has been registered for sale. Dual chamber packages typically have specifications for the active-containing chamber, the diluent-containing chamber, and the reconstituted product. Example specification assays for the lyophilized powder containing chamber might include identification, appearance, moisture, pH, potency assay, dose uniformity, and purity. Note that additional potency and purity assays will be required for biological type products. Example specification assays for the diluent chamber might include identification, appearance (including color, clarity, and particulates), preservative potency, pH, etc. Example assays for the reconstituted solution might include appearance (including color, clarity, and particulates), pH, reconstitution time, subvisible particulate matter dose uniformity, deliverable volume, purity, endotoxins, and sterility.

Additionally, functional testing according to ICH Q6A (30) may be included in specifications for dual chamber products. Potential functional testing may include pressure and seal integrity, resealability, tip-cap removal force, piston release force (break-loose force), piston travel force (glide force), and power injection function force. Note that data generated during product development may justify elimination of functional testing as part of the specifications.

STABILITY EVALUATION OF DUAL CHAMBER LYOPHILIZED PRODUCT PACKAGE SYSTEMS

A unique aspect for designing stability studies for dual chamber package systems is that stability of the active-containing chamber, diluent-containing chamber, and reconstituted solution must be monitored. This is similar to the three sets of specifications that must be set for a dual chamber lyophilized product. Specific guidances for stability testing of dual chamber package systems, in-use reconstituted solution stability testing, and sterility/container closure integrity are discussed below.

Active-Containing Chamber

At a minimum, potency assay, impurities, and water content are the chemical tests that should be performed on stability for the lyophile powder. Hydrolysis resulting from moisture content is a common cause of chemical instability and one that is potentially important in a lyophilized formulation. Temperature and pH can affect the rate of hydrolysis. Another potential cause of chemical instability, but not common for lyophilized products, is oxidation. If the formulation is susceptible to oxidation, typically a nitrogen overlay is used. Typical chemical tests for evaluating stability for small molecules are assay, degradation products, moisture content, and pH. Typical chemical tests for evaluating stability of biologicals include identity, protein concentration, purity (e.g., aggregates, etc.), biological potency, product-related substances (e.g., charge variants), oxidation for some cases, moisture content, and pH. If a functional excipient is part of the formulation, for example an antioxidant or preservative, then a test for the excipient is needed. Other tests that measure product quality may be included as warranted. Physical stability is also important for establishing product quality. The lyophile should preferably retain the original cake appearance, color, and morphology over its shelf life.

Diluent-Containing Chamber

The diluent-containing chamber should be tested for identification, appearance (including color, clarity, and particulates), preservative potency, pH, etc.

In-Use Reconstituted Solution Stability Studies

Reconstituted stability studies should be performed with product in full contact with the closure system (e.g., side-stored or inverted) to confirm compatibility. In-use test patterns applied to reconstituted small-molecule lyophiles typically include assay, degradation products, clarity of solution (USP and EP), pH, and visible and subvisible particulates. In-use test patterns applied to reconstituted biological lyophiles typically include identity, protein concentration, purity (e.g., monomer purity, IgG, etc.), biological potency, product-related substances (e.g., charge variants), oxidation for some cases, clarity of solution (USP and EP), pH, and visible and subvisible particulates. Functional excipient in-use testing of preservatives, antioxidants, etc. should be included when present in for-mulations. Other tests may also be added if critical to product quality.

Functional attributes of the syringe or cartridge should be monitored on stability because silicone disposition over the barrel may shift with time and could adversely impact compatibility with the device. Functional tests include pressure and seal integrity, tip-cap removal force, piston release force (break-loose force), piston travel force (glide force), and power injection function force.

Sterility/Container Closure Integrity

The requirement for sterility in parenteral products is absolute throughout the product shelf life at the intended storage condition. The sterility and container closure integrity should be monitored during the registration stability program. If the reconstituted liquid is to be multiused, or used over an extended period of time, an antimicrobial preservative may be part of the formulation. If a pre-servative is present, an antimicrobial effectiveness test should be part of testing. Preservative concentration that provides adequate antimicrobial protection should be determined in early development. The preservative stability in the reconstituted liquid needs to be assessed on stability; however, this is usually evaluated via a chemical test. During registration stability, AET is performed on the reconstituted product at initial and end of shelf life time points for the product. Unlike chemical and physical stability tests, microbiological tests are not normally performed at each time point, but instead are run on an annual basis. Usually only one lot per formulation is subjected to microbiological testing when multiple lots are placed on stability. CCIT is performed to evaluate the ability of the container closure system to provide protection and maintain integrity during the shelf life of the sterile drug product. According to a recent FDA guidance to industry, container closure system integrity testing can be used in lieu of sterility testing as a component of the stability protocol for sterile product (31). It should be noted that CCIT cannot replace sterility testing during product release. Before making the decision to use CCIT in lieu of sterility testing, the following should be considered: (*i*) confirm regulatory acceptance of this option in the country of filing and (*ii*) confirm that the methodology chosen is appropriate for the container closure system and is adequately validated.

COST ANALYSIS

The comparison of cost between a lyophilized product in a dual chamber package and a vial is complex due to the number of factors that need to be considered. Several of these key cost considerations are summarized in Table 2. It is difficult to generalize this topic since many of the factors are dependent on specifics of the product. One of the foremost cost differentiators is related to decreased overfill in a dual chamber system. Consider the example of a 1-mL fill in vial versus dual chamber system. The USP suggests a 0.15-mL overfill for 1-mL fill vial products due to loss during withdrawal and administration (32), which equates to a 15% overfill. However, the overfill for a dual chambered system is only dictated by hold-up volumes between chambers and that caused by the delivery needle. This is typically ≤0.1 mL (i.e., ≤10% overfill) for a 1-mL dual chamber system. This consideration is especially important for biologics drugs where the drug itself is expensive. Freeze-drying time and yield are specific to the formulation, primary package, and lyophilization parameters, but is generally less optimal for dual chambered systems. However, if a product in DCS has a relatively high collapse temperature and an efficient loading configuration in the lyophilizer, the overall difference in lyophilizer time and yield may not be much greater than for an identical product in vial. The number of accessories needed to reconstitute and deliver drug from a lyophilized drug in vial is much greater than for the lyophilized drug in a dual chamber system. This greater number of accessories must be taken into the overall cost of the final drug product. One exception is that drugs in DCCs must be delivered with a pen delivery system, which varies widely in cost. Finally, an overall cost differential that favors vials may not matter if a dual chamber system is required to compete in the market, which is the case for home administration of insulin and growth hormone.

SUMMARY

An increased use of lyophilization and the demand for more unit dose/unit of use packaging have created a requirement for new package forms. Lyophilizing in dual chamber packages eliminates the manual aseptic reconstitution step and greatly improves patient convenience and safety. However, lyophilizing a drug in dual chamber systems can be challenging compared with lyophilization in a standard vial due to differences in package configuration, heat flow, and lyophilizer trays. Appropriate considerations for the proper amount and fill for each chamber must be taken into account to produce a product designed to deliver the target label dose on use. Compatibility of the formulation with the package must be properly evaluated since these systems tend to be siliconized and possess large surface areas of the glass/rubber to volume, which can potentially impact product physical/chemical stability. Freeze-dry cycle development can be difficult due to poor heat transfer to the product during sublimation. Significant drying heterogeneity can occur within the dryer depending

TABLE 2 Cost Considerations for Dual Chamber Vs. Vial Packaging

	Dual chamber cartridge/syringe (DCC/DCS)	Vial	Comments
Drug product in primary package			
Primary package	Barrel, plungers, closure, finger rest, piston	Vial, closure, cap	Additional packaging needed to reconstitute and deliver dose from vial
Overfill	Minimal overfill due to elimination of product transfer step	Increased overfill due to loss in product transfer step	Cost of API must be considered, especially for a biologic drug
Freeze-drying cycle time	Longer due to poorer heat transfer	Shorter due to better heat transfer	Depends on specifics of the formulation, primary package and lyophilization parameters
Batch size	Generally larger due to smaller unit dimension	Generally smaller due to larger unit dimension	Depends on rack and loading configuration for DCC/DCS
Yield	Rejection rate potentially higher	Rejection rate potentially lower	Depends on specifics for the formulation, primary package and lyophilization cycle
Accessories			
Injection needle	1	1	
Reconstitution needle	N/a	1	
Plastic syringes	N/a	2	Syringes needed for reconstitution and delivery
Vial with diluent	N/a	1	
Secondary package	Blister + carton	Carton	Syringe packaged in blister
Disinfection swabs for reconstitution and loading of delivery needle	N/a	2	
Device for delivery	Pen injection device needed for DCC	N/a	
Other			
Competition in marketplace	Needed to compete in certain therapeutic areas, e.g., home administration of chronic therapeutics like insulin or growth hormone		Choice of package, and ability to sell product, is dependent on the standard-of-care for therapy

Abbreviations: DCC, dual chamber cartridge; DCS, dual chamber syringe; API, active pharmaceutical ingredient.

on the loading configurations and freeze-dry cycle utilized. Therefore, appropriate optimization studies are needed to develop a successful, robust freeze-dry cycle. Control of moisture for the product at the end of drying, after sealing, and absorption from the package (i.e., from the closure or diluent chamber) during the product shelf life must be evaluated and correlated to product stability. Specifications need to be established, which cover each chamber separately and the combined reconstituted product. The benefits of dual chamber lyophilized products compared to standard vial packages will need to be evaluated based on a variety of factors such as total production cost for the units required for dose administration, ease-of-use by the consumer, increased safety/sterility assurance, etc.

REFERENCES

1. Michaels TM Jr. Dual chamber prefilled syringes. J Parenter Sci Tech 1988; 42: 199–202.
2. Executive summary. In: Drug Delivery Pens and Autoinjectors: Devices, Diseases, and Delivery Strategies: Applied Data Research, 2007:6.
3. Pollin J. The ins and outs of prefilled syringes. Pharm Med Packaging News 2003; 11:40–44.
4. French D, Masters S. Current trends in the injection of biopharmaceuticals. Am Pharm Rev 2005; 8:30–37.
5. Executive summary. In: Pen Injectors: Worldwide Markets and Therapies. Amherst, New Hampshire: Greystone Associates, 2006:3.
6. Cotten J. Current market and key trends of devices used in self-injection. In: Furness G, ed. Prefilled Syringes: the Trend for Growth Strengthens. West Sussex: ONDrugDelivery Ltd, 2005:25–27.
7. Solomun L, Ibric S, Boltic Z, et al. The impact of primary packaging on the quality of parenteral products. J Pharm Biomed Anal 2008; 48:744–748.
8. Colandene J, Maldonado L, Creagh A, et al. Lyophilization cycle development for high-concentration monoclonal antibody formulation lacking a crystalline bulking agent. J Pharm Sci 2007; 96:1598–1608.
9. Rowles B, Sperandio GJ, Shaw SM. Effects of elastomer closures on the sorption of certain 14C-labeled drug and preservative combinations. Bull Parenter Drug Assoc 1971; 25:2–22.
10. Wang Y, Chien Y. Sterile pharmaceutical packaging: compatibility and stability. In: Technical Report No 5, Parenteral Drug Association, Inc., 1984:107–125.
11. Mundry T. Einbrennsilikonisierung Bei Pharmazeutischen Glaspackmitteln—Analytische Studien eines Produktionsprozesses. Berlin, Germany: Humoldt Universitaet, 1999.
12. Mundry T, Schurreit T, Surmann P. The fate of silicone oil during heat-curing glass siliconization-changes in molecular parameters analyzed by size exclusion and high temperature gas chromatography. PDA J Pharm Sci Technol 2000; 54:383–397.
13. Gabrielson J, Bates DG, Williams BM, et al. Silicone oil contamination of therapeutic protein formulations: surfactant and protein effects. Abstracts of Papers, 234th ACS National Meeting, Boston, MA, United States, August 19–23, 2007:BIOT-227.
14. Jones LS, Kaufmann A, Middaugh CR. Silicone oil induced aggregation of proteins. J Pharm Sci 2005; 94:918–927.
15. Chantelau E. Silicone oil contamination of insulin. Diabet Med 1989; 6:278.
16. Rosenburg A. Effects of protein aggregates: an immunologic perspective. AAPS J 2006; 8:E501–E507.
17. Schellekens H. Immunogenicity of therapeutic proteins: clinical implications and future prospects. Clin Ther 2002; 24:1720–1740; discussion 19.
18. Overcashier DE, Chan EK, Hsu CC. Technical considerations in the development of prefilled syringes for protein products. Am Pharm Rev 2006; 9:77–83.

19. Wen Z, Vance A, Vega F, et al. Distribution of Silicone Oil in Prefilled Glass Syringes Probed with Optical and Spectroscopic Methods. PDA J Pharm Sci Technol 2009; 63:149–158.
20. Hottot A, Andrieu J, Vessot S, et al. Experimental study and modeling of freeze-drying in syringe configuration. Part I: freezing step. Drying Technol 2009; 27:40–48.
21. Hottot A, Andrieu J, Hoang V, et al. Experimental study and modeling of freeze-drying in syringe configuration. Part II: mass and heat transfer parameters and sublimation end-points. Drying Technol 2009; 27:49–58.
22. Decision Trees for the Selection of Sterilization Methods (CPMP/QWP/054/98). Annex to note for guidance on development pharmaceutics (CPMP/QWP/155/96). London: The European Agency for Evaluation of Medicinal Products. Committee for Propriety of Medicinal Products (CPMP), April 5, 2000.
23. Shalaev E, Reddy R, Kimball R, et al. Protection of a protein against irradiation-induced degradation by additives in the solid-state. Radiat Phys Chem 2003; 66: 237–245.
24. Yamamoto O. Effect of radiation on protein stability. In: Ahern T, Manning M, eds. Stability of Protein Pharmaceuticals, Part A Chemical and Physical Pathways of Protein Degradation. New York and London: Plenum Press, 1992:361–421.
25. Kiang P, Ambrosio T, Buchanan R, et al. Technical report No. 16: effects of gamma irradiation on elastomeric closures. J Parenter Sci Technol 1992; 46(S2):S1–S13.
26. Shalaev EY, Zografi G. How does residual water affect the solid-state degradation of drugs in the amorphous state? J Pharm Sci 1996; 85:1137–1141.
27. Nail S, Gatlin S. Freeze-Drying: principles and practice. In: Avis K, Lieberman H, Lachman L, eds. Pharmaceutical Dosage Forms: Parenteral Medications. New York: Marcel Dekker, Inc., 1993:208.
28. DeGrazio F. Closure and container considerations in lyophilization. In: Rey L, May J, eds. Freeze-Drying/Lyophilization of Pharmaceuticals and Biological Products. New York: Marcel Dekker, 2004:290–292.
29. Hinneburg-Wolf B. Moisture uptake of freeze-dried product during storage: comparison of vials versus dual chamber syringe. In: Symposium on the Freeze-Drying of Pharmaceuticals and Biologicals. Presented by the Center for Pharmaceutical Processing Research Breckenridge, CO, 2004.
30. Section 3.3.2.3 Parenteral Drug Products. In: International Conference on Harmonization Topic Q6A, Specifications: Test Procedures and Acceptance Criteria for New Drug Substances and New Drug Products: Chemical substances, May 2000.
31. Guidance for Industry: Container and closure system integrity testing *in lieu* of sterility testing as a component of the stability protocol for sterile products. Rockville, MD, February 2008.
32. <1151> Pharmaceutical Dosage Forms-Injections. In: United States Pharmacopeia–National Formulary (USP 32–NF 27), 2009.

22 Lyophilization and Irradiation as an Integrated Process

Dalal Aoude-Werner and Florent Kuntz
Aerial-crt, Illkirch, France

INTRODUCTION

Many pharmaceutical preparations and especially injectables or parenterals are freeze-dried to maintain their physical and biological integrity after months or even years of storage. Sterility of these pharmaceutical preparations is often required.

The usual rule to achieve that goal is to start with a sterile filtered or heat-treated solution and, from there on, to carry out an entirely sterile process. Today all freeze-dryers have their cabinets opening within a sterile room while the machinery is sitting behind the wall in the engine room. Moreover, the drying chambers are all equipped with clean-in-place (CIP) systems and can be sterilized by pressure steam before each operation (SIP). Finally, those products that are prepared in vials are sealed directly within the chamber, thanks to moving pressure plates that drive the stoppers tight into the neck of the vials.

Preparation and handling of these products in a sterile environment imposes drastic production conditions leading to very important economic costs. Moreover, when facing complex mechanical setups such as semi-continuous or continuous freeze-drying processes (1), serious hazards in the sterility control of the operation can occur that definitely bear on the production cost.

Irradiation is an alternative technique for terminal sterilization of freeze-dried therapeutic products. Its effectiveness to eliminate microorganisms is no longer to prove (2).

According to this scheme, the whole freeze-drying operation would be carried out under clean hygienic conditions without any attempt to secure sterility. Then, when the freeze-dried products are ready and fully protected in their sealed vials, they are sterilized by ionizing *radiation*. Those radiation types are of sufficient energy to penetrate any standard packaging system. If the initial product is itself sterile and if the whole operation has been done carefully, the potential contamination will be quite low and a rather small irradiation dose might be required. As this is done for medical and surgical devices, sterility is then guaranteed by the process itself and the release of the product done on purely parametric criteria after they have been, of course, validated by applying the current international standards (3–5).

The various processes of interaction of these radiations with the matter lead throughout their way, with the release of electrons having a certain kinetic energy (secondary electrons). The loss of the kinetic energy of the secondary electrons in the product locally involves the ionization and the excitation of the matter crossed by the radiations. The absorption of this energy by the matter results in direct physicochemical effects to which can be added indirect effects in the event of presence of water in the medium (6).

These physicochemical effects induce biological effects in the particular case of the living organisms of which the structure is strongly hydrated and has a high level of organization. But, like all methods of sterilization, irradiation involves a compromise between inactivation of the contaminating micro-organisms and damage to the product being sterilized.

The imparted energy, in the form of γ-photons, X rays, or electrons, does not always differentiate between molecules of the contaminating microorganism and those of the pharmaceutical substrate.

The potential consequences of irradiation on the stability of the active principal ingredient (API) may be restricted when combined with the freeze-drying. Indeed, the latter, by eliminating water, limits the formation of radio-induced free radicals and their propagation responsible for the radiolysis reactions that can lead to the degradation of the of therapeutic preparations.

Various studies dealing with radiation sterilization of therapeutics have been reported in the literature. Some of them prove its effectiveness for freeze-dried pharmaceuticals radio sterilization and the absence of significant deleterious effect (7). The industrial applications of radio sterilization of therapeutic products are in strong progress.

IRRADIATION TECHNIQUES
Ionizing Radiation Used in Industry
Radiation sterilization appeared from the second half of the 20th century. In 1949, Charles Artlandi gave the bases of the electron beam (corpuscular radiation) sterilization using a 2 MeV Van de Graaff electron accelerator. This research allowed the installation of the first commercial irradiators in 1956 with accelerated electrons and in 1964 with γ rays (electromagnetic radiation).

The idea to use the X rays (electromagnetic radiation) to sterilize medical devices was born in the 1960s (8).

Radiation sterilization is a clean process that does not leave toxic residues in the products and makes it possible to sterilize products conditioned in their final packaging with a weak increase of temperature only. It is therefore known as the "the process of 'cold' sterilization."

The energy of the radiation usable in industry is limited to 5 MeV for the electromagnetic radiation and 10 MeV for the electron beams. However, higher energies can be implemented with the obligation to demonstrate the integrity of the treated product in terms of induced radioactivity. Taking into account their nature and these limitations, these radiations cannot disturb the nucleus of the atom and so can in no way induce in the irradiated matter a significant amount of radioactivity (9).

Electron Beams
The electron beams are generated by electric accelerators, machines, allowing by "construction" the control of the characteristics (energy and intensity) of the beam. The electron acceleration principles, to near the speed of light, are varied (high-frequency or electrostatic techniques), but the principal components of these machines remain always the same: source of electrons, accelerating tube or cavity, and scanning horn with electron beam shape device.

The sterilization industry employs mainly medium powerful beams (20–300 kW) from 5 to 10 MeV produced by high-performance accelerators.

Their principal interest lies in their short necessary processing time since the electron beam quickly provides an important dose to the product to be sterilized due to its very high dose rate. However, their very weak penetrating capacity constrained to irradiate limited surface mass (<8 g/cm^2, i.e., 8 cm in density 1 material) products packaged in cardboard boxes.

With the development of medium energy, self-shielded accelerators, electron beam sterilization can also be processed in a different way. Products can be treated piece by piece. This provides the advantage to require lower energy level (a few MeV) and as a matter of fact cheaper accelerators with cheaper shielding. The high-throughput rates are well suited to a just-in-time supply process or to full integration in the production line. In-line sterilization tunnels using electron beams have become a reality since the development of low and medium energy accelerators small enough to fit into self-shielded units for integration in production lines. They provide fast continuous room temperature sterilization that is simple to validate and traceable. The economies are apparent in terms of time, logistics, fixed assets costs, labor costs, etc. Environmental impact is considered low.

Electromagnetic Radiations
γ-Radiation

The very large majority of the treatments by electromagnetic radiations are thus carried out currently using γ rays emitted by radioactive sources, primarily cobalt-60. It is produced by neutron bombardment in a nuclear reactor of the metal cobalt-59, and then doubly encapsulated in stainless steel pencils to prevent any leakage during its use in an irradiator. This radioelement emits two γ rays (1.17 and 1.33 MeV) and has a period of 5.27 years (time after which the emitted radiation quantity decreases by half). The implementation of this type of treatment is based on the exposure of the product during a given time, with the radiation emitted by the radioactive source.

γ-Sources require important installations and moreover require managing the radioactive waste problem because the radioactive sources are changed before they are completely stable. However, because of its low attenuation inside the treated material, this type of radiation makes it possible to sterilize important volumes of products, packaged onto pallets.

X Rays

X-ray electromagnetic radiation (Bremsstrahlung radiation) is generated by the electron beams absorbed by a heavy metal target (generally out of tungsten or tantalum). Contrary to the γ rays, the X rays are mainly concentrated in the direction of the incidental electron beam, toward the products to be treated.

This type of treatment is currently little used in industry and is limited of principle by poor conversion yield of the electrons into X rays.

However, with the development of a new generation of high-energy, high-power electron accelerators (Rhodotron), X rays seem to be an interesting combination of the two other processes, that is, electron beam and γ.

X rays have a similar or even higher penetration capacity than γ rays while profiting from a sizeable dose rate inherited from the high-power accelerated electrons (10).

RADIATION STERILIZATION AND ITS CONSEQUENCE ON THE API
Radiation Sterilization (11)

The sterilization effect of the radiation rests on their biological effect.

When an ionizing radiation interacts with the matter, it releases an electron of their orbit that acquires a certain kinetic energy. The target atom loses an electron and becomes a positive ion. As for the released electrons, they strike other electrons on their pathway and are responsible for the formation of many other ions. In fact, the major part of ionizations is due to electrons whatever the incident radiation type may be (electron beam, γ or X ray).

The most critical damage caused by the ionizing radiations is made on the DNA molecule. The modifications of this macromolecule generate defects of genetic coding at the origin of changes or cellular death.

Actually, there are two processes of deterioration of the DNA molecule by the ionizing rays: either directly by ionization of the molecule and release of electrons as explained previously or by water radiolysis present in the microorganism's cell. This second phenomenon creates in the vicinity of the DNA molecule, powerful oxygen chemical reagents H_2O_2 or free radicals such as OH, H, HO_2, which will deteriorate the genetic material.

The ionizing rays deteriorate the vital structures of the living microorganisms whose consequences are generally lethal for them.

The inactivation of the microorganisms follows an exponential law and consequently there always exists a certain statistical probability that a microorganism can survive sterilization. The European and U.S. Pharmacopeia give a threshold value known as sterility assurance level (SAL) to reach so that sterilization is effective. Thus, the SAL for a given process is expressed like the probability of occurrence of a nonsterile article in a population of sterilized objects. For sterilization of drugs or medical devices, the SAL is set to 10^{-6} (4).

When using a radiation sterilization method, it is necessary to establish the dose received by the product to achieve sterility. The effectiveness of this dose must be shown and controlled over time. It thus requires validating the process that is the guarantee of the conformity of the product to the requirements retained by the legislation (3).

European Pharmacopeia indicates that the sterilization absorbed dose of reference is 25 kGy.[a] It also states that other values can be selected provided that they make it possible to obtain a SAL lower than 10^{-6}.

In the medical device industry, the validation of a radiation sterilization process is done on the ISO 11137 standard basis: *ISO standard 11137-2: 2006, Sterilization of health care products—Radiation—Part 2: Establishing the sterilization dose*. This text specifies methods to determine the minimum dose necessary to reach a specified requirement for sterility and methods to justify the use of the sterilization dose of 25 kGy or the sterilization dose of 15 kGy, to obtain a sterility insurance level of 10^{-6}.

---

[a]*Absorbed dose* is the mean energy imparted to a quantity of matter divided by the mass of that matter, i.e., energy per unit mass. Its unit is Gray (Gy). 1 Gy = 1 J/kg.

Inactivation of the API by Radiation Treatment

By using radiation as an effective method of pathogen inactivation in drugs one would deal with the main disadvantage of API stability.

Under irradiation treatment, the main problem results in unacceptable losses in functional activity of the APIs. There are two mechanisms by which irradiation can damage therapeutic molecules. One is the direct result of a radiation depositing energy into the target. The transfer of this energy results predominantly in the dislocation of outer electrons from molecules and breakage of covalent bonds. The second type of damage has been termed "indirect" and is the result of chemical attack by free radicals and reactive oxygen species typically generated by the interaction of radiation with water molecules and oxygen. Although these free radicals and reactive oxygen species are very short lived, they are responsible for the majority of the damage that occurs in therapeutics irradiated by conventional irradiation procedures.

During irradiation, the damaging secondary effects can be controlled by combinations with freeze-drying to minimize the potential for generating free radicals, protection from free radical damage through the use of free radical scavengers, and reduction of the free radical mobility by irradiation at cryogenic temperatures.

It is difficult to predict a general behavior of the active ingredients against irradiation. Each API is unique and may react differently depending on its own physicochemical characteristics and the characteristics of the environment in which it resides. Later in this chapter the impact of irradiation and its application conditions on the stability of a protein of human origin is presented as an example.

IMPACT OF IRRADIATION ON THERAPEUTIC PROTEINS

Proteins are composed of one or multiple linear amino acid sequences that interact with each other to form secondary structures (sheet, helix, etc.), which are then folded back in space by hydrophobic links to form tertiary structures. The protein may be subjected to several changes that could lead to different levels of degradation such as (*i*) important secondary changes of conformation that can lead to aggregations in liquid environment (irreversible modifications) (12), (*ii*) modification of the secondary structures not leading to aggregates formation in the liquid state but which can affect the stability of the protein during storage (13), and (*iii*) denaturations ending with the fragmentation of proteins.

The impact of irradiation on the therapeutic proteins is affected by a number of parameters such as the structure and state of the proteins (globular, native, or denatured), the medium composition (presence of other substances, wet, dry, in solution, or whether liquid or frozen), and the irradiation conditions (absorbed dose, dose rate, temperature, presence of oxygen, etc.)

The irradiation process initiates a series of reactions leading to ionic and free radical intermediates and ultimately to stable products that under certain conditions might lead to permanent modifications in proteins. The chemical changes in irradiated proteins include deamination, decarboxylation, reduction of disulfide linkages, oxydation of sulfhydryl groups, peptide-chain cleavage, aggregation, cross-linking, and aggregation (14–17). Hydroxyl radicals and superoxide anion radicals generated by radiation modify the primary structure of proteins, which results in the distortion of the secondary and tertiary structures.

The negative effect of irradiation on the stability and integrity of proteins can be widely limited if the irradiation conditions are well controlled to minimize the formation of radiolytic products and their diffusion.

More than 40% of currently marketed protein drug products are solids, primarily lyophilized (i.e., freeze-dried) powders for reconstitution. This form is often selected to preserve potency and prolong shelf life. In freeze-dried products, diffusion of radicals will be minimized and movement of large radicals will be virtually eliminated making the impact of irradiation on the proteins stability insignificant.

Impact of the Radiation Treatment Combined to Freeze-Drying on the Stability of a Therapeutic Protein: Human Serum Albumin

Human serum albumin (HSA) is a monomeric protein. It is a polypeptidic chain of 585 amino acids among which 35 cysteines forming 17 disulfures bonds (12). It has a molecular mass of 65 kDa and is mainly constituted by α-helix. It is a globular protein formed by three different but very similar domains in their peptide sequences and their three-dimensional conformations (18,19). Regarding its conformation, hydrophobic sites are internal and hydrophilic sites are external. HSA is the most abundant protein in human blood plasma. It is produced in the liver and is involved in many transports (hormones and fatty acids) and in the osmotic pressure maintenance (18,19). It may be deficient in the case of liver cirrhosis of nephritic syndrome or choleretic enteropathy. It is therefore a key protein that must be administered to deficient individuals.

In our work, the effect of radiation combined to the freeze-drying process is assessed for a pharmaceutical formulation containing 1% of HSA. This formulation is exposed before or after freeze-drying to 15 and 25 kGy (doses usually applied for sterilization) and 40 kGy excessive radiation dose. Radiation doses are delivered using a Van de Graaff electron beam accelerator (2.2 MeV, 150 μA).

Changes in the molecular properties of HSA due to irradiation and freeze-drying are evaluated, thanks to size exclusion HPLC (HP-SEC) (20), SDS-PAGE electrophoresis (21), and Fourier transform infrared (FTIR) spectroscopy (22,23).

HP-SEC and SDS-PAGE electrophoresis allow the separation of molecules according to their molecular weight and, thus, can highlight irreversible degradation such as aggregation and fragmentation. FTIR is based on the excitation of molecules by a laser and on the vibrations of intermolecular links. FTIR equipped with attenuated total reflection (ATR) is used to investigate alteration of the secondary structure of the protein and to learn about changes that may be reversible.

According to the HP-SEC results confirmed by electrophoresis on SDS-PAGE, the HSA protein within the framework of this study is present under monomer form (main form) and dimer and trimer forms (minors forms) (Figs. 1 and 2).

The freeze-drying of the HSA formulation does not seem to affect the stability of the protein (Fig. 3). The same chromatographic and electrophoretic profiles are obtained before and after freeze-drying (Fig. 3).

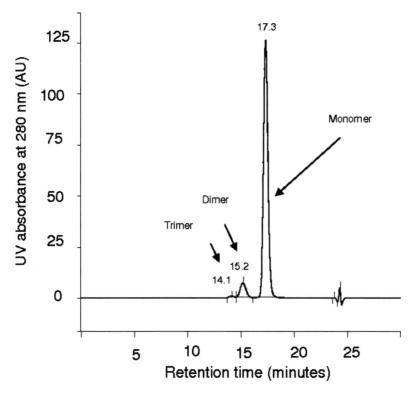

FIGURE 1 Chromatographic profile (on HP-SEC) of HSA. *Abbreviations*: HP-SEC, size exclusion high-performance liquid chromatography; HSA, human serum albumin.

However, variations are observed in these chromatographic and electrophoretic profiles as the formulation of HSA is irradiated under its liquid state (before freeze-drying) or solid one (after freeze-drying). Regarding the radiation damage to HSA, two types of damage are observed: fragmentation and aggregation (Fig. 4).

When irradiation is applied at 15, 25, and 40 kGy on the freeze-dried formulation, it induces partial breakdown of the polypeptide chain, resulting in the formation of degraded low molecular weight molecules revealed by the appearance of (*i*) small chromatographic peaks at retention times from 24 to 28 minutes (Fig. 4) and (*ii*) bands on SDS-PAGE corresponding to lower molecular weight molecules than HSA monomer (Fig. 5).

However, the degradation rate is lower than 18% (Table 1), even at the higher applied absorbed dose.

According to the literature, irradiation generally causes irreversible changes at the molecular level by the breakage of the covalent bonds of polypeptide chains. Exposure of proteins to oxygen radicals results in nonrandom and random fragmentations (24).

The protein fragmentation is affected by the local conformation of a particular amino acid inside the protein, its accessibility to water radiolysis

TABLE 1 Impact of Irradiation Applied Under Various Conditions on HSA Stability

	Part of HSA (%)	Part of HSA fragment (%)	Part of HSA aggregate (%)
Freeze-dried HSA formulation	100.00	–	–
HSA formulation irradiated at 15 kGy under liquid state	88.14	10.89	0.97
Freeze-dried HSA formulation irradiated at 15 kGy	89.94	9.92	0.14
HSA formulation irradiated at 25 kGy under liquid state	82.46	13.94	3.69
Freeze-dried HSA formulation irradiated at 25 kGy	86.18	13.57	0.25
HSA formulation irradiated at 40 kGy under liquid state	57.40	17.68	24.92
Freeze-dried HSA formulation irradiated at 40 kGy	81.58	17.97	0.45

The results are expressed as follows:
● Part of HSA (%) = (Peak area of HSA/total of peak areas) × 100.
● Part of HSA fragments (%) = (Peak area of HSA fragments/total of peak areas) × 100.
● Part of HSA aggregates (%) = (Peak area of HSA aggregates/total of peak areas) × 100.
Abbreviation: HSA, human serum albumin.

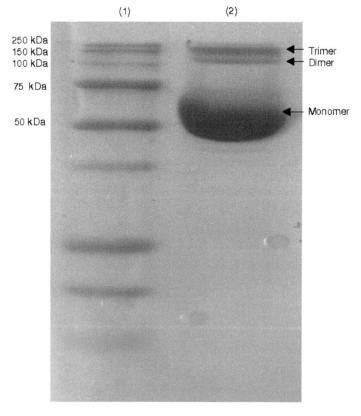

FIGURE 2 Electrophoresis profile (on SDS-PAGE) of HSA. (1) Protein markers and (2) HSA. *Abbreviation*: HSA, human serum albumin.

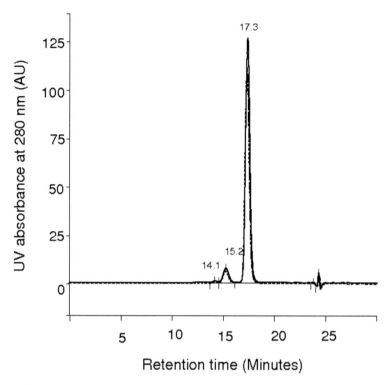

FIGURE 3 Chromatographic profile (on HP-SEC) of HSA formulation before freeze-drying (—) and after freeze-drying (– – – –). *Abbreviations*: HP-SEC, size exclusion high-performance liquid chromatography; HSA, human serum albumin.

products, and the primary amino acid sequence (25). In our case, this fragmentation is probably restricted because of the presence of few free radicals. Indeed, the low content of residual humidity in the freeze-dried formulation (2%) limits the radiolysis of water and thus the formation of free radicals (26); thus, the observed fragmentation is probably due to the direct effect of irradiation only.

Irradiation of the HSA formulation under its liquid state induces also a fragmentation of HSA materialized by the appearance of chromatographic peaks at the retention times between 24 and 27 minutes. However, this alteration remains low (less than 18% at 40 kGy) in comparison with the aggregation phenomenon that seems to be predominant (Table 1).

In fact, according to the results shown in Figures 4 and 5, the presence of chromatographic peaks between 12 and 14 minutes and a smear on SDS-PAGE with higher molecular weight than the HSA trimer form seems to be a signature of the formation of aggregates.

It is known that irradiation causes the aggregation and cross-linking of proteins. However, the nature of bonds involved in the protein aggregates is still an open question. Proteins may be converted to high molecular weight

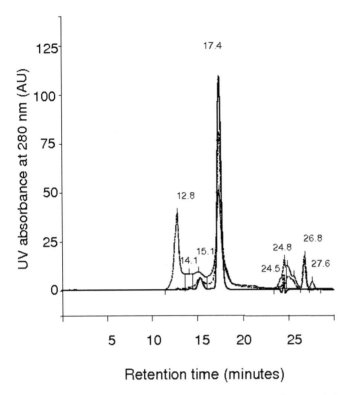

FIGURE 4 Chromatographic profile (on HP-SEC) of HSA formulation unirradiated (– – – –), irradiated at 40 kGy under liquid state (—), and after freeze-dried (– – – –). *Abbreviations*: HP-SEC, size exclusion high-performance liquid chromatography; HSA, human serum albumin.

aggregates due to the generation of interprotein cross-linking reactions, hydrophobic and electrostatic interactions, and the formation of disulfide bonds (27). Any amino acid radical formed within a peptide chain could cross-link with an amino acid radical from another protein. In the high-dose range, HSA proteins exposed to radiation underwent covalent cross-linking.

The formation of high molecular weight aggregates is negligible at 15 and 25 kGy and increases significantly with the higher absorbed dose of 40 kGy (up to 25%). However, no change in the secondary structure of HSA could be highlighted by FTIR.

In conclusion, irradiation of HSA formulation under its liquid state causes degradation, cross-linking, and aggregation of the polypeptide chains due to radicals generated by the water radiolysis. No aggregation is observed when HSA formulation is irradiated under its solid state (i.e., freeze-dried) and the amount of fragmentation due to irradiation remains low (less than 15%) at doses usually applied for sterilization. Removing water from the formulation by the freeze-drying process restricts the formation of radiolytic products and thus the occurrence of those alterations.

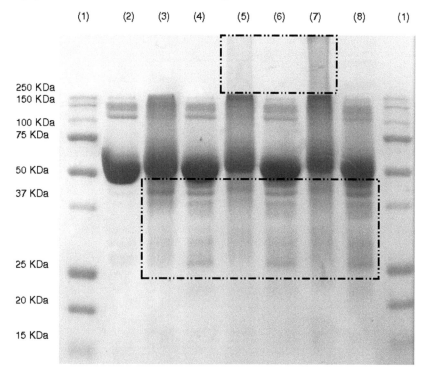

FIGURE 5 Electrophoretic profile of HSA formulation on SDS–PAGE gel when irradiated under various conditions. (1) Protein markers, (2) HSA formulation irradiated at 15 kGy under liquid state, (3) freeze-dried HSA formulation irradiated at 15 kGy, (4) HSA formulation irradiated at 25 kGy under liquid state, (5) freeze-dried HSA formulation irradiated at 25 kGy, (6) HSA formulation irradiated at 40 kGy under liquid state, (7) freeze-dried HSA formulation irradiated at 40 kGy. *Abbreviation*: HSA, human serum albumin.

CONCLUSION

Radiation processing is a rather efficient technique to achieve terminal sterilization of pharmaceuticals. However, its treatment conditions have to be well established and controlled to reduce as much as possible the impact on the API.

As it has been discussed and demonstrated in the presented example, the consequences of radiation on the API are mainly due to the radio-induced free radicals which are predominantly generated in the water environment of the therapeutic molecule.

As a matter of fact, the combined treatment of freeze-drying and radiation sterilization is an effective solution.

Irradiation in cryogenic temperature condition may improve the recovery and, of course, it is advisable for those products that cannot be freeze-dried, and, thus, should be processed/irradiated in frozen state to avoid the diffusion of the potentially damaging free radicals.

Whatever the chosen treatment conditions may be, it is necessary to examine each therapeutic product to assess its radiation stability.

REFERENCES

1. Rey L. Potential prospects in freeze-drying. In: Rey L, May JC, eds. Freeze-Drying/ Lyophilization of Pharmaceutical and Biological Products. New York: Marcel Dekker, Inc, 1999:465–471.
2. Swarbrick J, Boylan JC. Encyclopedia of Pharmaceutical Technology. New York: Marcel Dekker Inc., 1992; 5:105–136.
3. ISO 11137-1:2006 Sterilization of health care products—Radiation—Part 1: Requirements for development, validation, and routine control of a sterilization process for medical devices. Part 2: Establishing the sterilization dose.
4. ISO 11737-2:2009 Sterilization of medical devices—Microbiological methods—Part 2: Tests of sterility performed in the definition, validation and maintenance of a sterilization process.
5. ISO 14937:2009 Sterilization of health care products—General requirements for characterization of a sterilizing agent and the development, validation and routine control of a sterilization process for medical devices.
6. Grieb T, Forng RY, Brown R, et al. Effective use of gamma irradiation for pathogen inactivation of monoclonal antibody preparations. Biologicals 2002; 30:207–216.
7. May JC, Rey L, Lee C. Evaluation of some selected vaccines and other biological products irradiated by gamma rays, electron beams and X-Rays. Radiat Phys Chem 2002; 63:709–711.
8. Chmielewski AG, Haji-Saeid M. Radiation technologies: past, present and future. Radiat Phys Chem 2004; 71(1–2):17–21.
9. Les applications innovantes de l'ionisation dans l'industrie, Les guides de l'innovation NOVELECT, 1999, NOV.607.
10. Bol JL. IBA Industrial, Flexible Multipass X-ray configuration, International Meeting on Radiation Processing, London, September 21–26, 2008.
11. Trends in radiation sterilization of Health care products, International Atomic Energy Agency, Vienna, 2008, STI/PUB/1313, 261 pp.; 79 figures; ISBN 978-92-0-111007-7.
12. Lin JJ, Meyer JD, Carpenter JF, et al. Stability of human serum albumin in bioprocessing of albumin paste. Pharm Res 2000; 17(4):391–397.
13. Wei W. Lyophilization and development of solid protein pharmaceuticals. Int J Pharm 2000; 203:1–60.
14. Delincée H. Recent Advances in radiation Chemistry of proteins. In: Elias PS, Cohen AJ, eds. Recent Advances in Food Irradiation. Amsterdam: Elsevier Biomedical, 1983:129–147.
15. Garrison WM. Reaction mechanisms in the radiolysis of peptides, polypeptides and proteins. Chem Rev 1987; 87:381–398.
16. Davies KJA, Delsignore ME. Protein damage and degradation by oxygen radicals. III. Modification of secondary and tertiary structure. J Biol Chem 1987; 262:9908–9913.
17. Woods RJ, Pichaev AK. Applied radiation chemistry. New York: John Wiley and Sons, 1994.
18. Charbonneau D, Beauregard M, Tajmir-Riahi HA. Structural analysis of human serum albumin complexes with cationic lipids. J Phys Chem B 2009; 113(6):1777–1784.
19. Sugio S, Kashima A, Mochizuki S, et al. Crystal Structure of human serum albumin at 2,5 Angström resolution. Protein Eng 1999; 12(6):439–446.
20. Qian J, Tang Q, Cronin B, et al. Development of a high performance size exclusion chromatography method to determine the stability of human serum albumin in a lyophilized formulation of interferon alfa-2b. J Chromato A 2008; 1194:48–56.
21. Tamikazu K, Tsukasa M. Changes in structural and antigenic properties of proteins by radiation. Radiat Phys Chem 1995; 46(2):225–231.
22. Carpenter JF, Prestrelski SJ, Dong A. Application of infrared spectroscopy to development of stable lyophilized protein formulations. Eur J Pharm Biopharm 1998; 45:231–238.
23. Kong J, YU S. Fourier transform infrared spectroscopic analysis of protein secondary structures. Acta Biochimica Biophys Sin (Shanghai) 2007; 39(8):549–559.

24. Kume T, Matsuda T. Changes in the structural and antigenic properties of proteins by radiation. Radiat Phys Chem 1995; 46:225–231.
25. Filali-Mouhim A, Audette M, St-Louis M, et al. Lysozyme fragmentation induced by radiolysis. Int J Radiat Biol 1997; 72:63–70.
26. Wang SL, Lin SY, Li MJ, et al. Temperature effect on the structural stability, similarity, and reversibility of human serum albumin in different states. Biophys Chem 2005; 114:205–212.
27. Wei YS, Lin SY, Wang SL, et al. Fourier transform IR attenuated total reflectance spectroscopy studies of cysteine-induced changes in secondary conformations of bovine serum albumin after UV-B irradiation. Biopolymers (Biospectroscopy) 2003; 72:345–351.

APPENDIX

Table
Saturation Vapor Pressure Of Ice

SATURATION VAPOR PRESSURE OF ICE

Temperature	Saturation Vapor Pressure			
°C	mbar	mtorr	torr	Pa
−90	9.672E−05	0.07255	- -	9.672E−03
−89	1.160E−04	0.08701	- -	1.160E−02
−88	1.388E−04	0.1041	- -	1.388E−02
−87	1.658E−04	0.1244	- -	1.658E−02
−86	1.977E−04	0.1483	- -	1.977E−02
−85	2.353E−04	0.1765	- -	2.353E−02
−84	2.796E−04	0.2097	- -	2.796E−02
−83	3.316E−04	0.2487	- -	3.316E−02
−82	3.925E−04	0.2944	- -	3.925E−02
−81	4.638E−04	0.3479	- -	4.638E−02
−80	5.473E−04	0.4105	- -	5.473E−02
−79	6.444E−04	0.4833	- -	6.444E−02
−78	7.577E−04	0.5683	- -	7.577E−02
−77	8.894E−04	0.6671	- -	8.894E−02
−76	1.042E−03	0.7816	- -	1.042E−01
−75	1.220E−03	0.9151	- -	1.220E−01
−74	1.425E−03	1.069	- -	1.425E−01
−73	1.662E−03	1.247	- -	1.662E−01
−72	1.936E−03	1.452	- -	1.936E−01
−71	2.252E−03	1.689	- -	2.252E−01
−70	2.615E−03	1.961	- -	2.615E−01
−69	3.032E−03	2.274	- -	3.032E−01
−68	3.511E−03	2.633	- -	3.511E−01
−67	4.060E−03	3.045	- -	4.060E−01
−66	4.688E−03	3.516	- -	4.688E−01
−65	5.406E−03	4.055	- -	5.406E−01
−64	6.225E−03	4.669	- -	6.225E−01
−63	7.159E−03	5.370	- -	7.159E−01
−62	8.223E−03	6.168	- -	8.223E−01
−61	9.432E−03	7.075	- -	9.432E−01
−60	1.080E−02	8.101	- -	1.080E+00
−59	1.236E−02	9.271	- -	1.236E+00
−58	1.413E−02	10.60	- -	1.413E+00
−57	1.612E−02	12.09	- -	1.612E+00
−56	1.838E−02	13.79	- -	1.838E+00
−55	2.092E−02	15.69	- -	2.092E+00
−54	2.380E−02	17.85	- -	2.380E+00
−53	2.703E−02	20.27	- -	2.703E+00
−52	3.067E−02	23.00	- -	3.067E+00
−51	3.476E−02	26.07	- -	3.476E+00
−50	3.935E−02	29.51	- -	3.935E+00
−49	4.449E−02	33.37	- -	4.449E+00
−48	5.026E−02	37.70	- -	5.026E+00
−47	5.671E−02	42.54	- -	5.671E+00
−46	6.393E−02	47.95	- -	6.393E+00
−45	7.198E−02	53.99	- -	7.198E+00

Temperature	Saturation Vapor Pressure			
°C	mbar	mtorr	torr	Pa
−44	8.097E−02	60.73	- -	8.097E+00
−43	9.098E−02	68.24	- -	9.098E+00
−42	1.021E−01	76.58	- -	1.021E+01
−41	1.145E−01	85.88	- -	1.145E+01
−40	1.283E−01	96.23	- -	1.283E+01
−39	1.436E−01	107.7	- -	1.436E+01
−38	1.606E−01	120.5	- -	1.606E+01
−37	1.794E−01	134.6	- -	1.794E+01
−36	2.002E−01	150.2	- -	2.002E+01
−35	2.233E−01	167.5	- -	2.233E+01
−34	2.488E−01	186.6	- -	2.488E+01
−33	2.769E−01	207.7	- -	2.769E+01
−32	3.079E−01	230.9	- -	3.079E+01
−31	3.421E−01	256.6	- -	3.421E+01
−30	3.798E−01	284.9	- -	3.798E+01
−29	4.213E−01	316.0	- -	4.213E+01
−28	4.669E−01	350.2	- -	4.669E+01
−27	5.170E−01	387.8	- -	5.170E+01
−26	5.720E−01	429.0	- -	5.720E+01
−25	6.323E−01	474.3	- -	6.323E+01
−24	6.985E−01	523.9	- -	6.985E+01
−23	7.709E−01	578.2	- -	7.709E+01
−22	8.502E−01	637.7	- -	8.502E+01
−21	9.370E−01	702.8	- -	9.370E+01
−20	1.032	774.1	- -	1.032E+02
−19	1.135	851.3	- -	1.135E+02
−18	1.248	936.1	- -	1.248E+02
−17	1.371	- -	1.028	1.371E+02
−16	1.506	- -	1.130	1.506E+02
−15	1.652	- -	1.239	1.652E+02
−14	1.811	- -	1.358	1.811E+02
−13	1.984	- -	1.488	1.984E+02
−12	2.172	- -	1.629	2.172E+02
−11	2.376	- -	1.782	2.376E+02
−10	2.597	- -	1.948	2.597E+02
−9	2.837	- -	2.128	2.837E+02
−8	3.097	- -	2.323	3.097E+02
−7	3.379	- -	2.534	3.379E+02
−6	3.685	- -	2.764	3.685E+02
−5	4.015	- -	3.011	4.015E+02
−4	4.372	- -	3.279	4.372E+02
−3	4.757	- -	3.568	4.757E+02
−2	5.173	- -	3.880	5.173E+02
−1	5.623	- -	4.218	5.623E+02
0	6.108	- -	4.581	6.108E+02

Source: Smithsonian Meteorological Tables 6th ed. (1971) VDI vapor tables 6th ed. (1963)

Index